Introduction to Practice of Molecular Simulation

Introduction to Practice of Molecular Simulation

Molecular Dynamics, Monte Carlo, Brownian Dynamics, Lattice Boltzmann, Dissipative Particle Dynamics

Akira Satoh

Akita Prefectural University

Japan

ELSEVIER

AMSTERDAM • BOSTON • HEIDELBERG • LONDON • NEW YORK • OXFORD
PARIS • SAN DIEGO • SAN FRANCISCO • SINGAPORE • SYDNEY • TOKYO

Elsevier
32 Jamestown Road London NW1 7BY
30 Corporate Drive, Suite 400, Burlington, MA 01803, USA

First published 2011

Notices
Knowledge and best practice in this field are constantly changing. As new research and experience broaden our understanding, changes in research methods, professional practices, or medical treatment may become necessary.

Practitioners and researchers must always rely on their own experience and knowledge in evaluating and using any information, methods, compounds, or experiments described herein. In using such information or methods they should be mindful of their own safety and the safety of others, including parties for whom they have a professional responsibility.

To the fullest extent of the law, neither the Publisher nor the authors, contributors, or editors, assume any liability for any injury and/or damage to persons or property as a matter of products liability, negligence or otherwise, or from any use or operation of any methods, products, instructions, or ideas contained in the material herein.

British Library Cataloguing-in-Publication Data
A catalogue record for this book is available from the British Library

Library of Congress Cataloging-in-Publication Data
A catalog record for this book is available from the Library of Congress

ISBN: 978-0-323-16519-8

This book has been manufactured using Print On Demand technology. Each copy is produced to order and is limited to black ink. The online version of this book will show color figures where appropriate.

Contents

Preface

The control of internal structure during the fabrication of materials on the nanoscale may enable us to develop a new generation of materials. A deeper understanding of phenomena on the microscopic scale may lead to completely new fields of application. As a tool for microscopic analysis, molecular simulation methods—such as the molecular dynamics and the Monte Carlo methods—have currently been playing an extremely important role in numerous fields, ranging from pure science and engineering to the medical, pharmaceutical, and agricultural sciences. The importance of these methods is expected to increase significantly with the advance of science and technology.

Many physics textbooks address the molecular simulation method for pure liquid or solid systems. In contrast, textbooks concerning the simulation method for suspensions or dispersions are less common; this fact provided the motivation for my previous textbook. Moreover, students or nonexperts needing to apply the molecular simulation method to a physical problem have few tools for cultivating the skill of developing a simulation program that do not require training under a supervisor with expertise in simulation techniques. It became clear that students and nonexpert researchers would find useful a textbook that taught the important concepts of the simulation technique and honed programming skills by tackling practical physical problems with guidance from sample simulation programs. This book would need to be written carefully; it would not simply explain a sample simulation program, but also explains the analysis procedures and include the essence of the theory, the specification of the basic equations, the method of nondimensionalization, and appropriate discussion of results. A brief explanation of the essence of the grammar of programming languages also would be useful.

In order to apply the simulation methods to more complex systems, such as carbon-nanotubes, polymeric liquids, and DNA/protein systems, the present book addresses a range of practical methods, including molecular dynamics and Monte Carlo, for simulations of practical systems such as the spherocylinder and the disk-like particle suspension. Moreover, this book discusses the dissipative particle dynamics method and the lattice Boltzmann method, both currently being developed as simulation techniques for taking into account the multibody hydrodynamic interaction among dispersed particles in a particle suspension or among polymers in a polymeric liquid.

The resulting characteristics of the present book are as follows. The important and essential background relating to the theory of each simulation technique is explained, avoiding complex mathematical manipulation as much as possible. The equations that are included herein are all important expressions; an understanding of them is key to reading a specialized textbook that treats the more theoretical aspects of the simulation methods. Much of the methodology, such as the assignment of the initial position and velocity of particles, is explained in detail in order to be useful to the reader developing a practical simulation program.

In the chapters dedicated to advancing the reader's practical skill for developing a simulation program, the following methodology is adopted. First, the sample physical phenomenon is described in order to discuss the simulation method that will be addressed in the chapter. This is followed by a series of analyses (including the theoretical backgrounds) that are conducted mainly from the viewpoint of developing a simulation program. Then, the assignment of the important parameters and the assumptions that are required for conducting the simulation of the physical problem are described. Finally, results that have been obtained from the simulation are shown and discussed, with emphasis on the visualization of the results by snapshots. Each example is conducted with a sample copy of the simulation program from which the results were obtained, together with sufficient explanatory descriptions of the important features in the simulation program to aid to the reader's understanding.

Most of the sample simulation programs are written in the FORTRAN language, excepting the simulation program for the Brownian dynamics method. We take into account that some readers may be unfamiliar with programming languages, that is, the FORTRAN or the C language; therefore, an appendix explains the important features of these programming languages from the viewpoint of developing a scientific simulation program. These explanations are expected to significantly reduce the reader's effort of understanding the grammar of the programming languages when referring to a textbook of the FORTRAN or the C language.

The present book has been written in a self-learning mode as much as possible, and therefore readers are expected to derive the important expressions for themselves—that is the essence of each simulation demonstration. This approach should appeal to the reader who is more interested in the theoretical aspects of the simulation methods.

Finally, the author strongly hopes that this book will interest many students in molecular and microsimulation methods and direct them to the growing number of research fields in which these simulation methods are indispensable, and that one day they will be the preeminent researchers in those fields.

The author deeply acknowledges contribution of Dr. Geoff N. Coverdale, who volunteered valuable assistance during the development of the manuscript. The author also wishes to express his thanks to Ms. Aya Saitoh for her dedication and patience during the preparation of so many digital files derived from the handwritten manuscripts.

Akira Satoh
Kisarazu City, Chiba Prefecture, Japan
December 2010

1 Outline of Molecular Simulation and Microsimulation Methods

In the modern nanotechnology age, microscopic analysis methods are indispensable in order to generate new functional materials and investigate physical phenomena on a molecular level. These methods treat the constituent species of a system, such as molecules and fine particles. Macroscopic and microscopic quantities of interest are derived from analyzing the behavior of these species.

These approaches, called "molecular simulation methods," are represented by the Monte Carlo (MC) and molecular dynamics (MD) methods [1–3]. MC methods exhibit a powerful ability to analyze thermodynamic equilibrium, but are unsuitable for investigating dynamic phenomena. MD methods are useful for thermodynamic equilibrium but are more advantageous for investigating the dynamic properties of a system in a nonequilibrium situation. This book examines MD and MC methods of a nonspherical particle dispersion in a three-dimensional system, which may be directly applicable to such complicated dispersions as DNA and polymeric liquids. This book also addresses Brownian dynamics (BD) methods [1,4], which can simulate the Brownian motion of dispersed particles; dissipative particle dynamics (DPD) [5–8]; and lattice Boltzmann methods [9–12], in which a liquid system is regarded as composed of virtual fluid particles. Simulation methods using the concept of virtual fluid particles are generally used for pure liquid systems, but are useful for simulating particle dispersions.

1.1 Molecular Dynamics Method

A spherical particle dispersion can be treated straightforwardly in simulations because only the translational motion of particles is important, and the treatment of the rotational motion is basically unnecessary. In contrast, since the translational and rotational motion has to be simulated for an axisymmetric particle dispersion, MD simulations become much more complicated in comparison with the spherical particle system. Simulation techniques for a dispersion composed of nonspherical particles with a general shape may be obtained by generalizing the methods employed to an axisymmetric particle dispersion. It is, therefore, very important to understand the MD method for the axisymmetric particle system.

Introduction to Practice of Molecular Simulation. DOI: 10.1016/B978-0-12-385148-2.00001-X

1.1.1 Spherical Particle Systems

The concept of the MD method is rather straightforward and logical. The motion of molecules is generally governed by Newton's equations of motion in classical theory. In MD simulations, particle motion is simulated on a computer according to the equations of motion. If one molecule moves solely on a classical mechanics level, a computer is unnecessary because mathematical calculation with pencil and paper is sufficient to solve the motion of the molecule. However, since molecules in a real system are numerous and interact with each other, such mathematical analysis is impracticable. In this situation, therefore, computer simulations become a powerful tool for a microscopic analysis.

If the mass of molecule i is denoted by m_i, and the force acting on molecule i by the ambient molecules and an external field denoted by $\mathbf{f}_i$, then the motion of a particle is described by Newton's equation of motion:

$$m_i \frac{\mathrm{d}^2 \mathbf{r}_i}{\mathrm{d}t^2} = \mathbf{f}_i \tag{1.1}$$

If a system is composed of N molecules, there are N sets of similar equations, and the motion of N molecules interacts through forces acting among the molecules.

Differential equations such as Eq. (1.1) are unsuitable for solving the set of N equations of motion on a computer. Computers readily solve simple equations, such as algebraic ones, but are quite poor at intuitive solving procedures such as a trial-and-error approach to find solutions. Hence, Eq. (1.1) will be transformed into an algebraic equation. To do so, the second-order differential term in Eq. (1.1) must be expressed as an algebraic expression, using the following Taylor series expansion:

$$x(t+h) = x(t) + h\frac{\mathrm{d}x(t)}{\mathrm{d}t} + \frac{1}{2!}h^2\frac{\mathrm{d}^2x(t)}{\mathrm{d}t^2} + \frac{1}{3!}h^3\frac{\mathrm{d}^3x(t)}{\mathrm{d}t^3} + \cdots \tag{1.2}$$

Equation (1.2) implies that x at time $(t+h)$ can be expressed as the sum of x itself, the first-order differential, the second-order differential, and so on, multiplied by a constant for each term. If x does not significantly change with time, the higher-order differential terms can be neglected for a sufficiently small value of the time interval h. In order to approximate the second-order differential term in Eq. (1.1) as an algebraic expression, another form of the Taylor series expansion is necessary:

$$x(t-h) = x(t) - h\frac{\mathrm{d}x(t)}{\mathrm{d}t} + \frac{1}{2!}h^2\frac{\mathrm{d}^2x(t)}{\mathrm{d}t^2} - \frac{1}{3!}h^3\frac{\mathrm{d}^3x(t)}{\mathrm{d}t^3} + \cdots \tag{1.3}$$

If the first-order differential term is eliminated from Eqs. (1.2) and (1.3), the second-order differential term can be solved as

$$\frac{\mathrm{d}^2x(t)}{\mathrm{d}t^2} = \frac{x(t+h) - 2x(t) + x(t-h)}{h^2} + \mathrm{O}(h^2) \tag{1.4}$$

The last term on the right-hand side of this equation implies the accuracy of the approximation, and, in this case, terms higher than h^2 are neglected. If the second-order differential is approximated as

$$\frac{d^2x(t)}{dt^2} = \frac{x(t+h) - 2x(t) + x(t-h)}{h^2} \tag{1.5}$$

This expression is called the "central difference approximation." With this approximation and the notation $\mathbf{r}_i = (x_i, y_i, z_i)$ for the molecular position and $\mathbf{f}_i = (f_{xi}, f_{yi}, f_{zi})$ for the force acting on particle i, the equation of the x-component of Newton's equation of motion can be written as

$$x_i(t+h) = 2x_i(t) - x_i(t-h) + \frac{h^2}{m_i} f_{xi}(t) \tag{1.6}$$

Similar equations are satisfied for the other components. Since Eq. (1.6) is a simple algebraic equation, the molecular position at the next time step can be evaluated using the present and previous positions and the present force. If a system is composed of N molecules, there are $3N$ algebraic equations for specifying the motion of molecules; these numerous equations are solved on a computer, where the motion of the molecules in a system can be pursued with the time variable. Eq. (1.6) does not require the velocity terms for determining the molecular position at the next time step. This scheme is called the "Verlet method" [13]. The velocity, if required, can be evaluated from the central difference approximation as

$$\mathbf{v}_i(t) = \frac{\mathbf{r}_i(t+h) - \mathbf{r}_i(t-h)}{2h} \tag{1.7}$$

This approximation can be derived by eliminating the second-order differential terms in Eqs. (1.2) and (1.3). It has already been noted that the velocities are unnecessary for evaluating the position at the next time step; however, a scheme using the positions and velocities simultaneously may be more desirable in order to keep the system temperature constant. We show such a method in the following paragraphs.

If we take into account that the first- and second-order differentials of the position are equal to the velocity and acceleration, respectively, the neglect of differential terms equal to or higher than third-order in Eq. (1.2) leads to the following equation:

$$\mathbf{r}_i(t+h) = \mathbf{r}_i(t) + h\mathbf{v}_i(t) + \frac{h^2}{2m_i}\mathbf{f}_i(t) \tag{1.8}$$

This equation determines the position of the molecules, but the velocity term arises on the right-hand side, so that another equation is necessary for specifying the velocity. The first-order differential of the velocity is equal to the acceleration:

$$\mathbf{v}_i(t+h) = \mathbf{v}_i(t) + \frac{h}{m_i}\mathbf{f}_i(t) \tag{1.9}$$

In order to improve accuracy, the force term in Eq. (1.9) is slightly modified and the following equation obtained:

$$\mathbf{v}_i(t+h) = \mathbf{v}_i(t) + \frac{h}{2m_i}(\mathbf{f}_i(t) + \mathbf{f}_i(t+h)) \tag{1.10}$$

The scheme of using Eqs. (1.8) and (1.10) for determining the motion of molecules is called the "velocity Verlet method" [14]. It is well known that the velocity Verlet method is significantly superior in regard to the stability and accuracy of a simulation.

Consider another representative scheme. Noting that the first-order differential of the position is the velocity and that of the velocity is the acceleration, the application of the central difference approximation to these first-order differentials leads to the following equations:

$$\mathbf{r}_i(t+h) = \mathbf{r}_i(t) + h\mathbf{v}_i(t+h/2) \tag{1.11}$$

$$\mathbf{v}_i(t+h/2) = \mathbf{v}_i(t-h/2) + \frac{h}{m_i}\mathbf{f}_i(t) \tag{1.12}$$

The scheme of pursuing the positions and velocities of the molecules with Eqs. (1.11) and (1.12) is called the "leapfrog method" [15]. This name arises from the evaluation of the positions and forces, and then the velocities, by using time steps in a leapfrog manner. This method is also a significantly superior scheme in regard to stability and accuracy, comparable to the velocity Verlet method.

The MD method is applicable to both equilibrium and nonequilibrium physical phenomena, which makes it a powerful computational tool that can be used to simulate many physical phenomena (if computing power is sufficient).

We show the main procedure for conducting the MD simulation using the velocity Verlet method in the following steps:

1. Specify the initial position and velocity of all molecules.
2. Calculate the forces acting on molecules.
3. Evaluate the positions of all molecules at the next time step from Eq. (1.8).
4. Evaluate the velocities of all molecules at the next time step from Eq. (1.10).
5. Repeat the procedures from step 2.

In the above procedure, the positions and velocities will be evaluated at every time interval h in the MD simulation. The method of specifying the initial positions and velocities will be shown in Chapter 2.

Finally, we show the method of evaluating the system averages, which are necessary to make a comparison with experimental or theoretical values. Since microscopic quantities such as positions and velocities are evaluated at every time interval in MD simulations, a quantity evaluated from such microscopic values—for example, the pressure—will differ from that measured experimentally. In order to compare with experimental data, instant pressure is sampled at each time step, and these values are averaged during a short sampling time to yield a macroscopic pressure. This average can be expressed as

$$\overline{A} = \sum_{n=1}^{N} A_n/N \tag{1.13}$$

in which A_n is the nth sampled value of an arbitrary physical quantity A, and $\overline{A}$, called the "time average," is the mathematical average of N sampling data.

1.1.2 Nonspherical Particle Systems

1.1.2.1 Case of Taking into Account the Inertia Terms

For the case of nonspherical particles, we need to consider the translational motion of the center of mass of a particle and also the rotational motion about an axis through the center of mass. Axisymmetric particles are very useful as a particle model for simulations, so we will focus on the axisymmetric particle model in this section. As shown in Figure 1.1, the important rotational motion is to be treated about the short axis line. If the particle mass is denoted by m, the inertia moment by I, the position and velocity vectors of the center of mass of particle i by $\mathbf{r}_i$ and $\mathbf{v}_i$, respectively, the angular velocity vector about the short axis by $\boldsymbol{\omega}_i$, and the force and torque acting on the particle by $\mathbf{f}_i$ and $\mathbf{T}_i$, respectively, then the equations of motion concerning the translational and rotational motion can be written as

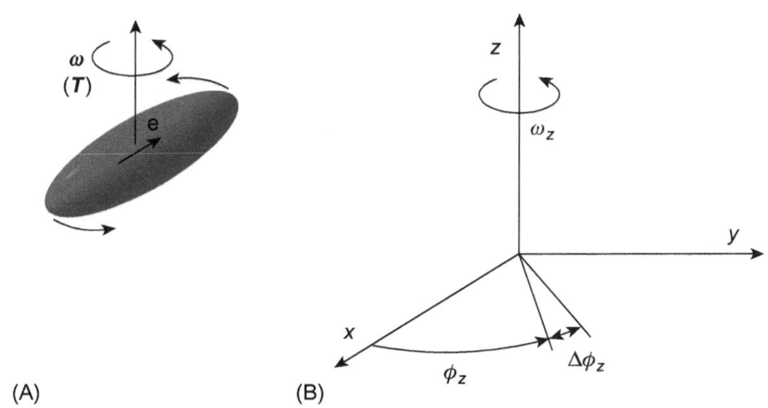

Figure 1.1 Linear particle and angular velocity: (A) the axisymmetric particle and (B) the coordinate system.

$$m\frac{d^2\mathbf{r}_i}{dt^2} = \mathbf{f}_i \tag{1.14}$$

$$I\frac{d\boldsymbol{\omega}_i}{dt} = \mathbf{T}_i \tag{1.15}$$

Since the translational velocity $\mathbf{v}_i$ is related to the position vector $\mathbf{r}_i$ as $\mathbf{v}_i = d\mathbf{r}_i/dt$, we now consider the meaning of a quantity $\boldsymbol{\phi}_i$, which is related to the angular velocity $\boldsymbol{\omega}_i$ as $\boldsymbol{\omega}_i = d\boldsymbol{\phi}_i/dt$. It is assumed that during a short time interval Δt, $\boldsymbol{\phi}_i$ changes into $(\boldsymbol{\phi}_i + \Delta\boldsymbol{\phi}_i)$ where $\Delta\boldsymbol{\phi}_i$ is expressed as $\Delta\boldsymbol{\phi}_i = (\Delta\phi_{ix}, \Delta\phi_{iy}, \Delta\phi_{iz})$. As shown in Figure 1.1B, ω_z is related to the rotational angle in the xy-plane about the z-axis, $\Delta\phi_z$. The other components have the same meanings, so that $\boldsymbol{\phi}_i$ and $\boldsymbol{\omega}_i$ for particle i can be related in the following expression:

$$\Delta\boldsymbol{\phi}_i = \boldsymbol{\phi}_i(t + \Delta t) - \boldsymbol{\phi}_i(t) = \Delta t\boldsymbol{\omega}_i(t) \tag{1.16}$$

Is the use of the quantity $\boldsymbol{\phi}_i$, corresponding to $\mathbf{r}_i$, general? It seems to be more direct and more intuitive to use the unit vector $\mathbf{e}_i$ denoting the particle direction rather than the quantity $\boldsymbol{\phi}_i$. The change in $\mathbf{e}_i$ during an infinitesimal time interval, $\Delta\mathbf{e}_i$, can be written using the angular velocity $\boldsymbol{\omega}_i$ as

$$\Delta\mathbf{e}_i(t) = \mathbf{e}_i(t + \Delta t) - \mathbf{e}_i(t) = \Delta t\boldsymbol{\omega}_i(t) \times \mathbf{e}_i(t) \tag{1.17}$$

From Eqs. (1.16) and (1.17), $\mathbf{e}_i$ can be related to $\boldsymbol{\phi}_i$ as

$$\Delta\mathbf{e}_i(t) = \Delta\boldsymbol{\phi}_i(t) \times \mathbf{e}_i(t) \tag{1.18}$$

Equation (1.17) leads to the governing equation specifying the change of the particle direction:

$$\frac{d\mathbf{e}_i(t)}{dt} = \boldsymbol{\omega}_i(t) \times \mathbf{e}_i(t) \tag{1.19}$$

Hence, Eq. (1.15) for the angular velocity and Eq. (1.19) for the particle direction govern the rotational motion of an axisymmetric particle.

In order to solve Eqs. (1.15) and (1.19) for the rotational motion on a computer, these equations have to be translated into finite difference equations. To do so, as already explained, the first- and second-order differentials have to be expressed as algebraic expressions using the finite difference approximations based on Taylor series expansions. General finite difference expressions are as follows:

$$\left.\begin{aligned} &\frac{dx(t)}{dt} = \frac{x(t+\Delta t) - x(t)}{\Delta t} + O(\Delta t), \quad \frac{dx(t)}{dt} = \frac{x(t) - x(t-\Delta t)}{\Delta t} + O(\Delta t) \\ &\frac{dx(t)}{dt} = \frac{x(t+\Delta t) - x(t-\Delta t)}{2\Delta t} + O((\Delta t)^2) \end{aligned}\right\} \tag{1.20}$$

$$\frac{d^2x(t)}{dt^2} = \frac{x(t+\Delta t) - 2x(t) + x(t-\Delta t)}{(\Delta t)^2} + O((\Delta t)^2) \tag{1.21}$$

The simplest algorithm can be obtained using the forward finite difference approximation in Eq. (1.20) as

$$\left.\begin{aligned} \mathbf{e}_i(t+\Delta t) &= \mathbf{e}_i(t) + \Delta t \omega_i(t) \times \mathbf{e}_i(t) \\ \omega_i(t+\Delta t) &= \omega_i(t) + \Delta t \frac{\mathbf{T}_i(t)}{I} \end{aligned}\right\} \tag{1.22}$$

This algorithm is quite straightforward and understandable, but in practice does not have sufficient accuracy, since the error of the forward finite difference approximation is of the order of Δt. In order to improve the accuracy, the following algorithm has already been presented.

If the new vector function $\mathbf{u}_i(t)$ such as $\mathbf{u}_i\ (t) = \omega_i\ (t) \times \mathbf{e}_i\ (t)$ is introduced, Eq. (1.19) can be written as

$$\frac{d\mathbf{e}_i(t)}{dt} = \mathbf{u}_i(t) \tag{1.23}$$

By conducting the operator $\times \mathbf{e}$ from the right side on the both sides of Eq. (1.15), the following equation is obtained:

$$\frac{d\omega_i(t)}{dt} \times \mathbf{e}_i(t) = \frac{1}{I}\mathbf{T}_i(t) \times \mathbf{e}_i(t) \tag{1.24}$$

The left-hand side of this equation leads to

$$\frac{d\omega_i}{dt} \times \mathbf{e}_i = \frac{d(\omega_i \times \mathbf{e}_i)}{dt} - \omega_i \times \frac{d\mathbf{e}_i}{dt} = \frac{d\mathbf{u}_i}{dt} - \omega_i \times \mathbf{u}_i \tag{1.25}$$

By substituting this equation into Eq. (1.24), the following equation can be obtained:

$$\begin{aligned} \frac{d\mathbf{u}_i(t)}{dt} &= \frac{1}{I}\mathbf{T}_i(t) \times \mathbf{e}_i(t) + \omega_i(t) \times \mathbf{u}_i(t) = \frac{1}{I}\mathbf{T}_i(t) \times \mathbf{e}_i(t) - \left|\omega_i(t)\right|^2 \mathbf{e}_i(t) \\ &= \frac{1}{I}\mathbf{T}_i(t) \times \mathbf{e}_i(t) + \lambda_i(t)\mathbf{e}_i(t) \end{aligned} \tag{1.26}$$

In the transformation from the first to the second expressions on the right-hand side, we have used the identity $\mathbf{a} \times (\mathbf{b} \times \mathbf{c}) = (\mathbf{a} \cdot \mathbf{c})\mathbf{b} - (\mathbf{a} \cdot \mathbf{b})\mathbf{c}$ in evaluating $\omega \times (\omega \times \mathbf{e})$. The quantity $\lambda_i\ (t)$ in the third expression has been introduced in order to satisfy the following relationship:

$$\mathbf{e}_i \cdot \mathbf{u}_i = \mathbf{e}_i \cdot (\omega_i \times \mathbf{e}_i) = 0 \tag{1.27}$$

We have now completed the transformation of the variables from $\mathbf{e}_i$ and ω_i to $\mathbf{e}_i$ and $\mathbf{u}_i$ for solving the rotational motion of particles.

According to the leapfrog algorithm [15], Eqs. (1.23) and (1.26) reduce to the following algebraic equations:

$$\mathbf{e}_i(t + \Delta t) = \mathbf{e}_i(t) + \Delta t \mathbf{u}_i(t + \Delta t/2) \tag{1.28}$$

$$\mathbf{u}_i(t + \Delta t/2) = \mathbf{u}_i(t - \Delta t/2) + \Delta t \frac{\mathbf{T}_i(t) \times \mathbf{e}_i(t)}{I} + \Delta t \lambda_i(t) \mathbf{e}_i(t) \tag{1.29}$$

Another equation is necessary for determining the value of λ_i (t). The velocity $\mathbf{u}_i(t)$ can be evaluated from the arithmetic average of $\mathbf{u}_i(t + \Delta t/2)$ and $\mathbf{u}_i(t + \Delta t/2)$, and the expression is finally written using Eq. (1.29) as

$$\begin{aligned} \mathbf{u}_i(t) &= \frac{\mathbf{u}_i(t + \Delta t/2) + \mathbf{u}_i(t - \Delta t/2)}{2} \\ &= \mathbf{u}_i(t - \Delta t/2) + \frac{\Delta t}{2} \cdot \frac{\mathbf{T}_i(t) \times \mathbf{e}_i(t)}{I} + \frac{\Delta t}{2} \lambda_i(t) \mathbf{e}_i(t) \end{aligned} \tag{1.30}$$

Since $\mathbf{u}_i(t)$ has to satisfy the orthogonality condition shown in Eq. (1.27), the substitution of Eq. (1.30) into Eq. (1.27) leads to the equation of $\lambda_i(t)$ as

$$\lambda_i(t) = -\frac{2}{\Delta t} \cdot \mathbf{e}_i(t) \cdot \mathbf{u}_i(t - \Delta t/2) \tag{1.31}$$

In obtaining this expression, the identity $\mathbf{a} \cdot (\mathbf{b} \times \mathbf{a}) = 0$ has been used to evaluate $\mathbf{e} \cdot (\mathbf{T} \times \mathbf{e})$.

Now all the equations have been derived for determining the rotational motion of axisymmetric particles. With the value $\lambda_i(t)$ in Eq. (1.31), $\mathbf{u}_i$ at $(t + \Delta t/2)$ is first evaluated from Eq. (1.29), and then $\mathbf{e}_i$ at $(t + \Delta t)$ is obtained from Eq. (1.28). This procedure shows that the solution of $\mathbf{u}_i$ $(t + \Delta t/2)$ gives rise to the values of $\mathbf{e}_i(t + \Delta t)$ and $\mathbf{T}_i(t + \Delta t)$, and these solutions lead to $\mathbf{u}_i(t + 3\Delta t/2)$, and so forth. This algorithm is therefore another example of a leapfrog algorithm.

For the translational motion, the velocity Verlet algorithm may be used, and the particle position $\mathbf{r}_i(t + \Delta t)$ and velocity $\mathbf{v}_i(t + \Delta t)$ can be evaluated as

$$\left.\begin{aligned} \mathbf{r}_i(t + \Delta t) &= \mathbf{r}_i(t) + \Delta t \mathbf{v}_i(t) + \frac{(\Delta t)^2}{2m} \mathbf{f}_i(t) \\ \mathbf{v}_i(t + \Delta t) &= \mathbf{v}_i(t) + \frac{\Delta t}{2m} \left\{ \mathbf{f}_i(t) + \mathbf{f}_i(t + \Delta t) \right\} \end{aligned}\right\} \tag{1.32}$$

These equations can be derived in a straightforward manner from the finite difference approximations in Eqs. (1.20) and (1.21).

We have shown all the equations for specifying the translational and rotational motion of axisymmetric particles for the case of taking into account the inertia terms. The main procedure for conducting the MD simulation is as follows:

1. Specify the initial configuration and velocity of the axisymmetric particles for the translational and rotational motion.
2. Calculate the forces and torques acting on particles.
3. Evaluate the positions and velocities of the translational motion at $(t + \Delta t)$ from Eq. (1.32).
4. Evaluate $\lambda_i(t)$ $(i = 1, 2, \ldots, N)$ from Eq. (1.31).
5. Evaluate $\mathbf{u}_i$ $(i = 1, 2, \ldots, N)$ at $(t + \Delta t/2)$ from Eq. (1.29).
6. Evaluate the unit vectors $\mathbf{e}_i$ $(i = 1, 2, \ldots, N)$ at $(t + \Delta t)$ from Eq. (1.28).
7. Advance one time step to repeat the procedures from step 2.

By following this procedure, the MD method for axisymmetric particles with the inertia terms can simulate the positions and velocities, and the directions and angular velocities, at every time interval Δt.

1.1.2.2 Case of Neglected Inertia Terms

When treating a colloidal dispersion or a polymeric solution, the Stokesian dynamics and BD methods are usually employed as a microscopic or mesoscopic analysis tool. In these methods, dispersed particles or polymers are modeled as idealized spherical or dumbbell particles, but the base liquid is usually assumed to be a continuum medium and its effect is included in the equations of motion of the particles or the polymers only as friction terms. If particle size approximates to or is smaller than micron-order, the inertia terms may be considered as negligible. In this section, we treat this type of small particles and neglect the inertia terms. For the case of axisymmetric particles moving in a quiescent fluid, the translational and angular velocities of particle i, $\mathbf{v}_i$ and $\boldsymbol{\omega}_i$, are written as

$$\mathbf{v}_i = \frac{1}{\eta}\left\{\frac{1}{X^A}\mathbf{e}_i\mathbf{e}_i + \frac{1}{Y^A}(\mathbf{I} - \mathbf{e}_i\mathbf{e}_i)\right\}\cdot\mathbf{F}_i \tag{1.33}$$

$$\boldsymbol{\omega}_i = \frac{1}{\eta}\left\{\frac{1}{X^C}\mathbf{e}_i\mathbf{e}_i + \frac{1}{Y^C}(\mathbf{I} - \mathbf{e}_i\mathbf{e}_i)\right\}\cdot\mathbf{T}_i \tag{1.34}$$

in which X^A, Y^A, X^C, and Y^C are the resistance functions specifying the particle shape. If the long- and short-axis lengths are denoted by $2a$ and $2b$, respectively, and the eccentricity is denoted by s $(=(a^2 - b^2)^{1/2}/a)$, the resistance functions for the spheroidal particle are written as [16–18]

$$X^A = 6\pi a\cdot\frac{8}{3}\cdot\frac{s^3}{-2s + (1 + s^2)L}, \qquad Y^A = 6\pi a\cdot\frac{16}{3}\cdot\frac{s^3}{2s + (3s^2 - 1)L} \tag{1.35}$$

$$X^C = 8\pi a^3 \cdot \frac{4}{3} \cdot \frac{s^3(1-s^2)}{2s-(1-s^2)L}, \quad Y^C = 8\pi a^3 \cdot \frac{4}{3} \cdot \frac{s^3(2-s^2)}{-2s+(1+s^2)L} \tag{1.36}$$

in which L is a function of the eccentricity and is expressed as

$$L = L(s) = \ln\frac{1+s}{1-s} \tag{1.37}$$

For the case of $s \ll 1$, Eqs. (1.35) and (1.36) are approximated using Taylor series expansions as

$$X^A = 6\pi a\left(1 - \frac{2}{5}s^2 + \cdots\right), \quad Y^A = 6\pi a\left(1 - \frac{3}{10}s^2 + \cdots\right) \tag{1.38}$$

$$X^C = 8\pi a^3\left(1 - \frac{6}{5}s^2 + \cdots\right), \quad Y^C = 8\pi a^3\left(1 - \frac{9}{10}s^2 + \cdots\right) \tag{1.39}$$

In the limit of $s \to 0$, the well-known Stokes drag formula for a spherical particle in a quiescent fluid can be obtained from Eqs. (1.33), (1.34), (1.38), and (1.39):

$$\mathbf{v}_i = \frac{1}{6\pi\eta a}\mathbf{F}_i, \quad \boldsymbol{\omega}_i = \frac{1}{8\pi\eta a^3}\mathbf{T}_i \tag{1.40}$$

It is possible to pursue the motion of an axisymmetric particle using Eqs. (1.33) and (1.34), but further simplified equations can be used for the present axisymmetric particle. For an axisymmetric particle, the translational motion can be decomposed into the motion in the long axis direction and that in a direction normal to the particle axis. Similarly, the rotational motion can be decomposed into the rotation about the particle axis and that about a line normal to the particle axis through the mass center. If the force $\mathbf{F}_i$ acting on the particle is expressed as the sum of the force $\mathbf{F}_i^{||}$ parallel to the particle axis and the force $\mathbf{F}_i^{\perp}$ normal to that axis, then these forces can be expressed using the particle direction vector $\mathbf{e}_i$ as

$$\mathbf{F}_i^{||} = \mathbf{e}_i(\mathbf{e}_i \cdot \mathbf{F}_i) = \mathbf{e}_i\mathbf{e}_i \cdot \mathbf{F}_i, \quad \mathbf{F}_i^{\perp} = \mathbf{F}_i - \mathbf{F}_i^{||} = (\mathbf{I} - \mathbf{e}_i\mathbf{e}_i) \cdot \mathbf{F}_i \tag{1.41}$$

With these expressions, the velocities $\mathbf{v}_i^{||}$ and $\mathbf{v}_i^{\perp}$ parallel and normal to the particle axis, respectively, can be written from Eq. (1.33) as

$$\mathbf{v}_i^{||} = \frac{1}{\eta X^A}\mathbf{F}_i^{||}, \quad \mathbf{v}_i^{\perp} = \frac{1}{\eta Y^A}\mathbf{F}_i^{\perp} \tag{1.42}$$

Similarly, the angular velocities $\boldsymbol{\omega}_i^{||}$ and $\boldsymbol{\omega}_i^{\perp}$ about the long and short axes, respectively, are written from Eq. (1.34) as

$$\boldsymbol{\omega}_i^{||} = \frac{1}{\eta X^C}\mathbf{T}_i^{||}, \quad \boldsymbol{\omega}_i^{\perp} = \frac{1}{\eta Y^C}\mathbf{T}_i^{\perp} \tag{1.43}$$

According to Eqs. (1.42) and (1.43), $\mathbf{v}_i^{||}$, $\mathbf{v}_i^{\perp}$, $\boldsymbol{\omega}_i^{||}$, and $\boldsymbol{\omega}_i^{\perp}$ can be evaluated from values of $\mathbf{F}_i^{||}$, $\mathbf{F}_i^{\perp}$, $\mathbf{T}_i^{||}$, and $\mathbf{T}_i^{\perp}$. The translational velocity $\mathbf{v}_i$ and angular velocity $\boldsymbol{\omega}_i$ are then obtained as

$$\mathbf{v}_i = \mathbf{v}_i^{||} + \mathbf{v}_i^{\perp}, \quad \boldsymbol{\omega}_i = \boldsymbol{\omega}_i^{||} + \boldsymbol{\omega}_i^{\perp} \tag{1.44}$$

With the solutions of the translational and angular velocities at the time step t shown in Eq. (1.44), the position vector $\mathbf{r}_i$ and the particle direction vector $\mathbf{e}_i$ at the next time step $(t + \Delta t)$ can finally be obtained as

$$\mathbf{r}_i(t + \Delta t) = \mathbf{r}_i(t) + \Delta t\mathbf{v}_i(t) \tag{1.45}$$

$$\mathbf{e}_i(t + \Delta t) = \mathbf{e}_i(t) + \Delta t\omega_i(t) \times \mathbf{e}_i(t) \tag{1.46}$$

Lastly, we show the main procedure for the simulation in the following steps:

1. Specify the initial configuration and velocity of all axisymmetric particles for the translational and rotational motion.
2. Calculate all the forces and torques acting on particles.
3. Evaluate $\mathbf{F}_i^{||}$, $\mathbf{F}_i^{\perp}$, $\mathbf{T}_i^{||}$, and $\mathbf{T}_i^{\perp}$ $(i = 1, 2, \ldots, N)$ from Eq. (1.41) and similar equations for the torques.
4. Calculate $\mathbf{v}_i^{||}$, $\mathbf{v}_i^{\perp}$, $\boldsymbol{\omega}_i^{||}$, and $\boldsymbol{\omega}_i^{\perp}$ $(i = 1, 2, \ldots, N)$ from Eqs. (1.42) and (1.43).
5. Calculate $\mathbf{v}_i$ and $\boldsymbol{\omega}_i$ $(i = 1, 2, \ldots, N)$ from Eq. (1.44).
6. Calculate $\mathbf{r}_i$ and $\mathbf{e}_i$ $(i = 1, 2, \ldots, N)$ at the next time step $(t + \Delta t)$ from Eqs. (1.45) and (1.46).
7. Advance one time step and repeat the procedures from step 2.

1.2 Monte Carlo Method

In the MD method, the motion of molecules (particles) is simulated according to the equations of motion and therefore it is applicable to both thermodynamic equilibrium and nonequilibrium phenomena. In contrast, the MC method generates a series of microscopic states under a certain stochastic law, irrespective of the equations of motion of particles. Since the MC method does not use the equations of motion, it cannot include the concept of explicit time, and thus is only a simulation technique for phenomena in thermodynamic equilibrium. Hence, it is unsuitable for the MC method to deal with the dynamic properties of a system, which are dependent on time. In the following paragraphs, we explain important points of the concept of the MC method.

Figure 1.2 Typical energy situations for a two particle system.

How do microscopic states arise for thermodynamic equilibrium in a practical situation? We discuss this problem by considering a two-particle attractive system using Figure 1.2. As shown in Figure 1.2A, if the two particles overlap, then a repulsive force or a significant interaction energy arises. As shown in Figure 1.2B, for the case of close proximity, the interaction energy becomes low and an attractive force acts on the particles. If the two particles are sufficiently distant, as shown in Figure 1.2C, the interactive force is negligible and the interaction energy can be regarded as zero. In actual phenomena, microscopic states which induce a significantly high energy, as shown in Figure 1.2A, seldom appear, but microscopic states which give rise to a low-energy system, as shown in Figure 1.2B, frequently arise. However, this does not mean that only microscopic states that induce a minimum-energy system appear. Consider the fact that oxygen and nitrogen molecules do not gather in a limited area, but distribute uniformly in a room. It is seen from this discussion that, for thermodynamic equilibrium, microscopic states do not give rise to a minimum of the total system energy, but to a minimum free energy of a system. For example, in the case of a system specified by the number of particles N, temperature T, and volume of the system V, microscopic states arise such that the following Helmholtz free energy F becomes a minimum:

$$F = E - TS \tag{1.47}$$

in which E is the potential energy of the system, and S is the entropy. In the preceding example, the reason why oxygen or nitrogen molecules do not gather in a limited area can be explained by taking into account the entropy term on the right-hand side in Eq. (1.47). That is, the situation in which molecules do not gather together and form flocks but expand to fill a room gives rise to a large value of the entropy. Hence, according to the counterbalance relationship of the energy and the entropy, real microscopic states arise such that the free energy of a system is at minimum.

Next, we consider how microscopic states arise stochastically. We here treat a system composed of N interacting spherical particles with temperature T and volume V of the system; these quantities are given values and assumed to be constant. If the position vector of an arbitrary particle i ($i = 1, 2, \ldots, N$) is denoted by $\mathbf{r}_i$, then the total interaction energy U of the system can be expressed as a function of the particle positions; that is, it can be expressed as $U = U(\mathbf{r}_1, \mathbf{r}_2, \ldots, \mathbf{r}_N)$. For the present system specified by given values of N, T, and V, the appearance of a microscopic state that the particle i ($i = 1, 2, \ldots, N$) exits within the small range of $\mathbf{r}_i \sim (\mathbf{r}_i + \Delta\mathbf{r}_i)$ is governed by the probability density function $\rho(\mathbf{r}_1, \mathbf{r}_2, \ldots, \mathbf{r}_N)$. This can be expressed from statistical mechanics [19,20] as

$$\rho(\mathbf{r}_1, \mathbf{r}_2, \ldots, \mathbf{r}_N) = \frac{\exp\{-U(\mathbf{r}_1, \mathbf{r}_2, \ldots, \mathbf{r}_N)/kT\}}{\int_V \cdots \int_V \exp\{-U(\mathbf{r}_1, \mathbf{r}_2, \ldots, \mathbf{r}_N)/kT\} d\mathbf{r}_1\, d\mathbf{r}_2 \ldots d\mathbf{r}_N} \tag{1.48}$$

If a series of microscopic states is generated with an occurrence according to this probability, a simulation may have physical meaning. However, this approach is impracticable, as it is extraordinarily difficult and almost impossible to evaluate analytically the definite integral of the denominator in Eq. (1.48). In fact, if we were able to evaluate this integral term analytically, we would not need a computer simulation because it would be possible to evaluate almost all physical quantities analytically.

The "Metropolis method" [21] overcomes this difficulty for MC simulations. In the Metropolis method, the transition probability from microscopic states i to j, p_{ij}, is expressed as

$$p_{ij} = \begin{cases} 1 & (\text{for } \rho_j/\rho_i \geq 1) \\ \dfrac{\rho_j}{\rho_i} & (\text{for } \rho_j/\rho_i < 1) \end{cases} \tag{1.49}$$

in which ρ_j and ρ_i are the probability density functions for microscopic states j and i appearing, respectively. The ratio of ρ_j/ρ_i is obtained from Eq. (1.48) as

$$\begin{aligned} \frac{\rho_j}{\rho_i} &= \exp\left\{-\frac{1}{kT}(U_j - U_i)\right\} \\ &= \exp\left[-\frac{1}{kT}\left\{U(\mathbf{r}_1{}^j, \mathbf{r}_2{}^j, \ldots, \mathbf{r}_N{}^j) - U(\mathbf{r}_1{}^i, \mathbf{r}_2{}^i, \ldots, \mathbf{r}_N{}^i)\right\}\right] \end{aligned} \tag{1.50}$$

In the above equations, U_i and U_j are the interaction energies of microscopic states i and j, respectively. The superscripts attached to the position vectors denote the same meanings concerning microscopic states. Eq. (1.49) implies that, in the transition from microscopic states i to j, new microscopic state j is adopted if the system energy decreases, with the probability ρ_j/ρ_i (<1) if the energy increases. As clearly demonstrated by Eq. (1.50), for ρ_j/ρ_i the denominator in Eq. (1.48) is not required in Eq. (1.50), because ρ_j is divided by ρ_i and the term is canceled through this operation. This is the main reason for the great success of the Metropolis method for MC simulations. That a new microscopic state is adopted with the probability ρ_j/ρ_i, even in the case of the increase in the interaction energy, verifies the accomplishment of the minimum free-energy condition for the system. In other words, the adoption of microscopic states, yielding an increase in the system energy, corresponds to an increase in the entropy.

The above discussion is directly applicable to a system composed of nonspherical particles. The situation of nonspherical particles in thermodynamic equilibrium can be specified by the particle position of the mass center, $\mathbf{r}_i(i=1, 2, \ldots, N)$, and the unit vector $\mathbf{e}_i(i=1, 2, \ldots, N)$ denoting the particle direction. The transition probability from microscopic states i to j, p_{ij} can be written in similar form to Eq. (1.49). The exact expression of ρ_j/ρ_i becomes

$$\frac{\rho_j}{\rho_i}=\exp\left\{-\frac{1}{kT}(U_j-U_i)\right\}=\exp\left[-\frac{1}{kT}\left\{U(\mathbf{r}_1^j,\mathbf{r}_2^j,\mathbf{r}_N^j,\mathbf{e}_1^j,\mathbf{e}_2^j,\ldots,\mathbf{e}_N^j)\right.\right.$$
$$\left.\left.-U(\mathbf{r}_1^j,\mathbf{r}_2^j,\mathbf{r}_N^j,\mathbf{e}_1^j,\mathbf{e}_2^j,\ldots,\mathbf{e}_N^j)\right\}\right] \tag{1.51}$$

The main procedure for the MC simulation of a nonspherical particle system is as follows:

1. Specify the initial position and direction of all particles.
2. Regard this state as microscopic state i, and calculate the interaction energy U_i.
3. Choose an arbitrary particle in order or randomly and call this particle "particle α."
4. Make particle α move translationally using random numbers and calculate the interaction energy U_j for this new configuration.
5. Adopt this new microscopic state for the case of $U_j \le U_i$ and go to step 7.
6. Calculate ρ_j/ρ_i in Eq. (1.51) for the case of $U_j > U_i$ and take a random number R_1 from a uniform random number sequence distributed from zero to unity.
   6.1. If $R_1 \le \rho_j/\rho_i$, adopt this microscopic state j and go to step 7.
   6.2. If $R_1 > \rho_j/\rho_i$, reject this microscopic state, regard previous state i as new microscopic state j, and go to step 7.
7. Change the direction of particle α using random numbers and calculate the interaction energy U_k for this new state.
8. If $U_k \le U_j$, adopt this new microscopic state and repeat from step 2.
9. If $U_k > U_j$, calculate ρ_k/ρ_j in Eq. (1.51) and take a random number R_2 from the uniform random number sequence.
   9.1. If $R_2 \le \rho_k/\rho_j$, adopt this new microscopic state k and repeat from step 2.
   9.2. If $R_2 > \rho_k/\rho_j$, reject this new state, regard previous state j as new microscopic state k, and repeat from step 2.

Although the treatment of the translational and rotational changes is carried out separately in the above algorithm, a simultaneous procedure is also possible in such a way that the position and direction of an arbitrary particle are simultaneously changed, and the new microscopic state is adopted according to the condition in Eq. (1.49). However, for a strongly interacting system, the separate treatment may be found to be more effective in many cases.

We will now briefly explain how the translational move is made using random numbers during a simulation. If the position vector of an arbitrary particle α in microscopic state i is denoted by $\mathbf{r}_\alpha=(x_\alpha, y_\alpha, z_\alpha)$, this particle is moved to a new position $\mathbf{r}'_\alpha=(x'_\alpha, y'_\alpha, z'_\alpha)$ by the following equations using random numbers R_1, R_2, and R_3, taken from a random number sequence ranged from zero to unity:

$$\left.\begin{aligned} x'_\alpha &= x_\alpha + R_1\delta r_{\max} \\ y'_\alpha &= y_\alpha + R_2\delta r_{\max} \\ z'_\alpha &= z_\alpha + R_3\delta r_{\max} \end{aligned}\right\} \tag{1.52}$$

These equations imply that the particle is moved to an arbitrary position, determined by random numbers, within a cube centered at the particle center with side length of $2\delta r_{\max}$. A series of microscopic states is generated by moving the particles according to the above-mentioned procedure.

Finally, we show the method of evaluating the average of a physical quantity in MC simulations. These averages, called "ensemble averages," are different from the time averages that are obtained from MD simulations. If a physical quantity A is a function of the microscopic states of a system, and A_n is the nth sampled value of this quantity in an MC simulation, then the ensemble average $\langle A\rangle$ can be evaluated from the equation

$$\langle A\rangle = \sum_{n=1}^{M} A_n/M \tag{1.53}$$

in which M is the total sampling number. In actual simulations, the sampling procedure is not conducted at each time step but at regular intervals. This may be more efficient because if the data have significant correlations they are less likely to be sampled by taking a longer interval for the sampling time. The ensemble averages obtained in this way may be compared with experimental data.

1.3 Brownian Dynamics Method

A dispersion or suspension composed of fine particles dispersed in a base liquid is a difficult case to be treated by simulations in terms of the MD method, because the characteristic time of the motion of the solvent molecules is considerably different from that of the dispersed particles. Simply speaking, if we observe such a dispersion based on the characteristic time of the solvent molecules, we can see only the active motion of solvent molecules around the quiescent dispersed particles. Clearly the MD method is quite unrealistic as a simulation technique for particle dispersions. One approach to overcome this difficulty is to not focus on the motion of each solvent molecule, but regard the solvent molecules as a continuum medium and consider the motion of dispersed particles in such a medium. In this approach, the influence of the solvent molecules is included into the equations of motion of the particles as random forces. We can observe such random motion when pollen moves at a liquid surface or when dispersed particles move in a functional fluid such as a ferrofluid. The BD method simulates the random motion of dispersed particles that is induced by the solvent molecules; thus, such particles are called "Brownian particles."

If a particle dispersion is so significantly dilute that each particle can be regarded as moving independently, the motion of this Brownian particle is governed by the following Langevin equation [22]:

$$m\frac{d\mathbf{v}}{dt} = \mathbf{f} - \xi\mathbf{v} + \mathbf{f}^{B} \tag{1.54}$$

This equation is valid for a spherical particle dispersion. In Eq. (1.54), m is the mass of a spherical particle, $\mathbf{v}$ is the velocity vector, ξ is the friction coefficient and is expressed as $\xi = 3\pi\eta d$ for the particle diameter d with the viscosity η of a base liquid, $\mathbf{f}$ is the force exerted by an external field, and $\mathbf{f}^{B}$ $(=(f_x^B, f_y^B, f_z^B))$ is the random force due to the motion of solvent molecules. This random force has the following stochastic properties:

$$\langle f_x^B(t)\rangle = \langle f_y^B(t)\rangle = \langle f_z^B(t)\rangle = 0 \tag{1.55}$$

$$\left\langle \{f_x^B(t)\}^2\right\rangle = \left\langle \{f_y^B(t)\}^2\right\rangle = \left\langle \{f_z^B(t)\}^2\right\rangle = 2\xi kT\delta(t - t') \tag{1.56}$$

in which $\delta(t - t')$ is the Dirac delta function. In Eq. (1.56) larger random forces act on Brownian particles at a higher temperature because the mean square average of each component of the random force is in proportion to the system temperature. At a higher temperature the solvent molecules move more actively and induce larger random forces.

In order to simulate the Brownian motion of particles, the basic equation in Eq. (1.54) has to be transformed into an algebraic equation, as in the MD method. If the time interval h is sufficiently short such that the change in the forces is negligible, Eq. (1.54) can be regarded as a simple first-order differential equation. Hence, Eq. (1.54) can be solved by standard textbook methods of differential equations [23], and algebraic equations can finally be obtained as

$$\begin{aligned}\mathbf{r}(t+h) = \mathbf{r}(t) &+ \frac{m}{\xi}\mathbf{v}(t)\left\{1 - \exp\left(-\frac{\xi}{m}h\right)\right\}\\ &+ \frac{1}{\xi}\mathbf{f}(t)\left\{h - \frac{m}{\xi}\left(1 - \exp\left(-\frac{\xi}{m}h\right)\right)\right\} + \Delta\mathbf{r}^{B}\end{aligned} \tag{1.57}$$

$$\mathbf{v}(t+h) = \mathbf{v}(t)\exp\left(-\frac{\xi}{m}h\right) + \frac{1}{\xi}\mathbf{f}(t)\left(1 - \exp\left(-\frac{\xi}{m}h\right)\right) + \Delta\mathbf{v}^{B} \tag{1.58}$$

in which $\Delta\mathbf{r}^{B}$ and $\Delta\mathbf{v}^{B}$ are a random displacement and velocity due to the motion of solvent molecules. The relationship of the x-components of $\Delta\mathbf{r}^{B}$ and $\Delta\mathbf{v}^{B}$ can be expressed as a two-dimensional normal distribution (similarly for the other components). We do not show such an expression here [4], but instead consider a method that is superior in regard to the extension of the BD method to the case with multibody hydrodynamic interactions. The BD method based on Eqs. (1.57) and (1.58) is applicable to physical phenomena in which the inertia term is a governing factor.

Since the BD method with multibody hydrodynamic interactions among the particles is very complicated, we here focus on an alternative method that treats the friction forces between the particles and a base liquid, and the nonhydrodynamic interactions between the particles. This simpler type of simulation method is sometimes used as a first-order approximation because of the complexity of treating hydrodynamic interactions. A representative nonhydrodynamic force is the magnetic force influencing the magnetic particles in a ferrofluid.

Although the BD method based on the Ermak–McCammon analysis [24] takes into account multibody hydrodynamic interactions among particles, we apply this analysis method to the present dilute dispersion without hydrodynamic interactions, and can derive the basic equation of the position vector $\mathbf{r}_i (i = 1, 2, \ldots, N)$ of Brownian particle i as

$$\mathbf{r}_i(t+h) = \mathbf{r}_i(t) + \frac{1}{\xi} h \mathbf{f}_i(t) + \Delta \mathbf{r}_i^{\mathrm{B}} \tag{1.59}$$

in which the components $(\Delta x_i^{\mathrm{B}}, \Delta y_i^{\mathrm{B}}, \Delta z_i^{\mathrm{B}})$ of the random displacement $\Delta \mathbf{r}_i^{\mathrm{B}}$ have to satisfy the following stochastic properties:

$$\langle \Delta x_i^{\mathrm{B}} \rangle = \langle \Delta y_i^{\mathrm{B}} \rangle = \langle \Delta z_i^{\mathrm{B}} \rangle = 0 \tag{1.60}$$

$$\left\langle (\Delta x_i^{\mathrm{B}})^2 \right\rangle = \left\langle (\Delta y_i^{\mathrm{B}})^2 \right\rangle = \left\langle (\Delta z_i^{\mathrm{B}})^2 \right\rangle = \frac{2kT}{\xi} h \tag{1.61}$$

Equations similar to Eq. (1.59) hold for every particle in the system. Interactions among particles arise through the force $\mathbf{f}_i (i = 1, 2, \ldots, N)$ acting on them.

If a Brownian particle exhibits magnetic properties and has, for example, a magnetic dipole moment at the particle center, it will have a tendency to incline in the direction of an applied magnetic field. Hence, even in the case of spherical particles, the rotational motion is influenced by an external field, so that both the translational and the rotational motion of a particle are treated simultaneously in simulations.

If the unit vector of the particle direction is denoted by $\mathbf{n}_i$, the equation of the change in $\mathbf{n}_i$ can be derived under the same conditions assumed in deriving Eq. (1.59) as

$$\mathbf{n}_i(t+h) = \mathbf{n}_i(t) + \frac{1}{\xi_R} h \mathbf{T}_i(t) \times \mathbf{n}_i(t) + \Delta \mathbf{n}_i^{\mathrm{B}} \tag{1.62}$$

in which ξ_R is the friction coefficient of the rotational motion, expressed as $\xi_R = \pi\eta d^3$, and $\mathbf{T}_i$ is the torque acting on particle i by nonhydrodynamic forces. Also, $\Delta\mathbf{n}_i^B$ is the rotational displacement due to random forces, expressed as

$$\Delta\mathbf{n}_i^B = \Delta\phi_{\perp 1}^B\mathbf{n}_{\perp 1} + \Delta\phi_{\perp 2}^B\mathbf{n}_{\perp 2} \tag{1.63}$$

in which $\mathbf{n}_{\perp 1}$ and $\mathbf{n}_{\perp 2}$ are a set of unit vectors normal to the direction of particle i, and $\Delta\phi_{\perp 1}^B$ and $\Delta\phi_{\perp 2}^B$ have the following stochastic properties:

$$\langle\Delta\phi_{\perp 1}^B\rangle = \langle\Delta\phi_{\perp 2}^B\rangle = 0 \tag{1.64}$$

$$\left\langle(\Delta\phi_{\perp 1}^B)^2\right\rangle = \left\langle(\Delta\phi_{\perp 2}^B)^2\right\rangle = \frac{2kT}{\xi_R}h \tag{1.65}$$

Now consider the correspondence of quantities in the translational and rotational motion. The velocity $\mathbf{v}_i$ in the translational motion corresponds to the angular velocity $\boldsymbol{\omega}_i$ in the rotational motion, and the position vector $\mathbf{r}_i$ corresponds to the quantity $\boldsymbol{\phi}_i$ defined as $d\boldsymbol{\phi}_i/dt = \boldsymbol{\omega}_i$. Obviously, due to the similarity of Eqs. (1.64) and (1.65) to Eqs. (1.60) and (1.61), the components $\Delta\phi_{\perp 1}^B$ and $\Delta\phi_{\perp 2}^B$ of the vector $\Delta\phi^B$ have to satisfy Eqs. (1.64) and (1.65).

The basic Eqs. (1.59) and (1.62) for governing the translational and rotational motion of particles have been derived under the assumptions that the momentum of particles is sufficiently relaxed during the time interval h and that the force acting on the particles is substantially constant during this infinitesimally short time. This is the essence of the Ermak–McCammon method for BD simulations.

Next, we show the method of generating random displacements according to Eqs. (1.60) and (1.61), but, before that, the normal probability distribution needs to be briefly described. If the behavior of a stochastic variable is described by the normal distribution $\rho_{\text{normal}}(x)$ with variance σ^2, $\rho_{\text{normal}}(x)$ is written as

$$\rho_{\text{normal}}(x) = \frac{1}{(2\pi\sigma^2)^{1/2}}\exp(-x^2/2\sigma^2) \tag{1.66}$$

in which the variance σ^2 is a measure of how wide the stochastic variable x is distributed around the mean value $\langle x\rangle$, which is taken as zero for this discussion. The variance σ^2 is mathematically defined as

$$\sigma^2 = \langle(x - \langle x\rangle)^2\rangle = \langle x^2\rangle - (\langle x\rangle)^2 \tag{1.67}$$

If Eq. (1.66) is applied to Eqs. (1.60) and (1.61), the random displacement Δx_i^B in the x-direction can be written in normal distribution form as

$$\rho_{\text{normal}}(\Delta x_i^B) = \left(\frac{\xi}{4\pi kTh}\right)^{1/2}\exp\left\{-\frac{\xi}{4kTh}(\Delta x_i^B)^2\right\} \tag{1.68}$$

The other components also obey a normal distribution. As seen in Eq. (1.68), larger random displacements tend to arise at a higher system temperature, which makes sense given that solvent molecules move more actively in the higher temperature case. The random displacements can therefore be generated by sampling according to the normal distributions shown in Eq. (1.68). An example of generating random displacements is shown in Appendix A2.

The main procedure for conducting the BD simulation based on Eqs. (1.59), (1.60), and (1.61) is:

1. Specify the initial position of all particles.
2. Calculate the forces acting on each particle.
3. Generate the random displacements $\Delta\mathbf{r}_i^B = (\Delta x_i^B, \Delta y_i^B, \Delta z_i^B)$ $(i = 1, 2, \ldots, N)$ using uniform random numbers: for example, Δx_i^B is sampled according to Eq. (1.68).
4. Calculate all the particle positions at the next time step from Eq. (1.59).
5. Return to step 2 and repeat.

The physical quantities of interest are evaluated by the time average, similar to the molecular dynamics method.

1.4 Dissipative Particle Dynamics Method

As already pointed out, it is not realistic to use the MD method to simulate the motion of solvent molecules and dispersed particles simultaneously, since the characteristic time of solvent molecules is much shorter than that of dispersed particles. Hence, in the BD method, the motion of solvent molecules is not treated, but a fluid is regarded as a continuum medium. The influence of the molecular motion is combined into the equations of motion of dispersed particles as stochastic random forces. Are there any simulation methods to simulate the motion of both the solvent molecules and the dispersed particles? As far as we treat the motion of real solvent molecules, the development of such simulation methods may be impractical. However, if groups or clusters of solvent molecules are regarded as virtual fluid particles, such that the characteristic time of the motion of such fluid particles is not so different from that of dispersed particles, then it is possible to simulate the motion of the dispersed and the fluid particles simultaneously. These virtual fluid particles are expected to exchange their momentum, exhibit a random motion similar to Brownian particles, and interact with each other by particle–particle potentials. We call these virtual fluid particles "dissipative particles," and the simulation technique of treating the motion of dissipative particles instead of the solvent molecules is called the "dissipative particle dynamics (DPD) method" [4–8].

The DPD method is principally applicable to simulations of colloidal dispersions that take into account the multibody hydrodynamic interactions among particles. For colloidal dispersions, the combination of the flow field solutions for a three- or four-particle system into a simulation technique enables us to address the physical situation of multibody hydrodynamic interactions as accurately as possible. However, it is extraordinarily difficult to solve analytically the flow field even for

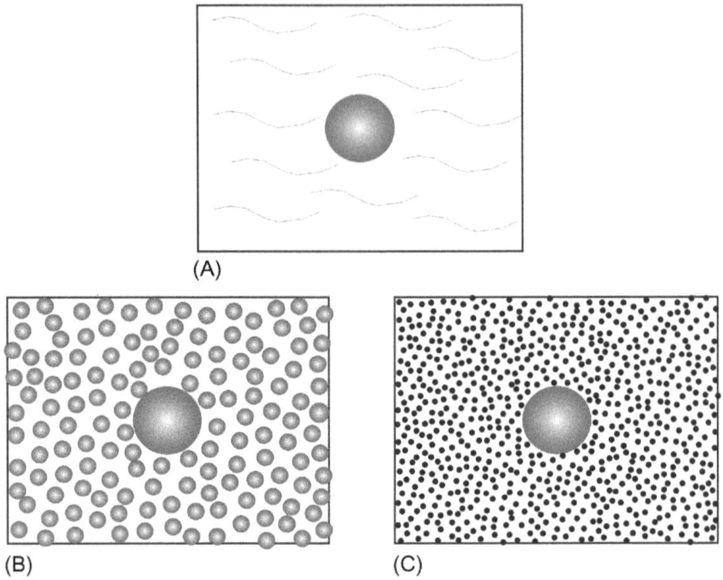

Figure 1.3 Modeling of a fluid: (A) the macroscopic model, (B) the mesoscopic model, and (C) the microscopic model.

a three-particle system, so a solution for a nonspherical particle system is futile to attempt. In contrast, the DPD method does not require this type of solution of the flow field in conducting simulations of colloidal dispersions that take into account multibody hydrodynamic effects. This is because they are automatically reproduced from consideration of the interactions between the dissipative and the colloidal particles. This approach to the hydrodynamic interactions is a great advantage of the DPD method. In addition, this method is applicable to nonspherical particle dispersions, and a good simulation technique for colloidal dispersions.

We will show the general categories of models employed in the modeling of a fluid for numerical simulations before proceeding to the explanation of the DPD method. Figure 1.3 schematically shows the classification of the modeling of a fluid. Figure 1.3A shows a continuum medium model for a fluid. In this case, a solution of a flow field can be obtained by solving the Navier–Stokes equations, which are the governing equations of the motion of a fluid. Figure 1.3C shows a microscopic model in which the solvent molecules are treated and a solution of the flow field can be obtained by pursuing the motion of the solvent molecules: this is the MD approach. Figure 1.3B shows a mesoscopic model in which a fluid is assumed to be composed of virtual fluid particles: the DPD method is classified within this category.

In the following paragraphs, we discuss the equations of motion of the dissipative particles for a system composed of dissipative particles alone, without colloidal particles. For simplification's sake, dissipative particles are simply called "particles" unless specifically identified.

In order that the solution of a flow field obtained from the particle motion agrees with that of the Navier–Stokes equations, the equations of motion of the particles have to be formalized in physically viable form. For example, as a physical restriction on the system behavior, the total momentum of a system should be conserved. The forces acting on particle i possibly seem to be a conservative force $\mathbf{F}_{ij}^{\mathrm{C}}$, exerted by other particles (particle j in this case); a dissipative force $\mathbf{F}_{ij}^{\mathrm{D}}$, due to the exchange of momentum; and a random force $\mathbf{F}_{ij}^{\mathrm{R}}$, inducing the random motion of particles. With the particle mass m and the particle velocity $\mathbf{v}_i$, the equation of motion can be written as

$$m\frac{\mathrm{d}\mathbf{v}_i}{\mathrm{d}t} = \sum_{j(\neq i)} \mathbf{F}_{ij}^{\mathrm{C}} + \sum_{j(\neq i)} \mathbf{F}_{ij}^{\mathrm{D}} + \sum_{j(\neq i)} \mathbf{F}_{ij}^{\mathrm{R}} \tag{1.69}$$

The subscripts in Eq. (1.69), for example in $\mathbf{F}_{ij}^{\mathrm{C}}$, represent the force acting on particle i by particle j. Now, we embody specific expressions for each force. Since $\mathbf{F}_{ij}^{\mathrm{C}}$ is a conservative force between particles i and j, it is assumed to be dependent on the relative position $\mathbf{r}_{ij}$ $(=\mathbf{r}_i - \mathbf{r}_j)$ alone, not on velocities. This specific expression will be shown later. $\mathbf{F}_{ij}^{\mathrm{D}}$ and $\mathbf{F}_{ij}^{\mathrm{R}}$ have to be conserved under a Galilean transformation (refer to a textbook of mechanics); thus, they must be independent of $\mathbf{r}_i$ and $\mathbf{v}_i$ in a given reference frame (quantities dependent on $\mathbf{r}_i$ and $\mathbf{v}_i$ are not conserved), but should be functions of the relative position vector $\mathbf{r}_{ij}$ and relative velocity vector $\mathbf{v}_{ij}$ $(=\mathbf{v}_i - \mathbf{v}_j)$. Furthermore, it is physically reasonable to assume that $\mathbf{F}_{ij}^{\mathrm{R}}$ is dependent only on the relative position $\mathbf{r}_{ij}$, and not on the relative velocity $\mathbf{v}_{ij}$. We also have to take into account that the particle motion is isotropic and the forces between particles decrease with the particle–particle separation. The following expressions for $\mathbf{F}_{ij}^{\mathrm{D}}$ and $\mathbf{F}_{ij}^{\mathrm{R}}$ satisfy all the above-mentioned requirements:

$$\mathbf{F}_{ij}^{\mathrm{D}} = -\gamma w_{\mathrm{D}}(r_{ij})(\mathbf{e}_{ij}\cdot\mathbf{v}_{ij})\mathbf{e}_{ij} \tag{1.70}$$

$$\mathbf{F}_{ij}^{\mathrm{R}} = \sigma w_{\mathrm{R}}(r_{ij})\mathbf{e}_{ij}\zeta_{ij} \tag{1.71}$$

in which $r_{ij} = |\mathbf{r}_{ij}|$, and $\mathbf{e}_{ij}$ is the unit vector denoting the direction of a line drawn from particles j to i, expressed as $\mathbf{e}_{ij} = \mathbf{r}_{ij}/r_{ij}$. The ζ_{ij} is the stochastic variable inducing the random motion of particles and has the following characteristics:

$$\langle \zeta_{ij} \rangle = 0, \quad \langle \zeta_{ij}(t)\zeta_{i'j'}(t') \rangle = (\delta_{ii'}\delta_{jj'} + \delta_{ij'}\delta_{ji'})\delta(t-t') \tag{1.72}$$

in which δ_{ij} is the Kronecker delta, and $\delta_{ij} = 1$ for $i = j$ and $\delta_{ij} = 0$ for the other cases. Since this variable satisfies the equation of $\zeta_{ij} = \zeta_{ji}$, the total momentum of a system is conserved. The $w_{\mathrm{D}}(r_{ij})$ and $w_{\mathrm{R}}(r_{ij})$ are weighting functions representing the characteristics of forces decreasing with the particle–particle separation, and γ and σ are constants specifying the strengths of the corresponding forces. As shown later, these constants are related to the system temperature and friction coefficients. The $\mathbf{F}_{ij}^{\mathrm{D}}$ acts such that the relative motion of particles i and j relaxes, and $\mathbf{F}_{ij}^{\mathrm{R}}$ functions such that the thermal motion is activated. Since the action–reaction law is satisfied by $\mathbf{F}_{ij}^{\mathrm{R}}$, the conservation of the total momentum is not violated by $\mathbf{F}_{ij}^{\mathrm{R}}$.

By substituting Eqs. (1.70) and (1.71) into Eq. (1.69), the equation of motion of particles can be written as

$$m\frac{\mathrm{d}\mathbf{v}_i}{\mathrm{d}t} = \sum_{j(\neq i)} \mathbf{F}_{ij}^{\mathrm{C}}(\mathbf{r}_{ij}) - \sum_{j(\neq i)} \gamma w_{\mathrm{D}}(r_{ij})(\mathbf{e}_{ij}\cdot\mathbf{v}_{ij})\mathbf{e}_{ij} + \sum_{j(\neq i)} \sigma w_{\mathrm{R}}(r_{ij})\mathbf{e}_{ij}\zeta_{ij} \tag{1.73}$$

The integral of this equation with respect to the time from t to $(t+\Delta t)$ leads to the finite difference equations specifying the motion of the simulation particles:

$$\Delta\mathbf{r}_i = \mathbf{v}_i\Delta t \tag{1.74}$$

$$\Delta\mathbf{v}_i = \frac{1}{m}\left(\sum_{j(\neq i)} \mathbf{F}_{ij}^{\mathrm{C}}(\mathbf{r}_{ij}) - \sum_{j(\neq i)} \gamma w_{\mathrm{D}}(r_{ij})(\mathbf{e}_{ij}\cdot\mathbf{v}_{ij})\mathbf{e}_{ij}\right)\Delta t + \frac{1}{m}\sum_{j(\neq i)} \sigma w_{\mathrm{R}}(r_{ij})\mathbf{e}_{ij}\Delta W_{ij} \tag{1.75}$$

The ΔW_{ij} has to satisfy the following stochastic properties, which can be obtained from Eq. (1.72):

$$\left.\begin{aligned} &\langle \Delta W_{ij}\rangle = 0 \\ &\langle \Delta W_{ij}\Delta W_{i'j'}\rangle = (\delta_{ii'}\delta_{jj'} + \delta_{ij'}\delta_{ji'})\Delta t \end{aligned}\right\} \tag{1.76}$$

If a new stochastic variable θ_{ij} is introduced from $\Delta W_{ij} = \theta_{ij}(\Delta t)^{1/2}$, the third term in Eq. (1.75) can be written as

$$\frac{1}{m}\sum_{j(\neq i)} \sigma w_{\mathrm{R}}(r_{ij})\mathbf{e}_{ij}\theta_{ij}\sqrt{\Delta t} \tag{1.77}$$

in which θ_{ij} has to satisfy the following stochastic characteristics:

$$\left.\begin{aligned} &\langle \theta_{ij}\rangle = 0 \\ &\langle \theta_{ij}\theta_{i'j'}\rangle = (\delta_{ii'}\delta_{jj'} + \delta_{ij'}\delta_{ji'}) \end{aligned}\right\} \tag{1.78}$$

In simulations, values of the stochastic variable are sampled from a normal distribution with zero-mean value and unit variance or from a uniform distribution.

The constants γ and σ and the weighting functions $w_{\mathrm{D}}(r_{ij})$ and $w_{\mathrm{R}}(r_{ij})$, which appeared in Eq. (1.75), must satisfy the following relationships:

$$\left.\begin{aligned} &w_{\mathrm{D}}(r_{ij}) = w_{\mathrm{R}}^2(r_{ij}) \\ &\sigma^2 = 2\gamma kT \end{aligned}\right\} \tag{1.79}$$

The second equation is called the "fluctuation–dissipation theorem." These relationships ensure a valid equilibrium distribution of particle velocities for thermodynamic equilibrium.

Next, we show expressions for the conservative force $\mathbf{F}_{ij}^{\mathrm{C}}$ and the weighting function $w_{\mathrm{R}}(r_{ij})$. The $\mathbf{F}_{ij}^{\mathrm{C}}$ functions as a tool for preventing particles from significantly overlapping, so that the value of $w_{\mathrm{R}}(r_{ij})$ has to increase with particles i and j approaching each other. Given this consideration, these expressions may be written as

$$\mathbf{F}_{ij}^{\mathrm{C}} = \alpha w_{\mathrm{R}}(r_{ij})\mathbf{e}_{ij} \tag{1.80}$$

$$w_{\mathrm{R}}(r_{ij}) = \begin{cases} 1 - \dfrac{r_{ij}}{r_c} & \text{for } r_{ij} \le r_c \\ 0 & \text{for } r_{ij} > r_c \end{cases} \tag{1.81}$$

in which α is a constant representing the strength of a repulsive force. By substituting the above-mentioned expressions into Eq. (1.75) and taking into account Eq. (1.77), the final expressions for the equations of motion of particles can be obtained as

$$\Delta \mathbf{r}_i = \mathbf{v}_i \Delta t \tag{1.82}$$

$$\begin{aligned} \Delta \mathbf{v}_i = {} & \frac{\alpha}{m}\sum_{j(\neq i)} w_{\mathrm{R}}(r_{ij})\mathbf{e}_{ij}\Delta t - \frac{\gamma}{m}\sum_{j(\neq i)} w_{\mathrm{R}}^2(r_{ij})(\mathbf{e}_{ij}\cdot\mathbf{v}_{ij})\mathbf{e}_{ij}\Delta t \\ & + \frac{(2\gamma kT)^{1/2}}{m}\sum_{j(\neq i)} w_{\mathrm{R}}(r_{ij})\mathbf{e}_{ij}\theta_{ij}\sqrt{\Delta t} \end{aligned} \tag{1.83}$$

As previously indicated, θ_{ij} satisfies the stochastic characteristics in Eq. (1.78) and is sampled from a normal distribution or from a uniform distribution. The DPD dynamics method simulates the motion of the dissipative particles according to Eqs. (1.82) and (1.83).

For actual simulations, we show the method of nondimensionalizing quantities. The following representative values are used for nondimensionalization: $(kT/m)^{1/2}$ for velocities, r_c for distances, $r_c(m/kT)^{1/2}$ for time, $(1/r_c^3)$ for number densities. Using these representative values, Eqs. (1.82) and (1.83) are nondimensionalized as

$$\Delta \mathbf{r}_i^* = \mathbf{v}_i^* \Delta t^* \tag{1.84}$$

$$\begin{aligned} \Delta \mathbf{v}_i^* = {} & \alpha^*\sum_{j(\neq i)} w_{\mathrm{R}}(r_{ij}^*)\mathbf{e}_{ij}\Delta t^* - \gamma^*\sum_{j(\neq i)} w_{\mathrm{R}}^2(r_{ij}^*)(\mathbf{e}_{ij}\cdot\mathbf{v}_{ij}^*)\mathbf{e}_{ij}\Delta t^* \\ & + (2\gamma^*)^{1/2}\sum_{j(\neq i)} w_{\mathrm{R}}(r_{ij}^*)\mathbf{e}_{ij}\theta_{ij}\sqrt{\Delta t^*} \end{aligned} \tag{1.85}$$

in which

$$w_{\mathrm{R}}(r_{ij}^*) = \begin{cases} 1 - r_{ij}^* & \text{for } r_{ij}^* \le 1 \\ 0 & \text{for } r_{ij}^* > 1 \end{cases} \tag{1.86}$$

$$\alpha^* = \alpha \frac{r_c}{kT}, \quad \gamma^* = \gamma \frac{r_c}{(mkT)^{1/2}} \tag{1.87}$$

Nondimensionalized quantities are distinguished by the superscript *. As seen in Eq. (1.85), the specification of the number density $n^*(=nr_c^3)$ and the number N of particles with appropriate values of α^*, γ^*, and Δt^* enables us to conduct DPD simulations. If we take into account that the time is nondimensionalized by the representative time based on the average velocity $\overline{v}$ ($\approx (kT/m)^{1/2}$) and distance r_c, the nondimensionalized time interval Δt^* has to be taken as $\Delta t^* \ll 1$.

The above-mentioned equations of motion retain a flexibility and are determined by our approach rather than the mathematical manipulation of certain basic key equations. These equations of motion are the revised version of the original equations, which were derived in order that the velocity distribution function of the particles converges to an equilibrium distribution for thermodynamic equilibrium. Hence, they are not the only valid equations of motion for the DPD method, and a new equation of motion may be proposed in order to enable us to conduct more accurate simulations.

The main procedure for conducting the DPD simulation is quite similar to the one we employed for BD simulations, so it is unnecessary to repeat the details here.

1.5 Lattice Boltzmann Method

Whether or not the lattice Boltzmann method is classified into the category of molecular simulation methods may depend on the researcher, but this method is expected to have a sufficient feasibility as a simulation technique for polymeric liquids and particle dispersions. We will therefore treat it in detail in this book. In the lattice Boltzmann method [4, 9–12], a fluid is assumed to be composed of virtual fluid particles, and such fluid particles move and collide with other fluid particles in a simulation region. A simulation area is regarded as a lattice system, and fluid particles move from site to site; that is, they do not move freely in a region. The most significant difference of this method in relation to the MD method is that the lattice Boltzmann method treats the particle distribution function of velocities rather than the positions and the velocities of the fluid particles.

Figure 1.4 illustrates the lattice Boltzmann method for a two-dimensional system. Figure 1.4A shows that a simulation region is divided into a lattice system. Figure 1.4B is a magnification of a unit square lattice cell. Virtual fluid particles, which are regarded as groups or clusters of solvent molecules, are permitted to move only to their neighboring sites, not to other, more distant sites. That is, the fluid particles at site 0 are permitted to stay there or to move to sites 1, 2,. . ., 8 at the next time step. This implies that fluid particles for moving to sites 1, 2, 3, and 4 have the velocity $c = (\Delta x/\Delta t)$, and those for moving to sites 5, 6, 7, and 8 have

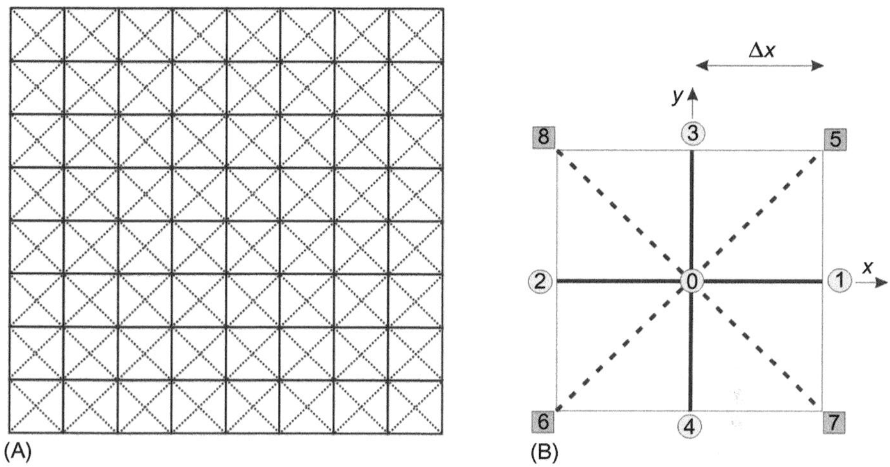

Figure 1.4 Two-dimensional lattice model for the lattice Boltzmann method (D2Q9 model).

the velocity $\sqrt{2}c$, in which Δx is the lattice separation of the nearest two sites and Δt is the time interval for simulations. Since the movement speeds of fluid particles are known as c or $\sqrt{2}c$, macroscopic velocities of a fluid can be calculated by evaluating the number of particles moving to each neighboring lattice site. In the usual lattice Boltzmann method, we treat the particle distribution function, which is defined as a quantity such that the above-mentioned number is divided by the volume and multiplied by the mass occupied by each lattice site. This is the concept of the lattice Boltzmann method. The two-dimensional lattice model shown in Figure 1.4 is called the "D2Q9" model because fluid particles have nine possibilities of velocities, including the quiescent state (staying at the original site).

Next, we explain the basic equations of the particle distribution function and the method of solving these equations. The detailed explanation will be shown in Chapter 8; here we outline the essence of the method. The velocity vector for fluid particles moving to their neighboring site is usually denoted by $\mathbf{c}_\alpha$ and, for the case of the D2Q9 model, there are nine possibilities, such as $\mathbf{c}_0$, $\mathbf{c}_1$, $\mathbf{c}_2$,. . ., $\mathbf{c}_8$. For example, the velocity of the movement in the left direction in Figure 1.4B is denoted by $\mathbf{c}_2$, and $\mathbf{c}_0$ is zero vector for the quiescent state ($\mathbf{c}_0 = \mathbf{0}$). We consider the particle distribution function $f_\alpha(\mathbf{r},t)$ at the position $\mathbf{r}$ (at point 0 in Figure 1.4B) at time t in the α-direction. Since $f_\alpha(\mathbf{r},t)$ is equal to the number density of fluid particles moving in the α-direction, multiplied by the mass of a fluid particle, the summation of the particle distribution function concerning all the directions ($\alpha = 0, 1,\ldots, 8$) leads to the macroscopic density $\rho(\mathbf{r},t)$:

$$\rho(\mathbf{r},t) = \sum_{\alpha=0}^{8} f_\alpha(\mathbf{r},t) \tag{1.88}$$

Similarly, the macroscopic velocity $\mathbf{u}(\mathbf{r},t)$ can be evaluated from the following relationship of the momentum per unit volume at the position $\mathbf{r}$:

$$\rho(\mathbf{r},t)\mathbf{u}(\mathbf{r},t) = \sum_{\alpha=0}^{8} f_\alpha(\mathbf{r},t)\,\mathbf{c}_\alpha \tag{1.89}$$

In Eqs. (1.88) and (1.89), the macroscopic density $\rho(\mathbf{r},t)$ and velocity $\mathbf{u}(\mathbf{r},t)$ can be evaluated if the particle distribution function is known. Since fluid particles collide with the other fluid particles at each site, the rate of the number of particles moving to their neighboring sites changes. In the rarefied gas dynamics, the well-known Boltzmann equation is the basic equation specifying the velocity distribution function while taking into account the collision term due to the interactions of gaseous molecules; this collision term is a complicated integral expression. The Boltzmann equation is quite difficult to solve analytically, so an attempt has been made to simplify the collision term. One such simplified model is the Bhatnagar-Gross-Krook (BGK) collision model. It is well known that the BGK Boltzmann method gives rise to reasonably accurate solutions, although this collision model is expressed in quite simple form. We here show the lattice Boltzmann equation based on the BGK model. According to this model, the particle distribution function $f_\alpha(\mathbf{r}+\mathbf{c}_\alpha\Delta t, t+\Delta t)$ in the α-direction at the position $(\mathbf{r}+\mathbf{c}_\alpha\Delta t)$ at time $(t+\Delta t)$ can be evaluated by the following equation:

$$f_\alpha(\mathbf{r}+\mathbf{c}_\alpha\Delta t, t+\Delta t) = f_\alpha(\mathbf{r},t) + \frac{1}{\tau}\{f_\alpha^{(0)}(\mathbf{r},t) - f_\alpha(\mathbf{r},t)\} \tag{1.90}$$

This equation is sometimes expressed in separate expressions indicating explicitly the two different processes of collision and transformation:

$$\left.\begin{aligned} f_\alpha(\mathbf{r}+\mathbf{c}_\alpha\Delta t, t+\Delta t) &= \tilde{f}_\alpha(\mathbf{r},t) \\ \tilde{f}_\alpha(\mathbf{r},t) &= f_\alpha(\mathbf{r},t) + \frac{1}{\tau}\{f_\alpha^{(0)}(\mathbf{r},t) - f_\alpha(\mathbf{r},t)\} \end{aligned}\right\} \tag{1.91}$$

in which τ is the relaxation time (dimensionless) and $f_\alpha^{(0)}$ is the equilibrium distribution, expressed for the D2Q9 model as

$$f_\alpha^{(0)} = \rho\, w_\alpha \left\{ 1 + 3\frac{\mathbf{c}_\alpha\cdot\mathbf{u}}{c^2} - \frac{3u^2}{2c^2} + \frac{9}{2}\cdot\frac{(\mathbf{c}_\alpha\cdot\mathbf{u})^2}{c^4} \right\} \tag{1.92}$$

$$w_\alpha = \begin{cases} 4/9 & \text{for } \alpha = 0 \\ 1/9 & \text{for } \alpha = 1,2,3,4 \\ 1/36 & \text{for } \alpha = 5,6,7,8 \end{cases} \qquad |\mathbf{c}_\alpha| = \begin{cases} 0 & \text{for } \alpha = 0 \\ c & \text{for } \alpha = 1,2,3,4 \\ \sqrt{2}c & \text{for } \alpha = 5,6,7,8 \end{cases} \tag{1.93}$$

In these equations ρ is the local density at the position of interest, $\mathbf{u}$ is the fluid velocity ($u = |\mathbf{u}|$), $c = \Delta x/\Delta t$, and w_α is the weighting constant.

The important feature of the BGK model shown in Eq. (1.91) is that the particle distribution function in the α-direction is independent of the other directions. The particle distributions in the other directions indirectly influence $f_\alpha(\mathbf{r} + \mathbf{c}_\alpha \Delta t, t + \Delta t)$ through the fluid velocity $\mathbf{u}$ and the density ρ. The second expression in Eq. (1.91) implies that the particle distribution $f_\alpha(\mathbf{r},t)$ at the position $\mathbf{r}$ changes into $\tilde{f}_\alpha(\mathbf{r}, t)$ after the collision at the site at time t, and the first expression implies that $\tilde{f}_\alpha(\mathbf{r}, t)$ becomes the distribution $f_\alpha(\mathbf{r} + \mathbf{c}_\alpha \Delta t, t + \Delta t)$ at $(\mathbf{r} + \mathbf{c}_\alpha \Delta t)$ after the time interval Δt.

The main procedure of the simulation is as follows:

1. Set appropriate fluid velocities and densities at each lattice site.
2. Calculate equilibrium particle densities $f_\alpha^{(0)}$ ($\alpha = 0, 1, \ldots, 8$) at each lattice site from Eq. (1.92) and regard these distributions as the initial distributions, $f_\alpha = f_\alpha^{(0)}$ ($\alpha = 0, 1, \ldots, 8$).
3. Calculate the collision terms $\tilde{f}_\alpha(\mathbf{r}, t)$ ($\alpha = 0, 1, \ldots, 8$) at all sites from the second expression of Eq. (1.91).
4. Evaluate the distribution at the neighboring site in the α-direction $f_\alpha(\mathbf{r} + \mathbf{c}_\alpha \Delta t, t + \Delta t)$ from the first expression in Eq. (1.91).
5. Calculate the macroscopic velocities and densities from Eqs. (1.88) and (1.89), and repeat the procedures from step 3.

In addition to the above-mentioned procedures, we need to handle the treatment at the boundaries of the simulation region. These procedures are relatively complex and are explained in detail in Chapter 8. For example, the periodic boundary condition, which is usually used in MD simulations, may be applicable.

For the D3Q19 model shown in figure 8.3, which is applicable for three-dimensional simulations, the equilibrium distribution function is written in the same expression of Eq. (1.92), but the weighting constants are different from Eq. (1.93) and are expressed in Eq. (8.69). The basic equations for $f_\alpha(\mathbf{r} + \mathbf{c}_\alpha \Delta t, t + \Delta t)$ are the same as Eq. (1.90) or (1.91), and the above-mentioned simulation procedure is also directly applicable to the D3Q19 model.

2 Outline of Methodology of Simulations

In order to develop a simulation program, it is necessary to have an overview of the general methodology, which should include the assignment of the initial configuration and velocities, the treatment of boundary conditions, and techniques for reducing computation time. An appropriate initial configuration has to be set with careful consideration given to the physical property of interest, so that the essential phenomena can be grasped. For example, if nonspherical molecules or particles are known to incline in a preferred direction, there may be some advantages to using a parallelepiped rectangular simulation region rather than a cubic one. The periodic boundary condition is a representative model to manage the boundary of a simulation region. It is almost always used for systems in thermodynamic equilibrium. On the other hand, for investigating the dynamic properties of a system, the simple shear flow is frequently treated and in this case the Lees–Edwards boundary condition is available. Techniques for reducing computation time become very important in large-scale three-dimensional simulations, and methods of tracking particle neighbors, such as the cell index method, are indispensable. The more important methods frequently employed in simulations are described in this chapter.

2.1 Initial Positions

2.1.1 Spherical Particle Systems

Setting an initial configuration of particles is an indispensable procedure for both MD and MC methods. Although it is possible to assign randomly the initial position of particles in a simulation region, a regular configuration, such as a simple cubic lattice or a face-centered cubic lattice, is handled in a more straightforward manner. The random allocation suffers from the problem of the undesirable overlap of particles and from possible difficulties in achieving high packing fractions. Lattice assignments are almost free from the overlap problem and can achieve high packing fractions. However, as will be shown later, the lattice packing may be too perfect for some simulations, requiring the adjustment of a small random perturbation. In the following paragraphs, we consider a system composed of spherical particles as an example to explain the method of setting the initial configuration in a

Introduction to Practice of Molecular Simulation. DOI: 10.1016/B978-0-12-385148-2.00002-1

regular lattice formation for a two-dimensional configuration. We then proceed to a three-dimensional configuration.

Figure 2.1 shows several lattice systems that may be used to assign an initial configuration for a two-dimensional system. A basic lattice form is expanded to fill the whole simulation region, and the particles are then located at each lattice point. Figure 2.1A, the simplest lattice model, may be suitable for a gaseous system. However, even if the particle–particle distance a is equal to the particle diameter, a high packing fraction cannot be obtained by using this simple lattice model. Hence, it is inappropriate for the simulations of a liquid or solid system. Since there is only one particle in the unit cell shown in Figure 2.1A, a system with total particle number N ($=Q^2$) can be generated by replicating the unit cell $(Q-1)$ times in each direction to make a square simulation region of side length $L=Qa$. So for the use of this lattice system as the initial configuration, the particle number N has to

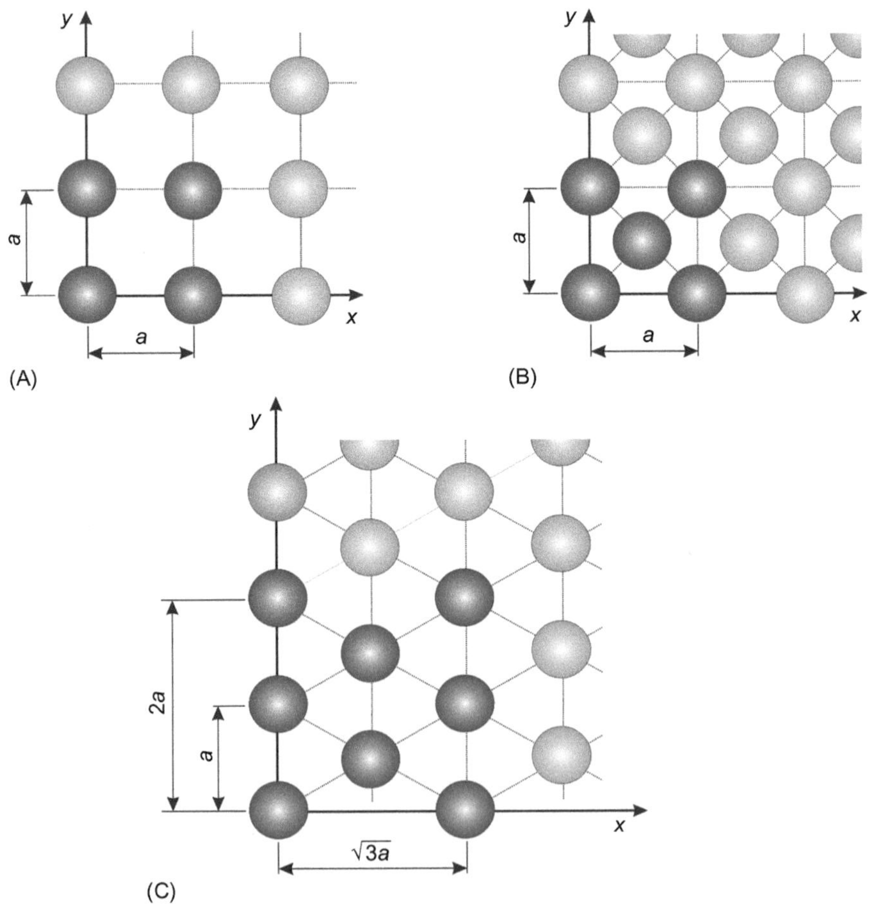

Figure 2.1 Initial conditions for a two-dimensional system.

be taken from $N = 1, 4, 9, 16, 25, 36, 49, 64, 81, 100$, and so on. The number density of particles n is given by $n = N/L^2$, and the area fraction is given by $\phi_s = \pi(d/2)^2 N/L^2$, where d is the particle diameter. In practice, the number of particles N and the area fraction ϕ_s are first chosen; then the values of Q and L are evaluated, from which the value of a can be determined. With these values, the initial configuration of particles can be assigned according to the simple lattice system shown in Figure 2.1A.

The lattice system shown in Figure 2.1B can yield a higher packing fraction and therefore may be applicable for an initial configuration of a gaseous or liquid state, but it has limited application to a solid state. Since there are two particles in the unit cell of this lattice, a system with total particle number $N = 2Q^2$ of particles can be generated by replicating the unit cell $(Q - 1)$ times in each direction. In this case, the simulation region is also a square of side length $L = Qa$, and the possible value of N is taken from 2, 8, 18, 32, 50, 72, 98, 128, 162, 200, and so on. The number density of particles n is given by $n = N/L^2$, and the area fraction ϕ_s is given by $\phi_s = \pi(d/2)^2 N/L^2$. Figure 2.1C shows the most compact lattice for a two-dimensional system. This lattice model may also be applicable to a solid system. If the dark particles are assumed to constitute the unit lattice, it follows that there are four particles in this unit lattice. Hence, by replicating the unit lattice $(Q - 1)$ times in each direction, the simulation region becomes a rectangle of side lengths $L_x = 3^{1/2}aQ$ and $L_y = 2aQ$, with a total number of particles $N = 4Q^2$, where the possible value of N is taken from 4, 16, 36, 64, 100, 144, 196, 256, 324, 400, and so on. The particle number density n is given by $n = N/L_xL_y$, and the area fraction ϕ_s is given by $\phi_s = \pi(d/2)^2 N/L_xL_y$. The actual assignment of the above-mentioned quantities for simulations is similar to that for Figure 2.1A.

Figure 2.2 shows several lattice models for a three-dimensional system. Figure 2.2A is the simple cubic lattice model, which is suitable as an initial configuration mainly for a gaseous or liquid system. Since there is only one particle in the unit cell, the number of particles in a system is given by $N = Q^3$ by replicating the unit cell $(Q - 1)$ times in each direction. In this case the possible value of N is taken from $N = 1, 8, 27, 64, 125, 216, 343, 512, 729, 1000$, and so on.

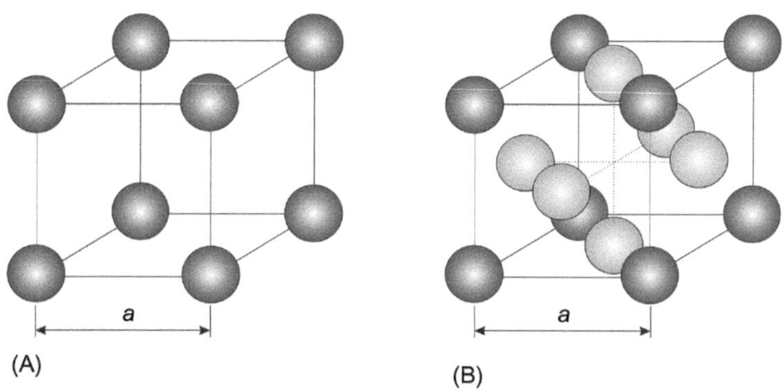

Figure 2.2 Initial conditions for a three-dimensional system.

The simulation region is a cube of side length $L = Qa$. The number density n and the volumetric fraction ϕ_V are given by $n = N/L^3$ and $\phi_V = 4\pi(d/2)^3N/3L^3$, respectively. The face-centered cubic lattice model shown in Figure 2.2B is one of the close-packed lattices, and therefore may be applicable as an initial configuration of a solid state. Since there are four particles in the unit cell, the total number of particles in the simulation region is given by $N = 4Q^3$ by replicating the unit cell $(Q - 1)$ times in each direction. In this case, the total number of particles is taken from $N = 4$, 32, 108, 256, 500, 864, 1372, and so on. The number density and the volumetric fraction are given by $n = N/L^3$ and $\phi_V = 4\pi(d/2)^3N/3L^3$, respectively. As in a two-dimensional system, for the actual assignment of the above-mentioned quantities, the particle number N and the volumetric fraction ϕ_V are first chosen, then Q and L are evaluated, and finally the lattice distance a is determined.

For a gaseous or liquid system, the simple lattice models shown in Figures 2.1A and 2.2A are applicable in a straightforward manner for developing a simulation program. In contrast, for the case of a solid system, the choice of an appropriate lattice used for the initial configuration of particles is usually determined by the known physical properties of the solid.

2.1.2 Nonspherical Particle Systems

The assignment of the initial configuration of particles for a spherical particle system, explained in the previous subsection, is quite clear because only the center of the particles needs to be considered. In this subsection we explain the method of setting the initial configuration for a system composed of nonspherical particles, using spherocylinders and disk-like particles as examples. For a nonspherical particle system, the orientation of the particles must be assigned in addition to their position, so that the technique for setting the initial configuration is a little more difficult than that for a spherical particle system. For this purpose, a versatile technique whereby a wide range of initial configurations can be assigned is desirable. If particle–particle interactions are large enough to induce the cluster formation of particles in a preferred direction, then an appropriately large initial configuration has to be adopted in order for the simulation to capture such characteristic aggregate structures.

We now consider the example of a system composed of spherocylinder particles with a magnetic moment at the particle center normal to the long particle axis. The spherocylinder is a cylinder with hemisphere caps at both the ends. An ensemble of these particles can be expected to aggregate to form raft-like clusters with the magnetic moments inclining in the applied magnetic field direction. Hence, a simulation region with sufficient length in the direction of the cluster formation has to be taken in order for the simulation particles to aggregate in a reasonable manner. We shall explain the technique for setting an initial configuration using Figure 2.3. The spherocylinder can be characterized by the ratio of the particle length l to the diameter d of the cylindrical part, known as the aspect ratio $r_p = l/d$. For the example in Figure 2.3 where $r_p = 3$, the particles are placed in contact with three and nine rows in the x- and y-directions, respectively, leading to a configuration of 27

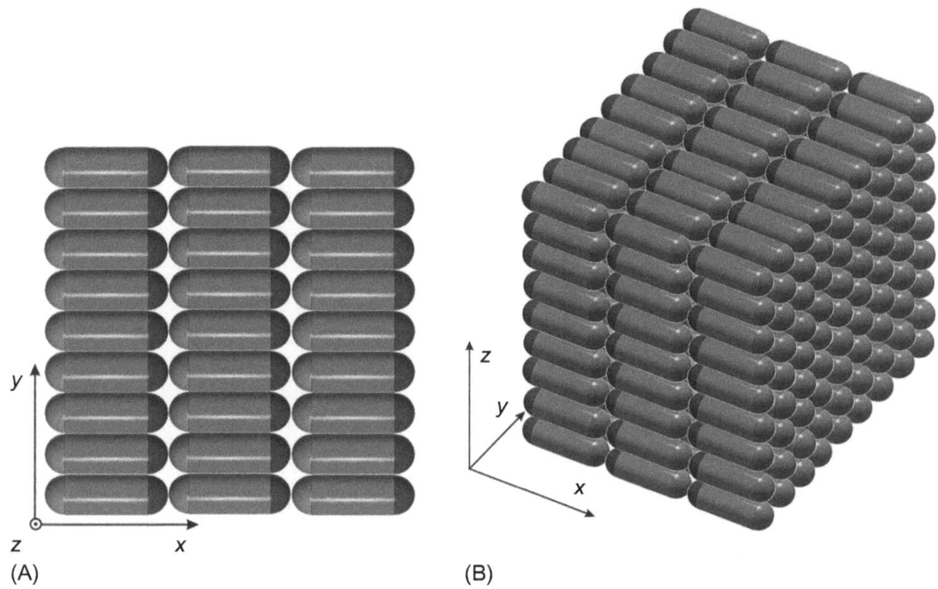

Figure 2.3 Initial conditions for spherocylinder particles.

particles in a square region in the xy-plane. Extending this configuration to 18 layers in the z-direction yields an initial configuration of spherocylinder particles with a simulation region $(L_x, L_y, L_z) = (3r_p d, 3r_p d, 6r_p d)$ with total number of particles $N = 486$; if four rows are arranged in the x-direction, then a simulation region larger than the present case can be adopted with a simulation region $(L_x, L_y, L_z) = (4r_p d, 4r_p d, 8r_p d)$.

If the particle–particle distances are expanded equally in each direction to yield a desired volumetric fraction of particles ϕ_V, then this expanded system may be used as an initial configuration for simulations. Such an expansion with a factor α of particle–particle distances gives rise to the system volume $V = 54r_p^3 d^3 \alpha^3$. The volumetric fraction ϕ_V is related to the system volume as $\phi_V = NV_p/V$, in which V_p is the volume of a spherocylinder particle, expressed as $V_p = \pi d^3(3r_p - 1)/12$. From these expressions, the expansion ratio α can be obtained as

$$\alpha = \frac{1}{r_p}\left(\frac{3\pi(3r_p - 1)}{4\phi_V}\right)^{1/3} \tag{2.1}$$

This initial configuration is applicable for a system in which particles are expected to aggregate in the direction of the particle short axis, as shown in Figure 2.4A. If particles are expected to aggregate in the direction of the particle long axis, as shown in Figure 2.4B, it is straightforward to follow a similar procedure with the spherocylinder particles aligned in the z-direction in Figure 2.3.

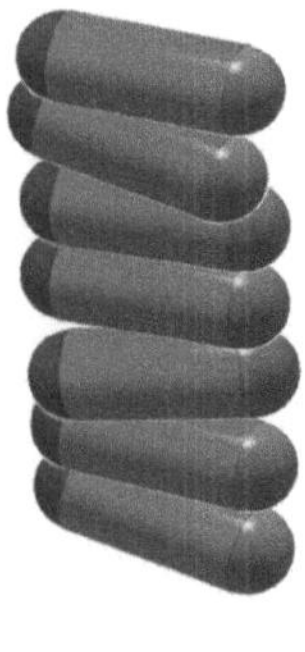

(A) Aggregation in the short axis direction

(B) Aggregation in the long axis direction

Figure 2.4 Aggregation for spherocylinder particles.

We now consider the method of setting an initial configuration of a disk-like particle system, in which particles are assumed to aggregate in a direction parallel to the disk plane surface, as shown in Figure 2.5B. Capturing such clusters in simulations requires a simulation region with suitable dimensions. As in the previous case of a spherocylinder particle system, a nearly close-packed configuration is first arranged. We here consider disk-like particles with particle aspect ratio r_p $(=d_1/b_1)=3$, in which the diameter of the circumference and the thickness are denoted by d_1 and b_1, respectively, as shown in Figure 4.12. If three and nine particles are arranged in the x- and y-directions, respectively, the subtotal number of $N=27$ particles can be located in the xy-plane, as shown in Figure 2.5A. Extending this configuration with 12 layers in the z-direction leads to an initial configuration of 324 particles with particle–particle contact in the simulation region of $(L_x, L_y, L_z) = (3r_pb_1, 3r_pb_1, 12r_pb_1)$. By expanding particle–particle distances equally in each direction by the expansion factor α, the volume of a system V becomes $V=108r_p^3b_1^3\alpha^3$. Given the volume of a disk-like particle, $V_p=(\pi/4)b_1^3(r_p-1)^2+(\pi^2/8)b_1^3(r_p-1)+(\pi/6)b_1^3$, the expansion factor α can be derived as

$$\alpha = \frac{1}{2r_p}\left[\frac{\pi}{\phi_V}\left\{6(r_p-1)^2+3\pi(r_p-1)+4\right\}\right]^{1/3} \tag{2.2}$$

In this derivation, the relationship of $\phi_V = NV_p/V$ has been used.

The main procedure for setting the initial configuration is summarized as follows:

1. Consider an appropriate initial configuration, with sufficient consideration given to the physical phenomenon of interest.
2. Set a nearly close-packed situation as an initial configuration.
3. Calculate the total number of particles N.
4. Evaluate the expansion ratio α from Eq. (2.1) or Eq. (2.2) to give rise to the desired volumetric fraction ϕ_V.
5. Expand particle–particle distances equally by the factor α.

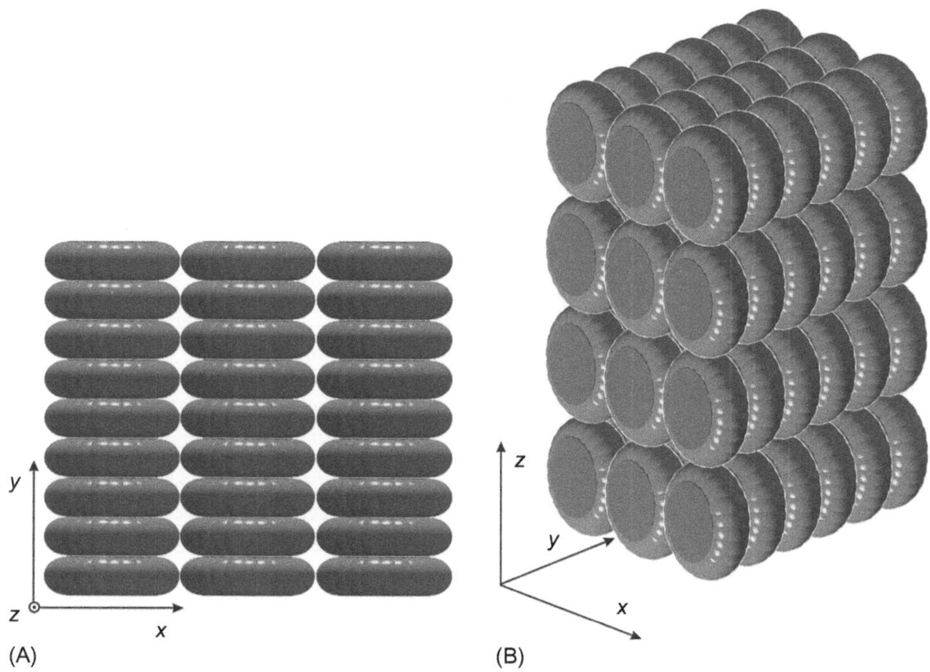

Figure 2.5 Initial conditions for disk-like particles.

6. Perturb the particle positions by small distances using random numbers in order to destroy the regularity of the initial configuration; otherwise, all particle–particle interactions may be zero and therefore the particles may not move with time.

2.2 Initial Velocities

2.2.1 Spherical Particle Systems

In the MD method, the motion of particles is described by pursuing their position and velocity over time, so these factors have to be specified as an initial condition. If the system of interest is in thermodynamic equilibrium with temperature T, the particle velocities are described by the following Maxwellian distribution [25]:

$$f(\mathbf{v}_i) = \left(\frac{m}{2\pi kT}\right)^{3/2} \exp\left\{-\frac{m}{2kT}\left(v_{ix}^2 + v_{iy}^2 + v_{iz}^2\right)\right\} \tag{2.3}$$

in which k is Boltzmann's constant, m is the mass of particles, and $\mathbf{v}_i = (v_{ix}, v_{iy}, v_{iz})$ is the velocity vector of particle i. Since the Maxwellian distribution f is the

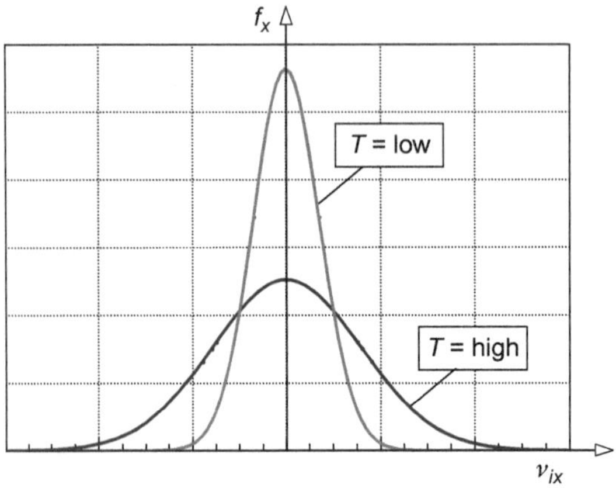

Figure 2.6 Velocity distributions in equilibrium.

probability density distribution function, the probability of particle i being found in the infinitesimal velocity range between $\mathbf{v}_i$ and $(\mathbf{v}_i + d\mathbf{v}_i)$ becomes $f(\mathbf{v}_i)d\mathbf{v}_i$. Characteristics of this function can be understood more straightforwardly by treating the distribution function f_x as the x-velocity component. Figure 2.6 clearly shows that a higher system temperature leads to an increase in the probability of particles appearing with a larger velocity component v_{ix}. If we focus on the magnitude of the particle velocities instead of the velocity components, clearer discussion concerning such characteristics becomes possible. The probability density distribution function $\chi(v_i)$ for the speed $v_i = ({v_{ix}}^2 + {v_{iy}}^2 + {v_{iz}}^2)$ of particle i can be derived from Eq. (2.3) as

$$\chi(v_i) = 4\pi\left(\frac{m}{2\pi kT}\right)^{3/2} v_i^2 \exp\left(-\frac{m}{2kT}v_i^2\right) \tag{2.4}$$

This equation is derived, first, by a transformation from orthogonal to spherical coordinates, that is, from (v_{ix}, v_{iy}, v_{iz}) to (v_i, θ, ϕ) with the relationship of $(v_{ix}, v_{iy}, v_{iz}) = (v_i \sin\theta \cos\phi,\ v_i \sin\theta \sin\phi,\ v_i \cos\theta)$, and second, from the integral with respect to θ and ϕ in the normalization equation of the Maxwellian distribution. The integrand in the normalization equation after this integral is the distribution function $\chi(v_i)$. Figure 2.7 shows the distribution χ as a function of the particle speed v_i for several system temperatures. Figure 2.7 shows that the curve of χ has a peak value position that moves further to the right with increasing value of the temperature. That is, the percentage of particles with larger velocities increases with the temperature. The particle speed v_{mp} yielding the peak value of the distribution can be derived from Eq. (2.4) as $v_{\mathrm{mp}} = (2kT/m)^{1/2}$, which is called the "most probable velocity." This means that particles with speed v_{mp} are likely to be the most numerous in the system. Note that the most probable speed is larger for a higher system temperature and a smaller mass.

For a given system temperature T, the initial velocities of particles for simulations can be assigned according to the probability density function in Eq. (2.3) or

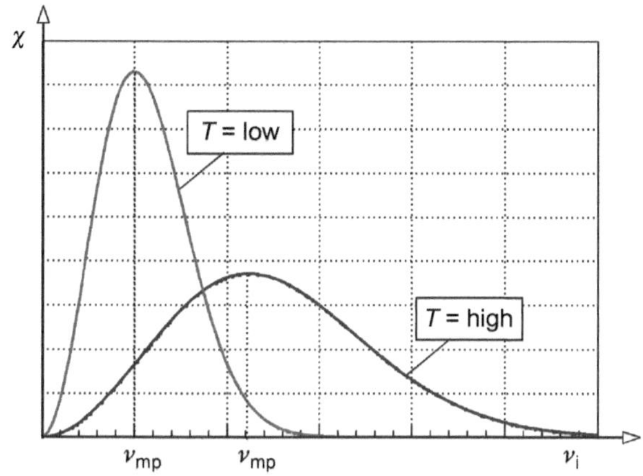

Figure 2.7 Particle speed distributions in equilibrium.

Eq. (2.4). The detailed explanation is given in Appendix A2, so here we only show the final technique. With six different uniform random numbers, $R_1, R_2, \ldots, R_6$, the initial velocity components (v_{ix}, v_{iy}, v_{iz}) of particle i can be set as

$$\left.\begin{aligned} v_{ix} &= \left(-2\frac{kT}{m}\ln R_1\right)^{1/2}\cos(2\pi R_2) \\ v_{iy} &= \left(-2\frac{kT}{m}\ln R_3\right)^{1/2}\cos(2\pi R_4) \\ v_{iz} &= \left(-2\frac{kT}{m}\ln R_5\right)^{1/2}\cos(2\pi R_6) \end{aligned}\right\} \tag{2.5}$$

Note that each particle requires a new, that is, a different, set of random numbers.

The temperature which is evaluated from the initial particle velocities assigned by the above-mentioned method is approximately equal to the desired system temperature, but may not necessarily be satisfactory. Hence, an equilibration procedure is usually necessary before starting the main loop in an actual simulation program. This will be explained in the next subsection.

2.2.2 Nonspherical Particle Systems

For a nonspherical particle system, the initial angular velocities need to be assigned in addition to the translational velocities. Similar to the translational velocity $\mathbf{v} = (v_x, v_y, v_z)$ discussed above, the angular velocity $\boldsymbol{\omega} = (\omega_x, \omega_y, \omega_z)$ is also governed by the Maxwellian distribution $f_\omega(\boldsymbol{\omega})$. The expression for $f_\omega(\boldsymbol{\omega})$ is

$$f_\omega(\boldsymbol{\omega}) = \left(\frac{I}{2\pi kT}\right)^{3/2} \exp\left\{-\frac{I}{2kT}\left(\omega_x^2 + \omega_y^2 + \omega_z^2\right)\right\} \tag{2.6}$$

in which I is the inertia moment of a particle. The characteristics of the exponential function in Eq. (2.3) or Eq. (2.6) demonstrate that the probability of particles appearing with larger translational and angular velocities increases with the system temperature. Similar to $v_{\rm mp}=(2kT/m)^{1/2}$, $\omega_{\rm mp}=(2kT/I)^{1/2}$ is the most probable angular velocity to yield the maximum value of the Maxwellian distribution f_ω. The method for setting the initial translational velocities using uniform random numbers, explained in the previous subsection, is applicable to the present angular velocity case. Here, m and (v_{ix}, v_{iy}, v_{iz}) in Eq. (2.5) are replaced by I and $(\omega_{ix}, \omega_{iy}, \omega_{iz})$; note that new uniform random numbers need to be used.

As already mentioned, an equilibration procedure may be necessary in order to obtain the desired system temperature T. In the example of a liquid, the temperature $T_{\rm cal}$, which is calculated from averaging the assigned velocities of particles, may differ significantly from the desired system temperature T. This may be due to the energy exchange between the kinetic and the potential energies. Hence, an equilibration procedure is frequently necessary before starting the main loop in a simulation program. The temperatures calculated from the translational and angular velocities of particles are denoted by $T_{\rm cal}^{(t)}$ and $T_{\rm cal}^{(r)}$, respectively, and written as

$$T_{\rm cal}^{(t)}=\frac{1}{3N}\sum_{i=1}^{N}\frac{mv_i^2}{k},\qquad T_{\rm cal}^{(r)}=\frac{1}{3N}\sum_{i=1}^{N}\frac{I\omega_i^2}{k} \tag{2.7}$$

in which N is the total number of particles, assumed to be $N \gg 1$. $T_{\rm cal}^{(t)}$ and $T_{\rm cal}^{(r)}$, calculated from $\mathbf{v}_i$ and $\boldsymbol{\omega}_i$ ($i=1, 2,\ldots, N$), are generally not equal to the desired temperature T. The equilibration procedure adjusts these temperatures to T during the simulation by using the method of scaling the translational and angular velocities of each particle. If $T_{\rm cal}^{(t)\rm ave}$ and $T_{\rm cal}^{(r)\rm ave}$ denote the averaged values of $T_{\rm cal}^{(t)}$ and $T_{\rm cal}^{(r)}$ taken, for example, over 50 time steps, then the scaling factors $c_0^{(t)}$ and $c_0^{(r)}$ are determined as

$$c_0^{(t)}=\sqrt{T/T_{\rm cal}^{(t)\rm ave}},\qquad c_0^{(r)}=\sqrt{T/T_{\rm cal}^{(r)\rm ave}} \tag{2.8}$$

With the scaling factors determined, the translational and angular velocities of all particles in a system are scaled as

$$\mathbf{v}_i'=c_0^{(t)}\mathbf{v}_i,\quad \boldsymbol{\omega}_i'=c_0^{(r)}\boldsymbol{\omega}_i\quad (i=1,2,\ldots,N) \tag{2.9}$$

This treatment yields the desired system temperature T. In this example the scaling procedure would be conducted at every 50 time steps, but in practice an appropriate time interval must be adopted for each simulation case. The above-mentioned equilibration procedure is repeated to give rise to the desired system temperature with sufficient accuracy. (Note that if a system has a macroscopic velocity, i.e., if it is not in a quiescent state, the scaling procedure has to be slightly modified.)

2.3 Reduction Methods of Computation Time

2.3.1 *Cutoff Distance*

Computation time is an important factor for successfully obtaining reasonable results from molecular simulations. In some cases, due to an excessive restriction of the required computation time, only a small or two-dimensional system is able to be treated. The most time-consuming procedure is the calculation of interaction energies between particles in the MC method and that of forces and torques in the MD method. Even with the action–reaction law taken into account, the $N(N-1)/2$ calculations of energies or forces are necessary per unit time step (or unit MC step) for an N-particle system. Therefore, if it is possible to restrict the pairs of particles for the calculation, the computation time may significantly decrease. Fortunately, many particle–particle potentials are of short-range order, so that the potential energy between particles rapidly decreases with the particle–particle separation over a distance only several times the particle diameter. Therefore we may be able to treat only interactions within this range. Although magnetic or electrostatic forces are of long-range order, the above-mentioned concept is applicable to these interactions if the criterion separation between particles is taken to be of sufficient length.

2.3.1.1 Spherical Particle Systems

An important concept in simulation methodology is that a significant limitation on the computation of interaction energies or forces between particles leads to an extraordinary reduction of the simulation time. To understand this concept, we consider the interaction energies between particles or potential curves. For example, the Lennard-Jones potential $U_{\rm LJ}$ is expressed as

$$U_{\rm LJ} = 4\varepsilon\left\{\left(\frac{\sigma}{r}\right)^{12} - \left(\frac{\sigma}{r}\right)^{6}\right\} \tag{2.10}$$

This potential is usually used as a model potential for rare gases such as Ar molecule; σ is the quantity corresponding to the particle diameter, and r is the separation between particles (molecules). Figure 2.8 shows the curve of the Lennard-Jones potential, in which $U_{\rm LJ}$ and r are nondimensionalized by ε and σ. Figure 2.8 illustrates a steep potential barrier in the range of $r \lesssim \sigma$, which induces such a significant repulsive interaction that particles are prevented from significantly overlapping, and an attractive interaction in the range of $r \gtrsim \sigma$, which rapidly decreases to zero. These characteristics of the potential curve indicate that the interaction energy after a distance of approximately $r = 3\sigma$ can be assumed to be negligible. Hence, particle interaction energies or forces do not need to be calculated in the range of $r > 3\sigma$ in actual simulations. The distance for cutting off the calculation of energies or forces is known as the cutoff distance or cutoff radius, denoted by $r_{\rm coff}$ in this book.

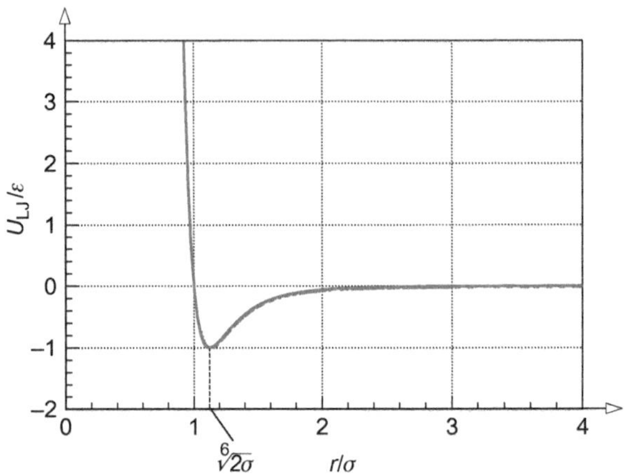

Figure 2.8 Lennard-Jones potential.

2.3.1.2 Nonspherical Particle Systems

The above cutoff procedure is directly applicable to a nonspherical particle system using the cutoff radius r_{coff} based on the particle center-to-center distance. That is, the calculation of energies or forces is unnecessary for $r_{ij} \geq r_{coff}$ in simulations. For example, this applies to the case of rod-like particles that have a magnetic dipole moment at their particle center, as shown in Figure 2.9A. Unfortunately, the method is not suitable for the case of rod-like particles with plus and minus (NS) magnetic charges at the centers of hemisphere caps, as shown in Figure 2.9B. For this case, the most direct criterion is to calculate the distance between each pair of magnetic charges of the two interacting spherocylinder particles and compare this separation with a suitable cutoff radius r_{coff}. This will require the distances of the four pairs of magnetic charges to be calculated. However, with prior knowledge of the arrangement of the two spherocylinder particles, it is possible to determine certain cases where we may know, without calculating the distances for all the four pairs of charges, that there are only two pairs of the distances satisfying the relationship of $r_{ij} \leq r_{coff}$. Referring to Figure 2.10, if the center-to-center distance between particles i and j is denoted by r_{ij} and the distance between the magnetic charges in the particle is denoted by l, then the following three cases have to be considered for this assessment:

1. For $r_{ij} \geq r_{coff} + l$
   No interactions.
2. For $r_{coff} + l > r_{ij} > r_{coff}$
   A possibility of two pairs of interactions at the most.
3. For $r_{ij} \leq r_{coff}$
   A possibility of all four pairs of interactions.

Figure 2.10A corresponds to case 1, in which the distance for any pair is beyond the cutoff radius. Figure 2.10B corresponds to case 2, in which there is a possibility of a certain magnetic charge interacting with both the magnetic charges in the other

(A) The model with a magnetic dipole at the center

(B) The model with magnetic charges at the centers of both the hemisphere caps

Figure 2.9 Treatment of the cutoff distance for different rod-like particle models.

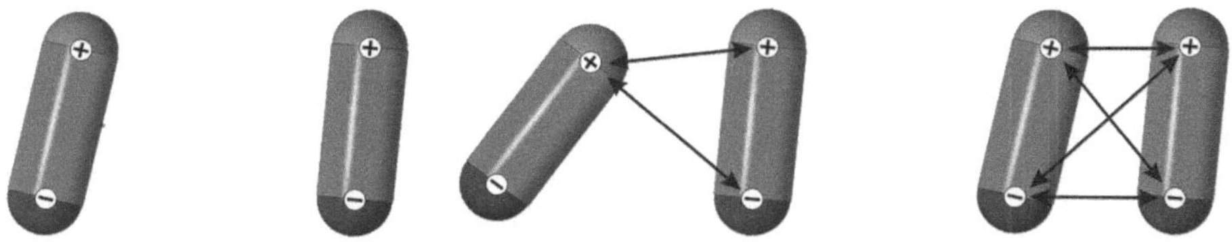

(A) No interactions

(B) Possibility of two pairs of interactions

(C) Possibility of four pairs of interactions

Figure 2.10 Check for interactions in the criterion of the particle distance r_{ij}.

particle within the cutoff radius. Figure 2.10C corresponds to case 3, in which the two particles are proximate enough to give rise to a possibility of four pairs of magnetic charges being within the cutoff range. Hence, if case 1 holds, the calculation of energies or forces between particles is unnecessary, and for case 2, if two pairs of magnetic charges are found to be within the cutoff range, the further calculation of energies or forces is unnecessary.

Finally, it should be noted that the introduction of the cutoff radius by itself does not necessarily lead to a significant reduction in the computational time, since the $N(N-1)/2$ calculations have to be conducted in order to evaluate the distances between particles. Hence, the following cell index method, or the Verlet neighbor list method, is used with the cutoff method to accomplish a significant reduction in the computation time.

2.3.2 Cell Index Method

If in some way we had already determined the names of the particles within the cutoff range from each particle, the calculation of the particle–particle distances for all pairs of particles at each time step would be unnecessary. Several methods have been developed for grasping such particle names. We first explain the cell index method [27,28] in this subsection. In order for the reader to understand the method straightforwardly, we treat a two-dimensional system composed of the spherocylinder particles shown in Figure 2.9A. With reference

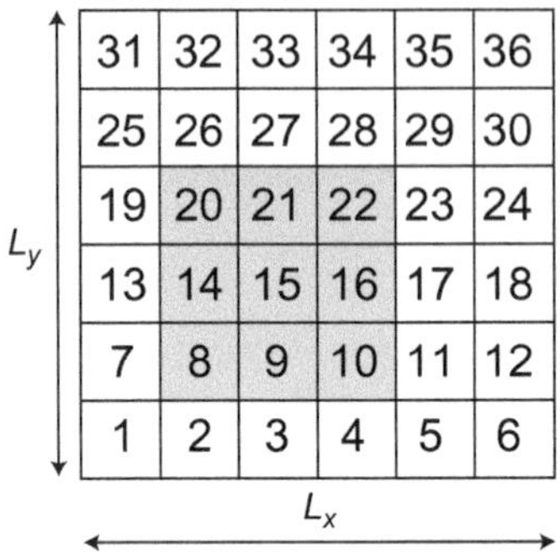

Figure 2.11 Cell index method for grasping neighboring particles.

to Figure 2.11, the simulation region with dimensions of (L_x, L_y) is divided into (Q_x, Q_y) equal parts in each direction $(Q_x = Q_y = 6)$ in order to divide the whole region into small cells. Each cell has the dimensions of $(L_x/Q_x, L_y/Q_y)$, in which (Q_x, Q_y) are maximized to satisfy the relationships $L_x/Q_x \geq r_{\text{coff}}$ and $L_y/Q_y \geq r_{\text{coff}}$. Since each side of a small cell is longer than the cutoff distance r_{coff}, the particles locating, for example, in the 15th cell in Figure 2.11, have a possibility to interact with those in their own cell 15 and those belonging to the neighboring cells, that is, in the 8th, 9th, 10th, 14th, 16th, 20th, 21st, and 22nd cells. Particles in other cells are beyond the cutoff area, so they are not used to calculate the distances between particles. Each cell needs to memorize the names of the particles which belong to it. As shown in Figure 2.11, the cell index method provides a significant reduction in the computation time for large values of (Q_x, Q_y). For the case of the particles shown in Figure 2.9B, the method is simply applied if the values of (Q_x, Q_y) are adopted in such a way to satisfy the relationships of $L_x/Q_x \geq r_{\text{coff}} + l$ and $L_y/Q_y \geq r_{\text{coff}} + l$.

2.3.3 Verlet Neighbor List Method

In the Verlet neighbor list method [29], a distance r_l, which is longer than the cutoff radius, is adopted, and each particle grasps the names of the particles within range of r_l from its center. Referring to Figure 2.12, it is clear that particles within range of $r < r_{\text{coff}}$ are certainly within range of $r < r_l$. Hence, if the list of particles within range of $r < r_l$ is renewed with such frequency that the particles outside the range of $r = r_l$ cannot attain to the cutoff area, then it is sufficient to calculate the distances between the particle of interest and its neighboring particles being within range of $r \leq r_l$. If r_l is sufficiently short compared with the dimensions of a simulation region, and the information concerning the names of the neighboring particles is renewed, for example, at every 10 time steps, then a significant reduction in the computation time can be expected. The Verlet neighbor list method is applicable to the MD method as well as to the MC method. Note that the cutoff distance is usually taken as $r_{\text{coff}} < L/2$ (L is the side length of a simulation region).

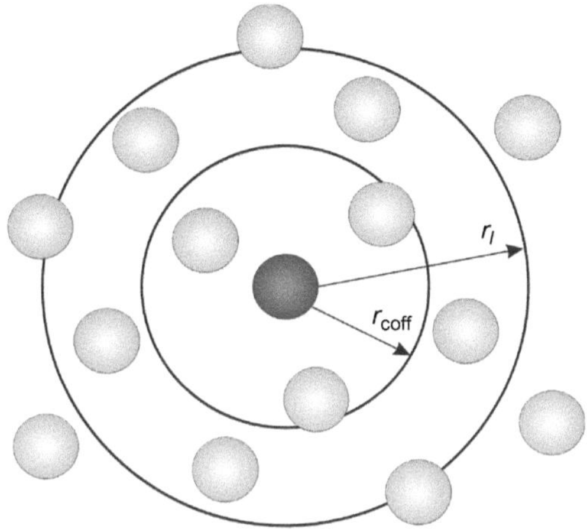

Figure 2.12 Verlet neighbor list method.

2.4 Boundary Conditions

2.4.1 Periodic Boundary Condition

Fortunately, a system of one-mol-order size, being composed of about 6×10^{23} particles, never needs to be directly treated in molecular simulations for thermodynamic equilibrium (actually, it is impossible). The use of the periodic boundary condition, explained below, enables us to treat only a relatively small system of about 100–10,000 particles in order to obtain such reasonable results as to explain the corresponding experimental data accurately.

Figure 2.13 schematically illustrates the concept of the periodic boundary condition for a two-dimensional system composed of spherocylinder particles. The central square is a simulation region and the surrounding squares are virtual simulation boxes, which are made by replicating the main simulation box. As Figure 2.13 shows, the origin of the xy-coordinate system is taken at the center of the simulation region, and the dimensions of the simulation region in the x- and y-directions are denoted by L_x and L_y. The two specific procedures are necessary in treating the periodic boundary condition, that is, first the treatment of outgoing particles crossing the boundary surfaces of the simulation region and second the calculation of interaction energies or forces with virtual particles being in the replicated simulation boxes.

As shown in Figure 2.13, a particle crossing and exiting the left boundary surface has to enter from the right virtual box. This treatment can be expressed using the FORTRAN language as

```
IF(RXI.GE.LX/2.D0) THEN
   RXI=RXI−LX
ELSE IF(RXI.LT.−LX/2.D0) THEN
   RXI=RXI+LX
END IF
```

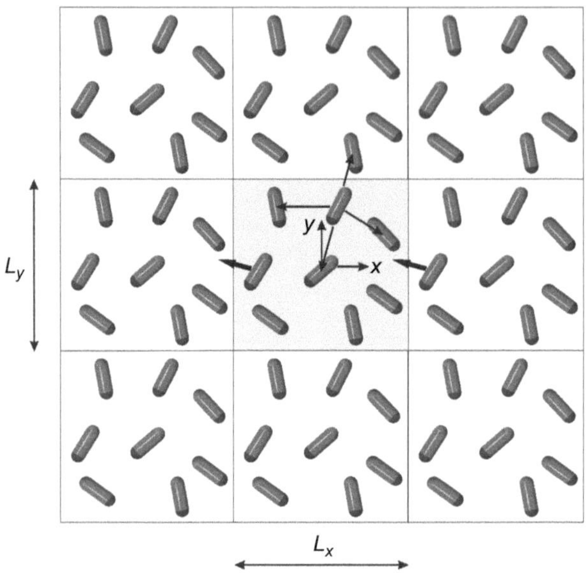

Figure 2.13 Periodic boundary condition.

The rounding-up function DNINT can yield a simple one-line expression as

```
RXI=RXI−DNINT(RXI/LX)*LX
```

Note that the position of particle i is denoted by (RXI, RYI). Similar procedures have to be conducted for the case of the y- and z-directions.

When the interaction energy or force of particle i with other particles, for example, particle j, is calculated, an appropriate particle j has to be chosen as an object from real and virtual particles j. This may be done in such a way that the distance between particle i and particle j is minimal. This treatment can be expressed using the FORTRAN language as

```
IF(RXIJ.GT.LX/2.D0) THEN
   RXIJ=RXIJ−LX
ELSE IF (RXIJ.LT.−LX/2.D0) THEN
   RXIJ=RXIJ+LX
END IF
```

The rounding-up function DNINT gives rise to a simple one-line expression as

```
RXIJ=RXIJ−DNINT(RXIJ/LX)*LX
```

in which RXIJ=RXI−RXJ, expressing the relative position of particles i to j. Similar procedures have to be conducted for the y- and z-directions. The abovementioned procedures are applicable directly to a system composed of rod-like particles, such as that shown in Figure 2.9A, in which the interaction energies or forces are dependent only on the particle center-to-center distance. If we treat the pairs of magnetic charges instead of particle center-to-center interactions, the above-mentioned procedures are also applicable, but in this case RXIJ and similar variables have to be taken as the distances between magnetic charges.

2.4.2 Lees–Edwards Boundary Condition

The periodic boundary condition is quite useful for molecular simulations of a system in thermodynamic equilibrium, but is this boundary condition still available for nonequilibrium situations? In treating the dynamic properties of a system in nonequilibrium, the most basic and important flow is a simple shear flow, as shown in Figure 2.14. The velocity profile, linearly varying from $-U$ at the lower surface to U at the upper one, can be generated by sliding the lower and upper walls in the left and right directions with the velocity U, respectively. This flow field is called the "simple shear flow." In generating such a simple shear flow in actual molecular simulations, the upper and lower replicated simulation boxes, shown in Figure 2.13, are made to slide in different directions with a certain constant speed. This sliding periodic boundary condition is called the "Lees–Edwards boundary condition" [30]. Figure 2.15 schematically depicts the concept of this boundary condition; the replicated boxes in the upper and lower layers slide in each direction by the distance ΔX. If particles move out of the simulation box by crossing the boundary surface normal to the x-axis, as shown in Figure 2.15, they are made to come into the simulation box through the opposite boundary surface, which is exactly the same procedure as the periodic boundary condition. The important treatment in the Lees–Edwards boundary condition concerns the particles crossing the boundary surfaces normal to the y-axis. The same treatment of the periodic boundary condition is applied to the y-coordinate of such particles, but the x-coordinate should be shifted from x to $(x - \Delta X)$ in the case of Figure 2.15. In addition, the x-component v_x of these particles needs to be modified to $(v_x - U)$, but the y-component v_y can be used without modification. The above-mentioned procedures concerning x and v_x can be expressed using the FORTRAN language as

```
IF (RYI.GE.LY/2.D0) THEN
   RXI=RXI−DX
   RXI=RXI−DNINT(RXI/LX)*LX
   VXI=VXI−U
ELSE (RYI.LT.−LY/2.D0) THEN
   RXI=RXI+DX
   RXI=RXI−DNINT(RXI/LX)*LX
   VXI=VXI+U
END IF
```

A slightly simplified expression can be written as

```
CORY=DNINT(RYI/LY)
RXI=RXI−CORY*DX
RXI=RXI−DNINT(RXI/LX)*LX
VXI=VXI−CORY*U
```

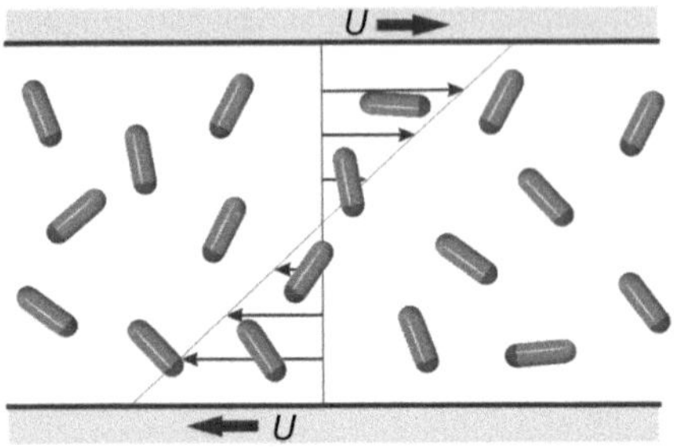

Figure 2.14 Simple shear flow.

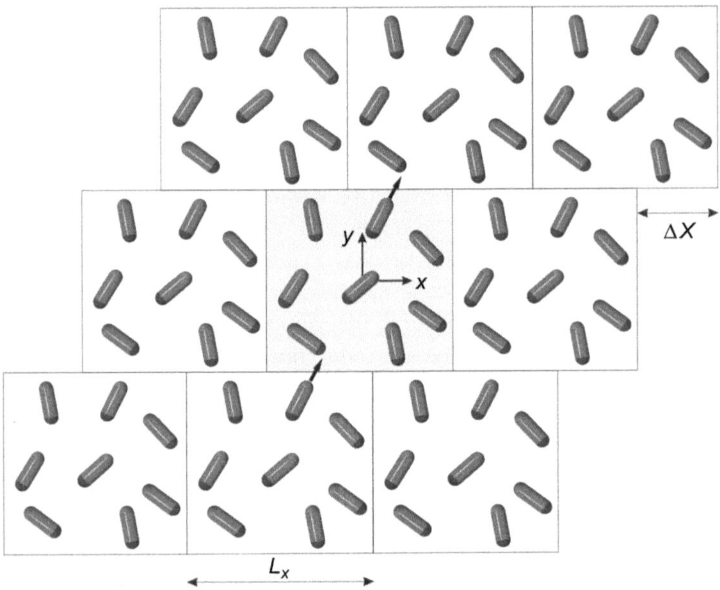

Figure 2.15 Lees–Edwards boundary condition.

The y- and z-coordinates are treated as in the periodic boundary condition, and the modification of v_y and v_z is unnecessary.

For the case of evaluating interaction energies or forces, the similar procedures have to be conducted for the particles interacting with virtual particles which are in the replicated simulation boxes in the upper or lower layers. This treatment can be expressed using the FORTRAN language as

```
IF (RYJI.GE.LY/2.D0) THEN
   RYJI=RYJI−LY
   RXJI=RXJI−DX
   RXJI=RXJI−DNINT(RXJI/LX)*LX
ELSE IF (RYJI.LT.−LY/2.D0) THEN
```

```
    RYJI=RYJI+LY
    RXJI=RXJI+DX
    RXJI=RXJI-DNINT(RXJI/LX)*LX
END IF
```

A slightly simplified expression can be written as

```
CORY=DNINT(RYJI/LY)
RYJI=RYJI-CORY*LY
RXJI=RXJI-CORY*DX
RXJI=RXJI-DNINT(RXJI/LX)*LX
```

The relative position RZJI in the z-direction is treated according to the periodic boundary condition.

The above-mentioned procedures are valid for the particle model shown in Figure 2.9A and also apply to the model shown in Figure 2.9B by focusing on the interactions between magnetic charges instead of the particle centers.

3 Practice of Molecular Dynamics Simulations

In the present and subsequent chapters, we consider examples of physical phenomena in order to explain a series of important procedures employed in conducting molecular simulations. In particular, we discuss the formalization of a problem and the method for nondimensionalizing quantities, and we make several analyses indispensable for developing a simulation program. These techniques are demonstrated in a sample simulation program with explanatory remarks included to clarify important features.

In this chapter, we consider two different physical phenomena as examples for the practice of molecular dynamics simulations. The first example discusses a diffusion problem with two kinds of molecules initially immersed in a small region in thermodynamic equilibrium. The simulation then follows the particle diffusion after the wall surrounding the region has been removed. For this case the Verlet algorithm is used for simulating the particle motion. The second example discusses the problem of the behavior of axisymmetric particles (spherocylinders in this case) in a simple shear flow. This case is an example of a more advanced type of molecular dynamics (MD) simulation where the translational and rotational motion of the particles is simulated simultaneously; therefore this exercise is considerably more advanced. The techniques demonstrated in this example are fundamental to many practical applications and may offer many valuable suggestions in developing a simulation program for systems such as DNA or polymer solutions.

3.1 Diffusion Phenomena in a System of Light and Heavy Molecules

In this section we demonstrate a MD simulation employing only the translational motion of spherical molecules. A spherical molecule system is a basic form employed in molecular simulations, and the diffusion problem in this system is a useful example because almost all the important methodology for developing a simulation program is included in this exercise. A system composed of two kinds of molecules has been chosen because the extra complexity renders the example more useful and practical.

Introduction to Practice of Molecular Simulation. DOI: 10.1016/B978-0-12-385148-2.00003-3

3.1.1 Physical Phenomena of Interest

The two kinds of molecules, that is, the N_A light molecules with mass m and N_B heavy molecules with mass M, are placed in a two-dimensional square cell with side length L in equilibrium with temperature T. Both kinds of molecules have the same diameter σ, and the interaction between molecules is assumed to be expressed by the Lennard-Jones potential. At the moment the wall surrounding the square retaining cell is removed, these molecules start to diffuse into the larger surrounding area. In this example, we will consider how this physical phenomenon depends on the system temperature and the mass ratio.

3.1.2 Specification of Problems in Equations

The starting point for the formalization of a problem is the development of the governing equation—in this case, the equation of motion of the molecules. The equation of motion of an arbitrary light molecule i and arbitrary heavy molecule j are written from Newton's equation of motion, respectively, as

$$m\frac{d^2\mathbf{r}_i}{dt^2} = \mathbf{f}_i = \sum_{p=1}^{N} \mathbf{f}_{ip} \tag{3.1}$$

$$M\frac{d^2\mathbf{r}_j}{dt^2} = \mathbf{f}_j = \sum_{p=1}^{N} \mathbf{f}_{jp} \tag{3.2}$$

in which $N = N_A + N_B$, $\mathbf{f}_{ip}$ is the force exerted by molecule p on molecule i, and $\mathbf{f}_i$ is the total force acting on molecule i from all the other molecules irrespective of the type of molecule. This notation is similarly applicable to a heavy molecule j. The force acting between molecules can be derived from the Lennard-Jones potential. With the aid of the basic formulae of vector analysis, the force $\mathbf{f}$ is derived from a potential U as

$$\mathbf{f} = -\nabla U = -\left(\mathbf{i}\frac{\partial U}{\partial x} + \mathbf{j}\frac{\partial U}{\partial y} + \mathbf{k}\frac{\partial U}{\partial z}\right) \tag{3.3}$$

The notation ∇ on the right-hand side is the nabla operator, which is defined by the last expression on the right-hand side, and ($\mathbf{i}$, $\mathbf{j}$, $\mathbf{k}$) are the unit vectors in the (x, y, z) directions, respectively. Equation (3.3) implies that the force acts in the direction of the interaction energy decreasing at the maximum. By substituting Eq. (2.10) into Eq. (3.3), the force $\mathbf{f}_{qp}$ exerted by molecule p on molecule q can be derived as

$$\mathbf{f}_{qp} = 24\varepsilon\left\{2\left(\frac{\sigma}{r_{qp}}\right)^{12} - \left(\frac{\sigma}{r_{qp}}\right)^{6}\right\}\frac{\mathbf{r}_{qp}}{r_{qp}^2} \tag{3.4}$$

in which $\mathbf{r}_{qp}$ is the relative position vector of molecule q to molecule p, expressed as $\mathbf{r}_{qp} = \mathbf{r}_q - \mathbf{r}_p$, and $r_{qp} = |\mathbf{r}_{qp}|$.

In practice, simulations usually treat a nondimensional system, in which the governing equations and all physical quantities are nondimensionalized by certain representative values. Therefore, in the following paragraphs, we show the method of nondimensionalizing the equations.

For a Lennard-Jones system, the following representative values are generally used for nondimensionalizing quantities: σ for distances, ε for energies, $(\varepsilon/m)^{1/2}$ for velocities, $\sigma(m/\varepsilon)^{1/2}$ for time, ε/σ for forces, ε/k for temperatures, $1/\sigma^3$ for number densities, and m/σ^3 for densities. Nondimensional quantities are expressed as the original quantities with superscript *. Each quantity is expressed as a nondimensional quantity multiplied by the corresponding representative value. The substitution of these quantities into the original dimensional equation yields the desired nondimensional equation. These procedures give rise to the nondimensional form of Eqs. (3.1) and (3.2) expressed, respectively, as

$$\frac{d^2\mathbf{r}_i^*}{dt^{*2}} = \mathbf{f}_i^* = \sum_{p=1}^{N} \mathbf{f}_{ip}^* \tag{3.5}$$

$$K\frac{d^2\mathbf{r}_j^*}{dt^{*2}} = \mathbf{f}_j^* = \sum_{p=1}^{N} \mathbf{f}_{jp}^* \tag{3.6}$$

in which the force is nondimensionalized from Eq. (3.4) as

$$\mathbf{f}_{qp}^* = 24\left\{2\left(\frac{1}{r_{qp}^*}\right)^{12} - \left(\frac{1}{r_{qp}^*}\right)^{6}\right\}\frac{\mathbf{r}_{qp}^*}{(r_{qp}^*)^2} \tag{3.7}$$

The parameter K, appearing in Eq. (3.6), is a nondimensional parameter expressing the mass ratio $K = M/m$, which arises due to the mass m being used as the representative mass. As in this example, it is usual for several additional nondimensional parameters to arise when equations and quantities are nondimensionalized. In order to compare the simulation with experimental results, appropriate values of these nondimensional parameters need to be adopted.

3.1.3 Verlet Algorithm

In this example we employ the Verlet algorithm in order to simulate the motion of the molecules. Referring to Eq. (1.6), the algebraic equations according to the Verlet algorithm can be expressed concerning a light molecule i and heavy molecule j as

$$\mathbf{r}_i^*(t^* + h^*) = 2\mathbf{r}_i^*(t^*) - \mathbf{r}_i^*(t^* - h^*) + h^{*2}\mathbf{f}_i^*(t^*) \tag{3.8}$$

$$\mathbf{r}_j^*(t^* + h^*) = 2\mathbf{r}_j^*(t^*) - \mathbf{r}_j^*(t^* - h^*) + \frac{h^{*2}}{K}\mathbf{f}_j^*(t^*) \tag{3.9}$$

As these equations indicate, in order to execute a simulation program, the Verlet algorithm needs the information of all the molecular positions at $t^* = 0$ and the first time step $t^* = h^*$. If the initial positions and velocities of molecules and the system temperature T are assigned at $t^* = 0$, then the molecular positions at $t^* = h^*$ may be evaluated from Eqs. (3.10) and (3.11).

For the given values of the molecular position $\mathbf{r}_i^*(0)$ and velocity $\mathbf{v}_i^*(0)$ at $t^* = 0$, the position $\mathbf{r}_i^*(h^*)$ at $t^* = h^*$ can be evaluated from Eq. (1.8) as

$$\mathbf{r}_i^*(h^*) = \mathbf{r}_i^*(0) + h^*\mathbf{v}_i^*(0) + \frac{h^{*2}}{2}\mathbf{f}_i^*(0) \tag{3.10}$$

Similarly, the equation for a heavy molecule j can be obtained as

$$\mathbf{r}_j^*(h^*) = \mathbf{r}_j^*(0) + h^*\mathbf{v}_j^*(0) + \frac{h^{*2}}{2K}\mathbf{f}_j^*(0) \tag{3.11}$$

Hence, if the initial position and velocity at $t^* = 0$ are assigned, the position at the next time step can be evaluated from Eqs. (3.10) and (3.11), and the simulation can commence according to Eqs. (3.8) and (3.9).

3.1.4 *Parameters for Simulations*

In addition to the above assignment of the initial positions and velocities, it is necessary to assign the number of molecules N, the system temperature T^*, and the mass ratio K. Setting these parameters corresponds to the specification of the physical system of interest. Moreover, an appropriate time interval h^* and the total number of time steps needed for one simulation run must also be carefully specified in order to conduct a simulation successfully without serious problems, such as a system divergence. Additionally, other specifications may be necessary to assist the postprocessing analysis and visualization. For example, in making an animation, it may be necessary to write out various types of data at specific time steps.

The initial positions are usually assigned by a method employing uniform random numbers. The Maxwellian distribution function, which is the velocity distribution for thermodynamic equilibrium, can be written in nondimensional form for a two-dimensional system from Eq. (2.3) as

$$f^*(\mathbf{v}_j^*) = \left(\frac{K}{2\pi T^*}\right)\exp\left\{-\frac{K}{2T^*}(v_{jx}^{*2} + v_{jy}^{*2})\right\} \tag{3.12}$$

This equation is for a heavy molecule j, but it also holds for a light molecule i by replacing subscript i and K by j and unity, respectively. The method of assigning the initial velocities according to this normal distribution function is explained in Appendix A2. Since the number of degrees of freedom for a two-dimensional system is different from that for a three-dimensional system, the relationship between the average velocity and the specified temperature has a slightly different expression from that in Eq. (2.7). If the square mean velocities of a light molecule i and heavy molecule j are denoted by $\overline{v_i^{*2}}$ and $\overline{v_j^{*2}}$, respectively, these quantities are related to the system temperature by

$$T^* = \frac{\overline{v_i^{*2}}}{2} = K\frac{\overline{v_j^{*2}}}{2} \tag{3.13}$$

The number of light molecules N_A is taken to be equal to that of heavy molecules N_B where $N_A = N_B = 20$ in this exercise. Note that in practice a personal computer can easily handle a much larger system, such as $N_A = 1000$ or 10,000. Generally speaking, it is desirable to run a set of simulations where each parameter is given at least three different values in order to grasp how it may influence the simulation results. If there are many parameters governing a phenomenon, it is advisable that important parameters are taken in several different cases, with a typical value set assigned to the other parameters. In the present exercise, therefore, the temperature T^* and mass ratio K are taken as $T^* = 1.5$, 5, and 10, and $K = 2$, 5, and 10, respectively. The number density $n^*(=n\sigma^2)$ is taken only for the single case of $n^* = 0.1$.

Finally, we discuss an appropriate value for the time interval, which has to be carefully determined because it has a significant influence on both the accuracy of the results and the stability of a simulation. If the mean speed of molecule i is assumed to be equal to the root mean square of velocity, the mean distance of travel for the translational motion during the time interval h is expected as $h(\overline{v_i^2})^{1/2}$. This distance is required to be much shorter than the characteristic distance of the Lennard-Jones potential. Referring to Figure 2.8, this can be expressed mathematically as

$$h(\overline{v_i^2})^{1/2} \ll 0.1 \times \sigma \tag{3.14}$$

Using Eq. (3.13) and expressing the average velocity as a function of T^*, it follows that Eq. (3.14) can be written in nondimensional form as

$$h^* \ll 0.1/\sqrt{2T^*} \tag{3.15}$$

As is clearly shown in Eq. (3.15), a shorter time interval is required for a higher temperature; for example, h^* is taken as $h^* = 0.005$, 0.001, and 0.0005 for $T^* = 1$, 5, and 10, respectively. Unless the time interval is sufficiently short, molecules will have a tendency to overlap in a manner that is physically unreasonable, which will induce divergence of the system. After determining an appropriate value of the time interval, one can determine the length of a simulation run, that is, the total number of time steps. For example, if T^* and h^* are adopted as $T^* = 10$ and

$h^* = 0.0005$, the mean travel distance of molecules per unit time step $h^*(2T^*)^{1/2}$ is approximately equal to 0.002. Hence, if the total number of time steps is set to be 50,000, the paths of the molecules will be of sufficient length to examine the diffusion phenomenon.

3.1.5 Results of Simulations

We show some results of the snapshots of molecules in Figures 3.1 and 3.2, which were obtained by conducting the simulation program that is shown in the next subsection. The figures were obtained for a molecular mass ratio of $K = 2$ and 10, respectively. Each figure shows two snapshots at $t^* = 8$ for the two cases of the temperature $T^* = 1.5$ and 5. These figures clearly show that both species of molecules move more actively and diffuse further in the higher-temperature case. If the snapshots for the same temperature are compared, the diffusion of heavy molecules is less active, and this situation is more significant for the larger mass ratio.

The sequence of snapshots in Figure 3.3 shows how molecules diffuse from the center toward the outer simulation boundaries with time for $K = 10$ and $T^* = 5$. This sequence clearly shows that light molecules start to diffuse from the central area in the outward directions more significantly than heavy molecules.

These results indicate the main qualitative features of the diffusion phenomenon of light and heavy molecules. However, the above discussion is too simple from an academic point of view, therefore quantitative considerations and discussion based on the theoretical background are indispensable. How can we theoretically explain the qualitative features that both the light and heavy molecules diffuse more significantly for a higher temperature, and also that heavy molecules are less able to move for larger values of the mass ratio? This may be explained theoretically by considering that Eq. (3.13) implies the mean velocity is larger for a higher

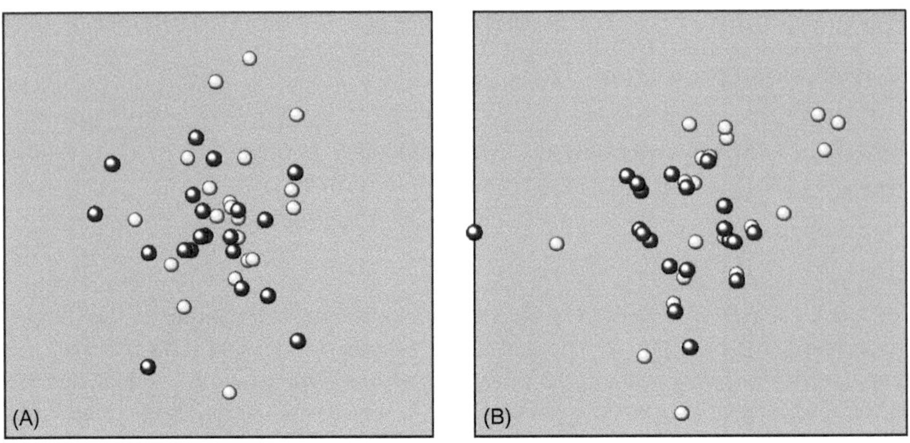

Figure 3.1 Diffusion phenomena of molecules for the mass ratio $K = 2$: (A) $T^* = 1.5$ and (B) $T^* = 5$ (white and black molecules denote light and heavy molecules, respectively).

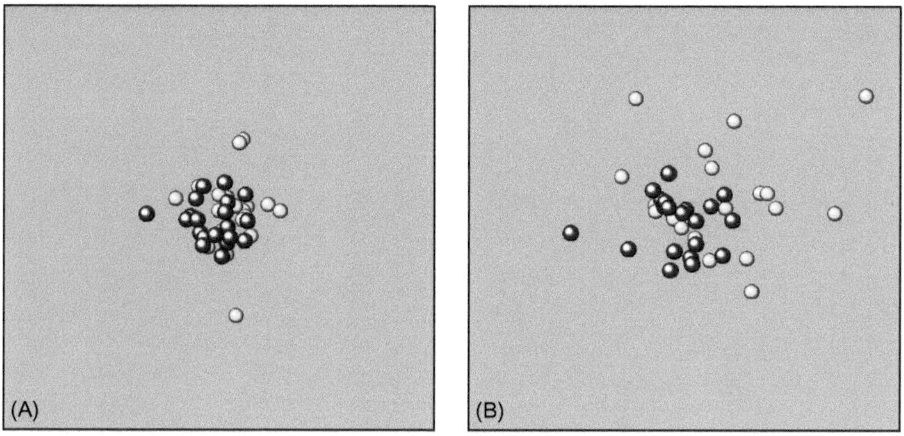

Figure 3.2 Diffusion phenomena of molecules for the mass ratio $K = 10$: (A) $T^* = 1.5$ and (B) $T^* = 5$ (white and black molecules denote light and heavy molecules, respectively).

temperature and that the mean velocity of the heavier molecules is smaller for a larger mass ratio. This may be one of the key theoretical considerations in fully understanding the present simulation results. In academic simulations, such theoretical considerations are conducted in more complex form by combining different threads in order to form a uniformed conclusion about the results. These sophisticated considerations help one to avoid presenting erroneous simulation results, which sometimes happens, for a variety of reasons. Although we here show only the results in the form of snapshots, academic research would require comprehensive quantitative results that might include the change in the internal structures and analysis of the transport coefficients. Furthermore, it will usually be necessary to check the influence of the time interval and the size of a system on the results.

3.1.6 Simulation Program

We here show a sample simulation program to simulate the present diffusion phenomenon. The program is written in the FORTRAN language. Since the main program is usually written in order to clarify a flow of procedure in a straightforward way, the assignment of the initial positions and velocities is treated in a subroutine subprogram. The reader is advised to develop a simulation program with a clear logical flow, thereby simplifying the debugging of a program under development and making it, on completion, a straightforward and useful resource.

For these reasons, the important variables in a program need to be explained in comments at the beginning of the program and each subroutine. These comments provide the user an image of a specific physical meaning from the variable name. In scientific numerical simulations, double-precision variables are usually used for real-type variables, but higher is sometimes more desirable in certain cases, such as solving the problem of an inverse matrix. The following simple simulation program

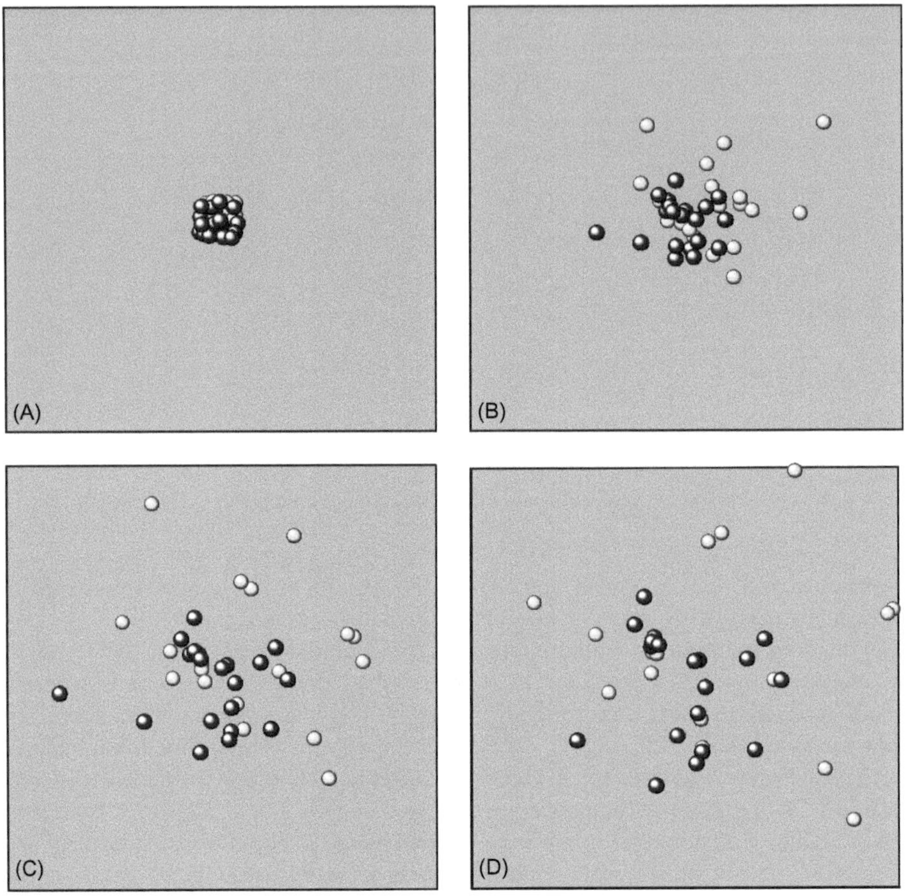

Figure 3.3 Movement of molecules with time for the mass ratio $K = 10$ and the temperature $T^* = 5$: (A) $t^* = 0$, (B) $t^* = 6$, (C) $t^* = 12$, and (D) $t^* = 18$.

has been developed according to these guidelines. The important variables are shown below to help the reader better understand the program.

`RX(I),RY(I)`	:	(x, y) coordinates of the position vector $\mathbf{r}_i^*$ of molecule i
`RX0(I),RY0(I)`	:	Position vector $\mathbf{r}_i^*$ at the previous time step
`FX(I),FY(I)`	:	Force $\mathbf{f}_i^*$ acting on molecule i
`N`	:	Number of molecules in the system
`NA,NB`	:	Numbers of light and heavy molecules, respectively
`K`	:	Mass ratio $K = M/m$
`T`	:	Desired temperature T^*
`H`	:	Time interval h^*
`NDENS`	:	Number density of molecules
`L`	:	Side length of the square simulation region

RAN(J) : Uniform random numbers ranging 0~1 (J=1~NRANMX)
NRAN : Number of used random numbers

Several remarks are attached to the more important statements in the program for the benefit of the reader. Note that the line numbers are for the sake of convenience only and are not necessary during the execution of a simulation program.

```
C*********************************************************************
C*                            diffuse.f                              *
C*                                                                   *
C*    MOLECULAR DYNAMICS METHOD FOR MOLECULAR DIFFUSION PROBLEM      *
C*                --- TWO-DIMENSIONAL CASE ---                       *
C*                                                                   *
C*     OPEN( 9,FILE= '@aaa1.data',STATUS='UNKNOWN')                  *
C*     OPEN(21,FILE='aaa001.data',STATUS='UNKNOWN') ; POSITION DATA  *
C*     OPEN(22,FILE='aaa011.data',STATUS='UNKNOWN') ; POSITION DATA  *
C*     OPEN(23,FILE='aaa021.data',STATUS='UNKNOWN') ; POSITION DATA  *
C*     OPEN(24,FILE='aaa031.data',STATUS='UNKNOWN') ; POSITION DATA  *
C*     OPEN(25,FILE='aaa041.data',STATUS='UNKNOWN') ; POSITION DATA  *
C*                                                                   *
C*                                                                   *
C*                             VER.4 , BY A.SATOH , '04 3/13         *
C*********************************************************************
C
C     RX(I) , RY(I)   : POSITION OF I-TH MOLECULE
C     RX0(I),RY0(I)   : POSITION OF I-TH MOLECULE AT PREVIOUS TIME
C     FX(I) , FY(I)   : FORCE ACTING ON I-TH MOLECULE
C     T        : TEMPERATURE
C     K        : MASS RATIO = MB/MA
C     NDENS    : NUMBER DENSITY OF MOLECULES
C     H        : TIME DIFFERENCE
C     RC       : CUTOFF RADIUS FOR FORCE
C     L        : MAGNITUDE OF CAGE
C     NA       : NUMBER OF MOLECULES OF SPECIES A
C     NB       : NUMBER OF MOLECULES OF SPECIES B
C     N        : TOTAL NUMBER OF MOLECULES
C            -L/2 < RX(I) < L/2   ,   -L/2 < RY(I) < L/2
C--------------------------------------------------------------------
C
      IMPLICIT REAL*8 (A-H,O-Z), INTEGER (I-N)
C
      COMMON /BLOCK1/ RX0 , RY0 , RX , RY
      COMMON /BLOCK2/ FX  , FY
      COMMON /BLOCK3/ VELX, VELY
      COMMON /BLOCK4/ H , RC , L , T , K , NDENS
      COMMON /BLOCK5/ NRAN  , RAN   , IX
C
      PARAMETER( NN=80, NRANMX=50000 )
      PARAMETER( PI=3.141592653589793D0 )
C
      REAL*8  RX0(NN), RY0(NN), RX(NN)  , RY(NN)
      REAL*8  FX(NN) , FY(NN) , VELX(NN), VELY(NN)
      REAL*8  H , RC , L , T , K , NDENS
C
      REAL    RAN(NRANMX)
      INTEGER NRAN  , IX
C
      REAL*8  RXI, RYI, TIME, HSQ, CC0, CC1
      INTEGER N, NA, NB
      INTEGER SWITCH, NTIME, NTIMEMX
      INTEGER NP, NOPT, NGRAPH, NPRINT
C
                 OPEN( 9,FILE= '@aaa1.data',STATUS='UNKNOWN')
                 OPEN(21,FILE='aaa001.data',STATUS='UNKNOWN')
                 OPEN(22,FILE='aaa011.data',STATUS='UNKNOWN')
                 OPEN(23,FILE='aaa021.data',STATUS='UNKNOWN')
```

• The given values and intermediate results are written out in @aaa1.data and the molecular positions are done in aaa001 to aaa041.

```
                    OPEN(24,FILE='aaa031.data',STATUS='UNKNOWN')
                    OPEN(25,FILE='aaa041.data',STATUS='UNKNOWN')
                                                                    NP=9
C                                                   ----- PARAMETER (1) -----
      T     = 5.0D0
      K     = 10.D0
      NA    = 20
      NB    = NA
      H     = 0.001D0
      RC    = 3.D0
      N     = NA + NB
      NDENS = 0.1D0
      L     = DSQRT( DBLE(N)/NDENS )
      HSQ = H*H
```

• Temperature $T^* = 5$, mass ratio $K = 10$, numbers of light and heavy molecules $N_A = N_B = 20$, time interval $h^* = 0.001$, cutoff distance $r^*_{coff} = 3$, number density $n^* = 0.1$, and simulation region size $L^* = (N/n^*)^{1/2}$.

```
C                                                   ----- PARAMETER (2) -----
      NTIMEMX= 10000
      NPRINT = 1000
      NGRAPH = 2000
      NOPT   = 20
```

• The total number of time steps is 10,000, and the molecular positions are written out at every 2000 time steps for the postprocessing analysis.

```
C                                                   ----- PARAMETER (3) -----
      IX     = 0
      CALL RANCAL(NRANMX,IX,RAN)
      NRAN   = 1
C
```

• A sequence of uniform random numbers is prepared in advance and when necessary, random numbers are taken out from the variable RAN(*).

```
C     -------------------------------------------------------------
C     -----------------    INITIAL CONFIGURATION    ---------------
C     -------------------------------------------------------------
C
C                                         --- SET INITIAL POSITIONS ---
      CALL INIPOSIT( N , L )
C                                          --- SET INITIAL VELOCITY ---
      CALL INIVEL( N, NA, NB, T, K, PI )
C                                    --- EVALUATE STARTING VALUE R1 ---
      DO 10 I=1,N
        RX(I) = RX0(I)
        RY(I) = RY0(I)
   10 CONTINUE
```

• The periodic BC is used for SWITCH=0 but not for the other cases.

```
C                                                    --- FORCE CAL. ---
      SWITCH = 0
      CALL FORCE( N, L, RC, SWITCH )
C
      CALL POSITR1( N, NA, H, K )
```

• The molecular positions are calculated at the next time step from Eqs. (3.10) and (3.11) in the subroutine POSITR1.

```
C
C                                          --- PRINT OUT CONSTANTS ---
      WRITE(NP,5) T , K , NDENS , NA , NB , L , H , RC
C                               --- PRINT OUT INITIAL CONFIGURATION ---
      CALL PRINTOUT( N, NA, TIME, NP )
C                                                --- INITIALIZATION ---
      TIME = 0.D0
C     ------------------------------------------------------------
C     ------------------    START OF MAIN LOOP    ----------------
C     ------------------------------------------------------------
C
      SWITCH = 10
C
      DO 100 NTIME=1, NTIMEMX
C
        CALL FORCE(N, L, RC,SWITCH)
        CC0 = 1.D0/K
        CC1 = 1.D0
C
        DO 50 I=1,N
C
          IF ( I .EQ. NA+1 ) CC1 = CC0
          RXI    = 2.D0*RX(I) - RX0(I) + FX(I)*HSQ*CC1
          RYI    = 2.D0*RY(I) - RY0(I) + FY(I)*HSQ*CC1
          RX0(I) = RX(I)
          RY0(I) = RY(I)
          RX(I)  = RXI
          RY(I)  = RYI
C
   50   CONTINUE
```

• The forces acting on each particle are calculated in the subroutine FORCE.
• The molecular positions are calculated from Eqs. (3.8) and (3.9). The previous molecular positions are saved in RX0(*) and RY0(*), and the present are saved in RX(*) and RY(*).

• The molecular positions are written out at every NPRINT time steps for subsequently checking the reliability of results.

```
C                                             --- PRINT OUT DATA ---
          IF ( MOD(NTIME,NPRINT) .EQ. 0 ) THEN
            TIME = H*DBLE(NTIME)
            CALL PRINTOUT( N, NA, TIME, NP )
          END IF
C                                             --- DATA OUTPUT FOR GRAPH ---
          IF ( MOD(NTIME,NGRAPH) .EQ. 0 ) THEN
            NOPT = NOPT + 1
            WRITE(NOPT,56)  N, NA, NB, L, REAL(H)*REAL(NTIME)
C
            DO 60 I =1,N
              IF( I .LE. NA ) THEN
                R = 1.D0
              ELSE
                R = 1.5D0
              END IF
              WRITE(NOPT,58) I, R, RX(I), RY(I)
   60       CONTINUE
            CLOSE(NOPT, STATUS='KEEP')
          END IF
C
C
  100 CONTINUE
C
C     ---------------------------------------------------------------
C     ---------------------   END OF MAIN LOOP   --------------------
C     ---------------------------------------------------------------
      CLOSE(NP, STATUS='KEEP')
C
C     ------------------------- FORMAT ------------------------------
    5 FORMAT(/1H ,'-----------------------------------------------'
     &       /1H ,'          MOLECULAR DYNAMICS SIMULATION         '
     &       /1H ,'FOR TWO-DIMENSIONAL MOLECULAR DIFFUSION PROBLEM  '
     &       /1H ,'-----------------------------------------------'
     &       /1H ,'TEMPERATURE=',F6.2 ,2X, 'MASS RATIO=',F6.2 ,2X,
     &            'NDENS=',F6.3
     &       /1H ,'NUMBER OF MOLECULES OF SPECIES A=',I4
     &       /1H ,'NUMBER OF MOLECULES OF SPECIES B=',I4
     &       /1H ,'MAGNITUDE OF CAGE=',F7.2 ,2X, 'TIME DIFF.=',
     &            F8.5 ,2X, 'CUTOFF RADIUS=',F6.2/)
   56 FORMAT( 3I6, 2E13.8 )
   58 FORMAT( I5, F8.3 , 2E26.18 )
                                                               STOP
                                                               END
C*****************************************************************
C*******************      SUBROUTINE     *************************
C*****************************************************************
C
C**** SUB PRINTOUT *****
      SUBROUTINE PRINTOUT( N, NA, TIME, NP )
C
      IMPLICIT REAL*8 (A-H,O-Z), INTEGER (I-N)
C
      COMMON /BLOCK1/ RX0, RY0, RX, RY
C
      PARAMETER( NN=80 )
C
      REAL*8  RX0(NN), RY0(NN), RX(NN), RY(NN)
      INTEGER N, NA, NP
C
      WRITE(NP,2) TIME
    2 FORMAT(/1H ,'--------------- TIME=',E13.5/)
      WRITE(NP,*)
      WRITE(NP,*)'MOLECULES OF SPECIES A'
      WRITE(NP,*)
      DO 10 I=1,NA
        WRITE(NP,5) I, RX(I), RY(I)
    5   FORMAT(1H ,'I=',I3 ,5X, 'RX=',F8.2 ,5X, 'RY=',F8.2)
   10 CONTINUE
      WRITE(NP,*)
```

• The molecular positions are written out at every NGRAPH time steps for the postprocessing analysis.

• The positions of light molecules are first written out and followed by those of the heavy molecules.

```
      WRITE(NP,*)'MOLECULES OF SPECIES B'
      WRITE(NP,*)
      DO 20 I=NA+1,N
        WRITE(NP,5) I,RX(I),RY(I)
   20 CONTINUE
      WRITE(NP,*)
                                                      RETURN
                                                      END
C**** SUB INIPOSIT *****
      SUBROUTINE INIPOSIT( N, L )
C
      IMPLICIT REAL*8 (A-H,O-Z), INTEGER (I-N)
C
      COMMON /BLOCK1/ RX0, RY0, RX, RY
      COMMON /BLOCK5/ NRAN   , RAN    , IX
C
      PARAMETER( NN=80, NRANMX=50000 )
C
      REAL*8  RX0(NN), RY0(NN), RX(NN), RY(NN), L
      REAL*8  RXIJ , RYIJ , RIJSQ , CRX0 , CRY0
      REAL    RAN(NRANMX)
      INTEGER N, NRAN
C
      DO 10 I=1,N
    2   NRAN = NRAN + 1
        CRX0 = L*( DBLE(RAN(NRAN))-0.5D0)
        NRAN = NRAN + 1
        CRY0 = L*( DBLE(RAN(NRAN))-0.5D0)
        IF( I .NE. 1 ) THEN
          DO 5 J=1,I-1
            RXIJ = CRX0 - RX0(J)
            RYIJ = CRY0 - RY0(J)
            RXIJ = RXIJ - DNINT( RXIJ/L )*L
            RYIJ = RYIJ - DNINT( RYIJ/L )*L
            RIJSQ = RXIJ*RXIJ + RYIJ*RYIJ
            IF ( RIJSQ .LT. 1.D0 )        GOTO 2
    5     CONTINUE
        END IF
        RX0(I) = CRX0
        RY0(I) = CRY0
C
   10 CONTINUE
                                                      RETURN
                                                      END
```

• A subroutine for setting the initial molecular positions.

• Dissimilar to the regular configuration explained in Section 2.1, the initial molecular positions are assigned using random numbers. If r^*_{ij}<1, such molecular positions are not employed because of an extraordinary overlap.

```
C**** SUB INIVEL ****
      SUBROUTINE INIVEL( N, NA, NB, T, K, PI )
C
      IMPLICIT REAL*8 (A-H,O-Z), INTEGER (I-N)
C
      COMMON /BLOCK3/ VELX , VELY
      COMMON /BLOCK5/ NRAN , RAN    , IX
C
      PARAMETER( NN=80, NRANMX=50000 )
C
      INTEGER N, NA, NB, NRAN
      REAL*8  VELX(NN) , VELY(NN) , T , K , PI
      REAL    RAN(NRANMX)
      REAL*8  MOMXA, MOMYA, MOMXB, MOMYB
      REAL*8  CC0, CC1, CC10, CC11
C
      CC0 = 1.D0/K
      CC1 = 1.D0
C
      DO 10 I=1,N
        IF ( I .EQ. NA+1 ) CC1 = CC0
        NRAN = NRAN + 1
        CC10 = DSQRT( -2.D0*T*CC1*DLOG( DBLE(RAN(NRAN)) ) )
        NRAN = NRAN + 1
        CC11 = 2.D0*PI*DBLE(RAN(NRAN))
        VELX(I) = CC10*DCOS(CC11)
        VELY(I) = CC10*DSIN(CC11)
```

• A subroutine for setting the initial molecular velocities.

• The initial velocities are set according to Eq. (2.5) based on Eq. (3.12) using random numbers.

```
   10 CONTINUE
C                                         --- SET TOTAL MOMENTUM ZERO ---
      MOMXA = 0.D0
      MOMYA = 0.D0
      MOMXB = 0.D0
      MOMYB = 0.D0
C
      DO 20 I=1,N
        IF ( I .LE. NA ) THEN
          MOMXA = MOMXA + VELX(I)
          MOMYA = MOMYA + VELY(I)
        ELSE
          MOMXB = MOMXB + VELX(I)
          MOMYB = MOMYB + VELY(I)
        END IF
   20 CONTINUE
C
```

• To make the system momentum zero, the total momentum is first calculated for each light and heavy molecule.

```
      MOMXA = MOMXA/DBLE(NA)
      MOMYA = MOMYA/DBLE(NA)
      MOMXB = MOMXB/DBLE(NB)
      MOMYB = MOMYB/DBLE(NB)
C
```

• The extra momentum permol-ecule is calculated from the total momentum.

```
C                                   --- CORRECT VELOCITIES TO SATISFY  ---
C                                   ---    ZERO TOTAL MOMENTUM         ---
C
      CC10  = MOMXA
      CC11  = MOMYA
      DO 30 I=1,N
        IF ( I .EQ. NA+1 ) THEN
          CC10 = MOMXB
          CC11 = MOMYB
        END IF
        VELX(I) = VELX(I)-CC10
        VELY(I) = VELY(I)-CC11
   30 CONTINUE
                                                              RETURN
                                                              END
```

• The total momentum is forced to be zero by subtracting the extra momentum per molecule from the velocity components of each molecule.

```
C**** SUB FORCE ****
      SUBROUTINE FORCE( N, L, RC, SWITCH )
C
      IMPLICIT REAL*8 (A-H,O-Z), INTEGER (I-N)
C
      COMMON /BLOCK1/ RX0, RY0, RX, RY
      COMMON /BLOCK2/ FX , FY
C
      PARAMETER( NN=80 )
C
      INTEGER N, SWITCH
      REAL*8  RX0(NN), RY0(NN), RX(NN), RY(NN)
      REAL*8  FX(NN) , FY(NN)
      REAL*8  L, RC
      REAL*8  RXI, RYI, RXIJ, RYIJ, RIJSQ
      REAL*8  FXI, FYI, FXIJ, FYIJ, FIJ
      REAL*8  RCSQ, LINV
      REAL*8  SR2, SR6, SR12
C
      RCSQ = RC*RC
      LINV = 1.D0/L
C
```

• A subroutine for calculating the forces acting on molecules.

```
      DO 5 I=1,N
        FX(I) = 0.D0
        FY(I) = 0.D0
    5 CONTINUE
C
```

• The force variables are initialized as zero before proceeding to the main loop.

```
      DO 20 I=1,N-1
C                                           --- FOR I-TH MOLECULE ---
        RXI = RX(I)
        RYI = RY(I)
        FXI = FX(I)
        FYI = FY(I)
```

• The action–reaction law enables us to calculate the forces of only pairs of particles satisfying $j > i$.

```
C
        DO 10 J=I+1,N
C                                              --- FOR I-TH AND J-TH ---
          RXIJ = RXI - RX(J)
          RYIJ = RYI - RY(J)
          IF ( SWITCH .EQ. 0 ) THEN
            RXIJ = RXIJ-DNINT( RXIJ*LINV )*L
            RYIJ = RYIJ-DNINT( RYIJ*LINV )*L
          END IF
          IF ( DABS(RXIJ) .GT. RC ) GOTO 10
          IF ( DABS(RYIJ) .GT. RC ) GOTO 10
          RIJSQ = RXIJ*RXIJ + RYIJ*RYIJ
          IF ( RIJSQ .GT. RCSQ )    GOTO 10
C
          SR2   = 1.D0/RIJSQ
          SR6   = SR2**3
          SR12  = SR6**2
          FIJ   = ( 2.D0*SR12-SR6 )/RIJSQ
          FXIJ  = FIJ*RXIJ
          FYIJ  = FIJ*RYIJ
          FXI   = FXI + FXIJ
          FYI   = FYI + FYIJ
          FX(J) = FX(J) - FXIJ
          FY(J) = FY(J) - FYIJ
   10   CONTINUE
C
        FX(I) = FXI
        FY(I) = FYI
C
   20 CONTINUE
C
      DO 30 I=1,N
        FX(I) = FX(I)*24.D0
        FY(I) = FY(I)*24.D0
   30 CONTINUE
                                                                 RETURN
                                                                 END
```

• The periodic BC is used for SWITCH = 0.
• The particles separating over the cutoff distance r^*_{coff} are passed without calculation of forces.

• The forces between molecules are calculated from Eq. (3.7).

• The factor 24 in Eq. (3.7) will be multiplied later.

• The action–reaction law can provide the force FX(J) and FY(J) as (–FXIJ) and (–FYIJ).

```
C**** SUB POSITR1 ****
      SUBROUTINE POSITR1( N, NA, H, K )
C
      IMPLICIT REAL*8 (A-H,O-Z), INTEGER (I-N)
C
      COMMON /BLOCK1/ RX0 , RY0,  RX,  RY
      COMMON /BLOCK2/ FX  , FY
      COMMON /BLOCK3/ VELX, VELY
C
      PARAMETER( NN=80 )
C
      REAL*8  RX0(NN), RY0(NN), RX(NN)  , RY(NN)
      REAL*8  FX(NN) , FY(NN) , VELX(NN), VELY(NN)
      REAL*8  H , K
      REAL*8  HSQ2, CC0, CC1
      INTEGER NA  , N
C
      HSQ2  = H*H/2.D0
      CC0   = 1.D0/K
      CC1   = 1.D0
C
      DO 10 I=1,N
        IF( I .EQ. NA+1 ) CC1 = CC0
        RX(I) = RX0(I) + H*VELX(I) + HSQ2*FX(I)*CC1
        RY(I) = RY0(I) + H*VELY(I) + HSQ2*FY(I)*CC1
   10 CONTINUE
                                                                 RETURN
                                                                 END
```

• The starting value of the molecular positions is calculated from Eqs. (3.10) and (3.11).

```
C**** SUB RANCAL ****
      SUBROUTINE RANCAL( N, IX, X )
C
      DIMENSION  X(N)
      DATA INTEGMX/2147483647/
      DATA INTEGST,INTEG/584287,48828125/
```

• A subroutine for generating a niform random number sequence.

```
C
      AINTEGMX = REAL( INTEGMX )
C
      IF ( IX.LT.0 ) PAUSE
      IF ( IX.EQ.0 ) IX = INTEGST
      DO 30 I=1,N
         IX = IX*INTEG
         IF (IX) 10, 20, 20
   10    IX   = (IX+INTEGMX)+1
   20    X(I) = REAL(IX)/AINTEGMX
   30 CONTINUE
      RETURN
      END
C*********************************************************************
C   THIS SUBROUTINE IS FOR GENERATING UNIFORM RANDOM NUMBERS         *
C   (SINGLE PRECISION) FOR 64-BIT COMPUTER.                          *
C      N       : NUMBER OF RANDOM NUMBERS TO GENERATE                *
C      IX      : INITIAL VALUE OF RANDOM NUMBERS (POSITIVE INTEGER)  *
C              : LAST GENERATED VALUE IS KEPT                        *
C      X(N)    : GENERATED RANDOM NUMBERS (0<X(N)<1)                 *
C*********************************************************************
C**** SUB RANCAL ****
ccc   SUBROUTINE RANCAL( N, IX, X )
C
ccc   IMPLICIT REAL*8(A-H,O-Z),INTEGER*8 (I-N)
C
ccc   REAL      X(N)
ccc   INTEGER*8 INTEGMX, INTEG64, INTEGST, INTEG
C
CCC   DATA INTEGMX/2147483647/
ccc   DATA INTEG64/2147483648/
ccc   DATA INTEGST,INTEG/584287,48828125/
C
CCC   AINTEGMX = REAL( INTEGMX )
ccc   AINTEGMX = REAL( INTEG64 )
C
ccc   IF ( IX.LT.0 ) PAUSE
ccc   IF ( IX.EQ.0 ) IX = INTEGST
ccc   DO 30 I=1,N
ccc      IX = IX*INTEG
ccc      IX = KMOD(IX,INTEG64)
CCC      IF (IX) 10, 20, 20
CCC10    IX   = (IX+INTEGMX)+1
ccc20    X(I) = REAL(IX)/AINTEGMX
ccc30 CONTINUE
ccc   RETURN
ccc   END
```

• This is for a 32-bit CPU based on the expression of two's complement.

• This is also a random number generating subroutine for a 64-bit CPU based on the expression of two's complement.

3.2 Behavior of Rod-like Particles in a Simple Shear Flow

In the present section, we consider the behavior of axisymmetric particles, known as spherocylinders, in a simple shear flow as the second demonstration of the MD method. MD simulations for rod-like particles are much more complex than those for a spherical particle system, since the translational and rotational motion of rod-like particles must be treated simultaneously. Hence, this exercise is of a considerably high level and may be applicable to a range of academic research fields. The present simulation method for a spherocylinder particle system is expected to offer many important suggestions in developing practical simulation programs, such as for the adsorption phenomenon between carbon-nanotubes and nonspherical molecules.

3.2.1 Physical Phenomena of Interest

The dispersion of interest in this exercise is composed of spherocylinder particles with mass m and inertia moment I, and it is subjected to a simple shear flow. The spherocylinder particle has a positive and negative magnetic charges (NS poles) at each center of the hemisphere cap situated at both ends of the cylindrical body. This magnetic particle is coated with a steric (surfactant) layer, which acts to prevent the particles from aggregating and thus sedimentation in the gravitational field. In this exercise we consider how such a dispersion behaves under the circumstance of an applied magnetic field in addition to the simple shear flow.

The main subjects for the formalization of this problem are explained in the following subsections. Essentially, they are the modeling of the particles, the formalization of the equation of motion, the derivation of forces and torques acting on particles, and the nondimensionalization of quantities.

3.2.2 Particle Model

As shown in Figure 3.4, a magnetic rod-like particle is modeled as a spherocylinder, with the magnetic charges $\pm q$ at the center of each hemisphere cap, which has a length l_0 and a cylindrical diameter d of the cylindrical part. The particle is covered with a uniform steric (surfactant) layer with thickness δ, and the overlap of these steric layers induces a repulsive interaction between the particles. In the following we show magnetic forces and torques acting on the magnetic particles.

If a magnetic charge q and a magnetic dipole moment $\mathbf{m}$ are acted upon by a uniform applied magnetic field $\mathbf{H}$, then the force $\mathbf{F}$ acting on the charge and the torque $\mathbf{T}$ acting on the dipole moment may be found from a standard textbook on magnetic material engineering as

$$\mathbf{F} = \mu_0 q\mathbf{H}, \quad \mathbf{T} = \mu_0 \mathbf{m} \times \mathbf{H} \tag{3.16}$$

The magnetic field $\mathbf{H}^{(\mathrm{ind})}$ at an arbitrary relative position $\mathbf{r}$ ($r = |\mathbf{r}|$) induced by the magnetic charge q is expressed as

$$\mathbf{H}^{(\mathrm{ind})} = \frac{q}{4\pi r^2} \cdot \frac{\mathbf{r}}{r} \tag{3.17}$$

Note that in this book we employ such a unit system concerning magnetic properties that the magnetization corresponds to the magnetic field, that is,

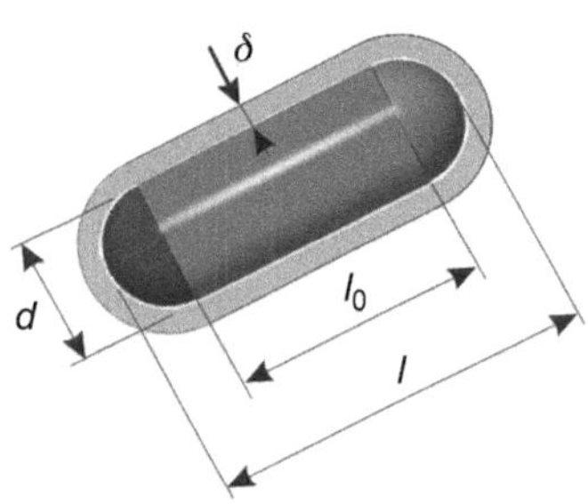

Figure 3.4 Particle model.

$\mathbf{B} = \mu_0(\mathbf{H} + \mathbf{M})$; the correspondence table between two representative unit systems is shown in Appendix A4. With these basic formulae, we derive the magnetic force and torque acting on the spherocylinder particle shown in Figure 3.4.

If the position vector of the center of particle i is denoted by $\mathbf{r}_i$ and the particle direction by $\mathbf{e}_i$, then the position vectors $\mathbf{r}_i^+$ and $\mathbf{r}_i^-$ of the magnetic charges q and $-q$ can be expressed as

$$\mathbf{r}_i^+ = \mathbf{r}_i + (l_0/2)\mathbf{e}_i, \quad \mathbf{r}_i^- = \mathbf{r}_i - (l_0/2)\mathbf{e}_i \tag{3.18}$$

The magnetic field $\mathbf{H}_{ij}^+$ at the position $\mathbf{r}_i^+$ induced by particle j can be written from Eq. (3.17) as

$$\begin{aligned}\mathbf{H}_{ij}^+ &= \frac{q}{4\pi}\cdot\frac{\mathbf{r}_i^+ - \mathbf{r}_j^+}{\left|\mathbf{r}_i^+ - \mathbf{r}_j^+\right|^3} - \frac{q}{4\pi}\cdot\frac{\mathbf{r}_i^+ - \mathbf{r}_j^-}{\left|\mathbf{r}_i^+ - \mathbf{r}_j^-\right|^3} \\ &= \frac{q}{4\pi}\left\{\frac{\mathbf{r}_{ij} + \dfrac{l_0}{2}(\mathbf{e}_i - \mathbf{e}_j)}{\left|\mathbf{r}_{ij} + \dfrac{l_0}{2}(\mathbf{e}_i - \mathbf{e}_j)\right|^3} - \frac{\mathbf{r}_{ij} + \dfrac{l_0}{2}(\mathbf{e}_i + \mathbf{e}_j)}{\left|\mathbf{r}_{ij} + \dfrac{l_0}{2}(\mathbf{e}_i + \mathbf{e}_j)\right|^3}\right\}\end{aligned} \tag{3.19}$$

Similarly, $\mathbf{H}_{ij}^-$ at $\mathbf{r}_i^-$ induced by particle j is written as

$$\mathbf{H}_{ij}^- = \frac{q}{4\pi}\left\{\frac{\mathbf{r}_{ij} - (l_0/2)(\mathbf{e}_i + \mathbf{e}_j)}{|\mathbf{r}_{ij} - (l_0/2)(\mathbf{e}_i + \mathbf{e}_j)|^3} - \frac{\mathbf{r}_{ij} - (l_0/2)(\mathbf{e}_i - \mathbf{e}_j)}{|\mathbf{r}_{ij} - (l_0/2)(\mathbf{e}_i - \mathbf{e}_j)|^3}\right\} \tag{3.20}$$

in which $\mathbf{r}_{ij} = \mathbf{r}_i - \mathbf{r}_j$. Hence, the magnetic forces acting on the positive and negative magnetic charges of particle i, $\mathbf{F}_{ij}^+$ and $\mathbf{F}_{ij}^-$, by the magnetic charges of particle j, are finally written as

$$\mathbf{F}_{ij}^+ = \mu_0 q\mathbf{H}_{ij}^+ = \frac{\mu_0 q^2}{4\pi}\left\{\frac{\mathbf{r}_{ij} + (l_0/2)(\mathbf{e}_i - \mathbf{e}_j)}{|\mathbf{r}_{ij} + (l_0/2)(\mathbf{e}_i - \mathbf{e}_j)|^3} - \frac{\mathbf{r}_{ij} + (l_0/2)(\mathbf{e}_i + \mathbf{e}_j)}{|\mathbf{r}_{ij} + (l_0/2)(\mathbf{e}_i + \mathbf{e}_j)|^3}\right\} \tag{3.21}$$

$$\mathbf{F}_{ij}^- = -\mu_0 q\mathbf{H}_{ij}^- = -\frac{\mu_0 q^2}{4\pi}\left\{\frac{\mathbf{r}_{ij} - (l_0/2)(\mathbf{e}_i + \mathbf{e}_j)}{|\mathbf{r}_{ij} - (l_0/2)(\mathbf{e}_i + \mathbf{e}_j)|^3} - \frac{\mathbf{r}_{ij} - (l_0/2)(\mathbf{e}_i - \mathbf{e}_j)}{|\mathbf{r}_{ij} - (l_0/2)(\mathbf{e}_i - \mathbf{e}_j)|^3}\right\} \tag{3.22}$$

Similarly, the magnetic torque about the particle axis of particle i, $\mathbf{T}_{ij}^+$, due to the force acting on the positive charge by the magnetic charges of particle j, is obtained as

$$\mathbf{T}_{ij}^+ = \frac{l_0}{2}\mathbf{e}_i \times \mathbf{F}_{ij}^+ = \frac{\mu_0 q^2 l_0}{8\pi}\left\{\frac{\mathbf{e}_i \times \mathbf{r}_{ij} + \dfrac{l_0}{2}(-\mathbf{e}_i \times \mathbf{e}_j)}{\left|\mathbf{r}_{ij} + \dfrac{l_0}{2}(\mathbf{e}_i - \mathbf{e}_j)\right|^3} - \frac{\mathbf{e}_i \times \mathbf{r}_{ij} + \dfrac{l_0}{2}(\mathbf{e}_i \times \mathbf{e}_j)}{\left|\mathbf{r}_{ij} + \dfrac{l_0}{2}(\mathbf{e}_i + \mathbf{e}_j)\right|^3}\right\} \tag{3.23}$$

Also, such a torque $\mathbf{T}_{ij}^{-}$ due to the force acting on the negative charge is as follows:

$$\mathbf{T}_{ij}^{-} = -\frac{l_0}{2}\mathbf{e}_i \times \mathbf{F}_{ij}^{-} = \frac{\mu_0 q^2 l_0}{8\pi}\left\{ \frac{\mathbf{e}_i \times \mathbf{r}_{ij} - \frac{l_0}{2}(\mathbf{e}_i \times \mathbf{e}_j)}{\left|\mathbf{r}_{ij} - \frac{l_0}{2}(\mathbf{e}_i + \mathbf{e}_j)\right|^3} - \frac{\mathbf{e}_i \times \mathbf{r}_{ij} - \frac{l_0}{2}(-\mathbf{e}_i \times \mathbf{e}_j)}{\left|\mathbf{r}_{ij} - \frac{l_0}{2}(\mathbf{e}_i - \mathbf{e}_j)\right|^3} \right\} \tag{3.24}$$

From these equations, the total magnetic force and torque acting on particle i by particle j are written as

$$\mathbf{F}_{ij}^{(\mathrm{m})} = \mathbf{F}_{ij}^{+} + \mathbf{F}_{ij}^{-}, \quad \mathbf{T}_{ij}^{(\mathrm{m})} = \mathbf{T}_{ij}^{+} + \mathbf{T}_{ij}^{-} \tag{3.25}$$

It is noted that $\mathbf{F}_{ji}^{(\mathrm{m})} = -\mathbf{F}_{ij}^{(\mathrm{m})}$ due to the action–reaction law.

A uniform applied magnetic field does not induce a force acting on a particle because there is no field gradient, but it does induce torque. Similar to the above derivation, the torque due to an applied magnetic field can be derived as

$$\mathbf{T}_i^{(\mathrm{H})} = \frac{l_0}{2}\mathbf{e}_i \times \mu_0 q\mathbf{H} - \frac{l_0}{2}\mathbf{e}_i \times (-\mu_0 q\mathbf{H}) = \mu_0(l_0 q\mathbf{e}_i) \times \mathbf{H} \tag{3.26}$$

Since the force and torque due to the overlap of the steric layers cannot be derived straightforwardly, we will discuss this interaction in detail later.

3.2.3 Equation of Motion and Molecular Dynamics Algorithm

The spherocylinder particle is axisymmetric and therefore we can employ the method shown in Section 1.1.2 for simulating the motion of particles. However, several modifications are necessary because we consider the behavior of the particles in a simple shear flow, not in a quiescent flow. If the particles are smaller than micron order, the inertia terms are negligible, which means that we can use the equations shown in Section 1.1.2. The equations of motion under the circumstance of a simple shear flow can be obtained by adding new terms due to the flow into Eqs. (1.42) and (1.43) as

$$\mathbf{v}_i^{||} = \mathbf{U}^{||}(\mathbf{r}_i) + \frac{1}{\eta X^A}\mathbf{F}_i^{||}, \quad \mathbf{v}_i^{\perp} = \mathbf{U}^{\perp}(\mathbf{r}_i) + \frac{1}{\eta Y^A}\mathbf{F}_i^{\perp} \tag{3.27}$$

$$\boldsymbol{\omega}_i^{||} = \boldsymbol{\Omega}^{||} + \frac{1}{\eta X^C}\mathbf{T}_i^{||}, \quad \boldsymbol{\omega}_i^{\perp} = \boldsymbol{\Omega}^{\perp} + \frac{1}{\eta Y^C}\mathbf{T}_i^{\perp} - \frac{Y^H}{Y^C}(\boldsymbol{\varepsilon}\cdot\mathbf{e}_i\mathbf{e}_i) : \mathbf{E} \tag{3.28}$$

in which an arbitrary vector is decomposed into the two vectors parallel and normal to the particle axis. These vectors are denoted by superscripts || and $\perp$, respectively: for example, $\mathbf{v}_i = \mathbf{v}_i^{||} + \mathbf{v}_i^{\perp}$. We here treat only the angular velocity $\boldsymbol{\omega}_i^{\perp}$ and neglect $\boldsymbol{\omega}_i^{||}$ because the rotational motion about the particle axis does not affect the particle orientation and the magnetic interactions. The velocity field $\mathbf{U}(\mathbf{r})$ for a simple shear flow is defined as

$$\mathbf{U}(\mathbf{r}) = \dot{\gamma}\begin{bmatrix} 0 & 1 & 0 \\ 0 & 0 & 0 \\ 0 & 0 & 0 \end{bmatrix}\begin{bmatrix} x \\ y \\ z \end{bmatrix} \tag{3.29}$$

in which $\mathbf{r}$ is the position vector from the origin of the coordinate system, expressed as $\mathbf{r} = (x,\ y,\ z)$. In this flow case, the rotational angular velocity $\boldsymbol{\Omega}$ and the rate-of-strain tensor $\mathbf{E}$ are derived from the definitions as

$$\boldsymbol{\Omega} = \frac{1}{2}\nabla \times \mathbf{U}(r) = -\frac{\dot{\gamma}}{2}\begin{bmatrix} 0 \\ 0 \\ 1 \end{bmatrix}, \quad \mathbf{E} = \frac{1}{2}(\nabla\mathbf{U} + (\nabla\mathbf{U})^t) = \frac{\dot{\gamma}}{2}\begin{bmatrix} 0 & 1 & 0 \\ 1 & 0 & 0 \\ 0 & 0 & 0 \end{bmatrix} \tag{3.30}$$

in which the superscript t denotes a transposed tensor, and $\dot{\gamma}$ is the shear rate and a constant representing the strength of a shear flow. Also, $\boldsymbol{\varepsilon}$ appeared as the last term in the second equation of Eq. (3.28) is a third-rank tensor called "Eddington's epsilon." The ijk-component of this tensor, ε_{ijk}, is expressed as

$$\varepsilon_{ijk} = \left\{ \begin{array}{lll} 1 & \text{for} & (i,j,k) = (x,y,z), (y,z,x), (z,x,y) \\ -1 & \text{for} & (i,j,k) = (z,y,x), (y,x,z), (x,z,y) \\ 0 & \text{for} & \text{the other cases} \end{array} \right\} \tag{3.31}$$

With these characteristics of ε_{ijk} and $\mathbf{E}$ in Eq. (3.30), the last term of the second equation in Eq. (3.28) can be simplified to

$$-\frac{Y^H}{Y^C}(\boldsymbol{\varepsilon}\cdot\mathbf{e}_i\mathbf{e}_i) : \mathbf{E} = -\frac{Y^H}{Y^C}\cdot\frac{\dot{\gamma}}{2}\begin{bmatrix} e_{iz}e_{ix} \\ -e_{iz}e_{iy} \\ {e_{iy}}^2 - {e_{ix}}^2 \end{bmatrix} \tag{3.32}$$

In obtaining Eq. (3.32), the following simple formulae have been used:

$$\mathbf{ab} = \begin{bmatrix} a_x b_x & a_x b_y & a_x b_z \\ a_y b_x & a_y b_y & a_y b_z \\ a_z b_x & a_z b_y & a_z b_z \end{bmatrix} \tag{3.33}$$

$$\begin{aligned} \mathbf{A}:\mathbf{B} &= A_{xx}B_{xx} + A_{xy}B_{yx} + A_{xz}B_{zx} + A_{yx}B_{xy} + A_{yy}B_{yy} + A_{yz}B_{zy} \\ &\quad + A_{zx}B_{xz} + A_{zy}B_{yz} + A_{zz}B_{zz} \end{aligned} \tag{3.34}$$

$$(\boldsymbol{\varepsilon}\cdot\mathbf{ab}) : \mathbf{A} = \begin{bmatrix} a_z(b_x A_{xy} + b_y A_{yy} + b_z A_{zy}) - a_y(b_x A_{xz} + b_y A_{yz} + b_z A_{zz}) \\ a_x(b_x A_{xz} + b_y A_{yz} + b_z A_{zz}) - a_z(b_x A_{xx} + b_y A_{yx} + b_z A_{zx}) \\ a_y(b_x A_{xx} + b_y A_{yx} + b_z A_{zx}) - a_x(b_x A_{xy} + b_y A_{yy} + b_z A_{zy}) \end{bmatrix} \tag{3.35}$$

in which **a** and **b** are arbitrary one-rank tensors, **A** and **B** are arbitrary two-rank tensors, and $\boldsymbol{\varepsilon}$ is the three-rank tensor previously defined.

The quantities used to determine the translational and angular velocities from Eqs. (3.27) and (3.28) can be obtained from the force $\mathbf{F}_i$ and torque $\mathbf{T}_i$ acting on particle i and also from the particle direction $\mathbf{e}_i$ as

$$\left.\begin{array}{lll} \mathbf{F}_i^{||} = (\mathbf{F}_i \cdot \mathbf{e}_i)\mathbf{e}_i, & \mathbf{F}_i^{\perp} = \mathbf{F}_i - \mathbf{F}_i^{||}, & \mathbf{T}_i^{||} = (\mathbf{T}_i \cdot \mathbf{e}_i)\mathbf{e}_i, \\ \mathbf{T}_i^{\perp} = \mathbf{T}_i - \mathbf{T}_i^{||}, & \boldsymbol{\Omega}_i^{||} = (\boldsymbol{\Omega}_i \cdot \mathbf{e}_i)\mathbf{e}_i, & \boldsymbol{\Omega}_i^{\perp} = \boldsymbol{\Omega}_i - \boldsymbol{\Omega}_i^{||} \end{array}\right\} \tag{3.36}$$

With the solutions of $\mathbf{v}_i(t)$ and $\boldsymbol{\omega}_i(t)$, the particle position $\mathbf{r}_i(t + \Delta t)$ and the particle direction $\mathbf{e}_i(t + \Delta t)$ at the next time step can be evaluated from Eqs. (1.45) and (1.46). That is,

$$\mathbf{r}_i(t + \Delta t) = \mathbf{r}_i(t) + \Delta t \mathbf{v}_i(t) \tag{3.37}$$

$$\mathbf{e}_i(t + \Delta t) = \mathbf{e}_i(t) + \Delta t \boldsymbol{\omega}_i^{\perp}(t) \times \mathbf{e}_i(t) \tag{3.38}$$

Finally, we discuss the resistance functions X^A, Y^A, X^C, Y^C, and Y^H [4,16–18]. There would be no difficulties for simulations if the solutions of these resistance functions were known for a spherocylinder particle. However, the solutions are known only for a cylindrical particle with sufficiently large aspect ratio, or for the spherical particle explained before. These solutions are for a solid particle, but in our case we are considering a solid particle coated with a soft steric layer, and the resistance functions have not yet been solved for this case.

Hence, in conducting MD simulations for the present particle dispersion, we have several options for overcoming the problem for the resistance functions. The first option is to tackle the difficult mathematical problem of solving these resistance functions. The second option is to apply the known solutions of a solid spheroidal particle as the first approximation. The third option is to introduce the modeling of the spherocylinder particle in order for the known solutions to be applied more accurately. Here we adopt the second option, that is, the solutions shown in Eqs. (1.35) and (1.36) for a solid spheroid are used for the resistance functions for the spherocylinder shown in Figure 3.4. In addition, the resistance function Y^H can be written as

$$Y^H = 8\pi a^3 \cdot \frac{4}{3} \cdot \frac{s^5}{-2s + (1 + s^2)L} \tag{3.39}$$

In the limiting case of $s \ll 1$, this can be approximated as

$$Y^H = 8\pi a^3 \left(\frac{1}{2} s^2 - \frac{1}{5} s^4 + \cdots \right) \tag{3.40}$$

in which a, b, and s are assumed to be expressed as $a = l/2 + \delta$, $b = d/2 + \delta$, and $s = \sqrt{(l/2 + \delta)^2 - (d/2 + \delta)^2} / (l/2 + \delta)$, respectively.

3.2.4 Modeling of Steric Repulsive Interaction

If the two spherocylinder particles coated with a surfactant layer, shown in Figure 3.4, overlap, how should we write this repulsive interaction as a mathematical expression? To answer this question, we first need to analyze the behavior of the surfactant molecules in detail in such a situation. However, it may be possible to develop a physically acceptable model as a first approximation by combining the known solutions in a sophisticated manner. For a spherical particle system, an expression for the repulsive interaction has already been obtained. Hence, the extension of this potential to the present spherocylinder particle system enables us to overcome the problem of the unknown potential for a spherocylinder coated with a soft surfactant layer.

We consider a spherical particle modeled as a solid sphere of diameter d coated by a uniform surfactant layer of thickness δ. An interaction energy arising from the overlap of these two particles has already been derived from the entropy calculation as [31,32]

$$u_{ij}^{(V)} = \frac{\pi d^2 n_s kT}{2}\left\{2 - \left(\frac{r_{ij}}{\delta}\right)\ln\left(\frac{d+2\delta}{r_{ij}}\right) - \frac{r_{ij}-d}{\delta}\right\} \tag{3.41}$$

in which n_s is the number of surfactant molecules per unit area on the particle surface, k is Boltzmann's constant, and T is the system temperature. The force acting on particle i, $\mathbf{F}_{ij}^{(V)}$, by particle j due to the overlap can be obtained from this equation as

$$\mathbf{F}_{ij}^{(V)} = -\frac{\partial}{\partial \mathbf{r}_i}u_{ij}^{(V)} = -\frac{\partial}{\partial \mathbf{r}_{ij}}u_{ij}^{(V)} = \frac{\pi d^2 n_s kT}{2\delta}\mathbf{t}_{ij}\ln\left(\frac{d+2\delta}{r_{ij}}\right) \quad (\text{for } d \le r_{ij} \le d+2\delta) \tag{3.42}$$

in which $\mathbf{t}_{ij}$ $(=\mathbf{r}_{ij}/r_{ij})$ is the unit vector. It is shown in Eq. (3.42) that this repulsive force acts along a line drawn between the two particles.

We now idealize the spherocylinder particle in order to apply Eq. (3.42). The most feasible model is a linear sphere-connected model shown in Figure 3.5. In this model, solid spheres are linearly connected in contact and covered by a uniform surfactant layer of thickness δ. If the constituent spherical particles are located at each fixed position in the rod-like particle, this model does not necessarily yield

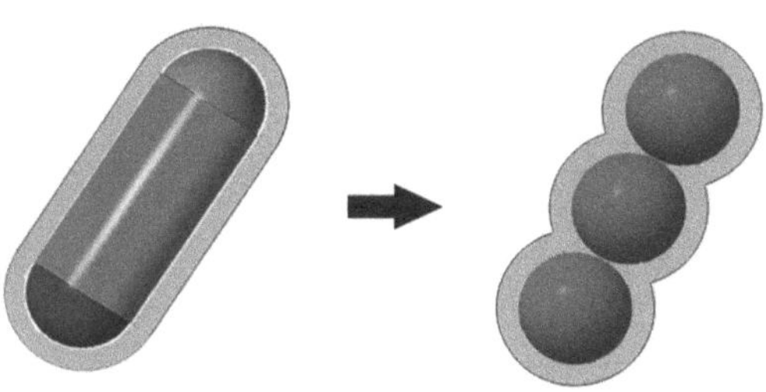

Figure 3.5 Sphere-connected model for calculating repulsive interactions.

a maximum repulsive interaction energy at a position where the maximum energy is provided from the overlap of the original spherocylinder particles. In order to overcome this shortcoming, the above model must be slightly modified to yield a maximum repulsive energy at a position of minimum separation between the two spherocylinder particles. To do so, two spheres are first located at the positions in each spherocylinder, where a maximum repulsive energy is yielded, and then other spheres are linearly added on each side of these two spheres on the original particle to produce a modified sphere-connected model. This is the particle model we use for evaluating interaction energies due to particle overlap.

In the following paragraphs, we show a method for calculating the force and torque acting between particles i and j based on the above-mentioned sphere-connected model. An important task for evaluating such a force and torque is to find the positions along each particle axis at which the separation between the two spherocylinder particles is minimized for the given position and orientation of these two particles. Hence, we focus on a method for finding this minimum separation, including a way of assessing the particle overlap.

The notation $\mathbf{r}_i$ is used for the center of spherocylinder particle i shown in Figure 3.4; similarly, $\mathbf{r}_j$ is used for particle j. Figure 3.6 demonstrates that there is a view angle from which the two particles can be seen as existing in two parallel planes. In Figure 3.6, two points P_i and P_j are taken on each particle axis line such that the line drawn between these points is normal to the two parallel planes. Consideration of the two points P_i and P_j enables us to make a systematic and sophisticated assessment of the particle overlap. If the points P_i and P_j are denoted by $(\mathbf{r}_i + k_i\mathbf{e}_i)$ and $(\mathbf{r}_j + k_j\mathbf{e}_j)$, respectively, the line $\overline{P_iP_j}$ has to satisfy the following equation from the orthogonality condition:

$$\mathbf{e}_i \cdot \{(\mathbf{r}_i + k_i\mathbf{e}_i) - (\mathbf{r}_j + k_j\mathbf{e}_j)\} = 0, \quad \mathbf{e}_j \cdot \{(\mathbf{r}_i + k_i\mathbf{e}_i) - (\mathbf{r}_j + k_j\mathbf{e}_j)\} = 0 \tag{3.43}$$

The solutions of k_i and k_j satisfying this relationship leads to the determination of the specific positions of P_i and P_j. Equation (3.43) yields the final results as

$$\begin{bmatrix} k_i \\ k_j \end{bmatrix} = \frac{1}{1 - (\mathbf{e}_i \cdot \mathbf{e}_j)^2} \begin{bmatrix} -1 & \mathbf{e}_i \cdot \mathbf{e}_j \\ -\mathbf{e}_i \cdot \mathbf{e}_j & 1 \end{bmatrix} \begin{bmatrix} \mathbf{e}_i \cdot \mathbf{r}_{ij} \\ \mathbf{e}_i \cdot \mathbf{r}_{ij} \end{bmatrix} \tag{3.44}$$

This equation has been derived under the assumption of $\mathbf{e}_i \cdot \mathbf{e}_j \neq \pm 1$. This condition is necessary for the existence of the solution because $\mathbf{e}_i \cdot \mathbf{e}_j = \pm 1$ implies a parallel

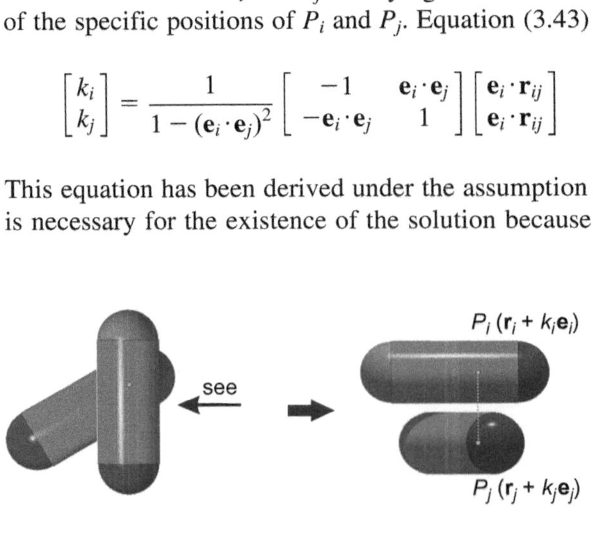

Figure 3.6 Assessment of the particle overlap.

or line configuration of the particles. If the line $\overline{P_iP_j}$ is longer than $(d + 2\delta)$, there is no particle overlap. Hence, we first consider the general case under the assumptions that $\mathbf{e}_i \cdot \mathbf{e}_j \neq \pm 1$ and the line $\overline{P_iP_j}$ is shorter than $(d + 2\delta)$.

There are three cases of overlap for the two spherocylinder particles: that is, hemisphere–hemisphere, hemisphere–cylinder, and cylinder–cylinder overlap. We first consider a cylinder–cylinder overlap between particles i and j. The condition for this overlap is derived as

$$|(\mathbf{r}_i + k_i\mathbf{e}_i) - (\mathbf{r}_j + k_j\mathbf{e}_j)| < d + 2\delta, \quad |k_i| < l_0/2, \quad |k_j| < l_0/2 \tag{3.45}$$

Next, we consider the criterion for the overlap between the cylindrical part of particle i and the hemisphere cap of particle j. In this case, the conditions of $|k_i| < l_0/2$ and $|k_j| \geq l_0/2$ are satisfied. A vertical line is drawn from the center of the hemisphere to the axis line of particle i, and the intersection point on this axis line of particle i is denoted by $Q_{i(j)}$, which is expressed as $(\mathbf{r}_i + k_i^s\mathbf{e}_i)$ with an unknown constant k_i^s. The determination of k_i^s yields explicit specification of the position $Q_{i(j)}$. If the center of hemisphere of particle j is denoted by $\mathbf{r}_j^s$ (similarly $\mathbf{r}_i^s$ for particle i), then k_i^s is solved from the orthogonality condition of $(\mathbf{r}_i + k_i^s\mathbf{e}_i - \mathbf{r}_j^s)$ and $\mathbf{e}_i$:

$$k_i^s = \mathbf{e}_i \cdot (\mathbf{r}_j^s - \mathbf{r}_i) \tag{3.46}$$

The use of this solution of k_i^s gives rise to the criterion condition for the overlap between the cylindrical part of particle i and the hemisphere cap of particle j as

$$|k_i^s| \leq l_0/2, \quad \left|(\mathbf{r}_i + k_i^s\mathbf{e}_i) - \mathbf{r}_j^s\right| < d + 2\delta \tag{3.47}$$

Finally, the overlap between the hemisphere caps between particles i and j arises when the following condition is satisfied:

$$|k_i^s| > l_0/2, \quad \left|\mathbf{r}_i^s - \mathbf{r}_j^s\right| < d + 2\delta \tag{3.48}$$

The above-mentioned criterion conditions are summarized as follows:

1. For $|(\mathbf{r}_i + k_i\mathbf{e}_i) - (\mathbf{r}_j + k_j\mathbf{e}_j)| \geq d + 2\delta$, there is no overlap.
2. For $|(\mathbf{r}_i + k_i\mathbf{e}_i) - (\mathbf{r}_j + k_j\mathbf{e}_j)| < d + 2\delta$, there is a possibility of overlap.
  2.1. For $|k_i| \leq l_0/2$ and $|k_j| \leq l_0/2$, an overlap occurs.
  2.2. For $|k_i| \leq l_0/2$ and $|k_j| > l_0/2$ and $|k_i^s| < l_0/2$, there is a possibility of overlap between the cylinder part of particle i and the hemisphere cap of particle j.
    2.2.1. $|(\mathbf{r}_i + k_i^s\mathbf{e}_i) - \mathbf{r}_j^s| \geq d + 2\delta$, there is no overlap.
    2.2.2. $|(\mathbf{r}_i + k_i^s\mathbf{e}_i) - \mathbf{r}_j^s| < d + 2\delta$, an overlap occurs.
  2.3. For $|k_i| \leq l_0/2$ and $|k_j| > l_0/2$ and $|k_i^s| \geq l_0/2$, there is a possibility of overlap between the hemisphere caps between particles i and j.
    2.3.1. For $|\mathbf{r}_i^s - \mathbf{r}_j^s| \geq d + 2\delta$, there is no overlap.
    2.3.2. For $|\mathbf{r}_i^s - \mathbf{r}_j^s| < d + 2\delta$, an overlap occurs.
  2.4. For $|k_j| > |k_i| > l_0/2$ and $|k_i^s| < l_0/2$, there is a possibility of overlap between the cylinder part of particle i and the hemisphere cap of particle j.
    2.4.1. For $|(\mathbf{r}_i + k_i^s\mathbf{e}_i) - \mathbf{r}_j^s| \geq d + 2\delta$, there is no overlap.
    2.4.2. For $|(\mathbf{r}_i + k_i^s\mathbf{e}_i) - \mathbf{r}_j^s| < d + 2\delta$, an overlap occurs.

2.5. For $|k_j| > |k_i| > l_0/2$ and $|k_i^s| \geq l_0/2$, there is a possibility of overlap between the hemisphere caps between particles i and j.
2.5.1. For $|\mathbf{r}_i^s - \mathbf{r}_j^s| \geq d + 2\delta$, there is no overlap.
2.5.2. For $|\mathbf{r}_i^s - \mathbf{r}_j^s| < d + 2\delta$, an overlap occurs.

These overlap criteria have been shown under the assumption of $|k_j| > |k_i|$. However, the above description is sufficient on the analysis level, because the exchange of subscripts i and j in a simulation program reduces to the same criterion procedure for particle overlap.

In addition to particle overlap in a general configuration, we need to consider several special cases, that is, particle overlap in a parallel or line configuration. The latter is straightforward to analyze and therefore we address the former case. According to the distance $|k_{ij}^c|$ $(= |\mathbf{r}_{ij} \cdot \mathbf{e}_i|)$ between the centers of particles i and j along the particle axis, whether or not particles i and j overlap can be determined by the following procedures:

1. For $|k_{ij}^c| \leq l_0$, an overlap occurs.
2. For $|k_{ij}^c| > l_0$,
2.1. For $|\mathbf{r}_i^s - \mathbf{r}_j^s| \geq d + 2\delta$, there is no overlap.
2.2. For $|\mathbf{r}_i^s - \mathbf{r}_j^s| < d + 2\delta$, an overlap occurs.

If the particle separation satisfies $(|\mathbf{r}_{ij}|^2 - |k_{ij}^c|^2)^{1/2} \geq d + 2\delta$, then overlap does not occur.

The above-assessing procedures concerning particle overlap enable us to recognize a specific configuration of the two particles in which the minimum distance can be obtained from the line of each particle axis. The notation $\mathbf{r}_i^{(\min)}$ and $\mathbf{r}_j^{(\min)}$ is used for expressing such positions on the axis lines. The present modified linear sphere-connected model for particle i can be constructed by placing other spheres on both sides of the sphere at $\mathbf{r}_i^{(\min)}$ repeatedly. According to this model, a force acting on particle i by particle j, arising from the overlap of the steric layers, can be obtained by evaluating the interaction forces between the constituent spherical particles and then by summing these interactions. Similarly, a torque acting on particle i by particle j can be evaluated by performing the vector product of each force vector of the constituent spheres and the corresponding relative position vectors from the center of particle i.

3.2.5 Nondimensionalization of Basic Equations

In actual simulations, it is usual to treat a nondimensional system in which quantities are nondimensionalized by the corresponding representative values. The present simulation employs the following representative values for nondimensionalization: d for distances, $1/\dot{\gamma}$ for time, $\dot{\gamma}d$ for velocities, $\dot{\gamma}$ for angular velocities, $3\pi\eta\dot{\gamma}d^2$ for forces, $\pi\eta\dot{\gamma}d^3$ for torques, and so on. With these representative values, the equations of motion in Eqs. (3.27) and (3.28) are nondimensionalized as

$$\mathbf{v}_i^{\|*} = \mathbf{U}^{\|*}(\mathbf{r}_i^*) + \frac{\mathbf{F}_i^{\|*}}{X^{A*}(l^* + 2\delta^*)}, \quad \mathbf{v}_i^{\perp*} = \mathbf{U}^{\perp*}(\mathbf{r}_i^*) + \frac{\mathbf{F}_i^{\perp*}}{Y^{A*}(l^* + 2\delta^*)} \tag{3.49}$$

$$\omega_i^{\perp *} = \Omega^{\perp *} + \frac{\mathbf{T}_i^{\perp *}}{Y^{C*}(l^* + 2\delta^*)^3} - \frac{Y^{H*}}{Y^{C*}}(\varepsilon \cdot \mathbf{e}_i\mathbf{e}_i) : \mathbf{E}^* \tag{3.50}$$

in which

$$\left.\begin{aligned} X^{A*} &= \frac{X^A}{6\pi(l/2+\delta)} = \frac{8}{3}\cdot\frac{s^3}{-2s+(1+s^2)L} \\ Y^{A*} &= \frac{Y^A}{6\pi(l/2+\delta)} = \frac{16}{3}\cdot\frac{s^3}{2s+(3s^2-1)L} \end{aligned}\right\} \tag{3.51}$$

$$Y^{C*} = \frac{Y^C}{8\pi(l/2+\delta)^3} = \frac{4}{3}\cdot\frac{s^3(2-s^2)}{-2s+(1+s^2)L} \tag{3.52}$$

$$Y^{H*} = \frac{Y^H}{8\pi(l/2+\delta)^3} = \frac{4}{3}\cdot\frac{s^5}{-2s+(1+s^2)L} \tag{3.53}$$

$$\mathbf{E}^* = \frac{1}{2}\begin{bmatrix} 0 & 1 & 0 \\ 1 & 0 & 0 \\ 0 & 0 & 0 \end{bmatrix}, \quad \Omega^* = -\frac{1}{2}\begin{bmatrix} 0 \\ 0 \\ 1 \end{bmatrix}, \quad (\varepsilon \cdot \mathbf{e}_i\mathbf{e}_i) : \mathbf{E}^* = \frac{1}{2}\begin{bmatrix} e_{iz}e_{ix} \\ -e_{iz}e_{iy} \\ e_{iy}^2 - e_{ix}^2 \end{bmatrix} \tag{3.54}$$

Also, Eqs. (3.37) and (3.38) can be written in nondimensionalized form as

$$\begin{aligned} \mathbf{r}_i^*(t^* + \Delta t^*) &= \mathbf{r}_i^*(t^*) + \Delta t^* \mathbf{v}_i^*(t^*), \\ \mathbf{e}_i(t^* + \Delta t^*) &= \mathbf{e}_i(t^*) + \Delta t^* \omega_i^{\perp *}(t^*) \times \mathbf{e}_i(t^*) \end{aligned} \tag{3.55}$$

The forces acting on the positive and negative magnetic charges of particle i in Eqs. (3.21) and (3.22) are nondimensionalized as

$$\mathbf{F}_{ij}^{+*} = \lambda_{\mathrm{m}}\left\{ \frac{\mathbf{r}_{ij}^* + (l_0^*/2)(\mathbf{e}_i - \mathbf{e}_j)}{\left|\mathbf{r}_{ij}^* + (l_0^*/2)(\mathbf{e}_i - \mathbf{e}_j)\right|^3} - \frac{\mathbf{r}_{ij}^* + (l_0^*/2)(\mathbf{e}_i + \mathbf{e}_j)}{\left|\mathbf{r}_{ij}^* + (l_0^*/2)(\mathbf{e}_i + \mathbf{e}_j)\right|^3} \right\} \tag{3.56}$$

$$\mathbf{F}_{ij}^{-*} = -\lambda_{\mathrm{m}}\left\{ \frac{\mathbf{r}_{ij}^* - (l_0^*/2)(\mathbf{e}_i + \mathbf{e}_j)}{\left|\mathbf{r}_{ij}^* - (l_0^*/2)(\mathbf{e}_i + \mathbf{e}_j)\right|^3} - \frac{\mathbf{r}_{ij}^* - (l_0^*/2)(\mathbf{e}_i - \mathbf{e}_j)}{\left|\mathbf{r}_{ij}^* - (l_0^*/2)(\mathbf{e}_i - \mathbf{e}_j)\right|^3} \right\} \tag{3.57}$$

in which ql_0 is the magnitude of a magnetic moment, expressed as $m = ql_0$, and λ_m is the nondimensional parameter representing the strength of magnetic forces relative to the shear force of a simple shear flow, expressed as

$$\lambda_{\mathrm{m}} = \frac{\mu_0 m^2}{12\pi^2 \eta \dot{\gamma} l_0^2 d^4} \tag{3.58}$$

The nondimensionalization procedure generally leads to the appearance of such nondimensional numbers; the most famous nondimensional number—the Reynolds number, in fluid mechanics—arises from a similar nondimensional procedure.

Similarly, the torque acting on particle i by particle j in Eqs. (3.23) and (3.24) is nondimensionalized as

$$\mathbf{T}_{ij}^{+*} = \frac{3l_0^*}{2}\lambda_{\rm m}\left\{ \frac{\mathbf{e}_i \times \mathbf{r}_{ij}^* - (l_0^*/2)(\mathbf{e}_i \times \mathbf{e}_j)}{\left|\mathbf{r}_{ij}^* + (l_0^*/2)(\mathbf{e}_i - \mathbf{e}_j)\right|^3} - \frac{\mathbf{e}_i \times \mathbf{r}_{ij}^* + (l_0^*/2)(\mathbf{e}_i \times \mathbf{e}_j)}{\left|\mathbf{r}_{ij}^* + (l_0^*/2)(\mathbf{e}_i + \mathbf{e}_j)\right|^3} \right\} \tag{3.59}$$

$$\mathbf{T}_{ij}^{-*} = \frac{3l_0^*}{2}\lambda_{\rm m}\left\{ \frac{\mathbf{e}_i \times \mathbf{r}_{ij}^* - (l_0^*/2)(\mathbf{e}_i \times \mathbf{e}_j)}{\left|\mathbf{r}_{ij}^* - (l_0^*/2)(\mathbf{e}_i + \mathbf{e}_j)\right|^3} - \frac{\mathbf{e}_i \times \mathbf{r}_{ij}^* + (l_0^*/2)(\mathbf{e}_i \times \mathbf{e}_j)}{\left|\mathbf{r}_{ij}^* - (l_0^*/2)(\mathbf{e}_i - \mathbf{e}_j)\right|^3} \right\} \tag{3.60}$$

The torque exerted by an applied magnetic field in Eq. (3.26) is written in nondimensional form:

$$\mathbf{T}_i^{(\mathrm{H})*} = \lambda_{\mathrm{H}}\mathbf{e}_i \times \mathbf{h} \tag{3.61}$$

in which $\mathbf{h}$ is a unit vector denoting the magnetic field direction, expressed as $\mathbf{h} = \mathbf{H}/H$. As before, λ_{H} is a nondimensional parameter representing the strength of magnetic particle–field interactions relative to the torque due to the shear flow force, expressed as

$$\lambda_{\mathrm{H}} = \frac{\mu_0 m H}{\pi\eta\dot{\gamma}d^3} \tag{3.62}$$

The repulsive force due to the overlap of the surfactant layers in Eq. (3.42) is nondimensionalized as

$$\mathbf{F}_{ij}^{(V)*} = \lambda_V \mathbf{t}_{ij} \ln\left(\frac{1+2\delta^*}{r_{ij}^*}\right) \quad (\text{for } 1 \le r_{ij}^* \le 1+2\delta^*) \tag{3.63}$$

in which λ_V is a nondimensional parameter representing the strength of such repulsive forces relative to the shear flow force.

We have finished nondimensionalizing almost all the quantities necessary for simulations. The nondimensional parameters characterizing the physical phenomenon are λ_{m} for magnetic particle–particle interactions, λ_{H} for magnetic particle–field interactions, and λ_{V} for steric repulsive interactions.

3.2.6 Treatment of the Criteria for Particle Overlap in Simulations

In the previous subsection on the modeling of steric repulsive interactions, we presented a mathematical discussion on the assessment for the overlap of the steric layers. In actual calculations in a simulation program, the systematic classification of the overlapping regimes enables us to quickly grasp a logical flow of the calculation procedures; this subject may be on a technical side rather than a mathematical one. As shown in Table 3.1, particle overlapping can be classified into four cases: that is, a general overlap (itree = 1), a linear overlap (itree = 0), a normal overlap (itree = 2), and a parallel overlap (itree = 3). Note that the variable names "itree" and "ipath" are commonly used in a simulation program, so that the overlap treatment is conducted for the cases specified by "itree" and "ipath" in a simulation program. The important point in a simulation program is that if $|k_j| < |k_i|$, then the overlap regimes shown in Table 3.1 are easily applicable after the replacement of indices i and j by j and i. Hence, the assumption of the condition $|k_j| \geq |k_i|$ for starting a mathematical analysis provides a relatively straightforward classification without losing our way in a mathematical labyrinth. The classification in the substage for each case depends on which hemisphere cap of particle j overlaps with particle i. That is, the directions of particles i and j are important for the successive treatment of repulsive interactions. For a linear overlapping case, the calculation of the repulsive force between only one pair of the spheres completes the overlapping treatment. On the other hand, for the other overlapping cases, two spheres are first placed at the nearest separation positions on each axis line, as previously explained, in order to calculate the force and torque for this pair of spheres. Then, other spheres are repeatedly added to the both ends of each sphere in linear formation to form the linear sphere-connected particles i and j. Finally, the interaction forces and torques are calculated for each pair of constituent spheres of particles i and j; the summation of these forces and torques for each pair of spheres yields the total force and torque acting on particle i by particle j. For example, we briefly consider the case of itree = 1 and ipath = 1 in Table 3.1. The positions of the two spheres are first determined on each axis line, and then the next spheres are placed at each neighboring position in the $(-\mathbf{n}_i)$ and $(-\mathbf{n}_j)$ directions; the repulsive forces and torques are calculated for each pair of these constituent spheres.

3.2.7 Parameters for Simulations

We set the following initial conditions for simulations. A magnetic field is applied in the y-axis direction, and a simple shear flow is applied in the x-direction. The spherocylinder particles are expected to aggregate in the magnetic field direction (y-axis direction) because they are magnetized in the particle axis direction. Hence we employ a rectangular-parallelepiped simulation box, with its longer axis along the field direction with a square base. We first place six rows of particles in the x-axis direction with their particle axis pointing to the y-axis direction, then repeat this procedure in the z-direction to obtain the initial configuration of 36 particles in the xz-plane. Finally, we expand this configuration in the y-axis direction to obtain the total six layers of these particles. The initial configuration of 216 particles, therefore, can be assigned from this procedure. A rectangular-parallelepiped simulation box needs to be set, with an appropriate aspect ratio dependent upon the particle aspect ratio. The present simulation uses a simulation box where the length in the y-axis direction is twice the length in the x-axis direction; note that the

Table 3.1 Regime of Overlap

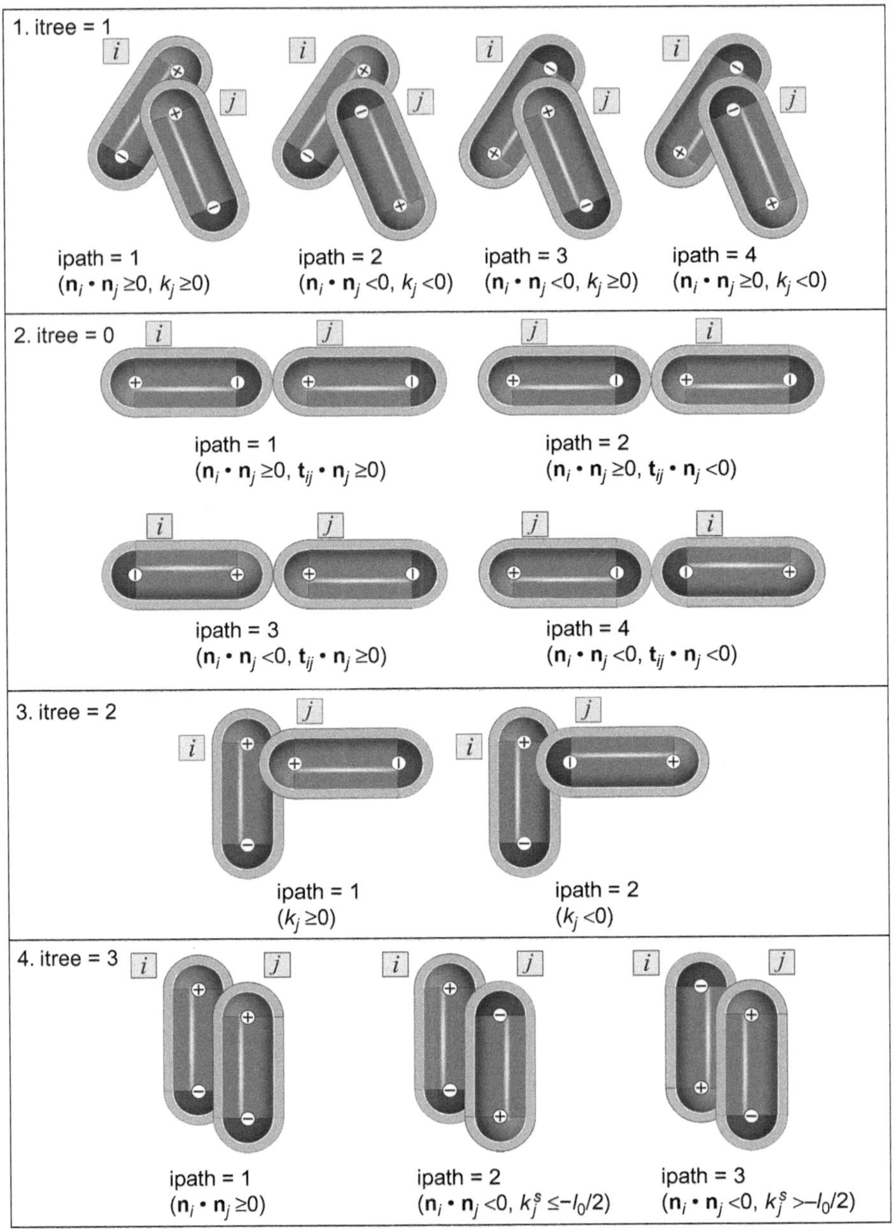

Note that $\mathbf{n}_i$ is used as $\mathbf{e}_i$.

above-mentioned setting procedure is slightly different from that explained in Section 2.1.2. In the present simulation, the particle aspect ratio r_p is taken as $r_p = 5$, the volumetric fraction as $\phi_V = 0.05$, and the thickness of a surfactant layer as $\delta^* = 0.15$.

A shear flow and a magnetic field have a tendency to make the spherocylinder particles incline in the flow direction and in the applied direction, respectively. The orientational behavior of the magnetic spherocylinder particles, therefore, depends in a complicated manner on the strength of magnetic interactions as well as the flow shear rate. The main objective of the present simulation is to discuss the influences of magnetic particle–field, magnetic particle–particle, and steric repulsive interactions on the behavior of spherocylinder particles in a simple shear flow. Hence, simulations are carried out for various cases of the nondimensional parameters λ_m and λ_H such as $\lambda_m = 0$, 10, 20, and 50 and $\lambda_H = 0$, 10, 20, 50, and 100. On the other hand, λ_V is taken to have the single value $\lambda_V = 150$; a larger value of λ_V induces a large repulsive force at the particle overlapping.

3.2.8 Results of Simulations

Figure 3.7 shows the change in aggregate structures with time for no applied magnetic field and no magnetic interactions between particles. The rod-like particles rotate in the *xy*-plane about the *z*-axis because there is no applied magnetic field. Describing in more detail, the particles incline in the flow direction (*x*-axis direction) during a long period as in Figures 3.7A and C. Once particles have been kicked below the *x*-axis, they quickly rotate toward the preferred direction, as shown in Figures 3.7A and C by way of a transient snapshot shown in Figure 3.7B. This is because much larger torques act on the rod-like particles when inclining in a direction normal to the flow.

Figure 3.8 shows a snapshot for no applied magnetic field under strong magnetic particle–particle interactions $\lambda_m = 10$. The figure on the left-hand side is a general snapshot viewed from a certain angle to grasp how nearly the particles incline in the flow direction. The figure on the right-hand side is an oblique view for grasping the formation of wall-like clusters along the flow direction, that is, it is viewed almost from the negative *x*-axis direction. In this case, even if no magnetic field is applied, rod-like particles seldom rotate from the situation in Figure 3.8 because magnetic particle–particle interactions become more dominant than viscous shear forces, and so the particles form complex three-dimensional aggregate structures. However, the individual particles have a tendency to incline in the shear flow direction.

Figure 3.9 shows a snapshot for a strong applied magnetic field $\lambda_H = 10$ and no magnetic interactions $\lambda_m = 0$. In this situation, the applied magnetic field makes rod-like particles incline in the magnetic field direction. The final particle orientation is determined by the balance of the torque due to the applied field and the torque due to a shear flow; in Figure 3.9 all rod-like particles tend to incline in the same direction (the direction of the flow) because there is no disturbance due to magnetic particle–particle interactions.

Figure 3.10 shows the result for magnetic interactions $\lambda_m = 10$ and for an external magnetic field $\lambda_H = 10$ as in Figure 3.9. A significant difference to the case of

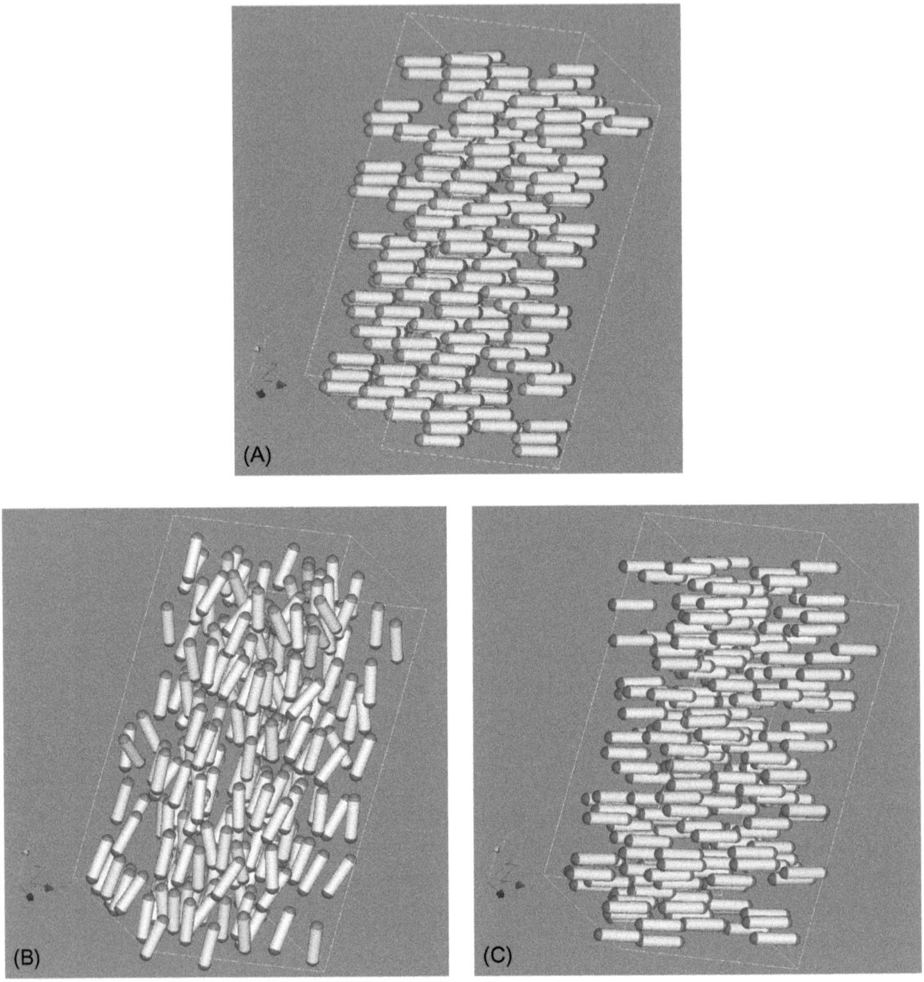

Figure 3.7 Time change in aggregate structures for $\lambda_H = 0$ and $\lambda_m = 0$: (A) $t = t_1$, (B) $t = t_2$, and (C) $t = t_3$.

Figure 3.9 is that, to a certain degree, aggregates have wall-like structures along the flow direction. The particle aggregation is due to magnetic interactions between particles, and the viscous forces and torques induce more complex aggregates, such as these wall-like structures. Wall-like clusters are also observed for the case of magnetic spherical particles in an applied magnetic field subject to a simple shear flow. Magnetic particle–particle interactions emphasize the tendency of particles to incline in the flow direction, which is clearly seen by comparing with the case in Figure 3.9. Note that the particles in Figure 3.10 do not orient toward the same preferred direction.

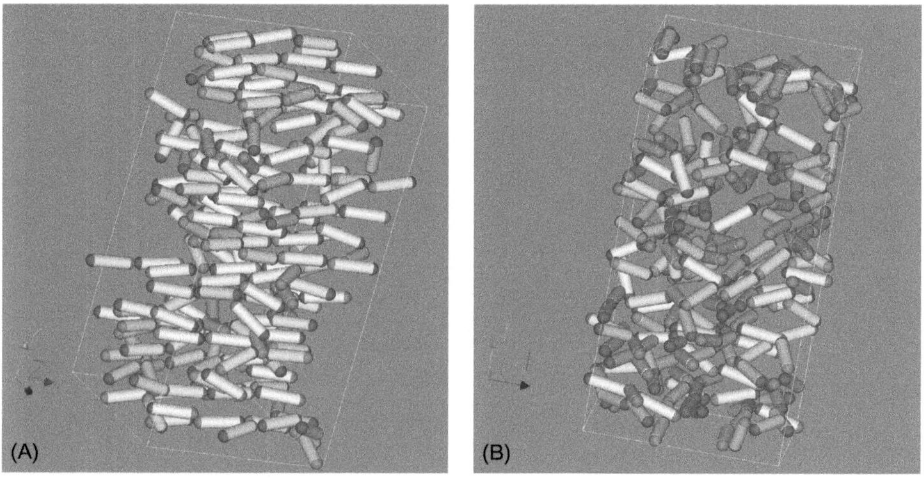

Figure 3.8 Aggregate structures for $\lambda_H = 0$ and $\lambda_m = 10$: (A) an oblique view and (B) viewed nearly from the negative x-axis.

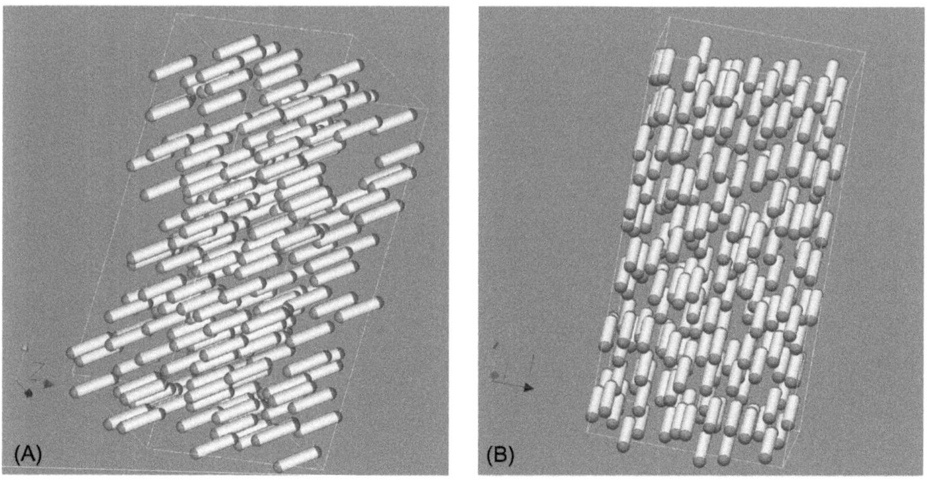

Figure 3.9 Aggregate structures for $\lambda_H = 10$ and $\lambda_m = 0$: (A) an oblique view and (B) viewed nearly from the negative x-axis.

Figure 3.11 is a snapshot for a significantly strong applied magnetic field $\lambda_H = 50$, but without magnetic particle–particle interactions. Since a magnetic field is significantly strong, each particle inclines to a higher degree in the magnetic field direction (y-axis direction) as compared with that in Figure 3.9. On the other hand, wall-like clusters are not formed in this case because there are no magnetic interactions.

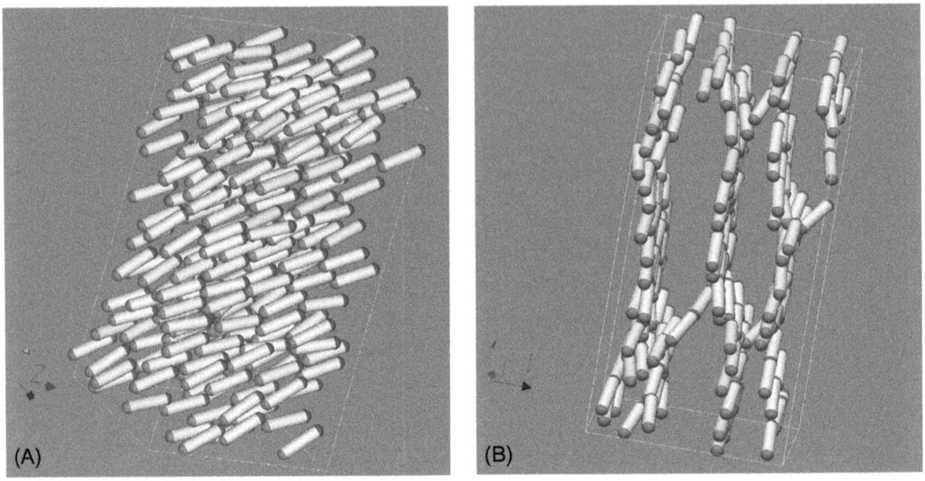

Figure 3.10 Aggregate structures for $\lambda_H = 10$ and $\lambda_m = 10$: (A) an oblique view and (B) viewed nearly from the negative *x*-axis.

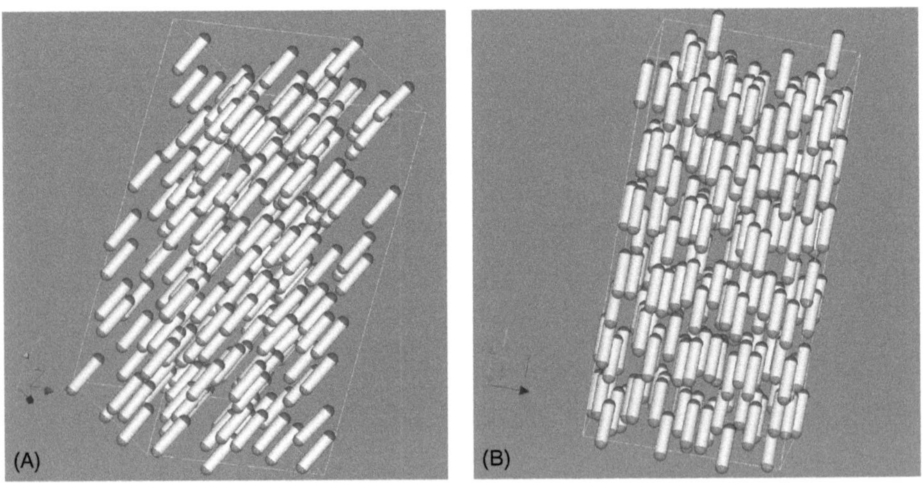

Figure 3.11 Aggregate structures for $\lambda_H = 50$ and $\lambda_m = 0$: (A) an oblique view and (B) viewed nearly from the negative *x*-axis.

Figure 3.12 also shows $\lambda_H = 50$, as in Figure 3.11, but magnetic interactions are $\lambda_m = 10$ in this case. Comparison with Figure 3.10, clearly reveals that wall-like clusters are formed along the flow direction. The detailed observation of the internal structures of wall-like clusters indicates that the rod-like particles aggregate to

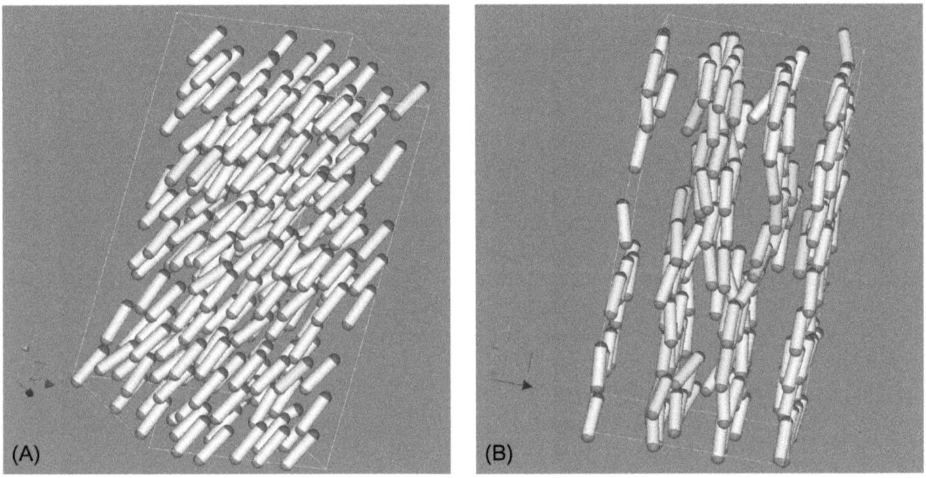

Figure 3.12 Aggregate structures for $\lambda_H = 50$ and $\lambda_m = 10$: (A) an oblique view and (B) viewed nearly from the negative x-axis.

form wall-like structures in such a way that one cluster is placed into two parallel clusters, with the plus magnetic charge of the center particle in contact with the minus magnetic charges of the two neighboring particles.

The above discussion has systematically used snapshots to present the properties of aggregates. However, this type of qualitative discussion is insufficient for an academic paper, and the addition of quantitative discussion is necessary. For this exercise, it would be suitable to discuss the radial, pair, and orientational distribution functions, whilst further investigation of the phenomena might necessitate Brownian dynamics in order to include random particle motion.

3.2.9 Simulation Program

The following sample simulation program has been written for the present simulation in FORTRAN. The important variables used in the simulation program are as follows:

`RX(I),RY(I),RZ(I)`	:	(x, y, z) components of the position vector $\mathbf{r}_i^*$ of particle i
`NX(I),NY(I),NZ(I)`	:	(x, y, z) components of the unit vector $\mathbf{n}_i(=\mathbf{e}_i)$ of particle i denoting the particle and magnetic moment direction
`FX(I),FY(I),FZ(I)`	:	(x, y, z) components of the force $\mathbf{F}_i^*$ acting on particle i
`TX(I),TY(I),TZ(I)`	:	(x, y, z) components of the torque $\mathbf{T}_i^*$ acting on particle i
`XL,YL,ZL`	:	Side lengths of the simulation box in the (x, y, z) directions
`L`	:	Length l^* of the solid part of the spherocylinder particle
`D`	:	Diameter d^* of the solid cylinder part of the spherocylinder
`DEL`	:	Thickness δ^* of the surfactant layer
`TD`	:	Ratio $2\delta^*(=2\delta/d)$ of the surfactant layer thickness to the particle radius

RP	:	Particle aspect ratio r_p $(=l/d)$
RP1	:	Particle aspect ratio $r_p'(=l_0/d=r_p-1)$
N	:	Number of particles
VDENS	:	Volumetric fraction of particles ϕ_V
NDENS	:	Number density of particles
HX,HY,HZ	:	(x,y,z) components of the unit vector denoting the magnetic field direction
RAM	:	Nondimensional parameter λ_m representing the strength of magnetic particle–particle interactions
RAH	:	Nondimensional parameter λ_H representing the strength of magnetic particle–field interactions
RAV	:	Nondimensional parameter λ_V representing the strength of repulsive interactions due to the overlap of steric layers
H	:	Time interval
RCOFF	:	Cutoff distance for calculations of forces and torques
XA,YA,YC,YH	:	Resistance functions
GAMDOT	:	Shear rate $\dot{\gamma}^*$
MOMX(*),MOMY(*), MOMZ(*)	:	Averaged values of the particle direction at each time step

As an aid for understanding the program, comments have been added to the important features. The line numbers shown at the beginning of each line are just for the reader's convenience and are unnecessary for executing the FORTRAN program.

We briefly explain quasi-random numbers, which are used in the subroutine "INITIAL" for setting an initial configuration. A quasi-random number is generated using an irrational. For example, if $\sqrt{2}$ is used, the fractional parts of $\sqrt{2}$, $2\sqrt{2}$, $3\sqrt{2}$, $4\sqrt{2}$,... provide a sequence of quasi-random numbers ranging from zero to unity.

```
C*********************************************************************
C*                            mdcylndr1.f                            *
C*                                                                   *
C*              OPEN(9, FILE='@bbb1.dat',  STATUS='UNKNOWN')          *
C*              OPEN(10,FILE='bbb11.dat',  STATUS='UNKNOWN')          *
C*              OPEN(13,FILE='bbb41.mgf',  STATUS='UNKNOWN')          *
C*              OPEN(21,FILE='bbb001.dat',STATUS='UNKNOWN')           *
C*              OPEN(22,FILE='bbb011.dat',STATUS='UNKNOWN')           *
C*              OPEN(23,FILE='bbb021.dat',STATUS='UNKNOWN')           *
C*              OPEN(24,FILE='bbb031.dat',STATUS='UNKNOWN')           *
C*              OPEN(25,FILE='bbb041.dat',STATUS='UNKNOWN')           *
C*              OPEN(26,FILE='bbb051.dat',STATUS='UNKNOWN')           *
C*              OPEN(27,FILE='bbb061.dat',STATUS='UNKNOWN')           *
C*              OPEN(28,FILE='bbb071.dat',STATUS='UNKNOWN')           *
C*              OPEN(29,FILE='bbb081.dat',STATUS='UNKNOWN')           *
C*              OPEN(30,FILE='bbb091.dat',STATUS='UNKNOWN')           *
C*                                                                   *
C*      ----------     MOLECULAR DYNAMICS SIMULATIONS    ---------    *
C*       THREE-DIMENSIONAL MOLECULAR DYNAMICS SIMULATIONS OF         *
C*       A DISPERSION COMPOSED OF MAGNETIC SPHEROCYLINDERS           *
C*       IN A SIMPLE SHEAR FLOW.                                     *
C*                                                                   *
C*       1. RODLIKE MODEL WITH ARBITRARY ASPECT RATIO.               *
C*       2. NO HYDRODYNAMIC INTERACTIONS AMONG PARTICLES.            *
C*                                                                   *
C*                                   VER.1  BY A.SATOH , '08  5/23   *
C*********************************************************************
```

```
C      N      : NUMBER OF PARTICLES
C      D      : DIAMETER OF SOLID HEMISPHERE PARTICLE (=1)
C      L      : LENGTH OF SOLID SPHEROCYLINDER
C      RP     : ASPECT RATIO (=L/D)
C      RP1    : ASPECT RATIO OF CYLINDER LENGTH TO D (=RP-1)
C      NDENS  : NUMBER DENSITY
C      VDENS  : VOLUMETRIC FRACTION
C      RAM    : NONDIMENSIONAL PARAMETER OF PARTICLE-PARTICLE INTERACT
C      RAH    : NONDIMENSIONAL PARAMETER OF PARTICLE-FIELD INTERACTION
C      RAV    : NONDIMENSIONAL PARAMETER OF STERIC REPULSION
C      RCOFF  : CUTOFF RADIUS FOR CALCULATION OF INTERACTION ENERGIES
C      XL,YL,ZL : DIMENSIONS OF SIMULATION REGION
C      BETA   : ASPECT RATIO OF SIMULATION BOX
C      (HX,HY,HZ) : UNIT VECTOR DENOTING MAGNETIC FIELD DIRECTION
C
C      XA,YA  : RESISTANCE FUNC. FOR TRANSLATIONAL MOTION
C      YC     : RESISTANCE FUNC. FOR ROTATIONAL MOTION
C      YH     : RESISTANCE FUNC. FOR SHEAR FLOW TERM
C      RX(I),RY(I),RZ(I)     : PARTICLE POSITION
C      NX(N),NY(N),NZ(N)     : DIRECTION OF PARTICLE MAJOR AXIS AND
C                              MAGNETIC MOMENT
C      FX(I),FY(I),FZ(I)     : FORCES ACTING ON PARTICLE I
C      TX(I),TY(I),TZ(I)     : TORQUES ACTING ON PARTICLE I
C      MOMX(**),MOMY(**)     : MAG. MOMENT OF SYSTEM AT EACH TIME STEP
C      MOMZ(**)
C
C      H        : INTERVAL OF TIME STEP FOR MOLE. DYNA. SIMULATIONS
C      GAMDOT   : SHEAR RATE (=1 FOR THIS CASE)
C      NTIMEMX  : MAXIMUM NUMBER OF TIME STEP
C
C         -XL/2 < RX < XL/2 ,  -YL/2 < RY < YL/2,  -ZL/2 < RZ < ZL/2
C-----------------------------------------------------------------------
C
       IMPLICIT REAL*8 (A-H,O-Z), INTEGER (I-N)
C
       COMMON /BLOCK1/  RX    , RY    , RZ
       COMMON /BLOCK2/  NX    , NY    , NZ
       COMMON /BLOCK3/  FX    , FY    , FZ
       COMMON /BLOCK4/  TX    , TY    , TZ
       COMMON /BLOCK5/  XL    , YL    , ZL
       COMMON /BLOCK6/  RP    , RP1   , D     ,  DEL   , TD
       COMMON /BLOCK7/  XA    , YA    , YC    , YH
       COMMON /BLOCK8/  N     , NDENS, VDENS
       COMMON /BLOCK9/  H     , RCOFF, GAMDOT, DX    , CORY
       COMMON /BLOCK10/ RAM   , RAH   , RAV
       COMMON /BLOCK11/ HX    , HY    , HZ
       COMMON /BLOCK12/ MOMX  , MOMY  , MOMZ
       COMMON /WORK20/  XRXI  , YRYI  , ZRZI  , XRXJ  , YRYJ  , ZRZJ
       COMMON /WORK21/  FXIJS, FYIJS, FZIJS, FXJIS, FYJIS, FZJIS
       COMMON /WORK22/  TXIJS, TYIJS, TZIJS, TXJIS, TYJIS, TZJIS
       COMMON /WORK23/  RCOFF2 , RP102 , D1     , D1SQ
       COMMON /WORK24/  CF0XA  , CF0YA , CT0YC , CE0YHYC
C
       PARAMETER( NN=1000 , NNS=500000 , PI=3.141592653589793D0 )
C
       REAL*8    NDENS
       REAL*8    RX(NN) , RY(NN) , RZ(NN) , NX(NN) , NY(NN) , NZ(NN)
       REAL*8    FX(NN) , FY(NN) , FZ(NN) , TX(NN) , TY(NN) , TZ(NN)
       REAL      MOMX(NNS) , MOMY(NNS) , MOMZ(NNS)
C
       REAL*8    BETA
       REAL*8    RXI , RYI , RZI , NXI , NYI , NZI , FXI , FYI , FZI
       REAL*8    TXI , TYI , TZI , WXI , WYI , WZI , WXIN, WYIN, WZIN
       REAL*8    FXIP , FYIP , FZIP , FXIN  , FYIN , FZIN
       REAL*8    TXIP , TYIP , TZIP , TXIN  , TYIN , TZIN
       REAL*8    OMEIPX , OMEIPY , OMEIPZ , OMEINX , OMEINY , OMEINZ
       REAL*8    C1   , C2   , C3   ,  C00
       REAL*8    C1X  , C1Y  , C1Z  ,  C2X  , C2Y  , C2Z
       REAL*8    C3X  , C3Y  , C3Z
       REAL*8    CCA1 , CCB1 , CCS1 ,  CCL1
       INTEGER   NTIME , NTIMEMX , NGRAPH , DNSMPL , NP   ,  NOPT
```

```
      INTEGER   NSMPL1  , NSMPL2
      INTEGER   NANIME  , NANMCTR, NOPT1
C
                 OPEN(9,FILE='@bbb1.dat' , STATUS='UNKNOWN')
                 OPEN(10,FILE='bbb11.dat', STATUS='UNKNOWN')
                 OPEN(13,FILE='bbb41.mgf', STATUS='UNKNOWN')
                 OPEN(21,FILE='bbb001.dat',STATUS='UNKNOWN')
                 OPEN(22,FILE='bbb011.dat',STATUS='UNKNOWN')
                 OPEN(23,FILE='bbb021.dat',STATUS='UNKNOWN')
                 OPEN(24,FILE='bbb031.dat',STATUS='UNKNOWN')
                 OPEN(25,FILE='bbb041.dat',STATUS='UNKNOWN')
                 OPEN(26,FILE='bbb051.dat',STATUS='UNKNOWN')
                 OPEN(27,FILE='bbb061.dat',STATUS='UNKNOWN')
                 OPEN(28,FILE='bbb071.dat',STATUS='UNKNOWN')
                 OPEN(29,FILE='bbb081.dat',STATUS='UNKNOWN')
                 OPEN(30,FILE='bbb091.dat',STATUS='UNKNOWN')
                                                              NP=9
```

• The given values and the magnetic moment results are written out in @bbb1 and bbb11, the data for MicroAVS are done in bbb41, and the intermediate positions and directions are done in bbb001–bbb091 in the time sequential order.

```
C
C                                         --- PARAMETER (1) ---
C                  ------------------------------------------------
C                    N=5**3(125), 6**3(216), 7**3(343), 8**3(512)
C                  ------------------------------------------------
C                     RAH   =   0.1     1.0      10.     100.
C                     H     =  0.001   0.001    0.001   0.0001
C                         ++++++++++++++++++++++++++++++++
C                     RAM   =   0.1     1.0      10.     100.
C                     H     =  0.001   0.001    0.001   0.0001
C
C                     THE MINIMUM VALUE ON THE ABOVE LIST MUST BE
C                     USED FOR THE TIME INTERVAL H.
C                  ------------------------------------------------
      N         = 216
      VDENS     = 0.05D0
      RAM       = 20.D0
      RAH       = 10.0D0
      RAV       = 150.D0
      RP        = 5.D0
      RP1       = RP - 1.D0
```

• The particle number N=216, volumetric fraction ϕ_V =0.05, λ_m=20, λ_H= 10, λ_V=150, and aspect ratio r_p=5.

```
C                                         --- PARAMETER (2) ---
      H         = 0.0001D0
      GAMDOT    = 1.D0
      TD        = 0.3D0
      RCOFF     = 5.D0*RP
      DEL       = TD/2.D0
      HX        = 0.D0
      HY        = 1.D0
      HZ        = 0.D0
```

• The time interval h^*=0.0001, t_δ=0.3, cutoff radius r^*_{coff}=5r_p, thickness of a surfactant layer δ^*=0.15, and magnetic field direction **h**=(0,1,0).

```
C
      BETA      = 2.D0
      DX        = 0.D0
      D         = 1.0D0
      NDENS     = (  12.D0/( PI*(3.D0*RP-1.D0) )  )*VDENS
```

• BETA is used in determining the simulation region size. NDENS is the number density of particles.

```
C
      NTIMEMX = 200000
      NGRAPH    = NTIMEMX/10
      NANIME    = NTIMEMX/200
      DNSMPL    = 2
      NOPT      = 20
```

• The main loop is finished when NTIME arrives at 200,000.
• The particle position and other data are written out at every NGRAPH time steps. 200 sets of data are written out for making an animation based on MicroAVS.

```
C                                         --- PARAMETER (5) ---
      CCA1      = RP/2.D0 + DEL
      CCB1      = 0.5D0    + DEL
      CCS1      = DSQRT( CCA1**2 - CCB1**2 ) / CCA1
      CCL1      = DLOG( (1.D0+CCS1)/(1.D0-CCS1) )
      XA        =  (8.D0/3.D0)*CCS1**3
     &                     / (  -2.D0*CCS1+(1.D0+CCS1**2)*CCL1   )
      YA        = (16.D0/3.D0)*CCS1**3
     &                     / (   2.D0*CCS1+(3.D0*CCS1**2-1.D0)*CCL1  )
      YC        = (4.D0/3.D0)*  ( CCS1**3*(2.D0-CCS1**2) )
     &                     / (  -2.D0*CCS1+(1.D0+CCS1**2)*CCL1   )
      YH        = (4.D0/3.D0)*CCS1**5
     &                     / (  -2.D0*CCS1+(1.D0+CCS1**2)*CCL1   )
```

• The resistance functions X^A, Y^A, Y^C, and Y^H are calculated in advance.

```
C
      RCOFF2   = RCOFF**2
      RP102    = RP1/2.D0
      D1       = 1.D0+TD
      D1SQ     = D1**2
      CF0XA    = 1.D0/( XA*(RP+2.D0*DEL ) )
      CF0YA    = 1.D0/( YA*(RP+2.D0*DEL ) )
      CT0YC    = 1.D0/( YC*(RP+2.D0*DEL )**3 )
      CE0YHYC = (YH/YC)*0.5D0
```

• CF0XA and CF0YA are the coefficients in the force term in Eq. (3.49), CT0YC is the coefficient in the torque term in Eq. (3.50), and CE0YHYC is a part of the coefficient of the shear rate in Eq. (3.50).

```
C
C     ----------------------------------------------------------------
C     -----------------   INITIAL CONFIGURATION   ------------------
C     ----------------------------------------------------------------
C
C                                              --- SET INITIAL CONFIG. ---
CCC   OPEN(19,FILE='qqq091.dat',STATUS='OLD')
CCC   READ(19,472)  N , XL , YL , ZL , D , TD , RP , RP1 , DX
CCC   READ(19,474) (RX(I),I=1,N),(RY(I),I=1,N),(RZ(I),I=1,N),
CCC  &             (NX(I),I=1,N),(NY(I),I=1,N),(NZ(I),I=1,N)
CCC   CLOSE(19,STATUS='KEEP')
CCC   GOTO 7
```

• These READ statements are for continuing the sequential simulation using the data saved previously.

```
C
      CALL INITIAL( BETA )
```

• The particle initial positions and velocities are assigned .

```
C
    7 IF( RCOFF .GE. XL/2.D0 ) THEN
        RCOFF = XL/2.D0 - 0.00001D0
      END IF
      RCOFF2 = RCOFF**2
C                                                       --- CAL FORCES ---
      NTIME = 0
      CALL FORCECAL( NP, NTIME )
```

• The forces and torques acting between particles are calculated.

```
C
C                                                        --- PRINT OUT ---
      WRITE(NP,12) N, VDENS, NDENS, RAM ,RAH ,RAV, RP, RP1, D, DEL,
     &             TD, XA, YA, YC, YH, H, RCOFF, GAMDOT, BETA,
     &             XL, YL, ZL
      WRITE(NP,13) RP102, D1, CF0XA, CF0YA, CT0YC, CE0YHYC
      WRITE(NP,14) NTIMEMX, NGRAPH, DNSMPL
C
C                                                   --- INITIALIZATION ---
      NANMCTR = 0
      NSMPL   = 0
C
C     ----------------------------------------------------------------
C     ------------------   START OF MAIN LOOP   --------------------
C     ----------------------------------------------------------------
C
      DO 1000 NTIME = 1,NTIMEMX
C
        DX = GAMDOT*YL*H*DBLE(NTIME)
        DX = DMOD( DX, XL )
```

• DX is ΔX in Fig. 2.15.

```
C
        DO 100 I = 1,N
C
          NXI    = NX(I)
          NYI    = NY(I)
          NZI    = NZ(I)
          FXI    = FX(I)
          FYI    = FY(I)
          FZI    = FZ(I)
          TXI    = TX(I)
          TYI    = TY(I)
          TZI    = TZ(I)
C                                       --- (1) TRANSLATIONAL MOTION ---
C
          C00    = FXI*NXI + FYI*NYI + FZI*NZI
          FXIP   = C00*NXI
          FYIP   = C00*NYI
          FZIP   = C00*NZI
          FXIN   = FXI - FXIP
          FYIN   = FYI - FYIP
```

• The force acting on particle *i* is decomposed into one in the particle direction and another in the direction normal to the particle axis according to Eq. (3.36).

```
          FZIN   = FZI - FZIP
C
          RXI    = RX(I) + H*( CF0XA*FXIP+CF0YA*FXIN ) + RY(I)*GAMDOT*H
          RYI    = RY(I) + H*( CF0XA*FYIP+CF0YA*FYIN )
          RZI    = RZ(I) + H*( CF0XA*FZIP+CF0YA*FZIN )
          CORY   = DNINT( RYI/YL )
          RXI    = RXI - CORY*DX
          RX(I)  = RXI - DNINT( RXI/XL )*XL
          RY(I)  = RYI - CORY*YL
          RZ(I)  = RZI - DNINT( RZI/ZL )*ZL
```

• This is the treatment of the Lees–Edwards BC explained in Section 2.4.2.

```
C
C                                   --- (2) ROTATIONAL MOTION ---
          C00    = TXI*NXI + TYI*NYI + TZI*NZI
          TXIP   = C00*NXI
          TYIP   = C00*NYI
          TZIP   = C00*NZI
          TXIN   = TXI - TXIP
          TYIN   = TYI - TYIP
          TZIN   = TZI - TZIP
```

• The torque acting on particle *i* is decomposed into one about the particle direction and another about a line normal to the particle direction through its center according to Eq. (3.36).

```
C
          C00     = -0.5D0*NZI
          OMEIPX = C00*NXI
          OMEIPY = C00*NYI
          OMEIPZ = C00*NZI
          OMEINX =          - OMEIPX
          OMEINY =          - OMEIPY
          OMEINZ = -0.5D0 - OMEIPZ
```

• The angular velocity is decomposed into two vectors in a similar way to the torque.

• The terms of the torque and the shear rate are calculated in the angular velocity in Eq. (3.50).

```
C
          C1X    = CT0YC*TXIN
          C1Y    = CT0YC*TYIN
          C1Z    = CT0YC*TZIN
          C2X    = -CE0YHYC* ( NZI*NXI )
          C2Y    = -CE0YHYC* (-NZI*NYI )
          C2Z    = -CE0YHYC* ( NYI**2 - NXI**2 )
C
          WXIN   = OMEINX + C1X + C2X
          WYIN   = OMEINY + C1Y + C2Y
          WZIN   = OMEINZ + C1Z + C2Z
```

• The angular velocity in Eq.(3.50) is calculated.

```
          C3X    = WYIN*NZI - WZIN*NYI
          C3Y    = WZIN*NXI - WXIN*NZI
          C3Z    = WXIN*NYI - WYIN*NXI
```

• To evaluate the particle direction from Eq.(3.55), the vector product of the angular velocity and the particle direction is first calculated.

```
C
          NXI    = NXI + H*C3X
          NYI    = NYI + H*C3Y
          NZI    = NZI + H*C3Z
```

• The particle direction is evaluated from Eq. (3.55).

```
          C00    = DSQRT( NXI**2 + NYI**2 + NZI**2 )
          NX(I) = NXI/C00
          NY(I) = NYI/C00
          NZ(I) = NZI/C00
```

• The modification is made to yield the unit vector.

```
C
  100   CONTINUE
C                                          --- CAL FORCES ---
        CALL FORCECAL( NP, NTIME )
```

• Calculation of the forces and torques.

```
C
C  -----------------------------------------------------------------
C                                      --- MOMENT OF SYSTEM ---
        IF( MOD(NTIME,DNSMPL) .EQ. 0 ) THEN
          NSMPL    = NSMPL + 1
          C1 = 0.D0
          C2 = 0.D0
          C3 = 0.D0
          DO 450 J=1,N
            C1 = C1 + NX(J)
            C2 = C2 + NY(J)
            C3 = C3 + NZ(J)
  450     CONTINUE
          MOMX(NSMPL) = REAL(C1)/REAL(N)
          MOMY(NSMPL) = REAL(C2)/REAL(N)
          MOMZ(NSMPL) = REAL(C3)/REAL(N)
        END IF
```

• To check the system convergence afterward, the average of the particle direction vector is calculated.

```
C
C                             --- DATA OUTPUT (1) FOR GRAPHICS ---
```

• The data of the particle positions and directions are written out at every NGRAPH time steps for the post processing analysis.

```
      IF( MOD(NTIME,NGRAPH) .EQ. 0 ) THEN
        NOPT = NOPT + 1
        WRITE(NOPT,472)  N , XL , YL , ZL , D , TD , RP , RP1 , DX
        WRITE(NOPT,474) (RX(I),I=1,N),(RY(I),I=1,N),(RZ(I),I=1,N),
     &                  (NX(I),I=1,N),(NY(I),I=1,N),(NZ(I),I=1,N)
                                   CLOSE(NOPT,STATUS='KEEP')
      END IF
C
C                                --- DATA OUTPUT FOR ANIMATION (2) ---
      IF( MOD(NTIME,NANIME) .EQ. 0 ) THEN
        NANMCTR = NANMCTR + 1
        NOPT1   = 13
        CALL DATAOPUT( NOPT1, NANMCTR, NTIMEMX, NANIME, N )
      END IF
C
C
 1000 CONTINUE
C
C     -----------------------------------------------------------------
C     -------------------  END OF MOLECULAR DYNAMICS   ---------------
C     -----------------------------------------------------------------
C
C                                               --- PRINT OUT (2) ---
      WRITE(NP,1011)
      NSMPL1 = 1
      NSMPL2 = NSMPL
      CALL PRNTDATA( NSMPL1 , NSMPL2 , NP )
      WRITE(NP,1013) NSMPL1 , NSMPL2
C                                  --- DATA OUTPUT (2) FOR GRAPHICS ---
      WRITE(10,1111) N, VDENS, NDENS, RAM, RAH, RAV
      WRITE(10,1113) RP, RP1, D, DEL, TD, XA, YA, YC, YH
      WRITE(10,1115) H, RCOFF, GAMDOT, BETA, XL, YL, ZL
      WRITE(10,1117) RP102, D1, CF0XA, CF0YA, CT0YC, CE0YHYC
      WRITE(10,1119) NTIMEMX, NGRAPH, DNSMPL
      WRITE(10,1121) ( MOMX(I),I=NSMPL1, NSMPL2)
     &              ,( MOMY(I),I=NSMPL1, NSMPL2)
     &              ,( MOMZ(I),I=NSMPL1, NSMPL2)
C
                                   CLOSE(9, STATUS='KEEP')
                                   CLOSE(10,STATUS='KEEP')
                                   CLOSE(13,STATUS='KEEP')
C     ------------------------- FORMAT ----------------------------
   12 FORMAT(/1H ,'-----------------------------------------------'
     &       /1H ,'-    MOLECULAR DYNAMICS SIMULATIONS OF SPHERO-   -'
     &       /1H ,'-    CYLINDER PARTICLES IN A SIMPLE SHEAR FLOW   -'
     &       /1H ,'-----------------------------------------------'
     &      //1H ,'N=',I6, 2X, 'VDENS=',F7.4, 2X ,'NDENS=',F9.6
     &       /1H ,'RAM=',F6.2, 2X, 'RAH=',F6.2, 2X ,'RAV=',F7.2
     &       /1H ,'RP=',F5.2, 2X ,'RP1=',F5.2, 2X ,'D=',F5.2, 2X ,
     &            'DEL=',F5.2, 2X,'TD=',F5.2
     &       /1H ,'XA=',E12.4,2X,'YA=',E12.4,2X,'YC=',E12.4,2X,
     &            'YH=',E12.4
     &       /1H ,'H=',E12.4,3X,'RCOFF=',F5.2,2X,'GAMDOT=',F5.2,2X,
     &            'BETA=', F5.2
     &       /1H ,'XL=',F6.2,2X,'YL=',F6.2,2X,'ZL=',F6.2)
   13 FORMAT( 1H ,'RP102=',F5.2,2X,'D1=',F5.2,2X,
     &            'CF0XA=',E11.3, 2X, 'CF0YA=',E11.3
     &       /1H ,'CT0YC=',E11.3, 2X, 'CE0YHYC=',E11.3)
   14 FORMAT( 1H ,'NTIMEMX=',I8, 2X,'NGRAPH=',I8,2X,'DNSMPL=',I8/)
  472 FORMAT( I5 , 3F9.4 , 4F8.4 , E16.8 )
  474 FORMAT( (5F16.10) )
 1011 FORMAT(/1H ,'+++++++++++++++++++++++++++++'
     &       /1H ,'        MD SIMULATIONS       '
     &       /1H ,'+++++++++++++++++++++++++++++'/)
 1013 FORMAT(///1H ,18X, 'START OF MD SAMPLING STEP=',I7
     &          /1H ,18X, 'END   OF MD SAMPLING STEP=',I7/)
 1111 FORMAT( I5 , 2F7.4 , 3F12.5 )
 1113 FORMAT( 3F6.2 , 2F7.3 , 4E12.4 )
 1115 FORMAT( E11.3 , F8.3 , 2F7.4 , 3F9.3 )
 1117 FORMAT( 2F6.2 , 4E12.4 )
 1119 FORMAT( 3I8 )
 1121 FORMAT( (10F8.5) )
```

• The data of the particle positions and directions are written out at every NANIME time steps for making an animation based on MicroAVS.

• To check the system convergence afterward, the data of the particle directions are written out.

```
      STOP
      END
C***********************************************************************
C**************************    SUBROUTINE    *************************
C***********************************************************************
C
C**** SUB PRNTDATA ****
      SUBROUTINE PRNTDATA( MCSST, MCSMX, NP )
C
      IMPLICIT REAL*8 (A-H,O-Z), INTEGER (I-N)
C
      COMMON /BLOCK12/ MOMX , MOMY , MOMZ
C
      PARAMETER( NN=1000 , NNS=500000 , PI=3.141592653589793D0 )
C
      INTEGER   MCSST      , MCSMX      , NP
      REAL      MOMX(NNS)  , MOMY(NNS)  , MOMZ(NNS)
C
      REAL      AMOMX(10) , AMOMY(10)  , AMOMZ(10)  , C0
      INTEGER   IC , IMC(0:10) , JS , JE
C
C                                             ----- KEIKA INSATU -----
      IC = ( MCSMX-MCSST+1 )/50
      DO 20 I= MCSST-1+IC , MCSMX , IC
        WRITE(NP,10) I, MOMX(I), MOMY(I), MOMZ(I)
   20 CONTINUE
C                                         ----- TIME STEP HEIKIN -----
      IC = ( MCSMX-MCSST+1 )/10
      DO 30 I=0,10
        IMC(I) = MCSST - 1 + IC*I
        IF( I .EQ. 10 ) IMC(I) =MCSMX
   30 CONTINUE
C
C
      DO 35 I=1,10
        AMOMX( I) = 0.
        AMOMY( I) = 0.
        AMOMZ( I) = 0.
   35 CONTINUE
C
      DO 50 I=1,10
        JS = IMC(I-1) + 1
        JE = IMC(I)
        DO 40 J=JS,JE
          AMOMX(I)  = AMOMX(I)   + MOMX( J)
          AMOMY(I)  = AMOMY(I)   + MOMY( J)
          AMOMZ(I)  = AMOMZ(I)   + MOMZ( J)
   40   CONTINUE
   50 CONTINUE
C
      DO 70 I=1,10
        C0          = REAL( IMC(I)-IMC(I-1) )
        AMOMX(I)    = AMOMX(I) /C0
        AMOMY(I)    = AMOMY(I) /C0
        AMOMZ(I)    = AMOMZ(I) /C0
   70 CONTINUE
C                                      ----- STEP HEIKIN INSATU -----
      WRITE(NP,75)
      DO 90 I=1,10
       WRITE(NP,80) I, IMC(I-1)+1, IMC(I), AMOMX(I),AMOMY(I),AMOMZ(I)
   90 CONTINUE
C     ----------------------------------------------------------------
   10 FORMAT(1H ,'SMPL=',I7, 1X ,'NX=',F6.3, 1X,'NY=',F6.3,
     &                                       1X,'NZ=',F6.3)
   75 FORMAT(//1H ,'-----------------------------------------------'
     &        /1H ,'                TIME AVERAGE                   '
     &        /)
   80 FORMAT(1H ,'I=',I2, 2X ,'SMPLMN=',I7, 2X ,'SMPLMX=',I7
     &       /1H ,5X ,'NX=',F6.3, 2X,'NY=',F6.3, 2X,'NZ=',F6.3/)
      RETURN
      END
C**** SUB INITIAL ****
```

• The total time steps are equally divided into 50 blocks, and the end value of each block is written out.

• The total time steps are equally divided into 10 blocks, and the subaverages are calculated for each block.

```
      SUBROUTINE INITIAL( BETA )
C
      IMPLICIT REAL*8 (A-H,O-Z), INTEGER (I-N)
C
      COMMON /BLOCK1/  RX    , RY    , RZ
      COMMON /BLOCK2/  NX    , NY    , NZ
      COMMON /BLOCK5/  XL    , YL    , ZL
      COMMON /BLOCK6/  RP    , RP1   , D      ,  DEL   , TD
      COMMON /BLOCK8/  N     , NDENS, VDENS
      COMMON /BLOCK9/  H     , RCOFF, GAMDOT, DX    , CORY
C
      PARAMETER( NN=1000 , PI=3.141592653589793D0 )
C
      REAL*8    NDENS
      REAL*8    RX(NN) , RY(NN) , RZ(NN) , NX(NN) , NY(NN) , NZ(NN)
C
      INTEGER   Q , PTCL
      REAL*8    A , XLUNT , YLUNT , ZLUNT,  RAN1 , RAN2 , RAN3
      REAL*8    C1 , C2 , C3
C
      A      = 1.D0/( (BETA*NDENS)**(1./3.) )
      Q      = NINT( (REAL(N+1))**(1./3.) )
      XL     = A*DBLE(Q)
      YL     = A*DBLE(Q)*BETA
      ZL     = A*DBLE(Q)
      XLUNT  = A
      YLUNT  = A*BETA
      ZLUNT  = A
```

• A subroutine for setting the initial positions and velocities of particles.

• The volume occupied by one particle is βa^{*3} and therefore $a^{*}=1/(\beta n^{*})^{1/3}$ because of $\beta a^{*3} n^{*}=1$ (n^{*} is the number density).
• The side lengths of the unit cell are (a^{*}, βa^{*}, a^{*}) in each direction.

```
C                                           ----- POSITION -----
      RAN1   = DSQRT( 2.D0 )
      RAN2   = DSQRT( 7.D0 )
      RAN3   = DSQRT( 11.D0 )
      PTCL   = 0
      DO 10 K=0,Q-1
      DO 10 J=0,Q-1
      DO 10 I=0,Q-1
        PTCL = PTCL + 1
        C1 = RAN1*DBLE(PTCL)
        C1 = C1 - DINT(C1)
        C1 = C1 - 0.5D0
        C2 = RAN2*DBLE(PTCL)
        C2 = C2 - DINT(C2)
        C2 = C2 - 0.5D0
        C3 = RAN3*DBLE(PTCL)
        C3 = C3 - DINT(C3)
        C3 = C3 - 0.5D0
        RX(PTCL) = DBLE(I)*XLUNT+XLUNT/3D0+C1*(XLUNT/8.D0)-XL/2.D0
        RY(PTCL) = DBLE(J)*YLUNT+YLUNT/3D0+C2*(YLUNT/8.D0)-YL/2.D0
        RZ(PTCL) = DBLE(K)*ZLUNT+ZLUNT/3D0+C3*(ZLUNT/8.D0)-ZL/2.D0
   10 CONTINUE
      N = PTCL
```

• RAN1, RAN2, and RAN3 are quasi-random numbers.
• Q particles are located in each axis direction.
• Each particle is moved in parallel by (XLUNT/3, YLUNT/3, ZLUNT/3) to remove subtle situations at outer boundary surfaces. Also, to remove the regularity of the initial configuration, each particle is moved randomly by the maximum displacement (1/2)×(XLUNT/8, YLUNT/8, ZLUNT/8) using quasi-random numbers.
• Each particle is moved in parallel by (1/2)×(-XL, -YL, -ZL), so that the simulation box center is the coordinate origin.

```
C                                           ----- MOMENT -----
      RAN1 = DSQRT( 2.D0 )
      RAN2 = DSQRT( 3.D0 )
      DO 20 I=1,N
        C1 = RAN1*DBLE(I)
        C1 = C1 - DINT(C1)
        C1 = C1 - 0.5D0
        C1 = PI/2.D0 + (5.D0/180.D0)*PI*C1
        C2 = RAN2*DBLE(I)
        C2 = C2 - DINT(C2)
        C2 = C2 - 0.5D0
        C2 = PI/2.D0 + (5.D0/180.D0)*PI*C2
        NX(I) = DSIN(C1)*DCOS(C2)
        NY(I) = DSIN(C1)*DSIN(C2)
        NZ(I) = DCOS(C1)
   20 CONTINUE
                                                        RETURN
                                                        END
C**** SUB DATAOPUT ****
      SUBROUTINE DATAOPUT( NOPT1, NANMCTR, NTIMEMX, NANIME, N )
C
```

• The initial direction of each particle is randomly assigned with a certain angle range about the y-direction using quasi-random numbers.

• A subroutine for writing out the data which can be used for making an animation based on the commercial software MicroAVS.

```
      IMPLICIT REAL*8 (A-H,O-Z), INTEGER (I-N)
C
      COMMON /BLOCK1/  RX    , RY    , RZ
      COMMON /BLOCK2/  NX    , NY    , NZ
      COMMON /BLOCK5/  XL    , YL    , ZL
      COMMON /BLOCK6/  RP    , RP1   , D      ,  DEL    , TD
      COMMON /BLOCK9/  H     , RCOFF, GAMDOT, DX     , CORY
C
      PARAMETER( NN=1000 , PI=3.141592653589793D0 )
C
      REAL*8    RX(NN) , RY(NN) , RZ(NN) , NX(NN) , NY(NN) , NZ(NN)
      REAL*8    RP102 , CRADIUS , CX1 , CY1 , CZ1 , CX2 , CY2 , CZ2
C
      RP102   = RP1/2.D0
      CRADIUS = ( D + TD )/2.D0
      XL2     = XL/2.D0
      YL2     = YL/2.D0
      ZL2     = ZL/2.D0
C
      IF( NANMCTR .EQ. 1 ) THEN
        WRITE(NOPT1,181) ( NTIMEMX/NANIME )
      END IF
C
      IF( (NANMCTR.GE.1) .AND. (NANMCTR.LE.9) ) THEN
        WRITE(NOPT1,183) NANMCTR
      ELSE IF( (NANMCTR.GE.10) .AND. (NANMCTR.LE.99) ) THEN
        WRITE(NOPT1,184) NANMCTR
      ELSE IF( (NANMCTR.GE.100) .AND. (NANMCTR.LE.999) ) THEN
        WRITE(NOPT1,185) NANMCTR
      ELSE IF( (NANMCTR.GE.1000) .AND. (NANMCTR.LE.9999) ) THEN
        WRITE(NOPT1,186) NANMCTR
      END IF
C
C     ----------------------------------------------- CYLINDER (1) ---
      WRITE(NOPT1,211)  N
      DO 250 I=1,N
        CX1 = RX(I) - NX(I)*RP102
        CY1 = RY(I) - NY(I)*RP102
        CZ1 = RZ(I) - NZ(I)*RP102
        CX2 = RX(I) + NX(I)*RP102
        CY2 = RY(I) + NY(I)*RP102
        CZ2 = RZ(I) + NZ(I)*RP102
        WRITE(NOPT1,248) CX1, CY1, CZ1, CX2, CY2, CZ2, (CRADIUS+1.D-5)
  250 CONTINUE
C
C     ------------------------------------------- SPHERE MINUS (2) ---
      WRITE(NOPT1,311)  N
      DO 350 I=1,N
        CX1 = RX(I) - NX(I)*RP102
        CY1 = RY(I) - NY(I)*RP102
        CZ1 = RZ(I) - NZ(I)*RP102
        WRITE(NOPT1,348) CX1, CY1, CZ1, CRADIUS, 0.0, 0.8, 1.0
  350 CONTINUE
C
C     -------------------------------------------- SPHERE PLUS (3) ---
      WRITE(NOPT1,311)  N
      DO 450 I=1,N
        CX1 = RX(I) + NX(I)*RP102
        CY1 = RY(I) + NY(I)*RP102
        CZ1 = RZ(I) + NZ(I)*RP102
        WRITE(NOPT1,348) CX1, CY1, CZ1, CRADIUS, 1.0, 0.0, 0.0
  450 CONTINUE
C
C     ------------------------------------- SIM.REGEON LINES  (4) ---
      WRITE(NOPT1,648)  17
      WRITE(NOPT1,649) -XL2, -YL2, -ZL2
      WRITE(NOPT1,649)  XL2, -YL2, -ZL2
      WRITE(NOPT1,649)  XL2,  YL2, -ZL2
      WRITE(NOPT1,649) -XL2,  YL2, -ZL2
      WRITE(NOPT1,649) -XL2, -YL2, -ZL2
      WRITE(NOPT1,649) -XL2, -YL2,  ZL2
      WRITE(NOPT1,649)  XL2, -YL2,  ZL2
```

• MicroAVS can make a visualization or animation by reading the data from bbb41.mgf.

• Drawing of the cylindrical part of particles.

• Drawing of the hemisphere of the negative charge.

• Drawing of the hemisphere of the positive charge.

• Drawing of the frame of the simulation box.

```
      WRITE(NOPT1,649)  XL2,  YL2,  ZL2
      WRITE(NOPT1,649) -XL2,  YL2,  ZL2
      WRITE(NOPT1,649) -XL2, -YL2,  ZL2
      WRITE(NOPT1,649) -XL2, -YL2, -ZL2
      WRITE(NOPT1,649) -XL2,  YL2, -ZL2
      WRITE(NOPT1,649) -XL2,  YL2,  ZL2
      WRITE(NOPT1,649)  XL2,  YL2,  ZL2
      WRITE(NOPT1,649)  XL2,  YL2, -ZL2
      WRITE(NOPT1,649)  XL2, -YL2, -ZL2
      WRITE(NOPT1,649)  XL2, -YL2,  ZL2
C
C     ---------------------------- FORMAT ----------------------------
  181 FORMAT('# Micro AVS Geom:2.00'
     &       /'# Animation of DPD simulation results'
     &       /I4)
  183 FORMAT('step',I1)
  184 FORMAT('step',I2)
  185 FORMAT('step',I3)
  186 FORMAT('step',I4)
  211 FORMAT( 'column'/'cylinder'/'dvertex'/'32'/I7 )
  248 FORMAT( 6F10.3 , F6.2 )
  311 FORMAT( 'sphere'/'sphere_sample'/'color'/I7 )
  348 FORMAT( 3F10.3 , F6.2 , 3F5.2 )
  648 FORMAT( 'polyline'/'pline_sample'/'vertex'/I3 )
  649 FORMAT( 3F10.3 )
                                                                 RETURN
                                                                 END
C**** SUB FORCECAL *****
      SUBROUTINE FORCECAL( NP, NTIME )
C
      IMPLICIT REAL*8 (A-H,O-Z), INTEGER (I-N)
C
      COMMON /BLOCK1/  RX    , RY    , RZ
      COMMON /BLOCK2/  NX    , NY    , NZ
      COMMON /BLOCK3/  FX    , FY    , FZ
      COMMON /BLOCK4/  TX    , TY    , TZ
      COMMON /BLOCK5/  XL    , YL    , ZL
      COMMON /BLOCK6/  RP    , RP1   , D      ,  DEL   , TD
      COMMON /BLOCK7/  XA    , YA    , YC     , YH
      COMMON /BLOCK8/  N     , NDENS, VDENS
      COMMON /BLOCK9/  H     , RCOFF, GAMDOT, DX     , CORY
      COMMON /BLOCK10/ RAM   , RAH   , RAV
      COMMON /BLOCK11/ HX    , HY    , HZ
      COMMON /WORK20/  XRXI , YRYI , ZRZI , XRXJ , YRYJ , ZRZJ
      COMMON /WORK21/  FXIJS, FYIJS, FZIJS, FXJIS, FYJIS, FZJIS
      COMMON /WORK22/  TXIJS, TYIJS, TZIJS, TXJIS, TYJIS, TZJIS
      COMMON /WORK23/  RCOFF2 , RP102 , D1     , D1SQ
      COMMON /WORK24/  CF0XA  , CF0YA , CT0YC , CE0YHYC
C
      PARAMETER( NN=1000 , PI=3.141592653589793D0 )
C
      REAL*8    NDENS
      REAL*8    RX(NN) , RY(NN) , RZ(NN) , NX(NN) , NY(NN) , NZ(NN)
      REAL*8    FX(NN) , FY(NN) , FZ(NN) , TX(NN) , TY(NN) , TZ(NN)
C
      REAL*8    RXI , RYI , RZI , RXIJ , RYIJ , RZIJ , RIJSQ , RIJ
      REAL*8    RXJ , RYJ , RZJ
      REAL*8    NXI , NYI , NZI , NXIJ , NYIJ , NZIJ
      REAL*8    NXJ , NYJ , NZJ , NXIJ2, NYIJ2, NZIJ2
      REAL*8    FXI , FYI , FZI , TXI  , TYI  , TZI
      REAL*8    FXIJP01 , FYIJP01, FZIJP01, FXIJP02, FYIJP02, FZIJP02
      REAL*8    FXIJM01 , FYIJM01, FZIJM01, FXIJM02, FYIJM02, FZIJM02
      REAL*8    FXIJP   , FYIJP  , FZIJP  , FXJIP  , FYJIP  , FZJIP
      REAL*8    TXIJP   , TYIJP  , TZIJP  , TXJIP  , TYJIP  , TZJIP
      REAL*8    FXIJM   , FYIJM  , FZIJM  , FXJIM  , FYJIM  , FZJIM
      REAL*8    TXIJM   , TYIJM  , TZIJM  , TXJIM  , TYJIM  , TZJIM
      REAL*8    TXIJ , TYIJ , TZIJ , TXIJ0 , TYIJ0 , TZIJ0
      REAL*8    XI   , YI   , ZI   , XJ    , YJ    , ZJ
      REAL*8    RRXI , RRYI , RRZI , RRXIJ , RRYIJ , RRZIJ , RRIJ
      REAL*8    RRXJ , RRYJ , RRZJ , RCHKSQ, RCHKSQ2
      REAL*8    RRXIC , RRYIC , RRZIC , RRXJC , RRYJC , RRZJC
      REAL*8    NNXI  , NNYI  , NNZI  , NNXJ  , NNYJ  , NNZJ
```

• A subroutine for calculating the forces and torques acting between particles.

```
      REAL*8    R11 , R12 , R21 , R22
      REAL*8    KI  , KJ  , KKI , KKJ , KIS , KJS , KKIS , KKIS2
      REAL*8    CKI , CKJ , CKKI, CKKJ
      REAL*8    KKIC, KKJC , CKKIC , CKKIC2
      REAL*8    CNINJ , CRIJNI , CRIJNJ , CRIJSQ , CWIDTH
      REAL*8    C01X, C01Y, C01Z
      REAL*8    C11X, C11Y, C11Z, C12X, C12Y, C12Z
      REAL*8    C21X, C21Y, C21Z, C22X, C22Y, C22Z
      REAL*8    C1R11, C1R12, C1R21, C1R22
      REAL*8    C2R11, C2R12, C2R21, C2R22
      REAL*8    C00 , C11 , C12 , C21 , C22
      INTEGER   ITREE , ISKIP , IPATH , ISUBTREE
      LOGICAL   KEEP
C
      DO 10 I=1,N
        FX(I) =  0.D0
        FY(I) =  0.D0
        FZ(I) =  0.D0
        TX(I) =  0.D0
        TY(I) =  0.D0
        TZ(I) =  0.D0
   10 CONTINUE
C
C     ---------------------------------------------- MAIN LOOP START
C
      DO 2000 I=1,N-1
C
```

• The treatment concerning particle *i* is conducted in the following.

```
        RXI  = RX(I)
        RYI  = RY(I)
        RZI  = RZ(I)
        NXI  = NX(I)
        NYI  = NY(I)
        NZI  = NZ(I)
        FXI  = FX(I)
        FYI  = FY(I)
        FZI  = FZ(I)
        TXI  = TX(I)
        TYI  = TY(I)
        TZI  = TZ(I)
C
      DO 1000 J=I+1,N
C
```

• The treatment concerning particles *i* and *j* is conducted in the following.

```
        RXJ  = RX(J)
        RYJ  = RY(J)
        RZJ  = RZ(J)
        NXJ  = NX(J)
        NYJ  = NY(J)
        NZJ  = NZ(J)
C
```

• The treatment of the periodic BC.

```
        RZIJ = RZI  - RZJ
        IF( RZIJ .GT. ZL/2.D0 ) THEN
          RZIJ = RZIJ - ZL
          RZJ  = RZJ  + ZL
        ELSE IF( RZIJ .LT. -ZL/2.D0 ) THEN
          RZIJ = RZIJ + ZL
          RZJ  = RZJ  - ZL
        END IF
        IF( DABS(RZIJ) .GE. RCOFF )       GOTO 1000
C
```

• The treatment of the Lees–Edwards BC.

```
        RYIJ   = RYI  - RYJ
        CORY   = - DNINT( RYIJ/YL)
        RYIJ   = RYIJ + CORY*YL
        RYJ    = RYJ  - CORY*YL
        IF( DABS(RYIJ) .GE. RCOFF )       GOTO 1000
C
        RXIJ   = RXI  - RXJ
        RXIJ   = RXIJ + CORY*DX
        RXJ    = RXJ  - CORY*DX
```

```
        IF( RXIJ .GT. XL/2.D0 ) THEN
          RXIJ = RXIJ - XL
          RXJ  = RXJ  + XL
        ELSE IF( RXIJ .LT. -XL/2.D0 ) THEN
          RXIJ = RXIJ + XL
          RXJ  = RXJ  - XL
        END IF
        IF( DABS(RXIJ) .GE. RCOFF )      GOTO 1000
C
        RIJSQ= RXIJ**2 + RYIJ**2 + RZIJ**2
        IF( RIJSQ .GE. RCOFF2 )          GOTO 1000
        RIJ  = DSQRT(RIJSQ)
C
```

• If the two particles are separated over the cutoff distance r^*_{coff}, the calculation of forces and torques is unnecessary.

```
C       -------------------------------- START OF MAGNETIC FORCES ---
        NXIJ    = NXI - NXJ
        NYIJ    = NYI - NYJ
        NZIJ    = NZI - NZJ
        NXIJ2   = NXI + NXJ
        NYIJ2   = NYI + NYJ
        NZIJ2   = NZI + NZJ
C                                          --- MAGNETIC FORCES ---
```

• The magnetic force acting between particles i and j is calculated.

```
        FXIJP01 = RXIJ + RP102*NXIJ
        FYIJP01 = RYIJ + RP102*NYIJ
        FZIJP01 = RZIJ + RP102*NZIJ
        FXIJP02 = RXIJ + RP102*NXIJ2
        FYIJP02 = RYIJ + RP102*NYIJ2
        FZIJP02 = RZIJ + RP102*NZIJ2
        FXIJM01 = RXIJ - RP102*NXIJ2
        FYIJM01 = RYIJ - RP102*NYIJ2
        FZIJM01 = RZIJ - RP102*NZIJ2
        FXIJM02 = RXIJ - RP102*NXIJ
        FYIJM02 = RYIJ - RP102*NYIJ
        FZIJM02 = RZIJ - RP102*NZIJ
C
```

• To calculate the first and second terms of Eq. (3.56) and also Eq. (3.57) separately, we calculate quantities needed in order.

```
        C2R11 =  FXIJP01**2 + FYIJP01**2 + FZIJP01**2
        C2R12 =  FXIJP02**2 + FYIJP02**2 + FZIJP02**2
        C2R21 =  FXIJM01**2 + FYIJM01**2 + FZIJM01**2
        C2R22 =  FXIJM02**2 + FYIJM02**2 + FZIJM02**2
        C1R11 = DSQRT( C2R11 )
        C1R12 = DSQRT( C2R12 )
        C1R21 = DSQRT( C2R21 )
        C1R22 = DSQRT( C2R22 )
        IF( C1R11 .GE. 1.D0 ) THEN
          R11 = C1R11*C2R11
        ELSE
          R11 = C1R11
        END IF
        IF( C1R12 .GE. 1.D0 ) THEN
          R12 = C1R12*C2R12
        ELSE
          R12 = C1R12
        END IF
        IF( C1R21 .GE. 1.D0 ) THEN
          R21 = C1R21*C2R21
        ELSE
          R21 = C1R21
        END IF
        IF( C1R22 .GE. 1.D0 ) THEN
          R22 = C1R22*C2R22
        ELSE
          R22 = C1R22
        END IF
C
```

• The denominators of the first and second terms in Eq. (3.56) are calculated and saved in R11 and R12.
• Similarly, those in Eq. (3.57) are calculated and saved in R21 and R22.

```
        C11X  =    FXIJP01/R11
        C11Y  =    FYIJP01/R11
        C11Z  =    FZIJP01/R11
        C12X  =    FXIJP02/R12
        C12Y  =    FYIJP02/R12
        C12Z  =    FZIJP02/R12
        FXIJP =    C11X - C12X
```

• The first and second terms in Eq. (3.56) are calculated and saved in (C11X, C11Y, C11Z) and (C12X, C12Y, C12Z).
• Eq. (3.56) is calculated, but λ_m is multiplied in the final stage.

```
      FYIJP =    C11Y - C12Y
      FZIJP =    C11Z - C12Z
      C21X  =    FXIJM01/R21
      C21Y  =    FYIJM01/R21
      C21Z  =    FZIJM01/R21
      C22X  =    FXIJM02/R22
      C22Y  =    FYIJM02/R22
      C22Z  =    FZIJM02/R22
      FXIJM =  - C21X + C22X
      FYIJM =  - C21Y + C22Y
      FZIJM =  - C21Z + C22Z
C
      FXJIP =  - C11X + C21X
      FYJIP =  - C11Y + C21Y
      FZJIP =  - C11Z + C21Z
      FXJIM =    C12X - C22X
      FYJIM =    C12Y - C22Y
      FZJIM =    C12Z - C22Z
C
      FXI   = FXI    + ( FXIJP + FXIJM )*RAM
      FYI   = FYI    + ( FYIJP + FYIJM )*RAM
      FZI   = FZI    + ( FZIJP + FZIJM )*RAM
      FX(J) = FX(J)  + ( FXJIP + FXJIM )*RAM
      FY(J) = FY(J)  + ( FYJIP + FYJIM )*RAM
      FZ(J) = FZ(J)  + ( FZJIP + FZJIM )*RAM
C                                               --- MAGNETIC TORQUES ---
      TXIJP =   ( NYI*FZIJP - NZI*FYIJP )
      TYIJP =   ( NZI*FXIJP - NXI*FZIJP )
      TZIJP =   ( NXI*FYIJP - NYI*FXIJP )
      TXIJM = -( NYI*FZIJM - NZI*FYIJM )
      TYIJM = -( NZI*FXIJM - NXI*FZIJM )
      TZIJM = -( NXI*FYIJM - NYI*FXIJM )
      TXI   = TXI + ( TXIJP + TXIJM )* (RP102*3.D0) * RAM
      TYI   = TYI + ( TYIJP + TYIJM )* (RP102*3.D0) * RAM
      TZI   = TZI + ( TZIJP + TZIJM )* (RP102*3.D0) * RAM
C
      TXJIP =   ( NYJ*FZJIP - NZJ*FYJIP )
      TYJIP =   ( NZJ*FXJIP - NXJ*FZJIP )
      TZJIP =   ( NXJ*FYJIP - NYJ*FXJIP )
      TXJIM = -( NYJ*FZJIM - NZJ*FYJIM )
      TYJIM = -( NZJ*FXJIM - NXJ*FZJIM )
      TZJIM = -( NXJ*FYJIM - NYJ*FXJIM )
      TX(J) = TX(J) + ( TXJIP + TXJIM )* (RP102*3.D0) * RAM
      TY(J) = TY(J) + ( TYJIP + TYJIM )* (RP102*3.D0) * RAM
      TZ(J) = TZ(J) + ( TZJIP + TZJIM )* (RP102*3.D0) * RAM
C
C     --------------------------------- END OF MAGNETIC FORCES ---
C
C     ----------------------------- FORCES DUE TO STERIC INER. ---
C
      CNINJ = NXI*NXJ +  NYI*NYJ +  NZI*NZJ
      TXIJ  = RXIJ/RIJ
      TYIJ  = RYIJ/RIJ
      TZIJ  = RZIJ/RIJ
      C11   = TXIJ*NXJ + TYIJ*NYJ + TZIJ*NZJ
C
      IF( DABS(CNINJ) .LT. 0.001D0 ) THEN
        ITREE = 2
      ELSE IF( DABS(CNINJ) .GT. 0.999D0) THEN
        IF(  DABS(C11) .GT. 0.999D0 )THEN
          ITREE = 0
        END IF
          ITREE = 3
      ELSE
        ITREE = 1
      END IF
C
C                                             --------------------
C                                             ITREE=0: LINEAR
C                                             ITREE=1: GENERAL
C                                             ITREE=2: NORMALL
C                                             ITREE=3: PARALLEL
```

• The first and second terms in Eq. (3.57) are calculated and saved in (C21X, C21Y, C21Z) and (C22X, C22Y, C22Z).
• Eq. (3.57) is calculated, but λ_m is multiplied in the final stage.

• The forces acting on the positive and negative charges of particle *j* can be obtained from the action–reaction law; λ_m is multiplied in the final stage.

• The force exerted by particle *j* is saved in the variable of particle *i*.
• Similarly, the force exerted by particle *i* is saved.

• The torque acting on particle *i* is calculated from Eqs. (3.59) and (3.60).

• The torque acting on particle *j* is calculated from Eqs. (3.59) and (3.60).

• The repulsive force due to the overlap of surfactant layers is calculated below.
• The variable ITREE implies the particle overlapping regime, and the procedures are performed according to ITREE.

• The regime in Table 3.1 is determined to proceed to appropriate treatment, and after the calculation of the repulsive forces, the calculation procedure returns to the main loop.

```
C                                                      --------------------
C
C     -------------------------------------------- (0) LINEAR ---
      IF( ITREE .EQ. 0 ) THEN
C
        IF( CNINJ .GE. 0 ) THEN
          IF( C11 .GE. 0 ) THEN
C                                                       --- IPATH=1
            XJ    = RXJ + NXJ*RP102
            YJ    = RYJ + NYJ*RP102
            ZJ    = RZJ + NZJ*RP102
            XI    = RXI - NXI*RP102
            YI    = RYI - NYI*RP102
            ZI    = RZI - NZI*RP102
          ELSE
C                                                       --- IPATH=2
            XJ    = RXJ - NXJ*RP102
            YJ    = RYJ - NYJ*RP102
            ZJ    = RZJ - NZJ*RP102
            XI    = RXI + NXI*RP102
            YI    = RYI + NYI*RP102
            ZI    = RZI + NZI*RP102
          END IF
        ELSE
          IF( C11 .GE. 0 ) THEN
C                                                       --- IPATH=3
            XJ    = RXJ + NXJ*RP102
            YJ    = RYJ + NYJ*RP102
            ZJ    = RZJ + NZJ*RP102
            XI    = RXI + NXI*RP102
            YI    = RYI + NYI*RP102
            ZI    = RZI + NZI*RP102
          ELSE
C                                                       --- IPATH=4
            XJ    = RXJ - NXJ*RP102
            YJ    = RYJ - NYJ*RP102
            ZJ    = RZJ - NZJ*RP102
            XI    = RXI - NXI*RP102
            YI    = RYI - NYI*RP102
            ZI    = RZI - NZI*RP102
          END IF
        END IF
C
        RRIJ = DSQRT( (XI-XJ)**2 + (YI-YJ)**2 + (ZI-ZJ)**2 )
        XRXI = XI - RXI
        YRYI = YI - RYI
        ZRZI = ZI - RZI
        XRXJ = XJ - RXJ
        YRYJ = YJ - RYJ
        ZRZJ = ZJ - RZJ
        ISKIP = 1
        CALL STEFORCE( RRIJ,RAV,ISKIP,TXIJ,TYIJ,TZIJ )
        FXI   = FXI   + FXIJS
        FYI   = FYI   + FYIJS
        FZI   = FZI   + FZIJS
        FX(J) = FX(J) + FXJIS
        FY(J) = FY(J) + FYJIS
        FZ(J) = FZ(J) + FZJIS
C
        GOTO 1000
C
      END IF
C     -------------------------------------------- END OF LINEAR --

      IF( (ITREE .EQ. 1) .OR. (ITREE .EQ. 2) ) THEN
C
        CRIJNI = NXI*RXIJ + NYI*RYIJ + NZI*RZIJ
        CRIJNJ = NXJ*RXIJ + NYJ*RYIJ + NZJ*RZIJ
        C00    = 1.D0 / (1.D0 - CNINJ**2)
        KI     = C00*( -CRIJNI + CNINJ*CRIJNJ )
        KJ     = C00*(  CRIJNJ - CNINJ*CRIJNI )
C                                               --- CHECK OVERLAP ---
```

• The treatment for the linear arrangement in Table 3.1.

• The positions (XI,YI,ZI) and (XJ,YJ,ZJ) of the magnetic charges of particles *i* and *j* are calculated.

• The calculation of torques is unnecessary, so ISKIP is set as ISKIP=1.

• The repulsive forces due to the overlap of the steric layers are calculated in the subroutine STEFORCE; the results concerning particles *i* and *j* are saved in (FXIJS,FYIJS,FZIJS) and (FXJIS, FYJIS, FZJIS), respectively.

• k_i and k_j are calculated from Eq. (3.44).

```
          CRIJSQ =    (RXIJ + KI*NXI - KJ*NXJ )**2
     &              + (RYIJ + KI*NYI - KJ*NYJ )**2
     &              + (RZIJ + KI*NZI - KJ*NZJ )**2
          IF( CRIJSQ .GE. D1SQ )  GOTO 1000
C
          IF( DABS(KJ) .GT. DABS(KI) ) THEN
            KEEP   = .TRUE.
            II     = I
            JJ     = J
            RRXI   = RXI
            RRYI   = RYI
            RRZI   = RZI
            RRXJ   = RXJ
            RRYJ   = RYJ
            RRZJ   = RZJ
            RRXIJ  = RXIJ
            RRYIJ  = RYIJ
            RRZIJ  = RZIJ
            NNXI   = NXI
            NNYI   = NYI
            NNZI   = NZI
            NNXJ   = NXJ
            NNYJ   = NYJ
            NNZJ   = NZJ
            KKI    = KI
            KKJ    = KJ
          ELSE
            KEEP   = .FALSE.
            II     = J
            JJ     = I
            RRXI   = RXJ
            RRYI   = RYJ
            RRZI   = RZJ
            RRXJ   = RXI
            RRYJ   = RYI
            RRZJ   = RZI
            RRXIJ  = -RXIJ
            RRYIJ  = -RYIJ
            RRZIJ  = -RZIJ
            NNXI   = NXJ
            NNYI   = NYJ
            NNZI   = NZJ
            NNXJ   = NXI
            NNYJ   = NYI
            NNZJ   = NZI
            KKI    = KJ
            KKJ    = KI
            TXIJ   = -TXIJ
            TYIJ   = -TYIJ
            TZIJ   = -TZIJ
          END IF
C                                       ------------------------------
C                                       ISUBTREE=1: i(sphe,cyl)-j(sphe)
C                                       ISUBTREE=2: i(cyl)      -j(cyl)
C                                       ------------------------------
          IF( DABS(KKJ) .GE. RP102 ) THEN
            ISUBTREE = 1
          ELSE
            ISUBTREE = 2
          END IF
C
        END IF
C                                                  --------------------
C                                                  ITREE=0: LINEAR
C                                                  ITREE=1: GENERAL
C                                                  ITREE=2: NORMALL
C                                                  ITREE=3: PARALLEL
C                                                  --------------------
        IF( ITREE .EQ. 1 ) GOTO 200
        IF( ITREE .EQ. 2 ) GOTO 400
        IF( ITREE .EQ. 3 ) GOTO 600
```

- The subscripts are exchanged between i and j so as to satisfy $|k_j|>|k_i|$.
- As a result, the particle names i and j in Table 3.1 are expressed as II and JJ in the program.

- Which part of particle j has a possibility of the overlap with particle i is grasped; there is an overlapping possibility of the hemisphere part for ISUBTREE=1 and of the cylindrical part for SUBTREE=2.

```
C
C     ------------------------------------------- (1) GENERAL ---
C                                          --- FOR II AND JJ ---
  200   CNINJ = NXI*NXJ + NYI*NYJ + NZI*NZJ
        IF( CNINJ .GT. 0.D0 ) THEN
          IF( KKJ .GE. 0.D0 ) THEN
            IPATH = 1
          ELSE
            IPATH = 4
          END IF
        ELSE
          IF( KKJ .GE. 0.D0 ) THEN
            IPATH = 3
          ELSE
            IPATH = 2
          END IF
        END IF
C
```

• The treatment for itree=1 of the general arrangement in Table 3.1.
• The treatment is conducted for the four cases depending on the position relationship of the positive and negative charges of particles i and j; IPATH is used for specifying the case chosen.

• k_i^s (KKIS) is calculated from Eq. (3.46). Similarly, $k_i^{s'}$ (KKIS2) concerning the negative magnetic charge of particle j is calculated.

```
        KKIS   =  CNINJ*RP102 - (RRXIJ*NNXI + RRYIJ*NNYI + RRZIJ*NNZI)
        KKIS2  = -CNINJ*RP102 - (RRXIJ*NNXI + RRYIJ*NNYI + RRZIJ*NNZI)
C
        C1 = RP102 - KKJ
        C1 = DINT( C1 )
        C2 = RP102 - DABS( KKJ )
        C2 = DINT( C2 )
C
        IF( IPATH .EQ. 1 ) THEN
C
          C12 =-1.D0
          C22 =-1.D0
          IF( ISUBTREE .EQ. 1 ) THEN
            C11 = RP102
            C21 = KKIS
            IF( KKIS .GT. RP102 )   C21 = RP102
            IF( KKIS .LT.-RP102 )   C21 =-RP102
          ELSE
            C11 = KKJ + C1
            C21 = KKI + C1
          END IF
C                                             --- PATH=2 ---
        ELSE IF( IPATH .EQ. 2 )  THEN
          C12 = 1.D0
          C22 =-1.D0
          IF( ISUBTREE .EQ. 1 ) THEN
            C11 =-RP102
            C21 = KKIS2
            IF( KKIS2 .GT. RP102 )  C21 = RP102
            IF( KKIS2 .LT.-RP102 )  C21 =-RP102
          ELSE
            C11 = KKJ - C2
            C21 = KKI + C2
          END IF
C                                             --- PATH=3 ---
        ELSE IF( IPATH .EQ. 3 )  THEN
          C12 =-1.D0
          C22 = 1.D0
          IF( ISUBTREE .EQ. 1 ) THEN
            C11 = RP102
            C21 = KKIS
            IF( KKIS .LT. -RP102 )  C21 = -RP102
            IF( KKIS .GT.  RP102 )  C21 =  RP102
          ELSE
            C11 = KKJ + C1
            C21 = KKI - C1
          END IF
C                                             --- PATH=4 ---
        ELSE
          C12 = 1.D0
          C22 = 1.D0
          IF( ISUBTREE .EQ. 1 ) THEN
            C11 =-RP102
            C21 = KKIS2
```

• According to the repulsive force model shown in Section 3.2.4, the position of the first constituent sphere to be placed is determined. The variables used to do so are C11 and C21 for particles j and i, respectively.

• The direction in which the next neighboring sphere is added to form the sphere-connected particle j is specified by C12; similarly, C22 is used for particle i. C12=1 means the particle axis direction. C12=–1 means the opposite direction to the particle axis.

```
            IF( KKIS2 .LT. -RP102 ) C21 = -RP102
            IF( KKIS2 .GT.  RP102 ) C21 =  RP102
          ELSE
            C11 = KKJ - C2
            C21 = KKI - C2
          END IF
        END IF
C
C
        JJJE = IDNINT(RP1)
        DO 250 JJJ= 0, JJJE
C
          CKKJ = C11 + C12*DBLE(JJJ)
          CKKI = C21 + C22*DBLE(JJJ)
          IF( ( DABS(CKKJ) .GT. RP102+1.D-10 ) .OR.
     &        ( DABS(CKKI) .GT. RP102+1.D-10 )         )    GOTO 250
C
          IF( ISUBTREE .EQ. 1 ) THEN
            IF( ( DABS(CKKI) .GT. RP102+1.D-10 ) .OR.
     &          ( DABS(CKKJ) .GT. RP102+1.D-10 )         )  GOTO 1000
          END IF
C
  245     XJ   = RRXJ + NNXJ*CKKJ
          YJ   = RRYJ + NNYJ*CKKJ
          ZJ   = RRZJ + NNZJ*CKKJ
          XI   = RRXI + NNXI*CKKI
          YI   = RRYI + NNYI*CKKI
          ZI   = RRZI + NNZI*CKKI
```

• The positions of the spheres of particle *i* and *j* are saved in (XI,YI,ZI) and (XJ,YJ, ZJ), respectively.

```
          RRIJ = DSQRT( (XI-XJ)**2 + (YI-YJ)**2 + (ZI-ZJ)**2 )
          IF( ISUBTREE .EQ. 1 ) THEN
            IF( RRIJ .GE. D1 )  GOTO 1000
          END IF
          XRXI = XI - RRXI
          YRYI = YI - RRYI
          ZRZI = ZI - RRZI
          XRXJ = XJ - RRXJ
          YRYJ = YJ - RRYJ
          ZRZJ = ZJ - RRZJ
```

• To evaluate the torque, the relative position of the sphere from the rod-like particle center is calculated.

```
          TXIJ0= (XI-XJ)/RRIJ
          TYIJ0= (YI-YJ)/RRIJ
          TZIJ0= (ZI-ZJ)/RRIJ
          ISKIP = 0
          CALL STEFORCE( RRIJ,RAV,ISKIP,TXIJ0,TYIJ0,TZIJ0 )
          IF( .NOT. KEEP ) THEN
```

• The posttreatment for the case of the particle names exchanged.

```
            C1    = FXIJS
            C2    = FYIJS
            C3    = FZIJS
            FXIJS = FXJIS
            FYIJS = FYJIS
            FZIJS = FZJIS
            FXJIS = C1
            FYJIS = C2
            FZJIS = C3
            C1    = TXIJS
            C2    = TYIJS
            C3    = TZIJS
            TXIJS = TXJIS
            TYIJS = TYJIS
            TZIJS = TZJIS
            TXJIS = C1
            TYJIS = C2
            TZJIS = C3
          END IF
          FXI   = FXI   + FXIJS
          FYI   = FYI   + FYIJS
          FZI   = FZI   + FZIJS
          FX(J) = FX(J) + FXJIS
          FY(J) = FY(J) + FYJIS
          FZ(J) = FZ(J) + FZJIS
          TXI   = TXI   + TXIJS
          TYI   = TYI   + TYIJS
          TZI   = TZI   + TZIJS
```

```
            TX(J) = TX(J) + TXJIS
            TY(J) = TY(J) + TYJIS
            TZ(J) = TZ(J) + TZJIS
C
  250     CONTINUE
C
          GOTO 1000
C         ---------------------------------------------- (2) NORMAL  ---
C                                                   --- FOR II AND JJ ---
C
  400     IF( KKJ .GE. 0.D0 ) THEN
            IPATH = 1
          ELSE
            IPATH = 2
          END IF
C
```

• The treatment for the normal arrangement in Table 3.1.

• k_i^s (KKIS) is calculated from Eq. (3.46). Similarly, $k_i^{s'}$ (KKIS2) concerning the negative magnetic charge of particle j is calculated

```
          CNINJ  =  NXI*NXJ + NYI*NYJ + NZI*NZJ
          KKIS   =  CNINJ*RP102 - (RRXIJ*NNXI + RRYIJ*NNYI + RRZIJ*NNZI)
          KKIS2  = -CNINJ*RP102 - (RRXIJ*NNXI + RRYIJ*NNYI + RRZIJ*NNZI)
C
          C11 = KKJ
          C21 = KKI
          IF( IPATH .EQ. 1 )  THEN
C                                                         --- PATH=1 ---
            IF( ISUBTREE .EQ. 1 ) THEN
              C11 = RP102
              C21 = KKIS
              IF( KKIS .GT. RP102 )   C21 = RP102
              IF( KKIS .LT.-RP102 )   C21 =-RP102
            END IF
          ELSE
C                                                         --- PATH=2 ---
            IF( ISUBTREE .EQ. 1 ) THEN
              C11 =-RP102
              C21 = KKIS2
              IF( KKIS2 .GT. RP102 )   C21 = RP102
              IF( KKIS2 .LT.-RP102 )   C21 =-RP102
            END IF
          END IF
C
          CKKJ = C11
          CKKI = C21
          XJ = RRXJ + CKKJ*NNXJ
          YJ = RRYJ + CKKJ*NNYJ
          ZJ = RRZJ + CKKJ*NNZJ
          XI = RRXI + CKKI*NNXI
          YI = RRYI + CKKI*NNYI
          ZI = RRZI + CKKI*NNZI
          RRIJ = DSQRT( (XI-XJ)**2 + (YI-YJ)**2 + (ZI-ZJ)**2 )
          IF( RRIJ .GE. D1 )  GOTO 1000
C
```

• According to the repulsive force model shown in Section 3.2.4, the position of the first constituent sphere to be placed is determined. The variables used to do so are C11 and C21 for particles j and i, respectively.

```
          XRXI = XI - RRXI
          YRYI = YI - RRYI
          ZRZI = ZI - RRZI
          XRXJ = XJ - RRXJ
          YRYJ = YJ - RRYJ
          ZRZJ = ZJ - RRZJ
          TXIJ0= (XI-XJ)/RRIJ
          TYIJ0= (YI-YJ)/RRIJ
          TZIJ0= (ZI-ZJ)/RRIJ
          ISKIP = 0
          CALL STEFORCE( RRIJ,RAV,ISKIP,TXIJ0,TYIJ0,TZIJ0 )
          IF( .NOT. KEEP ) THEN
            C1    = FXIJS
            C2    = FYIJS
            C3    = FZIJS
            FXIJS = FXJIS
            FYIJS = FYJIS
            FZIJS = FZJIS
            FXJIS = C1
            FYJIS = C2
```

• The positions of the spheres of particles i and j are saved in (XI,YI,ZI) and (XJ,YJ,ZJ), respectively.

• To evaluate the torque, the relative position of the sphere from the rod-like particle center is calculated.

• The posttreatment for the case of the particle names exchanged.

```
            FZJIS = C3
            C1    = TXIJS
            C2    = TYIJS
            C3    = TZIJS
            TXIJS = TXJIS
            TYIJS = TYJIS
            TZIJS = TZJIS
            TXJIS = C1
            TYJIS = C2
            TZJIS = C3
          END IF
          FXI   = FXI   + FXIJS
          FYI   = FYI   + FYIJS
          FZI   = FZI   + FZIJS
          FX(J) = FX(J) + FXJIS
          FY(J) = FY(J) + FYJIS
          FZ(J) = FZ(J) + FZJIS
          TXI   = TXI   + TXIJS
          TYI   = TYI   + TYIJS
          TZI   = TZI   + TZIJS
          TX(J) = TX(J) + TXJIS
          TY(J) = TY(J) + TYJIS
          TZ(J) = TZ(J) + TZJIS
C
          GOTO 1000
C         -------------------------------------------- (3) PARALLEL --
C                                                      --- FOR I AND J ---
C
   600    CNINJ =  NXI*NXJ + NYI*NYJ + NZI*NZJ
          KIS   =  CNINJ*RP102 - (RXIJ*NXI + RYIJ*NYI + RZIJ*NZI)
          KJS   =  CNINJ*RP102 + (RXIJ*NXJ + RYIJ*NYJ + RZIJ*NZJ)
C                                                   --- CHECK OVERLAP ---
          CWIDTH =    (RXIJ + KIS*NXI - RP102*NXJ )**2
     &              + (RYIJ + KIS*NYI - RP102*NYJ )**2
     &              + (RZIJ + KIS*NZI - RP102*NZJ )**2
          IF( CWIDTH .GE. D1SQ )  GOTO 1000
C
          IF( CNINJ .GE. 0.D0 ) THEN
            IPATH = 1
          ELSE
            IF( KIS .LE. -RP102 ) THEN
              IPATH = 2
            ELSE
              IPATH = 3
            END IF
          END IF
C
          KEEP   = .TRUE.
          II     = I
          JJ     = J
          RRXI   = RXI
          RRYI   = RYI
          RRZI   = RZI
          RRXJ   = RXJ
          RRYJ   = RYJ
          RRZJ   = RZJ
          RRXIJ  = RXIJ
          RRYIJ  = RYIJ
          RRZIJ  = RZIJ
          NNXI   = NXI
          NNYI   = NYI
          NNZI   = NZI
          NNXJ   = NXJ
          NNYJ   = NYJ
          NNZJ   = NZJ
          KKIS   = KIS
          IF( (IPATH .EQ. 1) .AND. (KIS .GT. KJS) ) THEN
            KEEP   = .FALSE.
            II     = J
            JJ     = I
            RRXI   = RXJ
            RRYI   = RYJ
```

• The treatment for the parallel arrangement in Table 3.1.

• The square distance between particles i and j is calculated and saved in CWIDTH. In this calculation, the length of the vertical line drawn from the positive magnetic charge of particle j to the axis line of particle i is evaluated.

```
          RRZI   = RZJ
          RRXJ   = RXI
          RRYJ   = RYI
          RRZJ   = RZI
          RRXIJ  = -RXIJ
          RRYIJ  = -RYIJ
          RRZIJ  = -RZIJ
          NNXI   = NXJ
          NNYI   = NYJ
          NNZI   = NZJ
          NNXJ   = NXI
          NNYJ   = NYI
          NNZJ   = NZI
          KKIS   = KJS
        END IF
```

• The point at which the vertical line drawn from the center of particle *j* intersects the axis line of particle *i* is assumed to be denoted by $\mathbf{r}_i+k_i^c e_i$, k_i^c, and a similar quantity k_j^c is evaluated.

```
C                                            --- FOR II AND JJ ---
C
        KKIC    = -( RRXIJ*NNXI+ RRYIJ*NNYI + RRZIJ*NNZI )
        KKJC    =  ( RRXIJ*NNXJ+ RRYIJ*NNYJ + RRZIJ*NNZJ )
        CKKIC   = DABS( KKIC )
        CKKIC2  = CKKIC/2.D0
C
        C11 = KKJC/2.D0
        C21 = KKIC/2.D0
        IF( IPATH .EQ. 1 ) THEN
```

• According to the repulsive force model in Section 3.2.4, the position of the first sphere to be placed is determined. The variables used to do so are C11 and C21 for particles *j* and *i*, respectively.

```
C                                                  --- PATH=1 ---
          C12 = 1.D0
          C22 = 1.D0
          IF( CKKIC2 .GT. RP102 ) THEN
            C11 =  RP102
            C21 = -RP102
          END IF
C                                                  --- PATH=2 ---
        ELSE IF( IPATH .EQ. 2 )  THEN
          C12 =-1.D0
          C22 = 1.D0
          IF( CKKIC2 .GT. RP102 ) THEN
            C11 = -RP102
            C21 = -RP102
          END IF
```

• The direction in which the next neighboring sphere is added to form the sphere-connected particle *j* is specified by C12.

```
C                                                  --- PATH=3 ---
        ELSE
          C12 = 1.D0
          C22 =-1.D0
          IF( CKKIC2 .GT. RP102 ) THEN
            C11 =  RP102
            C21 =  RP102
          END IF
        END IF
```

• Similarly, C22 is used for particle *i*. C12=1 means the particle axis direction; C12=–1 means the opposite direction to the particle axis.

```
C
        JJJE = IDNINT(RP102)
        DO 650 JJJ= 0, JJJE
C
          CKKJ = C11 + C12*DBLE(JJJ)
          CKKI = C21 + C22*DBLE(JJJ)
          IF( JJJ .EQ. 0 )  GOTO 645
          IF( ( DABS(CKKI) .GT. RP102+1.D-10)  .OR.
     &          ( DABS(CKKJ) .GT. RP102+1.D-10 )        )  GOTO 1000
C
  645     XJ    = RRXJ + NNXJ*CKKJ
          YJ    = RRYJ + NNYJ*CKKJ
          ZJ    = RRZJ + NNZJ*CKKJ
          XI    = RRXI + NNXI*CKKI
          YI    = RRYI + NNYI*CKKI
          ZI    = RRZI + NNZI*CKKI
          RRIJ = DSQRT( (XI-XJ)**2 + (YI-YJ)**2 + (ZI-ZJ)**2 )
          IF( RRIJ .GE. D1 )  GOTO 1000
          XRXI = XI - RRXI
          YRYI = YI - RRYI
          ZRZI = ZI - RRZI
          XRXJ = XJ - RRXJ
```

• The positions of the spheres of particles *i* and *j* are saved in (XI,YI,ZI) and (XJ,YJ,ZJ), respectively..

• To evaluate the torque, the relative position of the sphere from the rod-like particle center is calculated.

```
      YRYJ = YJ - RRYJ
      ZRZJ = ZJ - RRZJ
      TXIJ0= (XI-XJ)/RRIJ
      TYIJ0= (YI-YJ)/RRIJ
      TZIJ0= (ZI-ZJ)/RRIJ
      ISKIP = 0
      CALL STEFORCE( RRIJ,RAV,ISKIP,TXIJ0,TYIJ0,TZIJ0 )
      IF( .NOT. KEEP ) THEN
        C1    = FXIJS
        C2    = FYIJS
        C3    = FZIJS
```

• The posttreatment for the case of the particle names exchanged.

```
        FXIJS = FXJIS
        FYIJS = FYJIS
        FZIJS = FZJIS
        FXJIS = C1
        FYJIS = C2
        FZJIS = C3
        C1    = TXIJS
        C2    = TYIJS
        C3    = TZIJS
        TXIJS = TXJIS
        TYIJS = TYJIS
        TZIJS = TZJIS
        TXJIS = C1
        TYJIS = C2
        TZJIS = C3
      END IF
      FXI   = FXI   + FXIJS
      FYI   = FYI   + FYIJS
      FZI   = FZI   + FZIJS
      FX(J) = FX(J) + FXJIS
      FY(J) = FY(J) + FYJIS
      FZ(J) = FZ(J) + FZJIS
      TXI   = TXI   + TXIJS
      TYI   = TYI   + TYIJS
      TZI   = TZI   + TZIJS
      TX(J) = TX(J) + TXJIS
      TY(J) = TY(J) + TYJIS
      TZ(J) = TZ(J) + TZJIS
C                          --- COUNT JUST ONCE FOR CENTRAL PLACE ---
      IF( JJJ .EQ. 0 )  GOTO 650
C
```

• Because of the parallel arrangement, a similar calculation of the repulsive forces is carried out for the particles placed on the particle axis in the opposite direction.

```
      XJ   = RRXJ - NNXJ*CKKJ
      YJ   = RRYJ - NNYJ*CKKJ
      ZJ   = RRZJ - NNZJ*CKKJ
      XI   = RRXI - NNXI*CKKI
      YI   = RRYI - NNYI*CKKI
      ZI   = RRZI - NNZI*CKKI
      RRIJ = DSQRT( (XI-XJ)**2 + (YI-YJ)**2 + (ZI-ZJ)**2 )
      XRXI = XI - RRXI
      YRYI = YI - RRYI
      ZRZI = ZI - RRZI
      XRXJ = XJ - RRXJ
      YRYJ = YJ - RRYJ
      ZRZJ = ZJ - RRZJ
      TXIJ0= (XI-XJ)/RRIJ
      TYIJ0= (YI-YJ)/RRIJ
      TZIJ0= (ZI-ZJ)/RRIJ
      ISKIP = 0
      CALL STEFORCE( RRIJ,RAV,ISKIP,TXIJ0,TYIJ0,TZIJ0 )
      IF( .NOT. KEEP ) THEN
        C1    = FXIJS
        C2    = FYIJS
        C3    = FZIJS
        FXIJS = FXJIS
        FYIJS = FYJIS
        FZIJS = FZJIS
        FXJIS = C1
        FYJIS = C2
        FZJIS = C3
```

```
                C1     = TXIJS
                C2     = TYIJS
                C3     = TZIJS
                TXIJS = TXJIS
                TYIJS = TYJIS
                TZIJS = TZJIS
                TXJIS = C1
                TYJIS = C2
                TZJIS = C3
              END IF
              FXI    = FXI    + FXIJS
              FYI    = FYI    + FYIJS
              FZI    = FZI    + FZIJS
              FX(J)  = FX(J)  + FXJIS
              FY(J)  = FY(J)  + FYJIS
              FZ(J)  = FZ(J)  + FZJIS
              TXI    = TXI    + TXIJS
              TYI    = TYI    + TYIJS
              TZI    = TZI    + TZIJS
              TX(J)  = TX(J)  + TXJIS
              TY(J)  = TY(J)  + TYJIS
              TZ(J)  = TZ(J)  + TZJIS
C
  650       CONTINUE
C
            GOTO 1000
C
C           -----------------------   END OF ENERGY DUE TO STERIC INER. ---
C
 1000   CONTINUE
C
        FX(I) = FXI
        FY(I) = FYI
        FZ(I) = FZI
        TX(I) = TXI
        TY(I) = TYI
        TZ(I) = TZI
C
 2000   CONTINUE
C
C                                       --- TORQUES DUE TO MAG. FIELD ---
        DO 2010 I=1,N
          TX(I) = TX(I) + ( NY(I)*HZ - NZ(I)*HY )*RAH
          TY(I) = TY(I) + ( NZ(I)*HX - NX(I)*HZ )*RAH
          TZ(I) = TZ(I) + ( NX(I)*HY - NY(I)*HX )*RAH
 2010   CONTINUE
                                                                RETURN
                                                                END
C**** SUB STEFORCE ****
        SUBROUTINE STEFORCE( RRIJ,RAV,ISKIP,TXIJ,TYIJ,TZIJ )
C
        IMPLICIT REAL*8 (A-H,O-Z), INTEGER (I-N)
C
        COMMON /WORK20/  XRXI , YRYI , ZRZI , XRXJ , YRYJ , ZRZJ
        COMMON /WORK21/  FXIJS, FYIJS, FZIJS, FXJIS, FYJIS, FZJIS
        COMMON /WORK22/  TXIJS, TYIJS, TZIJS, TXJIS, TYJIS, TZJIS
        COMMON /WORK23/  RCOFF2 , RP102 , D1     , D1SQ
C
        REAL*8  FXIJ , FYIJ , FZIJ , C0
C                                                 --- STERIC REPULSION ---
        FXIJ   = 0.D0
        FYIJ   = 0.D0
        FZIJ   = 0.D0
C
        IF( RRIJ .LT. D1 ) THEN
          IF( RRIJ .LE. 1.D0 ) RRIJ = 1.0001D0
          C0   = DLOG( D1 / RRIJ )
          FXIJ = TXIJ*C0
          FYIJ = TYIJ*C0
          FZIJ = TZIJ*C0
        END IF
```

• The torque due to the external magnetic field is calculated and added to the corresponding variable.

• A subroutine for calculating the repulsive forces resulting from the overlap of the surfactant layers according to Eq. (3.63).

```
C
      FXIJS = FXIJ*RAV
      FYIJS = FYIJ*RAV
      FZIJS = FZIJ*RAV
      FXJIS = - FXIJS
      FYJIS = - FYIJS
      FZJIS = - FZIJS
      IF( ISKIP .EQ. 1 )  RETURN
C                                                    --- TORQUES ---
      TXIJS   =  YRYI*FZIJS - ZRZI*FYIJS
      TYIJS   =  ZRZI*FXIJS - XRXI*FZIJS
      TZIJS   =  XRXI*FYIJS - YRYI*FXIJS
      TXJIS   =  YRYJ*FZJIS - ZRZJ*FYJIS
      TYJIS   =  ZRZJ*FXJIS - XRXJ*FZJIS
      TZJIS   =  XRXJ*FYJIS - YRYJ*FXJIS
                                                              RETURN
                                                              END
```

• The torques acting on particles *i* and *j* due to the repulsive forces are calculated and saved in (TXIJS,TYIJS,TZIJS) and (TXJIS, TYJIS,TZJIS), respectively.

4 Practice of Monte Carlo Simulations

In the present chapter we demonstrate the two examples of an Monte Carlo (MC) simulation by considering the aggregation phenomena of magnetic particles in an applied magnetic field. The first exercise treats a two-dimensional suspension composed of magnetic spherocylinder particles with the purpose of discussing the dependence of the particle behavior on the magnetic particle–particle and the particle–field interactions. The second exercise treats a three-dimensional suspension composed of magnetic disk-like particles for discussing similar particle behavior in thermodynamic equilibrium. Understanding the MC method for simulations of these nonspherical systems is an important first step in treating a more complex system, such as DNA, polymeric liquids, or carbon-nanotubes. The sample simulation programs that follow each exercise have been taken from real-life academic-oriented research projects and are therefore realistic examples for guidance in writing an academic or commercial simulation program. In both examples demonstrated here, the canonical MC algorithm is used under the physical conditions of a given number of particles, temperature, and volume of the system.

4.1 Orientational Phenomena of Rod-like Particles in an Applied Magnetic Field

In the present section we consider a suspension composed of magnetic rod-like particles as a two-dimensional system that is in thermodynamic equilibrium under the conditions of a constant number of particles, temperature, and volume. A sample simulation program written in the FORTRAN language completes the exercise.

4.1.1 Physical Phenomena of Interest

The system, assumed to be in thermodynamic equilibrium, is composed of N ferromagnetic particles with diameter d and length l_0 $(=l+d)$ that are dispersed in a base liquid. Each magnetic rod-like particle is modeled as a spherocylinder, as already explained in Section 3.2, with magnetic plus and minus charges at the centers of each hemisphere cap; it is therefore magnetized in the particle axis direction. Each particle is coated with a surfactant layer for stabilization purposes. In this type of dispersion, the aggregation phenomenon of magnetic particles is strongly dependent on the magnetic field strength, magnetic interactions, and the number

Introduction to Practice of Molecular Simulation. DOI: 10.1016/B978-0-12-385148-2.00004-5

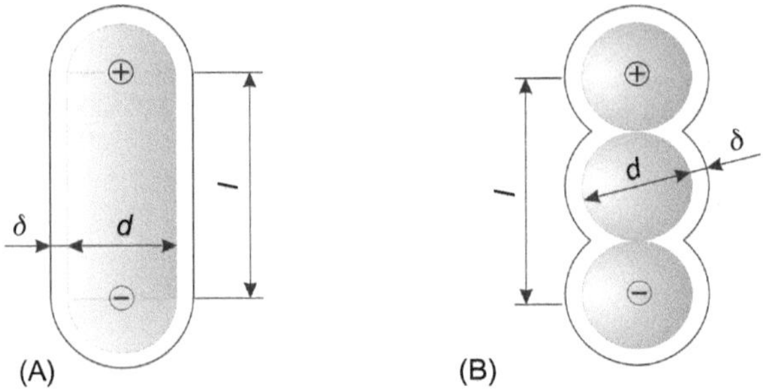

Figure 4.1 Rod-like particle model with a steric layer: (A) the spherocylinder model and (B) the sphere-connected model.

density. In this example we discuss the influence of these effects on particle aggregation by means of a canonical MC simulation.

4.1.2 Specification of Problems in Equations

The main consideration in formulating the present problem is the interaction energy between the particles. Similar to Section 3.2, it is necessary to take into account magnetic interactions and steric repulsive interactions acting between particles for the spherocylinder particle model shown in Figure 4.1A. The treatment of the steric interactions due to particle overlap is difficult even in the present two-dimensional case. Therefore, in evaluating the steric interactions, we employ the simple linear sphere-connected model shown in Figure 4.1B. In this model, each constituent sphere is covered by a uniform steric layer. Hence, a repulsive interaction energy due to the overlap of the two steric layers can be obtained by summing all interaction energies for each pair of spheres belonging to the two different rod-like particles. This is a characteristic feature of the sphere-connected model, which is different from the model employed in Section 3.2 in that the constituent spheres are in fixed positions in the present case.

It is difficult to treat the particle overlap in a manner that results in an efficient simulation program, even for the two-dimensional case, and therefore considerable effort is required to address this problem for a three-dimensional system. In many cases, rather than directly addressing the three-dimensional system, it is more effective to first develop a two-dimensional simulation program and then extend it to the three-dimensional case. The three-dimensional simulation program shown in Section 3.2 has been developed using this approach from the present two-dimensional program, which will be shown in Section 4.1.6.

We use the notation $\mathbf{r}_i$ for the position vector of the center of particle i ($i=1, 2,\ldots, N$), $\mathbf{e}_i$ for the particle axis direction vector, and $\pm q$ for the plus and minus magnetic charges at both hemispheres. The interaction energy with an applied magnetic field $\mathbf{H}$ is expressed similar to the spherical particles as

$$u_i = -\mu_0 \mathbf{m}_i \cdot \mathbf{H} \tag{4.1}$$

in which $\mathbf{m}_i$ is the magnetic moment, expressed as $\mathbf{m}_i = ql\mathbf{e}_i$ ($=m\mathbf{e}_i$). Eq. (4.1) implies that a rod-like particle tends to incline in the magnetic field direction, leading to a minimum interaction energy.

We first show an expression for the interaction energy u between magnetic charges q and q'. If the magnetic charges are separated by distance r, the interaction energy is expressed as

$$u = \frac{\mu_0 q q'}{4\pi r} \tag{4.2}$$

This equation is quite well known [31]. Eq. (4.2) is applied to the present magnetic rod-like particle shown in Figure 4.1. The interaction energy for the rod-like particles shown in Figure 4.1B can be obtained by summing the interaction energies for the four pairs of magnetic charges. If the position vectors of the plus and minus charges of an arbitrary particle i are denoted by $\mathbf{r}_i^+$ and $\mathbf{r}_i^-$, respectively, they are written as

$$\mathbf{r}_i^+ = \mathbf{r}_i + (l/2)\mathbf{e}_i, \quad \mathbf{r}_i^- = \mathbf{r}_i - (l/2)\mathbf{e}_i \tag{4.3}$$

With this notation, the magnetic interaction energy u_{ij} between rod-like particles i and j is expressed as

$$u_{ij} = \frac{\mu_0 q^2}{4\pi}\left\{\frac{1}{|\mathbf{r}_i^+ - \mathbf{r}_j^+|} - \frac{1}{|\mathbf{r}_i^+ - \mathbf{r}_j^-|} - \frac{1}{|\mathbf{r}_i^- - \mathbf{r}_j^+|} + \frac{1}{|\mathbf{r}_i^- - \mathbf{r}_j^-|}\right\} \tag{4.4}$$

The first term on the right-hand side is an interaction energy between the plus charges of particles i and j. The second term is an energy between the plus charge of particle i and the minus charge of particle j. The third term is an energy between the minus charge of particle i and the plus charge of particle j. The fourth term is an energy between the minus charges of particles i and j. Substitution of Eq. (4.3) into Eq. (4.4) leads to

$$u_{ij} = \frac{\mu_0 q^2}{4\pi}\left\{\frac{1}{|\mathbf{r}_{ij} + l\mathbf{e}_{ij}/2|} - \frac{1}{|\mathbf{r}_{ij} + l(\mathbf{e}_i + \mathbf{e}_j)/2|} - \frac{1}{|\mathbf{r}_{ij} - l(\mathbf{e}_i + \mathbf{e}_j)/2|} + \frac{1}{|\mathbf{r}_{ij} - l\mathbf{e}_{ij}/2|}\right\} \tag{4.5}$$

in which $\mathbf{r}_{ij} = \mathbf{r}_i - \mathbf{r}_j$ and $\mathbf{e}_{ij} = \mathbf{e}_i - \mathbf{e}_j$.

We now consider an interaction energy $u_{ij}^{(V)}$ arising from the overlap of the steric layers. For a spherical particle with diameter d covered by a uniform surfactant layer with thickness δ, an overlap of these two particles yields a repulsive interaction energy $u_{ij}^{(V)}$, as already shown in Eq. (3.41):

$$u_{ij}^{(V)} = kT\lambda_V\left\{2 - \frac{2r_{ij}/d}{t_\delta}\ln\left(\frac{d+2\delta}{r_{ij}}\right) - 2\frac{r_{ij}/d - 1}{t_\delta}\right\} \tag{4.6}$$

in which r_{ij} is the separation between particles i and j (center-to-center distance), t_δ is the ratio of the steric layer thickness to the particle radius expressed as $t_\delta = 2\delta/d$, λ_V is a nondimensional parameter representing the strength of steric repulsive interactions expressed as $\lambda_V = \pi d^2 n_s/2$, and n_s is the number of surfactant molecules per unit area on the particle surface. If the particle separation satisfies $r_{ij} < \mathrm{d} + 2\delta$, the two steric layers of particles i and j overlap. In the following, we apply this interaction energy to the two spherocylinder particles shown in Figure 4.1B.

The sphere-connected model enables us to employ the evaluation approach, which has been used for calculating magnetic interactions. That is, the net steric interaction energy between the two rod-like particles can be obtained by summing a steric interaction energy for each pair of constituent spherical particles belonging to the two different rod-like particles. However, this approach becomes inefficient, or requires enormous computation time, as the rod-like particle becomes longer (i.e., for an increase in the number of spherical particles). Since the steric layer is thin compared with the particle diameter, the pair-wise calculation of the repulsive interactions implies that, for some calculations, the result is negligible. We therefore need to develop an alternative technique for calculating the steric interactions. This kind of difficulty frequently appears in developing a simulation program, so the process of overcoming this problem provides a good opportunity for the development of a higher-level simulation program. Therefore, in the following we discuss this problem in more detail.

The spatial relationship of two rod-like particles i and j is a function of the particle position vectors $\mathbf{r}_i$ and $\mathbf{r}_j$ and the particle direction (unit) vectors $\mathbf{e}_i$ and $\mathbf{e}_j$. In practice, a two-dimensional system is considerably more straightforward than a three-dimensional system in treating the overlap assessment. Referring to Figure 4.2, we now discuss the overlap between particles i and j. Assessing how the two rod-like particles overlap first requires finding the intersection point of each particle axis. If the two axis lines intersect at the positions $(\mathbf{r}_i + k_i\mathbf{e}_i)$ and $(\mathbf{r}_j + k_j\mathbf{e}_j)$ of particles i and j, respectively, then the unknown constants k_i and k_j have to satisfy the following equation:

$$\mathbf{r}_i + k_i\mathbf{e}_i = \mathbf{r}_j + k_j\mathbf{e}_j \tag{4.7}$$

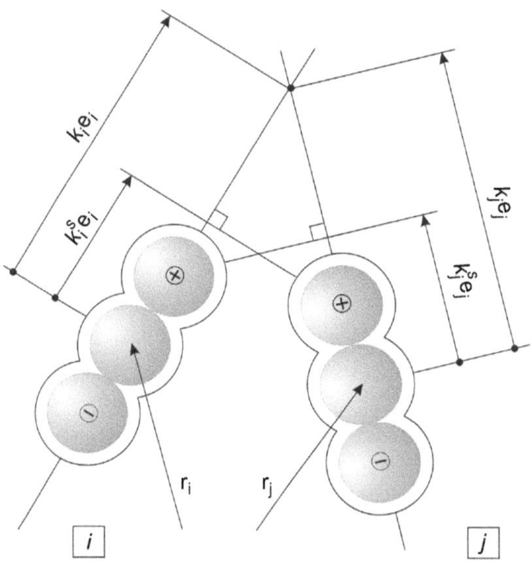

Figure 4.2 Analysis of the overlap condition of steric layers.

Vector product of $\mathbf{e}_j$ (or $\mathbf{e}_i$) on both sides of this equation yields $|k_i|$ (or $|k_j|$):

$$|k_i| = \frac{|\mathbf{r}_{ij} \times \mathbf{e}_j|}{|\mathbf{e}_i \times \mathbf{e}_j|}, \quad |k_j| = \frac{|\mathbf{r}_{ij} \times \mathbf{e}_i|}{|\mathbf{e}_i \times \mathbf{e}_j|} \tag{4.8}$$

These equations are valid for a nonparallel configuration. For parallel cases, the treatment of the particle overlap is quite straightforward and will be explained later.

Next we need to find the point $(\mathbf{r}_i + k_i^s \mathbf{e}_i)$ on the axis line of particle i, which is the intersection point of the line drawn from the position $\mathbf{r}_j^+$ of the plus magnetic charge of particle j that perpendicularly intersects the axis line of particle i. The orthogonality condition of this vertical line and the particle direction vector $\mathbf{e}_i$ provides the solution of the unknown constant k_i^s as

$$k_i^s = \frac{l}{2}\mathbf{e}_i \cdot \mathbf{e}_j - \mathbf{r}_{ij} \cdot \mathbf{e}_i \tag{4.9}$$

The solution of k_j^s can be obtained by exchanging the subscriptions i and j in this equation. Similarly, if a line drawn from the position $\mathbf{r}_j^-$ perpendicularly intersects the axis line of particle i at the position $(\mathbf{r}_i + k_i^{s\prime} \mathbf{e}_i)$, the above-mentioned mathematical procedure gives rise to the solution of $k_i^{s\prime}$ as

$$k_i^{s\prime} = -\frac{l}{2}\mathbf{e}_i \cdot \mathbf{e}_j - \mathbf{r}_{ij} \cdot \mathbf{e}_i \tag{4.10}$$

The use of these intersection points enables us to calculate effectively the repulsive interaction energy between particles i and j arising from the overlap of the steric layers. First, the solutions of k_i and k_j are obtained from Eqs. (4.7) and (4.8). From the large-or-small relationship and the positive-or-negative sign of k_i and k_j, we see which sphere of particle i has a possibility of interacting with which sphere of particle j. For example, since $k_j > k_i > 0$ in Figure 4.2, there is a possibility of the plus magnetic charged sphere of particle j interacting with any constituent spheres of particle i. Which sphere of particle i interacts with the plus charged sphere of particle j can be determined by the solution k_i^s in Eq. (4.9). Because $k_i^s > l/2$ in Figure 4.2, it has a possibility to interact with the plus magnetic charged sphere of particle i. At this stage, we have identified the first pair of constituent spheres of the particles i and j required for calculating the interaction energy due to the overlap of the steric layers.

After this calculation, we shift our attention to the next neighboring constituent spheres of each particle and calculate their interaction energy. Repeating this procedure finally yields the total interaction energy due to the particle overlap of particles i and j. An important advantage of this procedure is that the nonoverlap of the constituent spheres can be used to terminate the calculation. In other words, this method becomes much more efficient with an increasing particle length when compared to the simple calculation method, in which all possible pairs of constituent spheres are treated. Note that there may be situations where one constituent sphere of particle j may interact with two constituent spheres of particle i. For example, in Figure 4.2, this situation may arise if the two axis lines intersect under the condition of $-l/2 < k_i^s < l/2$; in this case, the sphere of particle j is located at a position between the two constituent spheres of particle i.

The parallel configuration and the linear configuration do not require values of k_i and k_j for the calculation of the steric interaction energy. The linear configuration satisfies the relationships of $|\mathbf{e}_i \cdot \mathbf{e}_j| = |\mathbf{e}_i \cdot \mathbf{t}_{ij}| = 1$, in which $\mathbf{t}_{ij}$ is the unit vector between particles i and j, expressed as $\mathbf{t}_{ij} = \mathbf{r}_{ij}/r_{ij}$. Only the two spheres at the end of each particle have a possibility to overlap for the linear configuration, so that just one calculation is sufficient for this case; these spheres can be straightforwardly specified by the signs of $\mathbf{e}_i \cdot \mathbf{e}_j$ and $\mathbf{e}_i \cdot \mathbf{t}_{ij}$. For the parallel configuration, a value of k_i^s in Eq. (4.9) provides information as to how the two particles are shifted in separation along the particle axis direction. The value of k_i^s or k_j^s indicates which sphere of particle j interacts with which sphere of particle i in the nearest configuration.

In the above discussion, we have explained the fundamental and mathematical aspects of evaluating the steric interaction between the particles. The technical aspect of this treatment, required for developing a simulation program, will be discussed in detail later in the next subsection on the MC algorithm.

Finally, we show the nondimensional expressions of the important physical quantities. If distances and energies are nondimensionalized by the particle diameter d and the thermal energy kT, respectively, Eqs. (4.1), (4.4), (4.5), and (4.6) are nondimensionalized as

$$u_i^* = u_i/kT = -\xi \mathbf{e}_i \cdot \mathbf{h} \tag{4.11}$$

$$u_{ij}^* = u_{ij}/kT = \lambda_0 \left\{ \frac{1}{|\mathbf{r}_i^{+*} - \mathbf{r}_j^{+*}|} - \frac{1}{|\mathbf{r}_i^{+*} - \mathbf{r}_j^{-*}|} - \frac{1}{|\mathbf{r}_i^{-*} - \mathbf{r}_j^{+*}|} + \frac{1}{|\mathbf{r}_i^{-*} - \mathbf{r}_j^{-*}|} \right\} \tag{4.12}$$

$$\begin{aligned} u_{ij}^* &= u_{ij}/kT \\ &= \lambda_0 \left\{ \frac{1}{|\mathbf{r}_{ij}^* + r_p \mathbf{e}_{ij}/2|} - \frac{1}{|\mathbf{r}_{ij}^* + r_p(\mathbf{e}_i + \mathbf{e}_j)/2|} - \frac{1}{|\mathbf{r}_{ij}^* - r_p(\mathbf{e}_i + \mathbf{e}_j)/2|} + \frac{1}{|\mathbf{r}_{ij}^* - r_p \mathbf{e}_{ij}/2|} \right\} \end{aligned} \tag{4.13}$$

$$u_{ij}^{(V)*} = u_{ij}^{(V)}/kT = \lambda_V \left\{ 2 - \frac{2r_{ij}^*}{t_\delta} \ln\left(\frac{1+t_\delta}{r_{ij}^*}\right) - 2\frac{r_{ij}^* - 1}{t_\delta} \right\} \tag{4.14}$$

in which r_p is the particle aspect ratio, defined as $r_p = l/d$. In addition, the nondimensional parameters ξ and λ_0 are expressed as

$$\xi = \mu_0 mH/kT, \quad \lambda_0 = \mu_0 (qd)^2/4\pi d^3 kT \tag{4.15}$$

in which $\mathbf{h} = \mathbf{H}/H(H = |\mathbf{H}|)$ and the quantities with superscript * are dimensionless quantities. As previously explained in Eqs. (3.62) and (3.58), the meanings of ξ and λ_0 are the strengths of magnetic particle–field and magnetic particle–particle interactions, respectively. A slightly different nondimensional parameter $\lambda = r_p^2 \lambda_0$ is introduced for discussion.

4.1.3 Canonical Monte Carlo Algorithm

The system is in thermodynamic equilibrium, composed of N rod-like particles with specified volume V (i.e., area in this two-dimensional case) and temperature T, and it is appropriate to use the canonical MC algorithm for the simulation. The total system potential energy is evaluated by summing the magnetic particle–field and the particle–particle interaction energy together with the steric repulsive interaction energy due to the overlap of the steric layers. That is,

$$U^* = \sum_{i=1}^{N} u_i^* + \sum_{i=1}^{N} \sum_{j=1(j>i)}^{N} \left(u_{ij}^* + u_{ij}^{(V)*} \right) \tag{4.16}$$

We now consider a transition from the present microscopic state k, which has a system potential energy U_k. A new microscopic state l is generated by selecting one particle and moving it to a new position by using random numbers, which yields a new system potential energy U_l. The transition probability from microscopic state k to l, p_{kl}, is given by Eq. (1.49), but in this case the probability density ratio is

$$\frac{\rho_l}{\rho_k} = \exp\left\{ -\frac{1}{kT}(U_l - U_k) \right\} = \exp\left\{ -\left(U_l^* - U_k^* \right) \right\} \tag{4.17}$$

After this treatment of the translational displacement of the particle, a similar procedure is conducted for the rotational displacement. A series of trials for the translational and rotational displacement, when applied to all the system particles, is called an "MC step," which corresponds to a time step in the molecular dynamics method.

From the viewpoint of developing a simulation program, we now show the scheme for calculating the interaction energy due to the overlap of the steric layers. Figure 4.3 shows the categories of overlap for the two particles. There are four typical overlap regimes: linear (itree = 0), general (itree = 1), perpendicular (itree = 2), and parallel (itree = 3). Any overlap of the steric layers can be classified into one of these four regimes. Note that the variables itree and ipath (appearing later) have no physical meaning but are used for the sake of convenience; these variables are used in the sample simulation program with consistent meaning. We explain the four overlap cases in more detail in the following paragraphs.

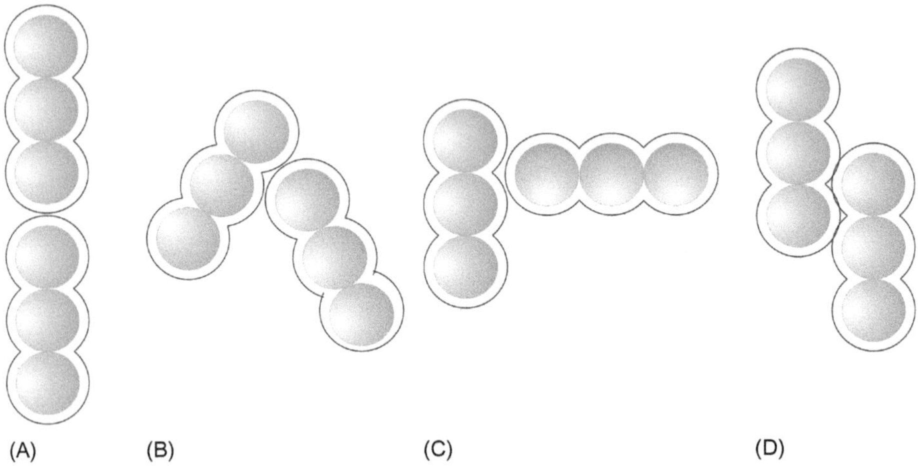

Figure 4.3 Typical overlap regime of the steric layers: (A) linear (itree = 0), (B) general (itree = 1), (C) perpendicular (itree = 2), and (D) parallel (itree = 3).

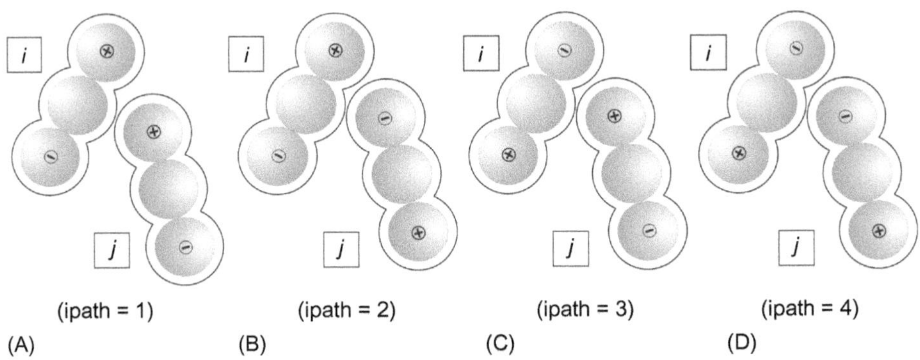

Figure 4.4 Overlap in the general situation (itree = 1).

4.1.3.1 General Overlap Case (itree = 1)

In this case, there are four types of overlap dependent upon the location of the plus and minus magnetic charges, which are schematically shown in Figure 4.4. In order to treat the particle overlap consistently in a simulation program, the names of the two particles may be exchanged in such a way so as to satisfy the relationship $|k_i| < |k_j|$. This condition is assumed to be satisfied in the following discussion. Figures 4.4A and C show the possibility of the plus magnetic charge of particle j overlapping with particle i. Figures 4.4B and D are for the overlap of the minus magnetic charge of particle j with particle i.

The four types of particle overlaps in Figure 4.4 can be identified in the following way. By reason of $|k_i| < |k_j|$, the particle on the left-hand side in Figure 4.4 is particle i, and the particle on the right-hand side is particle j. For the case of $\mathbf{e}_i \cdot \mathbf{e}_j \geq 0$, the overlap regime is ipath = 1 or ipath = 4, and for the case of $\mathbf{e}_i \cdot \mathbf{e}_j < 0$, it is ipath = 2 or ipath = 3. Furthermore, the sign of k_j enables us to identify whether ipath = 1 or ipath = 4 arises for the overlap, which is also applicable to the identification of ipath = 2 or ipath = 3. For example, for the case of $\mathbf{e}_i \cdot \mathbf{e}_j \geq 0$ and $k_j \geq 0$, there is a possibility of particle overlap in the situation ipath = 1.

We now discuss which constituent sphere of particle i interacts with the magnetic charged sphere of particle j. Since the principle is the same for all cases, we focus on the case of ipath = 1. The value of k_i^s can allow us to identify which sphere of particle i has the possibility to interact with the plus magnetic charged sphere of particle j. For simplification, we name the constituent spheres in the rod-like particle in such a way that the plus magnetic charged sphere is called "subparticle 1," the next neighboring sphere is called "subparticle 2," and so on. For $k_i^s \geq l/2$, subparticle 1 of particle j may overlap with subparticle 1 of particle i; similarly, $l/2 > k_i^s \geq (l/2 - d)$ overlaps with subparticle 1 or subparticle 2; and $(l/2 - d) > k_i^s \geq (l/2 - 2d)$ overlaps with subparticle 2 or subparticle 3. Even if the rod-like particle is composed of numerous subparticles, the above-mentioned procedure can provide us with a method to find which subparticle of particle i overlaps with particle j.

We now consider the case in which subparticle 1 of particle j overlaps with subparticle 2 or 3 of particle i. The total repulsive interaction energy between particles i and j can be obtained by calculating the interaction energy in Eq. (4.14) for this pair of subparticles and by repeating this calculation procedure for the neighboring subparticles for subparticle 2 of particle j and subparticle 3 or 4 (note that subparticle 4 does not exist for the present three-sphere-connected model) of particle i, and so on. The calculation procedure can be terminated when a pair of the subparticles is found to be separated by more than the distance $(d + 2\delta)$. In the case of Figure 4.4A, only the first two calculations are needed to obtain the total steric repulsive interaction energy between particles i and j. This discussion clearly suggests that the present method becomes much more effective for a longer rod-like particle. In the sample simulation program shown later, the above-mentioned procedures are employed for calculating the steric interaction energy together with the variables itree and ipath with the same meaning as above.

4.1.3.2 Normal Overlap Case (itree = 2)

Figure 4.5 shows the two categories of particle overlap in a normal orientation. As in the general overlap case, the subscripts i and j may be exchanged in order to satisfy $|k_i| < |k_j|$. Figure 4.5A shows an overlap between subparticle 1 of particle j and particle i, and Figure 4.5B is for the case of the other end subparticle of particle j overlapping with particle i. These two categories can be identified by the value of k_j; that is, there is a possibility of particle overlap in the situation ipath = 1 or ipath = 2 for $k_j > 0$ or $k_j < 0$, respectively.

We treat the case ipath = 1 shown in Figure 4.5 to consider which subparticle of particle i possibly overlaps with the subparticle of particle j. As in the general

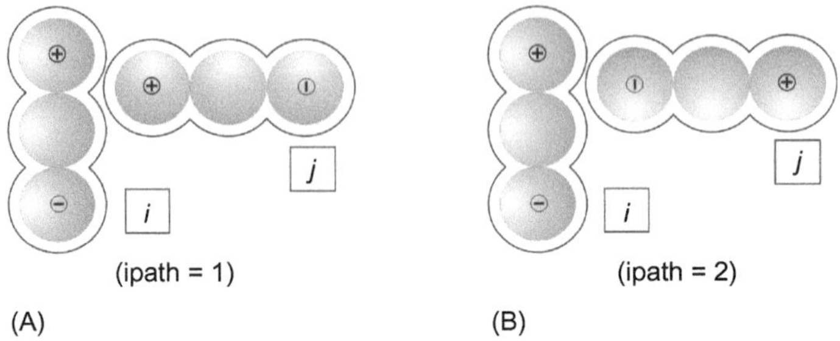

Figure 4.5 Overlap in the normal situation (itree = 2).

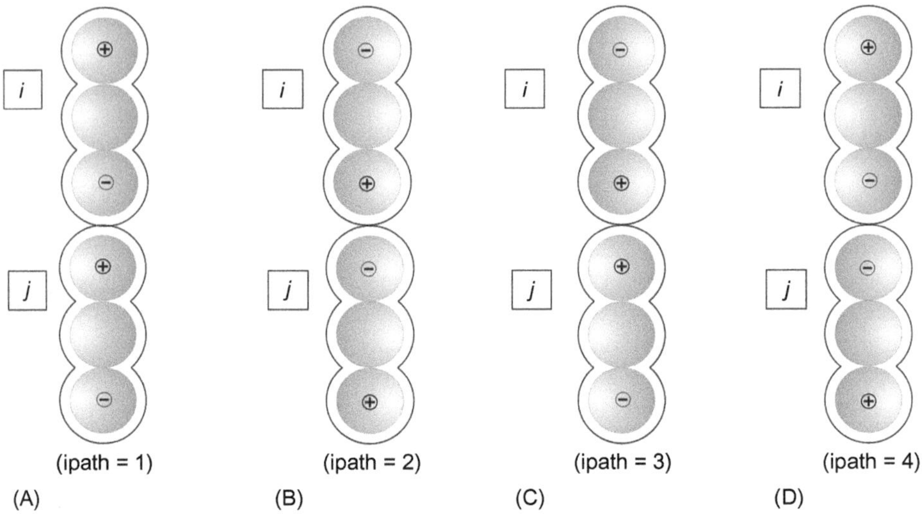

Figure 4.6 Overlap in the linear situation (itree = 0).

overlap situation, subparticle 1 possibly overlaps with subparticle 1 of particle j for $k_i^s \geq l/2$, subparticle 1 or 2 overlaps with particle j for $l/2 > k_i^s \geq (l/2 - d)$, and subparticle 2 or 3 does so for $(l/2 - d) > k_i^s \geq (l/2 - 2d)$. For the case where the rod-like particle is composed of numerous subparticles, the above-mentioned procedure is repeated to find a pair or two pairs of interacting subparticles.

4.1.3.3 *Linear Overlap Case (itree = 0)*

In the linear overlap case, there are four types of overlap possibility, as shown in Figure 4.6. The four categories can be identified by assessing the signs of $\mathbf{e}_i \cdot \mathbf{e}_j$ and $\mathbf{e}_j \cdot \mathbf{t}_{ij}$. That is, the relationship $\mathbf{e}_i \cdot \mathbf{e}_j > 0$ provides an overlap for ipath = 1 or ipath = 2, and $\mathbf{e}_i \cdot \mathbf{e}_j < 0$ provides an overlap for ipath = 3 or ipath = 4. For the case of $\mathbf{e}_i \cdot \mathbf{e}_j > 0$, the sign of $\mathbf{e}_j \cdot \mathbf{t}_{ij}$ can identify whether the overlap is for ipath = 1 or ipath = 2. Subsequently, there is a possibility of particle overlap in the situation

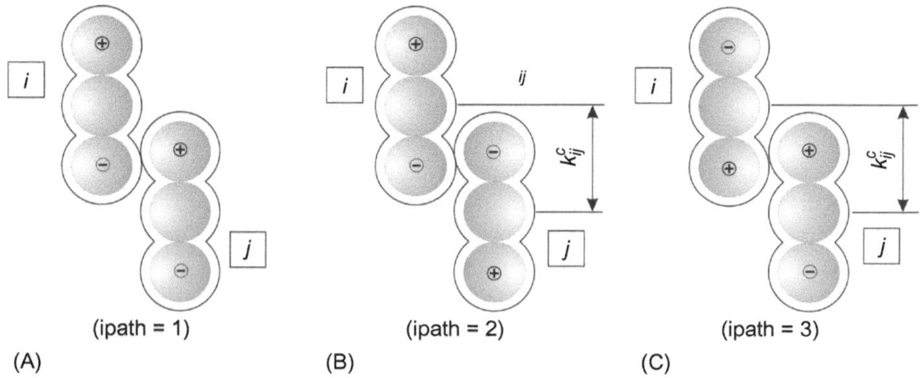

Figure 4.7 Overlap in the parallel situation (itree = 3).

where ipath = 1 for $\mathbf{e}_i \cdot \mathbf{e}_j > 0$ and $\mathbf{e}_j \cdot \mathbf{t}_{ij} > 0$, ipath = 2 for $\mathbf{e}_i \cdot \mathbf{e}_j > 0$ and $\mathbf{e}_j \cdot \mathbf{t}_{ij} < 0$, ipath = 3 for $\mathbf{e}_i \cdot \mathbf{e}_j < 0$ and $\mathbf{e}_j \cdot \mathbf{t}_{ij} > 0$, and ipath = 4 for $\mathbf{e}_i \cdot \mathbf{e}_j < 0$ and $\mathbf{e}_j \cdot \mathbf{t}_{ij} < 0$. Once the type of particle overlap is identified, the pair of the overlapping subparticles is readily identified in order to calculate the interaction energy.

4.1.3.4 Parallel Overlap Case (itree = 3)

For the parallel overlap case, there are three types of particle overlap, as shown in Figure 4.7. For the case of ipath = 1 in Figure 4.7A, the relationship $k_i^s \le k_j^s$ needs to be satisfied by exchanging the particle names. The overlap regime is identified by assessing the sign of $\mathbf{e}_i \cdot \mathbf{e}_j$ with a value of k_i^s. That is, the overlap regime is ipath = 1 for $\mathbf{e}_i \cdot \mathbf{e}_j > 0$, ipath = 2 for $\mathbf{e}_i \cdot \mathbf{e}_j < 0$ and $k_i^s \le -l/2$, and ipath = 3 for $\mathbf{e}_i \cdot \mathbf{e}_j < 0$ and $k_i^s > -l/2$.

We focus on the cases ipath = 2 and 3 for discussion, since the treatment for ipath = 1 is almost the same as in the general overlap case. For ipath = 2 and 3, the determination of the separation between the particle centers makes the subsequent treatment more straightforward. The separation between the particle centers along the particle axis, k_{ij}^c, is expressed as $k_{ij}^c = |k_i^s| - l/2$ for ipath = 2, and as $k_{ij}^c = k_i^s + l/2$ for ipath = 3. Because of the similarity in the treatment for ipath = 2 and 3, we explain only the case of ipath = 2. The value of k_{ij}^c allows us to find which subparticle of particle i overlaps with the minus magnetic charged sphere of particle j. There is a possibility of the overlap with subparticle 1 or 2 of particle i for $d \ge k_{ij}^c > 0$ and of the overlap with subparticle 2 or 3 for $2d \ge k_{ij}^c > d$. This calculation procedure is repeated until the end-sphere of particle i obtains the total steric interaction energy.

4.1.4 Parameters for Simulations

We employed the following parameters for conducting the simulations. It is presumed that the rod-like particles aggregate to form chain-like clusters along the applied field direction (i.e., y-axis direction). We therefore choose to employ a rectangular simulation region dependent upon the particle aspect ratio; we therefore adopt a rectangular region having a side length in the y-direction twice that of in the x-direction. The results shown in the next subsection were obtained under the assumption that a rod-like particle may be represented by three spherical subparticles. The area fraction $\phi_V = 0.2$, the nondimensional parameter λ_V, representing the strength of steric repulsive interactions, is set as $\lambda_V = 150$. The thickness of the steric layer is assumed as $t_\delta = 0.3$. The maximum distance δr^*_{max} and angle $\delta\theta_{max}$ per one trial in the MC algorithm are taken as $\delta r^*_{max} = 0.1$ and $\delta\theta_{max} = 5°$. The MC simulations were carried out for various cases of the magnetic particle–field and the particle–particle interactions, ξ and λ, respectively.

4.1.5 Results of Simulations

Figures 4.8–4.11 show the results relating to the aggregate structures, which were obtained by conducting the sample simulation program shown in the next subsection. Figure 4.8 was obtained for $\lambda_0 = 0.75$, Figure 4.9 for $\lambda_0 = 1.75$, Figure 4.10 for $\lambda_0 = 4$, and Figure 4.11 for $\lambda_0 = 7.5$. Each figure has two snapshots: one for the case of no external field, and the other for the case of a strong applied magnetic field.

For the case of $\lambda_0 = 0.75$, shown in Figure 4.8, the magnetic interaction between particles is of the same order of the thermal energy and therefore no aggregates are observed in Figures 4.8A and B. Figure 4.8A is for the case of no external field and therefore the rod-like particles have no specifically favored directional characteristic. On the other hand, the rod-like particles tend to incline in the magnetic field direction in Figure 4.8B because $\xi = 20$ represents a significantly strong magnetic field.

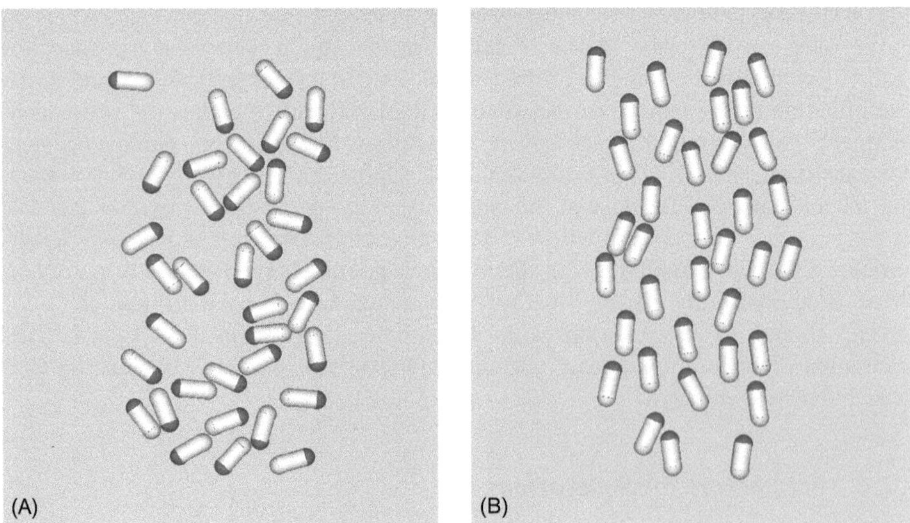

Figure 4.8 Snapshots of aggregate structures for $\lambda = 3$ ($\lambda_0 = 0.75$): (A) $\xi = 0$ and (B) $\xi = 20$.

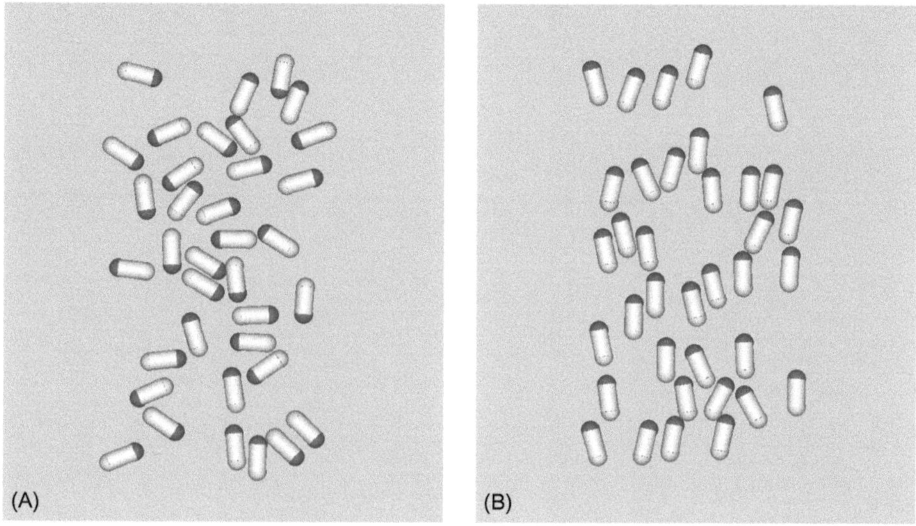

Figure 4.9 Snapshots of aggregate structures for $\lambda = 7$ ($\lambda_0 = 1.75$): (A) $\xi = 0$ and (B) $\xi = 20$.

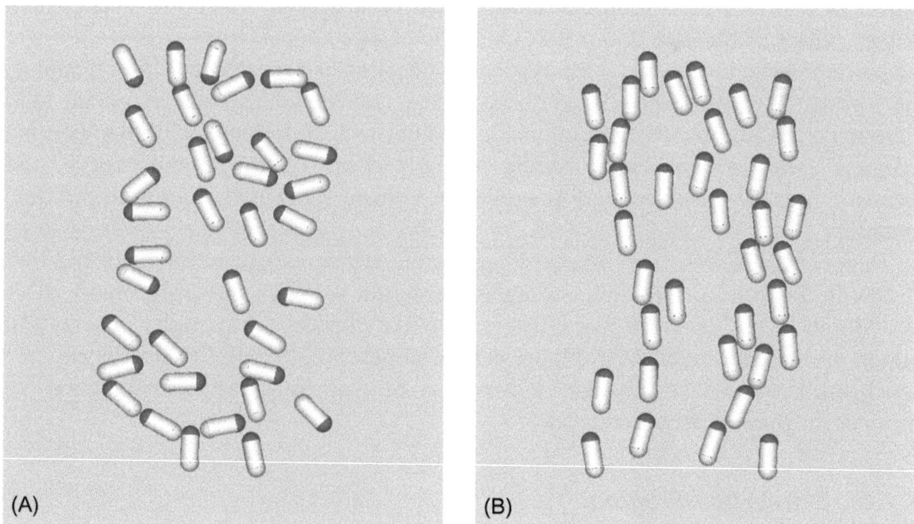

Figure 4.10 Snapshots of aggregate structures for $\lambda = 16$ ($\lambda_0 = 4$): (A) $\xi = 0$ and (B) $\xi = 20$.

Figure 4.9 shows snapshots for the slightly stronger interaction $\lambda_0 = 1.75$. These snapshots are similar to Figure 4.8, because $\lambda_0 = 1.75$ is not significantly larger than the thermal energy.

For the stronger case of $\lambda_0 = 4$, shown in Figure 4.10, the magnetic interaction between particles is now more dominant than the thermal energy, and thus significant aggregate structures are observed. In the case of no applied magnetic field,

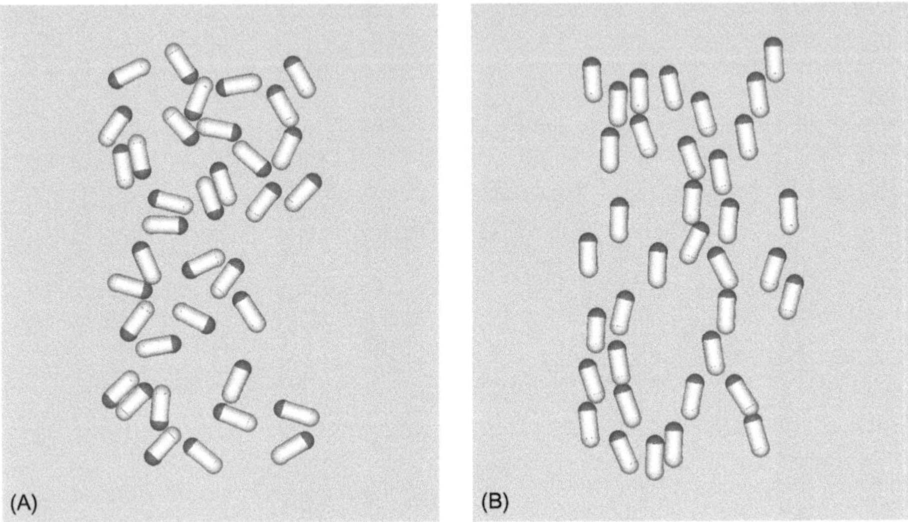

Figure 4.11 Snapshots of aggregate structures for $\lambda = 30$ ($\lambda_0 = 7.5$): (A) $\xi = 0$ and (B) $\xi = 20$.

shown in Figure 4.10A, loop-like clusters can be observed. Since the arrangement of the contact of the plus and minus magnetic charged spheres gives rise to a lower magnetic interaction energy, this type of connection is repeated and may result in the formation of necklace-like clusters. In the case of no external magnetic field there is no mechanism for forming chain-like clusters. In Figure 4.10B, the external magnetic field is significantly strong in comparison to the thermal energy, and therefore rod-like particles tend to aggregate to form chain-like clusters in the field direction.

These characteristics exhibited by aggregate structures can be recognized more clearly in the case of the much stronger interaction $\lambda_0 = 7.5$ shown in Figure 4.11. In addition to the necklace-like clusters, star-like clusters are partially observed in Figure 4.11A. The snapshot in Figure 4.11B suggests the possibility that large-scale network-like or thick chain-like clusters may be formed in the field direction for stronger magnetic interaction cases.

4.1.6 Simulation Program

We now show a sample simulation program written in the FORTRAN language employing the simulation techniques described above in the present demonstration of the MC method.

The important variables used in the program are described below.

`RX(I),RY(I)`	:	(x,y) components of the position vector $\mathbf{r}_i^*$ of particle i
`NX(I),NY(I)`	:	(x,y) components of the unit vector $\mathbf{e}_i$ of particle i denoting the particle direction

XL, YL : Side lengths of the simulation box in the (x,y) directions
N : Number of particles
D : Particle diameter (D = 1 in this case)
VDENS : Area fraction of particles ϕ_V
RA : Nondimensional parameter λ representing the strength of magnetic particle–particle interactions
KU : Nondimensional parameter ξ representing the strength of magnetic particle–field interactions
RV : Nondimensional parameter λ_V representing the strength of repulsive interactions due to the overlap of the steric layers
RCOFF : Cutoff distance for calculations of interaction energies
DELR : δr^*_{max}
DELT : $\delta\theta_{max}$
RAN(J) : Uniform random numbers ranging 0 ~ 1 (J = 1 ~ NRANMX)
NRAN : Number of used random numbers
E(I) : Energy of particle i interacting with other particles
MOMX(*), MOMY(*) : Mean value of the particle direction at each MC step
MEANENE(*) : Mean value of the system energy at each MC step

As an aid for the reader, comments have been placed beside important programming features. The line numbers are added for convenience and are unnecessary for the execution of the simulation program.

Finally, note that the cluster-moving method [4] may not be required for the case of a rod-like particle suspension, although it is indispensable for a spherical particle system in order to obtain physically reasonable aggregate structures in a strongly interacting system.

```
C***********************************************************************
C*                              mccylin3.f                             *
C*                                                                     *
C*          --------        MONTE CARLO SIMULATIONS      -------       *
C*            TWO-DIMENSIONAL MONTE CARLO SIMULATION OF                *
C*            FERROMAGNETIC COLLOIDAL DISPERSIONS COMPOSED OF          *
C*            RODLIKE PARTICLES                                        *
C*                                                                     *
C*      OPEN(9, FILE='@daa1.data', STATUS='UNKNOWN'); parameters       *
C*      OPEN(10,FILE='daa11.data', STATUS='UNKNOWN'); para. & data     *
C*      OPEN(21,FILE='daa001.data',STATUS='UNKNOWN'); particle pos.    *
C*      OPEN(22,FILE='daa011.data',STATUS='UNKNOWN'); particle pos.    *
C*      OPEN(23,FILE='daa021.data',STATUS='UNKNOWN'); particle pos.    *
C*      OPEN(24,FILE='daa031.data',STATUS='UNKNOWN'); particle pos.    *
C*      OPEN(25,FILE='daa041.data',STATUS='UNKNOWN'); particle pos.    *
C*      OPEN(26,FILE='daa051.data',STATUS='UNKNOWN'); particle pos.    *
C*      OPEN(27,FILE='daa061.data',STATUS='UNKNOWN'); particle pos.    *
C*      OPEN(28,FILE='daa071.data',STATUS='UNKNOWN'); particle pos.    *
C*      OPEN(29,FILE='daa081.data',STATUS='UNKNOWN'); particle pos.    *
C*      OPEN(30,FILE='daa091.data',STATUS='UNKNOWN'); particle pos.    *
C*                                                                     *
C*          1. WITHOUT CLUSTER MOVEMENT.                               *
C*          2. RODLIKE MODEL COMPOSED OF ARBITRARY NUMBER              *
C*             OF PARTICLES.                                           *
C*                                                                     *
C*                                                                     *
C*                                   VER.1  BY A.SATOH , '03 11/20     *
C***********************************************************************
```

```
C      N     : NUMBER OF PARTICLES
C      D     : DIAMETER OF PARTICLE ( =1 FOR THIS CASE )
C      VDENS : VOLUMETRIC FRACTION OF PARTICLES
C      RA    : NONDIMENSIONAL PARAMETER OF PARTICLE-PARTICLE INTERACT
C      KU    : NONDIMENSIONAL PARAMETER OF PARTICLE-FIELD INTERACTION
C      RV    : NONDIMENSIONAL PARAMETER OF STERIC REPULSION
C      RCOFF : CUTOFF RADIUS FOR CALCULATION OF INTERACTION ENERGIES
C      XL,YL : DIMENSIONS OF SIMULATION REGION
C
C      RX(N),RY(N) : PARTICLE POSITION
C      NX(N),NY(N) : DIRECTION OF MAGNETIC MOMENT
C      E(I)        : INTERACTION ENERGY OF PARTICLE I WITH THE OTHERS
C      MOMX(**),MOMY(**)  : MAGNETIC MOMENT OF SYSTEM AT EACH MC STEP
C      MEANENE(**)        : MEAN ENERGY OF SYSTEM AT EACH MC STEP
C
C      DELR    : MAXIMUM MOVEMENT DISTANCE
C      DELT    : MAXIMUM MOVEMENT IN ORIENTATION
C
C             -XL/2 < RX(*) < XL/2 ,   -YL/2 < RY(*) < YL/2
C---------------------------------------------------------------------
C
      IMPLICIT REAL*8 (A-H,O-Z), INTEGER (I-N)
C
      COMMON /BLOCK1/  RX    , RY
      COMMON /BLOCK2/  NX    , NY
      COMMON /BLOCK3/  XL    , YL
      COMMON /BLOCK4/  RA    , KU    , RV    , TD    ,  RP
      COMMON /BLOCK5/  VDENS, N     , NPTC  , RCOFF ,  D    , NPTCHF
      COMMON /BLOCK6/  E     , ENEW  , EOLD
      COMMON /BLOCK7/  NRAN  , RAN   , IX
      COMMON /BLOCK8/  DELR  , DELT
      COMMON /BLOCK9/  MOMX  , MOMY  , MEANENE
C
      PARAMETER( NN=1000  , NNS=200000 )
      PARAMETER( NRANMX=500000  , PI=3.141592653589793D0 )
C
      REAL*8    RX(NN) , RY(NN) , NX(NN) , NY(NN) , E(NN)
      REAL*8    VDENS  , KU
      REAL      MOMX(NNS), MOMY(NNS), MEANENE(NNS)
      INTEGER   N , NPTC , NDNSMX , NPTCHF
C
      REAL      RAN(NRANMX)
      INTEGER   NRAN , IX , NRANCHK
C
      REAL*8    RXCAN   , RYCAN   , NXCAN   , NYCAN
      REAL*8    RXI     , RYI     , NXI     , NYI
      REAL*8    RXIJ    , RYIJ    , RIJ     , RIJSQ   , RCOFF2
      REAL*8    ECAN    , C1      , C2      , C3      , CX      , CY
      INTEGER   MCSMPL  , MCSMPLMX , MCSMPL1 , MCSMPL2
      INTEGER   NGRAPH  , NOPT
      LOGICAL   OVRLAP
C
             OPEN(9, FILE='@daa1.data', STATUS='UNKNOWN')
             OPEN(10,FILE='daa11.data', STATUS='UNKNOWN')
             OPEN(21,FILE='daa001.data',STATUS='UNKNOWN')
             OPEN(22,FILE='daa011.data',STATUS='UNKNOWN')
             OPEN(23,FILE='daa021.data',STATUS='UNKNOWN')
             OPEN(24,FILE='daa031.data',STATUS='UNKNOWN')
             OPEN(25,FILE='daa041.data',STATUS='UNKNOWN')
             OPEN(26,FILE='daa051.data',STATUS='UNKNOWN')
             OPEN(27,FILE='daa061.data',STATUS='UNKNOWN')
             OPEN(28,FILE='daa071.data',STATUS='UNKNOWN')
             OPEN(29,FILE='daa081.data',STATUS='UNKNOWN')
             OPEN(30,FILE='daa091.data',STATUS='UNKNOWN')
C
```

```
                                                        NP=9
C
C
      N       = 36
      VDENS   = 0.2D0
      RA      = 5.0D0
      KU      = 8.0D0
      RV      = 150.D0
      D       = 1.0D0
      TD      = 0.3D0
      RP      = 2.0D0
      NPTC    = 3
      RCOFF   = 5.D0*RP
      NPTCHF  = (NPTC-1)/2
C
      DELR    =   0.1D0
      DELT    =   (5.D0/180.D0 )*PI
C
                                             --- PARAMETER (3) ---
      MCSMPLMX = 10000
      NGRAPH   = MCSMPLMX/10
      NOPT     = 20
      RCOFF2   = RCOFF**2
C
      IX = 0
      CALL RANCAL( NRANMX, IX, RAN )
      NRAN     = 1
      NRANCHK = NRANMX - 10*N
C
C     ----------------------------------------------------------------
C     -----------------   INITIAL CONFIGURATION   -----------------
C     ----------------------------------------------------------------
C
C                                          --- SET INITIAL CONFIG. ---
CCC   OPEN(19,FILE='daa091.dat',STATUS='OLD')
CCC     READ(19,462)  N , XL, YL, D , DT , NPTC
CCC     READ(19,464) (RX(I),I=1,N) , (RY(I),I=1,N) ,
CCC  &                (NX(I),I=1,N) , (NY(I),I=1,N)
CCC   CLOSE(19,STATUS='KEEP')
CCC   GOTO 7
C
      CALL INITIAL( VDENS , N , NPTC )
C
C                                                 --- PRINT OUT ---
    7 WRITE(NP,12) N, VDENS, RA, KU, RV, D, TD, XL, YL, RCOFF,
     &              RP, NPTC, DELR, DELT
      WRITE(NP,14) MCSMPLMX, NGRAPH
C
C                                              --- INITIALIZATION ---
C
C     ----------------------------------------------------------------
C     -----------------   START OF MONTE CARLO PART   ---------------
C     ----------------------------------------------------------------
C
C
      DO 500 MCSMPL = 1 , MCSMPLMX
C
        DO 400 I=1,N
C                                        ------------------ POSITION --------
C                                                   --- OLD ENERGY ---
          RXI = RX(I)
          RYI = RY(I)
          NXI = NX(I)
          NYI = NY(I)
          CALL ENECAL( I , RXI, RYI, NXI, NYI, RCOFF2 , ECAN, OVRLAP )
          EOLD = ECAN
C
C
          NRAN  = NRAN + 1
          RXCAN = RX(I) + DELR*( 1.D0 - 2.D0*DBLE(RAN(NRAN)) )
          RXCAN = RXCAN - DNINT(RXCAN/XL)*XL
          NRAN  = NRAN + 1
          RYCAN = RY(I) + DELR*( 1.D0 - 2.D0*DBLE(RAN(NRAN)) )
```

• The given values and subaveraged values are written out in @daa1.data and daa11.data; @daa1 is for confirming the values assigned for simulations and the results calculated, and daa11.data is for the postprocessing analysis.
• The particle positions and directions are written out in daa001 – daa091 for the postprocessing analysis.

• The particle number $N = 36$, area fraction $\phi_V = 0.2$, $\lambda = 5$, $\xi = 8$, $\lambda_V = 150$, $t_\delta = 0.3$, aspect ratio $r_p = 2$, number of constituent spheres forming the sphere-connected model NPTC, cutoff distance $r^*_{coff} = 5r_p$, $\delta r^*_{max} = 0.1$, and $\delta\theta_{max} = (5/180)\pi$.

• The total number of MC steps is 10,000, and the particle positions are written out at every NGRAPH steps for the postprocessing analysis.

• A sequence of uniform random numbers are prepared in advance and, when necessary, random numbers are taken out from the variable RAN(*)

• The READ statements are for continuing the sequential simulation using the data saved previously.

• The interaction energies are calculated between particle i and its interacting particles.

• After particle i is slightly moved according to Eq. (1.52), the interaction energy is calculated for this new microscopic state.

```
      RYCAN = RYCAN - DNINT(RYCAN/YL)*YL
C                                            --- NEW ENERGY ---
      CALL ENECAL(I, RXCAN, RYCAN, NXI, NYI, RCOFF2, ECAN, OVRLAP)
      IF( OVRLAP ) THEN
        ENEW = EOLD
        GOTO 150
      END IF
C                                   --------  (2) ENERGY HANDAN
C
      C3 = ECAN - EOLD
      IF( C3 .GE. 0.D0 )THEN
        NRAN = NRAN + 1
        IF( DBLE(RAN(NRAN)) .GE. DEXP(-C3) )THEN
          ENEW = EOLD
          GOTO 150
        END IF
      END IF
```

• The adoption of the new state is determined according to the transition probability in Eq. (1.49).

```
C                                     ++++++++++++++++++++++
C                                     CANDIDATES ARE ACCEPTED
C                                     ++++++++++++++++++++++
      RX(I) = RXCAN
      RY(I) = RYCAN
      ENEW  = ECAN
      E(I)  = ECAN
```

• The procedure after the acceptance of the new state.

```
C
C                        --------------------- MOMENT --------
  150 RXI = RX(I)
      RYI = RY(I)
      NXI = NX(I)
      NYI = NY(I)
C                                            --- OLD ENERGY ---
      EOLD = ENEW
C                                   ----------  (3) CANDIDATE
C
      NRAN  = NRAN + 1
      C1    = DELT*DBLE(RAN(NRAN))
      NRAN  = NRAN + 1
      C1    = DSIGN( C1 , DBLE(RAN(NRAN)-0.5) )
      CX    = DSIN(C1)
      CY    = DCOS(C1)
      NXCAN = NXI*CY + NYI*CX
      NYCAN = NYI*CY - NXI*CX
```

• After the direction of particle *i* is slightly changed according to a similar equation to Eq. (1.52), the interaction energy is calculated for this new microscopic state.

```
C
C                                            --- NEW ENERGY ---
      CALL ENECAL(I, RXI, RYI, NXCAN, NYCAN, RCOFF2, ECAN, OVRLAP)
      IF( OVRLAP ) GOTO 400
C
C                          --------  (4) ENERGY HANDAN --------
C
      C3 = ECAN - EOLD
      IF( C3 .GE. 0.D0 )THEN
        NRAN = NRAN + 1
        IF( DBLE(RAN(NRAN)) .GE. DEXP(-C3) )THEN
          GOTO 400
        END IF
      END IF
```

• The adoption of the new state is determined according to the transition probability in Eq. (1.49).

```
C                                     ++++++++++++++++++++++
C                                     CANDIDATES ARE ACCEPTED
C                                     ++++++++++++++++++++++
      NX(I) = NXCAN
      NY(I) = NYCAN
      E(I)  = ECAN
```

• The procedure after the acceptance of the new state.

```
C
ccc             if( i.eq.1) then
ccc               write(6,*) 'smpl,rx,ry',mcsmpl, rx(1), ry(1)
ccc             end if
C
  400 CONTINUE
```

• The average of the components of the vector denoting the particle direction is calculated.

```
C
C
C                      ----- MOMENT AND ENERGY OF SYSTEM -----
      C1 = 0.D0
      C2 = 0.D0
```

• The system energy can be obtained by summing the energy of each particle.

```
        C3 = 0.D0
        DO 450 J=1,N
          C1 = C1 + NY(J)
          C2 = C2 + NX(J)
          C3 = C3 + E(J)
  450   CONTINUE
        MOMY(MCSMPL)    = REAL(C1)/REAL(N)
        MOMX(MCSMPL)    = REAL(C2)/REAL(N)
        MEANENE(MCSMPL) = REAL(C3-KU*C1)/REAL(2*N)
C
C                               --- DATA OUTPUT FOR GRAPHICS (1) ---
C
        IF( MOD(MCSMPL,NGRAPH) .EQ. 0 ) THEN
          NOPT = NOPT + 1
          WRITE(NOPT,462)  N , XL , YL , D , DT , NPTC
          WRITE(NOPT,464)  (RX(I),I=1,N) , (RY(I),I=1,N) ,
     &                     (NX(I),I=1,N) , (NY(I),I=1,N)
                                         CLOSE(NOPT,STATUS='KEEP')
        END IF
C
C                           --- CHECK OF THE SUM OF RANDOM NUMBERS ---
C
        IF( NRAN .GE. NRANCHK )THEN
          CALL RANCAL( NRANMX, IX, RAN )
          NRAN = 1
        END IF
C
C
  500 CONTINUE
C
C     ---------------------------------------------------------------
C     -------------------  END OF MONTE CARLO PART  ----------------
C     ---------------------------------------------------------------
C
      WRITE(NP,592)
      MCSMPL1 = 1
      MCSMPL2 = MCSMPLMX
      CALL PRNTDATA( MCSMPL1 , MCSMPL2 , NP )
      WRITE(NP,612)  MCSMPL1 , MCSMPL2
C
C                               --- DATA OUTPUT FOR GRAPHICS (2) ---
      WRITE(10,1012) N, VDENS, RA, KU, RV, D, TD, XL, YL
      WRITE(10,1013) RCOFF, RP, NPTC, DELR, DELT
      WRITE(10,1014) MCSMPLMX, NGRAPH
      WRITE(10,1016) ( MEANENE(I),I=MCSMPL1, MCSMPL2)
     &              ,( MOMX(I),    I=MCSMPL1, MCSMPL2)
     &              ,( MOMY(I),    I=MCSMPL1, MCSMPL2)
C
                                         CLOSE(9, STATUS='KEEP')
                                         CLOSE(10,STATUS='KEEP')
C     ------------------------- FORMAT ----------------------------
   12 FORMAT(/1H ,'-----------------------------------------------'
     &        /1H ,'-              MONTE CARLO METHOD             -'
     &        /1H ,'-----------------------------------------------'
     &       //1H ,'N=',I4, 2X ,'VDENS=',F5.2, 2X ,
     &             'RA=',F5.2, 2X ,'KU=',F6.2, 2X ,'RV=',F6.2, 2X,
     &             'D=',F5.2, 2X ,'TD=',F5.2
     &        /1H ,'XL=',F6.2, 2X,'YL=',F6.2, 2X, 'RCOFF=',F6.2, 2X,
     &             'RP=',F7.4, 2X,'NPTC=',I3
     &        /1H ,'DELR=',F7.4, 2X ,'DELT=',F7.4)
   14 FORMAT( 1H ,'MCSMPMX=',I8, 2X, 'NGRAPH=',I8/)
  462 FORMAT( I5 , 4F9.4 , I5 )
  464 FORMAT( (8F10.5) )
  592 FORMAT(/1H ,'++++++++++++++++++++++++++++++'
     &        /1H ,'   WITHOUT CLUSTER MOVEMENT   '
     &        /1H ,'++++++++++++++++++++++++++++++'/)
  612 FORMAT(///1H ,18X, 'START OF MC SAMPLING STEP=',I7
     &          /1H ,18X, 'END   OF MC SAMPLING STEP=',I7/)
 1007 FORMAT(/1H ,'*********** NUMBER DENSITY OF CLUSTERS ***********'
     &        /1H ,'Q (MEAN LENGTH OF CLUSTERS)=',F10.5, 5X ,
     &             'NDNSMX=',I8
     &       //1H ,'NDNSCLS(1), NDNSCLS(2), NDNSCLS(3), ..............'
     &        /(1H , 6E13.6) )
 1012 FORMAT( I7 , 8F9.4 )
 1013 FORMAT( 2F9.5 , I4, 2F8.5 )
```

• Since each interaction energy is counted twice, the magnetic particle–field interaction is also added twice. The system energy can finally be obtained by dividing the result by two.

• The number of the random numbers used is checked. If over NRANCHK, a uniform random number sequence is renewed.

```
 1014 FORMAT( 2I8 )
 1016 FORMAT( (5E16.9) )
                                                                      STOP
                                                                      END
C*********************************************************************
C*************************    SUBROUTINE    *************************
C*********************************************************************
C
C**** SUB PRNTDATA ****
      SUBROUTINE PRNTDATA( MCSST, MCSMX, NP )
C
      IMPLICIT REAL*8 (A-H,O-Z), INTEGER (I-N)
C
      COMMON /BLOCK9/  MOMX , MOMY , MEANENE
C
      PARAMETER( NN=1000 , NNS=200000 )
      PARAMETER( NRANMX=500000 , PI=3.141592653589793D0 )
C
      INTEGER   MCSST      , MCSMX      , NP
      REAL      MOMX(NNS) , MOMY(NNS) , MEANENE(NNS)
C
      REAL      AMOMX(10)   , AMOMY(10)   , AMEANENE(10)  , C0
      INTEGER   IC , IMC(0:10) , JS , JE
C
C
      IC = ( MCSMX-MCSST+1 )/50
      DO 20 I= MCSST-1+IC , MCSMX , IC
        WRITE(NP,10) I ,MOMX(I) ,MOMY(I) ,MEANENE(I)
   20 CONTINUE
```

- The total MC steps are equally divided into 50 blocks, and the end value of each block is written out.

```
C                                     ----- MONTE CARLO STEP HEIKIN -----
      IC = ( MCSMX-MCSST+1 )/10
      DO 30 I=0,10
        IMC(I) = MCSST - 1 + IC*I
        IF( I .EQ. 10 ) IMC(I) =MCSMX
   30 CONTINUE
C
C
      DO 35 I=1,10
        AMOMY(I)    = 0.
        AMOMX(I)    = 0.
        AMEANENE(I) = 0.
   35 CONTINUE
```

- The particle direction and the averaged energy are written out.

- The total MC steps are equally divided into10 blocks, and the subaverages are calculated for each block.

```
C
      DO 50 I=1,10
        JS = IMC(I-1) + 1
        JE = IMC(I)
        DO 40 J=JS,JE
          AMOMY(I)    = AMOMY(I)    + MOMY(J)
          AMOMX(I)    = AMOMX(I)    + MOMX(J)
          AMEANENE(I) = AMEANENE(I) + MEANENE(J)
   40   CONTINUE
   50 CONTINUE
C
      DO 70 I=1,10
        C0          = REAL( IMC(I)-IMC(I-1) )
        AMOMY(I)    = AMOMY(I)    /C0
        AMOMX(I)    = AMOMX(I)    /C0
        AMEANENE(I) = AMEANENE(I)/C0
   70 CONTINUE
C                                          ----- STEP HEIKIN INSATU -----
      WRITE(NP,75)
      DO 90 I=1,10
       WRITE(NP,80)I,IMC(I-1)+1,IMC(I),AMOMX(I),AMOMY(I),AMEANENE(I)
   90 CONTINUE
C     -----------------------------------------------------------------
   10 FORMAT(1H ,'MCSMPL=',I5, 3X ,'MOMENT(X)=',F7.4, 3X ,
     &           'MOMENT(Y)=',F7.4, 3X ,'MEAN ENERGY=',E12.5)
   75 FORMAT(//1H ,'-----------------------------------------------'
     &        /1H ,'                 MONTE CARLO HEIKIN            '
     &        /)
   80 FORMAT(1H ,'I=',I2, 2X ,'SMPLMN=',I5, 2X ,'SMPLMX=',I5
     &       /1H ,15X ,'MOMENT(X)=',F7.4, 3X ,
     &                 'MOMENT(Y)=',F7.4, 3X ,'MEAN ENERGY=',E12.5/)
                                                                    RETURN
                                                                    END
```

```
C**** SUB INITIAL ****
      SUBROUTINE INITIAL( VDENS , N , NPTC )
C
      IMPLICIT REAL*8 (A-H,O-Z), INTEGER (I-N)
C
      COMMON /BLOCK1/  RX    , RY
      COMMON /BLOCK2/  NX    , NY
      COMMON /BLOCK3/  XL    , YL
C
      PARAMETER( NN=1000 )
      PARAMETER( NRANMX=500000 , PI=3.141592653589793D0 )
C
      REAL*8    RX(NN) , RY(NN) , NX(NN) , NY(NN)
      REAL*8    VDENS
C
      INTEGER   Q , PTCL
      REAL*8    A , XLUNT , YLUNT , RAN , RAN1 , RAN2 , C1 , C2
C
      A  = DSQRT( DBLE(NPTC)*PI/(8.D0*VDENS) )
      Q  = NINT( SQRT(REAL(N+1)) )
      XL = A*DBLE(Q)
      YL = A*DBLE(2*Q)
      XLUNT = A
      YLUNT = A*DBLE(2)
C                                             ----- POSITION -----
      RAN1 = DSQRT( 2.D0 )
      RAN2 = DSQRT( 7.D0 )
      PTCL=0
      DO 10 J=0,Q-1
        DO 10 I=0,Q-1
          PTCL = PTCL + 1
          C1 = RAN1*DBLE(PTCL)
          C1 = C1 - DINT(C1)
          C1 = C1 - 0.5D0
          C2 = RAN2*DBLE(PTCL)
          C2 = C2 - DINT(C2)
          C2 = C2 - 0.5D0
          RX(PTCL) = DBLE(I)*XLUNT+XLUNT/2.D0+C1*(XLUNT/6.D0)-XL/2.D0
          RY(PTCL) = DBLE(J)*YLUNT+YLUNT/2.D0+C2*(YLUNT/6.D0)-YL/2.D0
   10 CONTINUE
      N = PTCL
C                                             ----- MOMENT -----
      RAN = DSQRT( 2.D0 )
      DO 20 I=1,N
        C1 = RAN*DBLE(I)
        C1 = C1 - DINT(C1)
        C1 = C1 - 0.5D0
        C1 = (5.D0/180.D0)*PI*C1
        NX(I) = DSIN( C1 )
        NY(I) = DCOS( C1 )
   20 CONTINUE
                                                       RETURN
                                                       END
C**** SUB ENECAL *****
      SUBROUTINE ENECAL(I, RXI, RYI, NXI, NYI, RCOFF2 ,ECAN, OVRLAP)
C
      IMPLICIT REAL*8 (A-H,O-Z), INTEGER (I-N)
C
      COMMON /BLOCK1/  RX    , RY
      COMMON /BLOCK2/  NX    , NY
      COMMON /BLOCK3/  XL    , YL
      COMMON /BLOCK4/  RA    , KU    , RV    , TD    , RP
      COMMON /BLOCK5/  VDENS, N     , NPTC  , RCOFF ,  D    , NPTCHF
      COMMON /BLOCK6/  E     , ENEW  , EOLD
C
      PARAMETER( NN=1000 , PI=3.141592653589793D0 )
C
      REAL*8    RX(NN) , RY(NN) , NX(NN) , NY(NN) , E(NN)
      REAL*8    VDENS , KU
      LOGICAL   OVRLAP
C
      REAL*8    RXI , RYI , RXJ , RYJ , RXIJ , RYIJ , RIJ , RIJSQ
      REAL*8    NXI , NYI , NXJ , NYJ , NXIJ , NYIJ , NXIJ2 , NYIJ2
      REAL*8    RRXI , RRYI , RRXJ , RRYJ , RRXIJ , RRYIJ
      REAL*8    NNXI , NNYI , NNXJ , NNYJ
```

• A subroutine for setting the initial position and velocity of each particle.

• The area occupied by one particle is ($a^* \times 2a^*$) and therefore the relationship between the area fraction ϕ_V and a^* is expressed as ϕ_V =(NPTC)*$\pi/8a^{*2}$.

• The particles are initially set in the simple lattice unit formation in Figure 2.1A; the side lengths of the unit cell are (a^*, $2a^*$) in each direction.

• Each particle is moved in parallel by (XLUNT/2, YLUNT/2) to remove subtle situations at outer boundary surfaces. Also, to remove the regularity of the initial configuration, each particle is moved randomly by the maximum displacement 0.5×(XLUNT/6, YLUNT/6) using quasi-random numbers.

• To save pseudo-random numbers, quasi-random numbers based on irrational numbers are used for randomly setting the particle direction within a small angle range.

• A subroutine for calculating the interaction energies between particles.

```
      REAL*8     TXIJ , TYIJ , R00  , R01  , R10   , R11
      REAL*8     C11 , C12 , C21 , C22 , C31 , C32 , C41 , C42
      REAL*8     C00  , C01  , C02
      REAL*8     CNINJ , CNINJ2 , CRIJNI2 , CRIJNJ2 , CKI , CKJ
      REAL*8     KI , KJ , KKI , KKJ , KIS , KJS , KKIS , KKIS2 , KKIJC
      REAL*8     DSQ, RCHKSQ   , RCHKSQ2
      REAL*8     XI , YI , XJ  , YJ  , ENESTER
      INTEGER    ITREE , IPATH , II , JJ , JJS , JJE , IIDEF, IINUMBR
C
      OVRLAP = .FALSE.
      ECAN   = - KU*NYI
      DSQ    = ( 1.D0 + TD )**2
C
      DO 1000 J=1,N
C
        IF( J .EQ. I )            GOTO 1000
C
        RXJ = RX(J)
        RYJ = RY(J)
        RXIJ = RXI - RXJ
        RXIJ = RXIJ - DNINT(RXIJ/XL)*XL
        IF( DABS(RXIJ) .GE. RCOFF )  GOTO 1000
        RYIJ = RYI - RYJ
        RYIJ = RYIJ - DNINT(RYIJ/YL)*YL
        IF( DABS(RYIJ) .GE. RCOFF )  GOTO 1000
        RIJSQ= RXIJ**2 + RYIJ**2
        IF( RIJSQ .GE. RCOFF2 )      GOTO 1000
        RIJ  = DSQRT(RIJSQ)
C
        IF( DABS(RXIJ) .GT. XL/2.D0 ) THEN
          IF( RXIJ .GT. 0.D0 ) RXJ = RXJ + XL
          IF( RXIJ .LE. 0.D0 ) RXJ = RXJ - XL
        END IF
        IF( DABS(RYIJ) .GT. YL/2.D0 ) THEN
          IF( RYIJ .GT. 0.D0 ) RYJ = RYJ + YL
          IF( RYIJ .LE. 0.D0 ) RYJ = RYJ - YL
        END IF
        NXJ   = NX(J)
        NYJ   = NY(J)
        NXIJ  = NXI - NXJ
        NYIJ  = NYI - NYJ
        NXIJ2 = NXI + NXJ
        NYIJ2 = NYI + NYJ
C
        C11 = RXIJ*NXIJ  + RYIJ*NYIJ
        C21 = RXIJ*NXIJ2 + RYIJ*NYIJ2
        C12 = 1.D0 - ( NXI*NXJ + NYI*NYJ )
        C22 = 1.D0 + ( NXI*NXJ + NYI*NYJ )
        C00 = RA/(RP**2)
        C01 = RP/RIJSQ
        C02 = RP**2/(2.D0*RIJSQ)
C                                      --- MAGNETIC ENERGY ---
        R00 = RIJ*(1.D0 + C01*C11 + C02*C12)**0.5
        R11 = RIJ*(1.D0 - C01*C11 + C02*C12)**0.5
        R01 = RIJ*(1.D0 + C01*C21 + C02*C22)**0.5
        R10 = RIJ*(1.D0 - C01*C21 + C02*C22)**0.5
        IF(     (R00 .LT. 1.D0) .OR. (R11 .LT. 1.D0)
     &      .OR. (R01 .LT. 1.D0) .OR. (R10 .LT. 1.D0)     ) THEN
          OVRLAP = .TRUE.
          RETURN
        END IF
C
        ECAN = ECAN + C00*( 1.D0/R00 + 1.D0/R11 - 1.D0/R01 - 1.D0/R10 )
C
C
C       ------------------------------ ENERGY DUE TO STERIC INER. ---
C
        CNINJ = NXI*NXJ + NYI*NYJ
        IF( DABS(CNINJ) .LT. 0.2D0 ) THEN
          ITREE = 2
        ELSE IF( DABS(CNINJ) .GT. 0.9999D0) THEN
          ITREE = 3
```

• The treatment concerning particle *i*.

• The treatment of the periodic BC.
• If the two particles are separated over the cutoff distance $r^*_{\rm coff}$, the calculation is unnecessary.

• The position of the partner particle *j* is modified according to the periodic BC.

• The magnetic interaction energy is calculated from Eq. (4.13).
• The distance between the magnetic charges is first calculated.

• The interaction energy is summed for the four pairs of magnetic charges.

• The interaction energy due to the overlap of the steric layers is calculated in the following.

• The regime shown in Figure 4.3 is determined to proceed to the appropriate treatment, and after the calculation of the interaction energy, the calculation procedure returns to the main program.

```
          ELSE
            ITREE = 1
          END IF
C
          TXIJ = RXIJ/RIJ
          TYIJ = RYIJ/RIJ
          C11  = TXIJ*NXJ  + TYIJ*NYJ
          IF( (DABS(CNINJ).GT.0.9999D0).AND.(DABS(C11).GT.0.9999D0) )THEN
            ITREE=0
          END IF
C                                                  --------------------
C                                                   ITREE=0: LINEAR
C                                                   ITREE=1: GENERAL
C                                                   ITREE=2: NORMALL
C                                                   ITREE=3: PARALLEL
C                                                  --------------------
C
C         -------------------------------------------- (0) LINEAR ---
          IF( ITREE .EQ. 0 ) THEN
C
            IF( CNINJ .GE. 0 ) THEN
              IF( C11 .GE. 0 ) THEN
C                                                         --- IPATH=1
                XJ = RXJ + NXJ*DBLE(NPTCHF)
                YJ = RYJ + NYJ*DBLE(NPTCHF)
                XI = RXI - NXI*DBLE(NPTCHF)
                YI = RYI - NYI*DBLE(NPTCHF)
                ECAN = ECAN +  ENESTER( XI, YI, XJ, YJ, TD, RV, OVRLAP )
                IF ( OVRLAP ) RETURN
              ELSE
C                                                         --- IPATH=2
                XJ = RXJ - NXJ*DBLE(NPTCHF)
                YJ = RYJ - NYJ*DBLE(NPTCHF)
                XI = RXI + NXI*DBLE(NPTCHF)
                YI = RYI + NYI*DBLE(NPTCHF)
                ECAN = ECAN +  ENESTER( XI, YI, XJ, YJ, TD, RV, OVRLAP )
                IF ( OVRLAP ) RETURN
              END IF
            ELSE
              IF( C11 .GE. 0 ) THEN
C                                                         --- IPATH=3
                XJ = RXJ + NXJ*DBLE(NPTCHF)
                YJ = RYJ + NYJ*DBLE(NPTCHF)
                XI = RXI + NXI*DBLE(NPTCHF)
                YI = RYI + NYI*DBLE(NPTCHF)
                ECAN = ECAN +  ENESTER( XI, YI, XJ, YJ, TD, RV, OVRLAP )
                IF ( OVRLAP ) RETURN
              ELSE
C                                                         --- IPATH=4
                XJ = RXJ - NXJ*DBLE(NPTCHF)
                YJ = RYJ - NYJ*DBLE(NPTCHF)
                XI = RXI - NXI*DBLE(NPTCHF)
                YI = RYI - NYI*DBLE(NPTCHF)
                ECAN = ECAN +  ENESTER( XI, YI, XJ, YJ, TD, RV, OVRLAP )
                IF ( OVRLAP ) RETURN
              END IF
            END IF
C
            GOTO 1000
C
          END IF
C         ----------------------------------------------- END OF LINEAR --

          IF( (ITREE .EQ. 1) .OR. (ITREE .EQ. 2) ) THEN
C
            CNINJ2  = NXJ*NYI  - NYJ*NXI
            CRIJNI2 = RXIJ*NYI - RYIJ*NXI
            CRIJNJ2 = RXIJ*NYJ - RYIJ*NXJ
            CKJ     = DABS( CRIJNI2/CNINJ2 )
            CKI     = DABS( CRIJNJ2/CNINJ2 )
C
            C11 = RXIJ + CKI*NXI - CKJ*NXJ
            C12 = RYIJ + CKI*NYI - CKJ*NYJ
            C21 = RXIJ - CKI*NXI - CKJ*NXJ
            C22 = RYIJ - CKI*NYI - CKJ*NYJ
```

• The treatment for a linear arrangement in Figure 4.6.

• The position (XI,YI) and (XJ,YJ) of the spheres of particles *i* and *j* are calculated.

• The absolute values (CKI, CKJ) of (k_i, k_j) are calculated from Eq. (4.8).

```
          C31 = RXIJ + CKI*NXI + CKJ*NXJ
          C32 = RYIJ + CKI*NYI + CKJ*NYJ
          C41 = RXIJ - CKI*NXI + CKJ*NXJ
          C42 = RYIJ - CKI*NYI + CKJ*NYJ
          C00 = 1.0D-8
          IF( (DABS(C11).LT. C00) .AND. (DABS(C12).LT. C00) )THEN
            KI =  CKI
            KJ =  CKJ
            GOTO 110
          END IF
          IF( (DABS(C21).LT. C00) .AND. (DABS(C22).LT. C00) )THEN
            KI = -CKI
            KJ =  CKJ
            GOTO 110
          END IF
          IF( (DABS(C31).LT. C00) .AND. (DABS(C32).LT. C00) )THEN
            KI =  CKI
            KJ = -CKJ
            GOTO 110
          END IF
          IF( (DABS(C41).LT. C00) .AND. (DABS(C42).LT. C00) )THEN
            KI = -CKI
            KJ = -CKJ
            GOTO 110
          END IF
C
  110     IF( CKJ .GT. CKI ) THEN
            II     = I
            JJ     = J
            RRXI   = RXI
            RRYI   = RYI
            RRXJ   = RXJ
            RRYJ   = RYJ
            RRXIJ  = RXIJ
            RRYIJ  = RYIJ
            NNXI   = NXI
            NNYI   = NYI
            NNXJ   = NXJ
            NNYJ   = NYJ
            KKI    = KI
            KKJ    = KJ
          ELSE
            II     = J
            JJ     = I
            RRXI   = RXJ
            RRYI   = RYJ
            RRXJ   = RXI
            RRYJ   = RYI
            RRXIJ  = -RXIJ
            RRYIJ  = -RYIJ
            NNXI   = NXJ
            NNYI   = NYJ
            NNXJ   = NXI
            NNYJ   = NYI
            KKI    = KJ
            KKJ    = KI
          END IF
C
        END IF
C
C                                                    --------------------
C                                                     ITREE=0: LINEAR
C                                                     ITREE=1: GENERAL
C                                                     ITREE=2: NORMALL
C                                                     ITREE=3: PARALLEL
C                                                    --------------------
        IF( ITREE .EQ. 1 ) GOTO 200
        IF( ITREE .EQ. 2 ) GOTO 400
        IF( ITREE .EQ. 3 ) GOTO 600
C
C       ------------------------------------------ (1) GENERAL ---
  200   CNINJ = NXI*NXJ + NYI*NYJ
        IF( CNINJ .GT. 0.D0 ) THEN
          IF( KKJ .GE. 0.D0 ) THEN
            IPATH = 1
          ELSE
```

• The final results of k_i and k_j are obtained by checking the sign of k_i and k_j.

• The subscripts are exchanged between i and j so as to satisfy $|k_j|>|k_i|$.
• As a result, the particle names i and j in Figure 4.2 are expressed as II and JJ in the program.

• The treatment for a general arrangement in Figure 4.4.

```
            IPATH = 4
          END IF
        ELSE
          IF( KKJ .GE. 0.D0 ) THEN
            IPATH = 3
          ELSE
            IPATH = 2
          END IF
        END IF
C
        KKIS  =   CNINJ*DBLE(NPTCHF)   - ( RRXIJ*NNXI + RRYIJ*NNYI )
        KKIS2 = - CNINJ*DBLE(NPTCHF)   - ( RRXIJ*NNXI + RRYIJ*NNYI )
        RCHKSQ =( RRXIJ + KKIS *NNXI - NNXJ*DBLE(NPTCHF) )**2
     &         +( RRYIJ + KKIS *NNYI - NNYJ*DBLE(NPTCHF) )**2
        RCHKSQ2=( RRXIJ + KKIS2*NNXI + NNXJ*DBLE(NPTCHF) )**2
     &         +( RRYIJ + KKIS2*NNYI + NNYJ*DBLE(NPTCHF) )**2
C
        IF( IPATH .EQ. 1 ) THEN
C                                                  --- PATH=1 ---
          IF( RCHKSQ .GE. DSQ ) GOTO 1000
C
          IF( KKIS .GE. 0.D0 ) THEN
            IKKIS = IDINT(KKIS) + 1
          ELSE
            IKKIS = IDINT(KKIS)
          END IF
          IF( IKKIS .GT. NPTCHF )  IKKIS = NPTCHF
          JJS   = NPTCHF
          IIDEF = NPTCHF - IKKIS
          JJE   = -NPTCHF + IIDEF
C
          DO 250 JJ= JJS, JJE, -1
            XJ = RRXJ + DBLE(JJ)*NNXJ
            YJ = RRYJ + DBLE(JJ)*NNYJ
          DO 250 II= JJ-IIDEF, JJ-IIDEF-1, -1
            IF( II .LT. -NPTCHF )    GOTO 250
            XI = RRXI + DBLE(II)*NNXI
            YI = RRYI + DBLE(II)*NNYI
            ECAN = ECAN +  ENESTER( XI, YI, XJ, YJ, TD, RV, OVRLAP )
            IF ( OVRLAP ) RETURN
  250     CONTINUE
C
        ELSE IF( IPATH .EQ. 2 ) THEN
C                                                  --- PATH=2 ---
          IF( RCHKSQ2 .GE. DSQ ) GOTO 1000
C
          IF( KKIS2 .GE. 0.D0 ) THEN
            IKKIS2 = IDINT(KKIS2) + 1
          ELSE
            IKKIS2 = IDINT(KKIS2)
          END IF
          IF( IKKIS2 .GT. NPTCHF )  IKKIS2 = NPTCHF
          JJS   = NPTCHF
          IIDEF = NPTCHF - IKKIS2
          JJE   = -NPTCHF + IIDEF
C
          DO 252 JJ= JJS, JJE, -1
            JJJ= -JJ
            XJ = RRXJ + DBLE(JJJ)*NNXJ
            YJ = RRYJ + DBLE(JJJ)*NNYJ
          DO 252 II= JJ-IIDEF, JJ-IIDEF-1, -1
            IF( II .LT. -NPTCHF )    GOTO 252
            XI = RRXI + DBLE(II)*NNXI
            YI = RRYI + DBLE(II)*NNYI
            ECAN = ECAN +  ENESTER( XI, YI, XJ, YJ, TD, RV, OVRLAP )
            IF ( OVRLAP ) RETURN
  252     CONTINUE
C
        ELSE IF( IPATH .EQ. 3 ) THEN
C                                                  --- PATH=3 ---
          IF( RCHKSQ .GE. DSQ ) GOTO 1000
C
          IF( -KKIS .GE. 0.D0 ) THEN
            IKKIS = IDINT(-KKIS) + 1
          ELSE
            IKKIS = IDINT(-KKIS)
```

- After the assessment of the particle overlap regime, k_i^s (KKIS) and $k_i^{s'}$ (KKIS2) are calculated from Eqs. (4.9) and (4.10).

- The constituent spheres in the rod-like particle are named in such a way that the central sphere is 0, the neighboring spheres are 1,2,..., in the particle direction, and –1,–2,..., in the opposite direction.

- The interaction energy between the sphere IKKIS of particle *i* and the sphere JJS of particle *j* is checked.
- The two spheres of particle *i* are checked as an object interacting with the sphere of particle *j*.

- The center of the sphere of particle *i* is denoted by (XI,YI) and, similarly, (XJ,YJ) for the sphere of particle *j*.

```
              END IF
              IF( IKKIS .GT. NPTCHF )  IKKIS = NPTCHF
              JJS    = NPTCHF
              IIDEF = NPTCHF - IKKIS
              JJE    = -NPTCHF + IIDEF
C
              DO 254 JJ= JJS, JJE, -1
                XJ = RRXJ + DBLE(JJ)*NNXJ
                YJ = RRYJ + DBLE(JJ)*NNYJ
              DO 254 II= JJ-IIDEF, JJ-IIDEF-1, -1
                IF( II .LT. -NPTCHF )     GOTO 254
                III = -II
                XI = RRXI + DBLE(III)*NNXI
                YI = RRYI + DBLE(III)*NNYI
                ECAN = ECAN +  ENESTER( XI, YI, XJ, YJ, TD, RV, OVRLAP )
                IF ( OVRLAP ) RETURN
  254         CONTINUE
C
            ELSE IF( IPATH .EQ. 4 ) THEN
C                                                          --- PATH=4 ---
              IF( RCHKSQ2 .GE. DSQ ) GOTO 1000
C
              IF( -KKIS2 .GE. 0.D0 ) THEN
                IKKIS2 = IDINT(-KKIS2) + 1
              ELSE
                IKKIS2 = IDINT(-KKIS2)
              END IF
              IF( IKKIS2 .GT. NPTCHF )  IKKIS2 = NPTCHF
              JJS    = NPTCHF
              IIDEF = NPTCHF - IKKIS2
              JJE    = -NPTCHF + IIDEF
C
              DO 256 JJ= JJS, JJE, -1
                JJJ = -JJ
                XJ = RRXJ + DBLE(JJJ)*NNXJ
                YJ = RRYJ + DBLE(JJJ)*NNYJ
              DO 256 II= JJ-IIDEF, JJ-IIDEF-1, -1
                IF( II .LT. -NPTCHF )     GOTO 256
                III = -II
                XI = RRXI + DBLE(III)*NNXI
                YI = RRYI + DBLE(III)*NNYI
                ECAN = ECAN +  ENESTER( XI, YI, XJ, YJ, TD, RV, OVRLAP )
                IF ( OVRLAP ) RETURN
  256         CONTINUE
C
            END IF
C
            GOTO 1000
C           -------------------------------------------- (2) NORMAL  ---
C
  400       IF( KKJ .GE. 0.D0 ) THEN
              IPATH = 1
            ELSE
              IPATH = 2
            END IF
C
            CNINJ = NXI*NXJ + NYI*NYJ
            KKIS  =   CNINJ*DBLE(NPTCHF) - ( RRXIJ*NNXI + RRYIJ*NNYI )
            KKIS2 = - CNINJ*DBLE(NPTCHF) - ( RRXIJ*NNXI + RRYIJ*NNYI )
            RCHKSQ =( RRXIJ + KKIS *NNXI - NNXJ*DBLE(NPTCHF) )**2
     &             +( RRYIJ + KKIS *NNYI - NNYJ*DBLE(NPTCHF) )**2
            RCHKSQ2=( RRXIJ + KKIS2*NNXI + NNXJ*DBLE(NPTCHF) )**2
     &             +( RRYIJ + KKIS2*NNYI + NNYJ*DBLE(NPTCHF) )**2
C
            IF( IPATH .EQ. 1 ) THEN
              IF( RCHKSQ  .GE. DSQ ) GOTO 1000
            ELSE
              IF( RCHKSQ2 .GE. DSQ ) GOTO 1000
            END IF
C
            IF( IPATH .EQ. 2 )  KKIS = KKIS2
C
            IF( KKIS .GE. 0.D0 ) THEN
              IKKIS = IDINT(KKIS) + 1
            ELSE
```

• The treatment for a normal arrangement in Figure 4.5.

• After the assessment of the particle overlap regime, k_i^s (KKIS) and $k_i^{s'}$ (KKIS2) are calculated from Eqs. (4.9) and (4.10).

• The constituent spheres in the rod-like particle are named in such a way that the central sphere is 0, the neighboring spheres are 1,2,..., in the particle direction, and −1, −2,..., in the opposite direction.

```
            IKKIS = IDINT(KKIS)
          END IF
          IF( IKKIS .GT. NPTCHF )  IKKIS = NPTCHF + 1
          IIDEF = NPTCHF - IKKIS
C
          JJJ = NPTCHF
          IF( IPATH .EQ. 1 ) THEN
            JJ = JJJ
          ELSE
            JJ = -JJJ
          END IF
C
```

- The interaction energy between the sphere IKKIS of particle *i* and the sphere JJ of particle *j* is treated.
- The two spheres of particle *i* are checked as an object interacting with the sphere of particle *j*.

```

          XJ = RRXJ + DBLE(JJ)*NNXJ
          YJ = RRYJ + DBLE(JJ)*NNYJ
           DO 450 II= JJJ-IIDEF, JJJ-IIDEF-1, -1
             IF( II .GT.  NPTCHF )     GOTO 450
             IF( II .LT. -NPTCHF )     GOTO 450
            XI = RRXI + DBLE(II)*NNXI
            YI = RRYI + DBLE(II)*NNYI
            ECAN = ECAN +  ENESTER( XI, YI, XJ, YJ, TD, RV, OVRLAP )
            IF ( OVRLAP ) RETURN
    450   CONTINUE
C
          GOTO 1000
```

- The center of the sphere of particle *i* is denoted by (XI,YI) and, similarly, (XJ,YJ) for the sphere of particle *j*

- The treatment for a parallel arrangement in Figure 4.7.

```
C         --------------------------------------------- (3) PARALLEL --
C
    600   CNINJ = NXI*NXJ + NYI*NYJ
          KIS = CNINJ*DBLE(NPTCHF) - ( RXIJ*NXI + RYIJ*NYI )
          KJS = CNINJ*DBLE(NPTCHF) + ( RXIJ*NXJ + RYIJ*NYJ )
          IF( CNINJ .GE. 0.D0 ) THEN
            IPATH = 1
          ELSE
            IF( KIS .LE. -DBLE(NPTCHF) ) THEN
              IPATH = 2
            ELSE
              IPATH = 3
            END IF
          END IF
C
```

- After the assessment of the particle overlap regime, k_i^s (KIS) and k_j^s (KJS) are calculated from Eq. (4.9).

- The subscripts are exchanged between *i* and *j* so as to satisfy $k_j^s > k_i^s$.
- As a result, the particle names *i* and *j* in Figure 4.7 are expressed as II and JJ in the program.

```
          II     = I
          JJ     = J
          RRXI   = RXI
          RRYI   = RYI
          RRXJ   = RXJ
          RRYJ   = RYJ
          RRXIJ  = RXIJ
          RRYIJ  = RYIJ
          NNXI   = NXI
          NNYI   = NYI
          NNXJ   = NXJ
          NNYJ   = NYJ
          KKIS   = KIS
          IF( (IPATH .EQ. 1) .AND. (KIS .GT. KJS) ) THEN
            II     = J
            JJ     = I
            RRXI   = RXJ
            RRYI   = RYJ
            RRXJ   = RXI
            RRYJ   = RYI
            RRXIJ  = -RXIJ
            RRYIJ  = -RYIJ
            NNXI   = NXJ
            NNYI   = NYJ
            NNXJ   = NXI
            NNYJ   = NYI
            KKIS   = KJS
          END IF
C
          RCHKSQ = ( RRXIJ + KKIS *NNXI - NNXJ*DBLE(NPTCHF) )**2
     &            +( RRYIJ + KKIS *NNYI - NNYJ*DBLE(NPTCHF) )**2
          IF( RCHKSQ  .GE. DSQ ) GOTO 1000
```

- The constituent spheres in the rod-like particle are named in such a way that the central sphere is 0, the neighboring spheres are 1,2,..., in the particle direction, and –1, –2,..., in the opposite direction.

- The interaction energy between the sphere of the positive magnetic charge of particle j and the sphere IKKIS (and (IKKIS−1)) of particle i is calculated. There are IINUMBER pairs of particles.

- The center of the sphere of particle i is denoted by (XI,YI) and, similarly, (XJ,YJ) for particle j.

```
C
      IF( IPATH .EQ. 1 ) THEN
C
        IF( KKIS .GE. 0.D0 ) THEN
          IKKIS = IDINT(KKIS) + 1
        ELSE
          IKKIS = IDINT(KKIS)
        END IF
        IIDEF = NPTCHF - IKKIS
C
        XJ = RRXJ + DBLE(NPTCHF)*NNXJ
        YJ = RRYJ + DBLE(NPTCHF)*NNYJ
        IINUMBR = NPTC + 1 - IIDEF
        DO 650 II= NPTCHF-IIDEF, NPTCHF-IIDEF-1, -1
          IF( II .LT. -NPTCHF )    GOTO 650
          XI = RRXI + DBLE(II)*NNXI
          YI = RRYI + DBLE(II)*NNYI
          IINUMBR = IINUMBR - 1
          ECAN = ECAN + DBLE(IINUMBR)*
     &                 ENESTER( XI, YI, XJ, YJ, TD, RV, OVRLAP )
          IF ( OVRLAP ) RETURN
  650   CONTINUE
      END IF
C
      KKIJC = DABS( KKIS ) - DBLE(NPTCHF)
C
```

- The separation between the central spheres of particles i and j along the particle axis, k_{ij}^{c}(KKIJC), is calculated.

- The sphere of particle i is determined as an object according to Section 4.1.3.4. There are IINUMBER pairs of particles yielding such an arrangement.

```
      IF( IPATH .EQ. 2 ) THEN
C                                                  --- PATH=2 ---
        IKKIJC = IDINT( KKIJC )
        IIDEF  = IKKIJC
        XJ = RRXJ - DBLE(NPTCHF)*NNXJ
        YJ = RRYJ - DBLE(NPTCHF)*NNYJ
        IINUMBR = NPTC + 1 - IIDEF
        DO 652 II= NPTCHF-IIDEF, NPTCHF-IIDEF-1, -1
          IF( II .LT. -NPTCHF )    GOTO 652
          XI = RRXI + DBLE(II)*NNXI
          YI = RRYI + DBLE(II)*NNYI
          IINUMBR = IINUMBR - 1
          ECAN = ECAN + DBLE(IINUMBR)*
     &                 ENESTER( XI, YI, XJ, YJ, TD, RV, OVRLAP )
          IF ( OVRLAP ) RETURN
  652   CONTINUE
      END IF
C
      KKIJC = KKIS + DBLE(NPTCHF)
C
      IF( IPATH .EQ. 3 ) THEN
C                                                  --- PATH=3 ---
        IKKIJC = IDINT( KKIJC )
        IIDEF  = IKKIJC
        XJ = RRXJ + DBLE(NPTCHF)*NNXJ
        YJ = RRYJ + DBLE(NPTCHF)*NNYJ
        IINUMBR = NPTC + 1 - IIDEF
        DO 654 II= NPTCHF-IIDEF, NPTCHF-IIDEF-1, -1
          IF( II .LT. -NPTCHF )    GOTO 654
          III = -II
          XI = RRXI + DBLE(III)*NNXI
          YI = RRYI + DBLE(III)*NNYI
          IINUMBR = IINUMBR - 1
          ECAN = ECAN + DBLE(IINUMBR)*
     &                 ENESTER( XI, YI, XJ, YJ, TD, RV, OVRLAP )
          IF ( OVRLAP ) RETURN
  654   CONTINUE
      END IF
C
      GOTO 1000
C
C     ----------------------- END OF ENERGY DUE TO STERIC INER. ---
C
 1000 CONTINUE
                                                           RETURN
                                                           END
C#### FUN ENESTER ####
      DOUBLE PRECISION FUNCTION ENESTER(XI, YI, XJ, YJ, TD, RV, OVRLAP)
C
      IMPLICIT REAL*8 (A-H,O-Z), INTEGER (I-N)
```

• A function subprogram for calculating the interaction energy due to the overlap of the surfactant layers according to Eq. (4.14).

```
C
      LOGICAL   OVRLAP
C
      RIJ = DSQRT( (XI-XJ)**2 + (YI-YJ)**2 )
      X   = 2.D0*(RIJ-1.D0)
      IF( X .LT. 0.D0 ) THEN
        OVRLAP = .TRUE.
        ENESTER = 1.D9
        RETURN
      END IF
ccc      write(6,*)'xi,yi,xj,yj', xi,yi,xj,yj
C
      IF( RIJ .LE. (TD+1.D0) ) THEN
        C1 = (X+2.D0)/TD
        C2 = DLOG( (TD+1.D0)/(X/2.D0+1.D0) )
        C3 = X/TD
        ENESTER = RV*( 2.D0 - C1*C2 - C3 )
        RETURN
      ELSE
        ENESTER = 0.D0
        RETURN
      END IF
                                                      RETURN
                                                      END
```

• A subroutine for generating a uniform random number sequence.

• This is for a 32-bit CPU based on the expression of two's complement.

```
C**** SUB RANCAL ****
      SUBROUTINE RANCAL( N, IX, X )
C
      DIMENSION  X(N)
      DATA INTEGMX/2147483647/
      DATA INTEGST,INTEG/584287,48828125/
C
      AINTEGMX = REAL( INTEGMX )
C
      IF ( IX.LT.0 ) PAUSE
      IF ( IX.EQ.0 ) IX = INTEGST
      DO 30 I=1,N
         IX = IX*INTEG
         IF (IX) 10, 20, 20
   10    IX   = (IX+INTEGMX)+1
   20    X(I) = REAL(IX)/AINTEGMX
   30 CONTINUE
      RETURN
      END
C*********************************************************************
C   THIS SUBROUTINE IS FOR GENERATING UNIFORM RANDOM NUMBERS         *
C   (SINGLE PRECISION) FOR 64-BIT COMPUTER.                          *
C      N      : NUMBER OF RANDOM NUMBERS TO GENERATE                 *
C      IX     : INITIAL VALUE OF RANDOM NUMBERS (POSITIVE INTEGER)   *
C             : LAST GENERATED VALUE IS KEPT                         *
C      X(N)   : GENERATED RANDOM NUMBERS (0<X(N)<1)                  *
C*********************************************************************
C**** SUB RANCAL999 ****
ccc   SUBROUTINE RANCAL999( N, IX, X )
C
ccc   IMPLICIT REAL*8 (A-H,O-Z), INTEGER*8 (I-N)
C
ccc   REAL      X(N)
ccc   INTEGER*8 INTEGMX, INTEG64, INTEGST, INTEG
C
CCC   DATA INTEGMX/2147483647/
ccc   DATA INTEG64/2147483648/
ccc   DATA INTEGST,INTEG/584287,48828125/
C
CCC   AINTEGMX = REAL( INTEGMX )
ccc   AINTEGMX = REAL( INTEG64 )
C
ccc   IF ( IX.LT.0 ) PAUSE
ccc   IF ( IX.EQ.0 ) IX = INTEGST
ccc   DO 30 I=1,N
ccc      IX = IX*INTEG
ccc      IX = KMOD(IX,INTEG64)
CCC      IF (IX) 10, 20, 20
CCC10    IX   = (IX+INTEGMX)+1
ccc20    X(I) = REAL(IX)/AINTEGMX
ccc30 CONTINUE
ccc   RETURN
ccc   END
```

4.2 Aggregation Phenomena in a Dispersion of Plate-like Particles

In this section, we consider aggregation phenomena in a suspension composed of disk-like particles. As seen in the rod-like particle system, there are several obstacles to developing a simulation program employing a nonspherical particle system. That is, we need to first make a mathematical analysis of particle overlap and then express the overlap criterion in the language of a simulation program. Hence, in this section we show the mathematical analysis from the viewpoint of developing a simulation program. The exercise of interest is a circular disk-like particle with a magnetic dipole moment at the particle center. We discuss the influences of magnetic particle—particle interactions and the magnetic field strength on aggregation phenomena. The subject of the present exercise is partly under our research group's study, and therefore the sample simulation program has an academic emphasis. The system of interest is in thermodynamic equilibrium and has a given number of particles, temperature, and volume; therefore, the canonical MC algorithm is employed.

4.2.1 Physical Phenomena of Interest

It is assumed that the system composed of disk-like particles with a magnetic moment at the particle center is in thermodynamic equilibrium. In the present exercise, we discuss aggregation phenomena in this type of dispersion under the influence of an applied magnetic field by means of an MC simulation.

The main points in formalizing this demonstration are to develop the particle model, to express the potential energy between particles, and to analyze the criterion for particle overlap. We explain these important subjects in detail below.

4.2.2 Particle Model

As shown in Figure 4.12, we here employ a disk-like particle with a magnetic moment **m** (along the disk surface) normal to the particle axis at the particle center with the section shape of a spherocylinder. The central part of this disk-like particle is a short cylinder with diameter d and thickness b_1. The side of the cylinder is surrounded by the semi-shape of a torus shape, resulting in a particle circumcircle with dimension d_1 $(=d+b_1)$, as shown in Figure 4.12. The configurational state of a single axisymmetric particle i is specified by the position of the particle center $\mathbf{r}_i$, the particle direction (normal to the disk surface) $\mathbf{e}_i$, and the magnetic moment direction $\mathbf{n}_i$ where $\mathbf{e}_i$ and $\mathbf{n}_i$ are the unit vectors. In the MC method, knowledge of only the position and direction of each particle is sufficient to advance an MC step, while both the translational and angular velocities need to be treated in the MD method. The magnetic moment is assumed to be fixed in the particle body, so that only the rotation of the particle can provide a change in the magnetic moment direction.

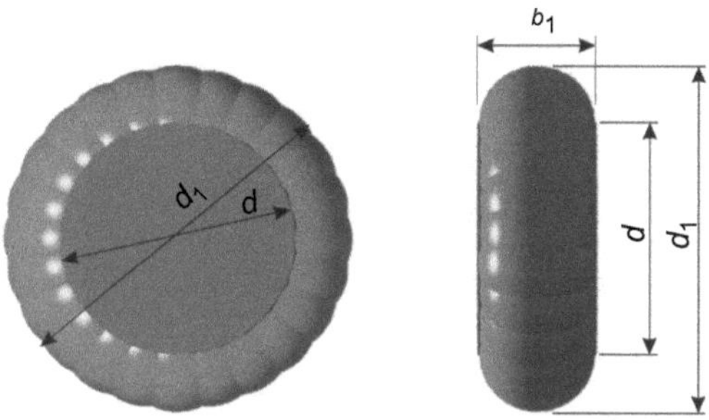

(A) (B)

Figure 4.12 Particle model: (A) plane view and (B) side view.

The interaction energy u_i between the magnetic moment $\mathbf{m}_i$ and an applied magnetic field $\mathbf{H}$ is expressed as

$$u_i = -\mu_0 \mathbf{m}_i \cdot \mathbf{H} \tag{4.18}$$

in which μ_0 is the permeability of free space. This expression clearly implies that the inclination of the magnetic moment along the field direction yields a minimum interaction energy; that is, the particle has a tendency to orient in such a way that the magnetic moment will incline in the field direction.

The magnetic interaction energy u_{ij} between particles i and j is expressed as [31]

$$u_{ij} = \frac{\mu_0}{4\pi r_{ij}^3}\left\{\mathbf{m}_i \cdot \mathbf{m}_j - \frac{3}{r_{ij}^2}(\mathbf{m}_i \cdot \mathbf{r}_{ij})(\mathbf{m}_j \cdot \mathbf{r}_{ij})\right\} \tag{4.19}$$

in which $\mathbf{r}_i$ is the position vector of particle i ($i = 1, 2, \ldots, N$), $\mathbf{r}_{ij} = \mathbf{r}_i - \mathbf{r}_j$, and $r_{ij} = |\mathbf{r}_{ij}|$. Eq. (4.19) implies that a minimum interaction energy can be obtained when both magnetic moments incline in the same direction along a line drawn between the particle centers. However, note that a thermodynamic equilibrium state will be determined by the balance of the decrease in the system energy and the increase in the system entropy; that is, the entropy should be treated in addition to the energy in order to discuss the thermodynamic equilibrium state. This approach may provide an important facility to molecular simulation methods as a tool for analyzing physical phenomena at the microscopic level. In addition to magnetic forces, the interactions due to electric double layers and steric layers are important considerations, but in this example we have chosen to neglect these interactions for simplification and clarification of the method.

In our approach, by treating a nondimensional form of the system, we are able to discuss the physical phenomenon of interest in a much more reasonable manner, since several important factors governing the physical phenomenon appear as explicit terms in the nondimensional equations. In the nondimensionalization procedure the representative values used are particle thickness b_1 for distances and thermal energy kT for energies. With these representative values, Eqs. (4.18) and (4.19) are written as

$$u_i^* = u_i/kT = -\xi \mathbf{n}_i \cdot \mathbf{h} \tag{4.20}$$

$$u_{ij}^* = u_{ij}/kT = \lambda \frac{1}{r_{ij}^{*3}} \{\mathbf{n}_i \cdot \mathbf{n}_j - 3(\mathbf{n}_i \cdot \mathbf{t}_{ij})(\mathbf{n}_j \cdot \mathbf{t}_{ij})\} \tag{4.21}$$

in which $\mathbf{n}_i = \mathbf{m}_i/m$, $m = |\mathbf{m}_i|$, $\mathbf{h} = \mathbf{H}/H$, $H = |\mathbf{H}|$, $\mathbf{t}_{ij} = \mathbf{r}_{ij}/r_{ij}$, and the superscript * implies nondimensionalized quantities; $\mathbf{n}_i$ and $\mathbf{h}$ are the unit vectors denoting the magnetic moment direction and the magnetic field direction, respectively. The procedure gives rise to the nondimensional parameters ξ and λ that are defined as

$$\xi = \mu_0 mH/kT, \quad \lambda = \mu_0 m^2/4\pi b_1^3 kT \tag{4.22}$$

This is a typical example of the nondimensionalizing procedure giving rise to the appearance of nondimensional parameters or nondimensional numbers. In the present exercise, the physical phenomenon is governed by the magnetic particle–field interactions, the particle–particle interactions, and the random forces and torques acting on each particle. It is therefore reasonable that the ratios of these factors appear in the basic equations as nondimensional parameters ξ and λ in Eq. (4.22). These parameters imply the strengths of the magnetic particle–field and the particle–particle interactions relative to the thermal energy, respectively.

4.2.3 Criterion of the Particle Overlap

Assessing the overlap of the two disk-like particles shown in Figure 4.12 is significantly different from that of a pair of spherical particles. Both the torus parts may overlap, or the torus part and the disk part may overlap. Taking into account all the possible overlap regimes during a simulation requires probing into the essence of the overlap and then making a systematic analysis based on the insight gained from a careful investigation of the problem. This is usually undertaken in advance as part of the preparation required in writing a computer simulation program. In the previous case of the spherocylinder, systematic analysis on the particle overlap criterion was achieved by viewing a pair from such a direction that the planes including the corresponding particles are seen to be parallel. For our disk-like particle, a systematic analysis may be possible by focusing on the line of intersection generated by the two corresponding planes. Hence, we first consider the case of nonparallel planes, in which the intersection line can certainly be defined. The use of the maximum section circle of diameter d_1 of the disk-like particle enables us to indicate the typical overlap patterns schematically in Figure 4.13. Figure 4.13A is for the case of the intersection line penetrating each particle (circle), Figures 4.13B and C are for the intersection line penetrating only one particle, and Figure 4.13D is for the intersection line located outside both particles. Since the present disk-like particles have a definite thickness, the above-mentioned regimes of the particle

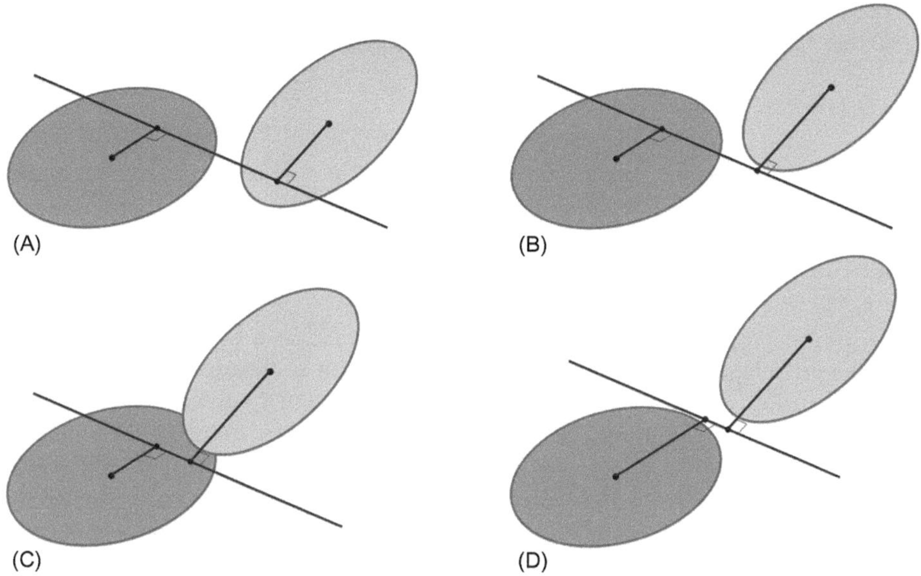

Figure 4.13 Overlap of circular disk-like particles with infinitesimal thin thickness.

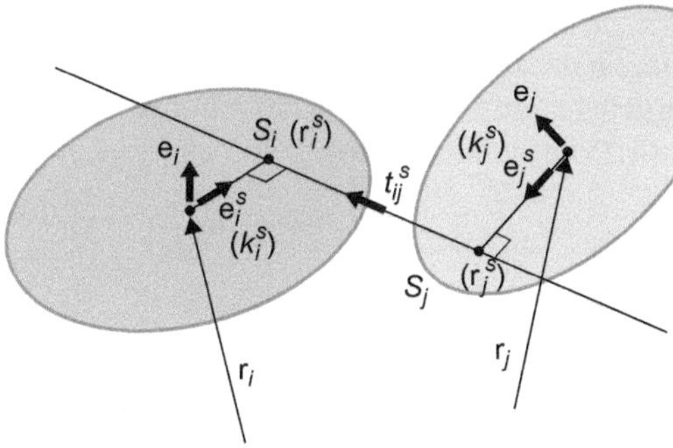

Figure 4.14 Analysis of circles with radius r_0 (=$d/2$).

overlap need to be slightly modified. That is, in each regime, the particle overlap is assessed by calculating the minimum separation between the two particles. We discuss this method of assessing an overlap in detail below.

Advancing our analysis, we now consider the configuration of particles i and j shown in Figure 4.14, with the notion $\mathbf{r}_i$ for the particle center position, the unit vector $\mathbf{e}_i$ for denoting the particle direction (normal to the disk surface), the point S_i for the intersection point of the vertical line drawn from $\mathbf{r}_i$ to the intersection line, the position vector $\mathbf{r}_i^s$ for point S_i, and the unit vector $\mathbf{e}_i^s$ for denoting the direction of $(\mathbf{r}_i^s - \mathbf{r}_i)$, with similar notation for particle j. In addition, the notation $\mathbf{t}_{ij}^s$ is used as the unit vector denoting the direction of the line drawn from points S_j to S_i. In the following paragraphs, these quantities are first evaluated for a pair of particles and then they are used to discuss the criterion for particle overlap.

The unit vector $\mathbf{t}_{ij}^s$ along the intersection line is normal to both the vectors $\mathbf{e}_i$ and $\mathbf{e}_j$, so that $\mathbf{t}_{ij}^s$ can be expressed from the formula of vector product as

$$\mathbf{t}_{ij}^s = \mathbf{e}_j \times \mathbf{e}_i / |\mathbf{e}_j \times \mathbf{e}_i| \tag{4.23}$$

in which $\mathbf{t}_{ij}^s$ is necessarily taken from S_j toward S_i. Since $\mathbf{e}_i^s$ is normal to both $\mathbf{e}_i$ and $\mathbf{t}_{ij}^s$ and, similarly, $\mathbf{e}_j^s$ is normal to $\mathbf{e}_j$ and $\mathbf{t}_{ij}^s$, the use of the vector $\mathbf{t}_{ij}^s$ provides the solutions of $\mathbf{e}_i^s$ and $\mathbf{e}_j^s$ as

$$\mathbf{e}_i^s = -\mathbf{e}_i \times \mathbf{t}_{ij}^s, \quad \mathbf{e}_j^s = \mathbf{e}_j \times \mathbf{t}_{ij}^s \tag{4.24}$$

In the particle configuration shown in Figure 4.14, it is clear that the unit vectors $\mathbf{e}_i^s$ and $\mathbf{e}_j^s$ in Eq. (4.24) point toward the intersection line from the center of each particle. In certain situations, however, these vectors may point in the opposite direction. The treatment of ensuring that $\mathbf{e}_i^s$ and $\mathbf{e}_j^s$ point toward the intersection line will be discussed in detail in Section 4.2.5. If the distance between the center of particle i and point S_i is denoted by k_i^s (similarly, k_j^s for particle j), and the separation between points S_i and S_j is denoted by k_{ij}^s, the expression of point S_i in the two different forms yields the following equation:

$$\mathbf{r}_i + k_i^s \mathbf{e}_i^s = \mathbf{r}_j + k_j^s \mathbf{e}_j^s + k_{ij}^s \mathbf{t}_{ij}^s \tag{4.25}$$

The left- and right-hand sides in this equation are related to the same position vector $\mathbf{r}_i^s$, which is traced from the center of particles i and j, respectively. With the orthogonality condition of the unit vectors, Eq. (4.25) provides the following expressions:

$$k_i^s = -\frac{\mathbf{e}_j \cdot \mathbf{r}_{ij}}{\mathbf{e}_j \cdot \mathbf{e}_i^s}, \quad k_j^s = \frac{\mathbf{e}_i \cdot \mathbf{r}_{ij}}{\mathbf{e}_i \cdot \mathbf{e}_j^s}, \quad k_{ij}^s = \mathbf{r}_{ij} \cdot \mathbf{t}_{ij}^s \tag{4.26}$$

Another preliminary discussion is necessary before proceeding to the analysis of the particle overlap. Figure 4.15 shows the possibility of the torus part of particle j overlapping with the disk surface part of particle i, where the angle between the two planes including each particle is denoted by θ_0. A line is drawn from the nearest point Q_j at the torus center circle of particle j so that it is perpendicular to the plane of particle i, and this line will intersect the plane at a point denoted by $Q_{i(j)}$, as shown in Figure 4.15. The length of the vertical line $k_{i(j)}^Q$ can be straightforwardly obtained from a simple geometric relationship as

$$k_{i(j)}^Q = (k_j^s - d/2)|\mathbf{e}_j^s \cdot \mathbf{e}_i| \tag{4.27}$$

The position vector $\mathbf{r}_{i(j)}^Q$ of point $Q_{i(j)}$ can therefore be written as

$$\mathbf{r}_{i(j)}^Q = \mathbf{r}_j + (d/2)\ \mathbf{e}_j^s - k_{i(j)}^Q \mathbf{e}_i \tag{4.28}$$

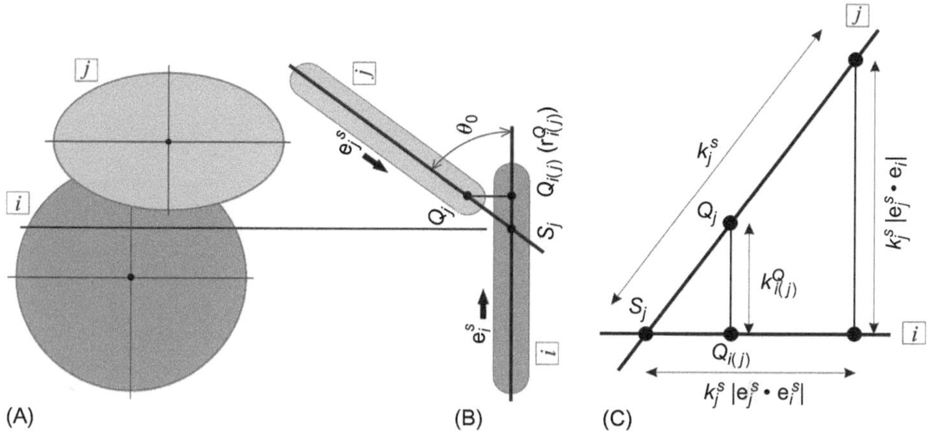

Figure 4.15 Analysis of the overlap of the flat part of particle i and the circumference of particle j: (A) plane view, (B) side view, and (C) vector expression.

Note that Eqs. (4.27) and (4.28) are valid for $k_j^S \geq d/2$, as shown in Figure 4.15. In the case of $k_j^S < d/2$, the following expressions are used instead of Eqs. (4.27) and (4.28):

$$k_{i(j)}^Q = (d/2 - k_j^s)|\mathbf{e}_j^s \cdot \mathbf{e}_i| \tag{4.29}$$

$$\mathbf{r}_{i(j)}^Q = \mathbf{r}_j + (d/2)\ \mathbf{e}_j^s + k_{i(j)}^Q \mathbf{e}_i \tag{4.30}$$

We have now completed the preparatory analysis and are able to begin discussion of the particle overlap conditions. For simplicity, the condition $k_i^S \leq k_j^S$ is assumed to be satisfied in the following. It is reasonable to discuss the particle overlap condition for the three different cases with regard to the directions of $\mathbf{e}_i$ and $\mathbf{e}_j$:

1. Case of $\mathbf{e}_i \neq \pm\mathbf{e}_j$ (general overlap).
2. Case of $\mathbf{e}_i = \pm\mathbf{e}_j$ and $\mathbf{e}_i \cdot \mathbf{r}_{ij} = 0$ (two particles being in the same plane).
3. Case of $\mathbf{e}_i = \pm\mathbf{e}_j$ and $\mathbf{e}_i \cdot \mathbf{r}_{ij} \neq 0$ (two particles being in the two parallel planes).

The procedure for assessing the particle overlap with regard to particles i and j is as follows:

1. For $\mathbf{e}_i = \pm\mathbf{e}_j$ and $\mathbf{e}_i \cdot \mathbf{r}_{ij} = 0$ (both particles being in one plane).
   1.1. For $|\mathbf{r}_i - \mathbf{r}_j| \geq d_1$, no overlap.
   1.2. For $|\mathbf{r}_i - \mathbf{r}_j| < d_1$, an overlap.
2. For $\mathbf{e}_i = \pm\mathbf{e}_j$ and $\mathbf{e}_i \cdot \mathbf{r}_{ij} \neq 0$ (particles i and j being in two parallel planes).
   2.1. For $|\mathbf{e}_i \cdot \mathbf{r}_{ij}| \geq \mathrm{b}_1$, no overlap.
   2.2. For $|\mathbf{e}_i \cdot \mathbf{r}_{ij}| < \mathrm{b}_1$, a possibility of overlap.

The line drawn between $\mathbf{r}_i$ and $\mathbf{r}_j$ is projected onto each plane. The projected lines will intersect the corresponding torus center circles at points P_i and P_j, respectively. Then the unit vector $\mathbf{e}_i^P$ denoting the direction from the particle center to point P_i (similarly $\mathbf{e}_j^P$) can be expressed as

$$\mathbf{e}_j^P = \frac{(\mathbf{e}_i \times \mathbf{r}_{ij}) \times \mathbf{e}_i}{|(\mathbf{e}_i \times \mathbf{r}_{ij}) \times \mathbf{e}_i|} = \frac{\mathbf{r}_{ij} - (\mathbf{e}_i \cdot \mathbf{r}_{ij})\mathbf{e}_i}{|\mathbf{r}_{ij} - (\mathbf{e}_i \cdot \mathbf{r}_{ij})\mathbf{e}_i|}, \quad \mathbf{e}_i^P = -\mathbf{e}_j^P \tag{4.31}$$

$$\mathbf{r}_{ij}^P = \mathbf{r}_{ij} - (\mathbf{e}_i \cdot \mathbf{r}_{ij})\mathbf{e}_i \tag{4.32}$$

With these vectors,

2.2.1. For $|\mathbf{r}_{ij}^P| < d$, an overlap.
2.2.2. For $|\mathbf{r}_{ij}^P| \geq d_1$, no overlap.
2.2.3. For $|\mathbf{r}_{ij}^P| \geq d$ and $|(\mathbf{r}_i + (d/2)\mathbf{e}_i^P) - (\mathbf{r}_j + (d/2)\mathbf{e}_j^P)| < b_1$, an overlap.
2.2.4. For $|\mathbf{r}_{ij}^P| \geq d$ and $|(\mathbf{r}_i + (d/2)\mathbf{e}_i^P) - (\mathbf{r}_j + (d/2)\mathbf{e}_j^P)| \geq b_1$, no overlap.

3. For $\mathbf{e}_i \neq \pm\mathbf{e}_j$ (general overlap situations)

3.1. For $k_j^s > d/2$,
3.1.1. For $k_{i(j)}{}^Q \geq b_1$, no overlap irrespective of values of $|\mathbf{r}_{i(j)}{}^Q - \mathbf{r}_i|$.
3.1.2. For $k_{i(j)}{}^Q < b_1$, a possibility of overlap.
a. For $|\mathbf{r}_{i(j)}{}^Q - \mathbf{r}_i| < d/2$, an overlap.
b. For $|\mathbf{r}_{i(j)}{}^Q - \mathbf{r}_i| \geq d/2$, a possibility of overlap.
b.1. For $r_{ij}{}^{(\min)} \geq b_1$, no overlap.
b.2. For $r_{ij}{}^{(\min)} < b_1$, an overlap.

3.2. For $k_i^s < d/2$ and $k_j^s \leq d/2$, depending on the value of $r_{ij}{}^{(\min)}$ (defined later)
3.2.1. For $|\mathbf{r}_{i(j)}{}^Q - \mathbf{r}_i| < d/2$, an overlap.
3.2.2. For $|\mathbf{r}_{i(j)}{}^Q - \mathbf{r}_i| \geq d/2$, a possibility of overlap.
a. For $r_{ij}{}^{(\min)} \geq b_1$, no overlap.
b. For $r_{ij}{}^{(\min)} < b_1$, an overlap.

The above-mentioned analysis has effectively generated an algorithm for assessing the particle overlap. Notice that the algorithm has been organized from the viewpoint of developing a simulation program, so it can be readily translated into a programming language.

Figure 4.16 shows a method of evaluating the minimum distance $r_{ij}{}^{(\min)}$, which has already been used in the analysis but not yet given an exact definition. The particle coordinate system XYZ is fixed at the center of the torus circle of particle i,

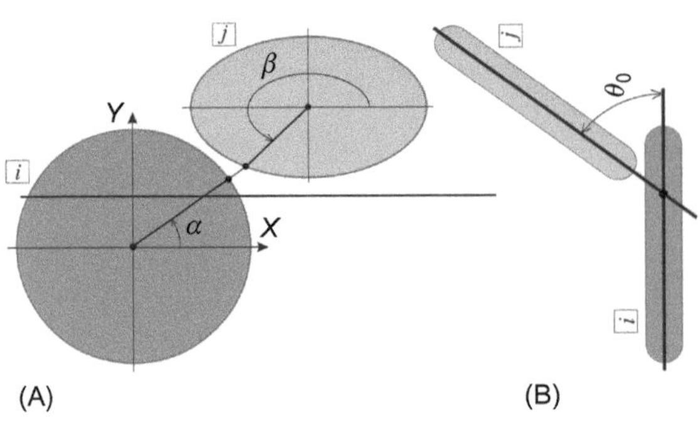

Figure 4.16 Evaluation of the minimum distance of particles i and j using the particle-fixed coordinate system XYZ: (A) plane view and (B) side view.

and the center of particle j is assumed to be expressed as (x_0,y_0,z_0) in this coordinate system, where the X-axis is taken parallel to the intersection line. The angle between the two planes that include particles i and j is denoted by the angle θ_0, as shown in Figure 4.16B. An arbitrary position vector $\mathbf{x}_1 = (x_1,y_1,z_1)$ on the torus center circle line of particle i is taken in the counterclockwise direction by the angle α. Similarly, an arbitrary position vector $\mathbf{x}_2 = (x_2,y_2,z_2)$ on the torus center circle line of particle j is taken in a similar way by the angle β, as shown in Figure 4.16A. Then $\mathbf{x}_1$ and $\mathbf{x}_2$ are expressed as

$$\mathbf{x}_1 = (r_0 \cos \alpha, r_0 \sin \alpha, 0) \tag{4.33}$$

$$\mathbf{x}_2 = (r_0 \cos \beta + x_0, r_0 \sin \beta \cos \theta_0 + y_0, r_0 \sin \beta \sin \theta_0 + z_0) \tag{4.34}$$

in which $r_0 = d/2$. The square separation between $\mathbf{x}_1$ and $\mathbf{x}_2$ is a function of the angles α and β, expressed as

$$\begin{aligned} g(\alpha,\beta) &= (x_2 - x_1)^2 + (y_2 - y_1)^2 + (z_2 - z_1)^2 \\ &= (r_0 \cos \beta + x_0 - r_0 \cos \alpha)^2 + (r_0 \sin \beta \cos \theta_0 + y_0 - r_0 \sin \alpha)^2 \\ &\quad + (r_0 \sin \beta \sin \theta_0 + z_0)^2 \end{aligned} \tag{4.35}$$

Certain values of α and β give rise to a minimum value of $g(\alpha, \beta)$. It is clear that the positions $\mathbf{x}_1$ and $\mathbf{x}_2$ on the different torus center circles specified by the angles α and β that minimize the function g yield their minimum separation distance. The values of α and β to satisfy a minimum $g(\alpha, \beta)$ can be obtained by solving the equations of $\partial g/\partial\alpha = \partial g/\partial\beta = 0$. The equation $\partial g/\partial\alpha = 0$ yields the following relationship:

$$\tan \alpha = \frac{y_2}{x_2} \tag{4.36}$$

Furthermore, the expression $\partial g/\partial\beta = 0$ gives rise to the following relationship:

$$\tan \beta = \frac{y_0 - y_1}{x_0 - x_1} \cos \theta_0 + \frac{z_0}{x_0 - x_1} \sin \theta_0 \tag{4.37}$$

The solutions of α and β can be obtained by solving Eqs. (4.36) and (4.37). However, because of the difficulty of an analytical approach, we here employ Newton's iteration method [33] for numerically solving these equations. From the particle configuration in Figure 4.16, we reasonably expect that Newton's iteration method will effectively provide a converged solution after several iterations, because $g(\alpha, \beta)$ has a relatively simple form. We show the algorithm of Newton's iteration method in the following steps:

1. Suppose a starting value β_n, around an expected solution, for β.
2. Calculate (x_2,y_2,z_2).

3. Calculate $(x_1, y_1, z_1) = \left(r_0 x_2/\sqrt{x_2^2 + y_2^2}, r_0 y_2/\sqrt{x_2^2 + y_2^2}, 0\right)$ from Eqs. (4.36) and (4.33).
4. Evaluate $f(\beta_n)$ from Eq. (4.37).

$$f(\beta_n) = \tan \beta_n - \frac{y_0 - y_1}{x_0 - x_1} \cos \theta_0 - \frac{z_0}{x_0 - x_1} \sin \theta_0 \tag{4.38}$$

5. Evaluate the derivative of $f(\beta)$ with respect to β.

$$f'(\beta_n) = \frac{1}{\cos^2 \beta_n} - \frac{\cos \theta_0}{(x_0 - x_1)^2} \left\{ -\frac{\partial y_1}{\partial \beta}(x_0 - x_1) + \frac{\partial x_1}{\partial \beta}(y_0 - y_1) \right\} - z_0 \sin \theta_0 \frac{\partial x_1/\partial \beta}{(x_0 - x_1)^2} \tag{4.39}$$

in which

$$\left.\begin{aligned} \frac{\partial x_1}{\partial \beta} &= r_0 \cdot \frac{\frac{\partial x_2}{\partial \beta}\sqrt{x_2^2 + y_2^2} - \left(x_2 \frac{\partial x_2}{\partial \beta} + y_2 \frac{\partial y_2}{\partial \beta}\right) x_2/\sqrt{x_2^2 + y_2^2}}{x_2^2 + y_2^2} \\ \frac{\partial y_1}{\partial \beta} &= r_0 \cdot \frac{\frac{\partial y_2}{\partial \beta}\sqrt{x_2^2 + y_2^2} - \left(x_2 \frac{\partial x_2}{\partial \beta} + y_2 \frac{\partial y_2}{\partial \beta}\right) y_2/\sqrt{x_2^2 + y_2^2}}{x_2^2 + y_2^2} \end{aligned}\right\} \tag{4.40}$$

The right-hand sides are evaluated by setting $\beta = \beta_n$.

6. Evaluate the next approximation β_{n+1} from Newton's method:

$$\beta_{n+1} = \beta_n - \frac{f(\beta_n)(x_0 - x_1)^2}{f'(\beta_n)(x_0 - x_1)^2} \tag{4.41}$$

7. Go to step 8 in the case of sufficiently convergence such as $|\beta_{n+1} - \beta_n| < \varepsilon$ (ε is infinitesimal small), otherwise repeat from step 2 by regarding β_{n+1} as β_n.
8. Calculate α_{n+1} from Eq. (4.36) with the converged value of β_{n+1}, and evaluate $g(\alpha_{n+1}, \beta_{n+1})$ from Eq. (4.35), yielding the desired minimum distance $r_{ij}^{(\min)} = \sqrt{g(\alpha_{n+1}, \beta_{n+1})}$.

We here employ a value satisfying $x_2 = x_0/2$ as a starting value of β. With this value, β can be obtained from Eq. (4.34) as $\beta = \cos^{-1}(-x_0/2r_0)$: although there are two solutions of the equation of $\cos \beta = -x_0/2r_0$, such a solution as satisfying $z_2 < z_0$ is adopted for β. This solution β provides the values of y_2 and z_2 from Eq. (4.34).

4.2.4 Canonical Monte Carlo Algorithm

As already indicated, we consider a system composed of N magnetic particles in an applied magnetic field in thermodynamic equilibrium. The canonical MC method is therefore adopted for a given system temperature T, volume V, and number of particles N. The system potential energy U^* can be expressed as the summation of the magnetic particle–particle interaction energy u_{ij}^* and the magnetic particle–field interaction energy u_i^* as

$$U^* = \sum_{i=1}^{N} u_i^* + \sum_{i=1}^{N} \sum_{j=1(j>i)}^{N} u_{ij}^* \tag{4.42}$$

in which u_i^* and u_{ij}^* have already been shown in Eqs. (4.20) and (4.21).

The canonical MC algorithm has been explained in Chapter 1 for a nonspherical particle system. According to Eq. (1.52), an arbitrary particle is translated into an adjacent position using random numbers. If the energy U^* decreases, the movement is accepted, but if it increases, it is employed according to the probability shown in Eq. (1.49). The rotational movement is first attempted and then accepted or rejected in a similar procedure. Although the simultaneous attempt of the translational and rotational movements is possible, the above-mentioned separate attempts will become more effective in the case of a strongly interacting system.

4.2.5 Treatment of the Criterion of the Particle Overlap in Simulations

The criterion of the particle overlap has already been discussed in detail from a mathematical point of view. In this subsection, we address important points to be noted with regard to the actual treatment of particle overlap in the simulation.

1. Exchange of the particle names i and j:
   The particle subscriptions i and j are exchanged in such a way to satisfy $k_i^s \le k_j^s$. That is, in the case of $k_i^s > k_j^s$, the subscriptions i and j are replaced with j and i, respectively; therefore the criterion for particle overlap in Section 4.2.3 is directly applicable.
2. Reversal of the directions of the unit vectors $\mathbf{e}_i$ and $\mathbf{e}_j$:
   As shown in Figures (4.14) and (4.15), the unit vectors $\mathbf{e}_i$ and $\mathbf{e}_j$ are temporarily reversed in such a way that the angle θ_0 will satisfy $0 \le \theta_0 \le \pi/2$. In the case of $\mathbf{r}_{ji} \cdot \mathbf{e}_i \ge 0$, $\mathbf{e}_i$ is unchanged, otherwise $\mathbf{e}_i$ is temporarily reversed in direction as $\mathbf{e}_i \rightarrow -\mathbf{e}_i$. For this new $\mathbf{e}_i$, $\mathbf{e}_j$ is unchanged for $\mathbf{e}_i \cdot \mathbf{e}_j \ge 0$; otherwise $\mathbf{e}_j$ is temporarily reversed for the successive procedure. These treatments confirm that θ_0 becomes an acute angle, as shown in Figure 4.14. Note that the exchange of the subscriptions i and j may be necessary in the following procedures.
3. Reversal of the direction of the unit vector $\mathbf{t}_{ij}^s$:
   The unit vector $\mathbf{t}_{ij}^s$ is taken in the direction from point S_j to S_i. For $\mathbf{t}_{ij}^s$ evaluated from Eq. (4.23), if $\mathbf{t}_{ij}^s \cdot \mathbf{r}_{ij} \ge 0$, $\mathbf{t}_{ij}^s$ is unchanged, otherwise $\mathbf{t}_{ij}^s$ is temporarily reversed as $\mathbf{t}_{ij}^s \rightarrow -\mathbf{t}_{ij}^s$. This treatment ensures that $\mathbf{t}_{ij}^s$ is from point S_j toward point S_i even if particle j is on the left-hand side.
4. Reversal of the unit vectors $\mathbf{e}_i$ and $\mathbf{e}_j$:
   With the unit vectors $\mathbf{e}_i^s$ and $\mathbf{e}_j^s$ evaluated from Eq. (4.24), the solutions k_i^s and k_j^s can be obtained from Eq. (4.26). However, note that the definition of these unit vectors

pointing toward the intersection line from each particle center is not necessarily satisfied but depends on the interaction position. In other words, since the sign of k_i^s or k_j^s is not necessarily positive, $\mathbf{e}_i^s$ or $\mathbf{e}_j^s$ may be reversed in this situation. In the case of $k_i^s \geq 0$, $\mathbf{e}_i^s$ is unchanged, and in the case of $k_i^s < 0$, $\mathbf{e}_i^s$ is reversed as $\mathbf{e}_i^s \rightarrow -\mathbf{e}_i^s$, making k_i^s positive. Similar treatment is made for k_j^s and $\mathbf{e}_j^s$.

These procedures ensure that the previous algorithm for assessing the particle overlap and Newton's iteration method for finding the minimum separation are directly applicable without any changes.

4.2.6 Particle-Fixed Coordinate System and the Absolute Coordinate System

We here explain the particle-fixed coordinate system and the absolute coordinate system, which are necessary for a rotation of the particle and a rotation of the magnetic moment. As previously defined, we use the notation **e** for the particle direction and **n** for the magnetic moment direction, as shown in Figure 4.17. We call the coordinate system fixed at the particle the "particle-fixed coordinate system," simply expressed as the *XYZ*-coordinate system, centered at the particle center with the *Z*-axis along the particle axis direction. On the other hand, the coordinate system fixed, for example, on the computational cell is called the "absolute coordinate system," simply expressed as the *xyz*-coordinate system. Note that each particle has its own particle-fixed coordinate system centered at its particle center.

We briefly consider the rotation of the *xyz*-coordinate system about the *z*-axis by an angle ϕ, and then the rotation of the rotated *xyz*-coordinate system about the *y*-axis by an angle θ to generate the *XYZ*-coordinate system. For these rotations, the rotational matrix **R** can be written as

$$\begin{aligned}\mathbf{R} &= \begin{pmatrix} \cos\theta & 0 & -\sin\theta \\ 0 & 1 & 0 \\ \sin\theta & 0 & \cos\theta \end{pmatrix} \begin{pmatrix} \cos\phi & \sin\phi & 0 \\ -\sin\phi & \cos\phi & 0 \\ 0 & 0 & 1 \end{pmatrix} \\ &= \begin{pmatrix} \cos\theta\cos\phi & \cos\theta\sin\phi & -\sin\theta \\ -\sin\phi & \cos\phi & 0 \\ \sin\theta\cos\phi & \sin\theta\sin\phi & \cos\theta \end{pmatrix}\end{aligned} \tag{4.43}$$

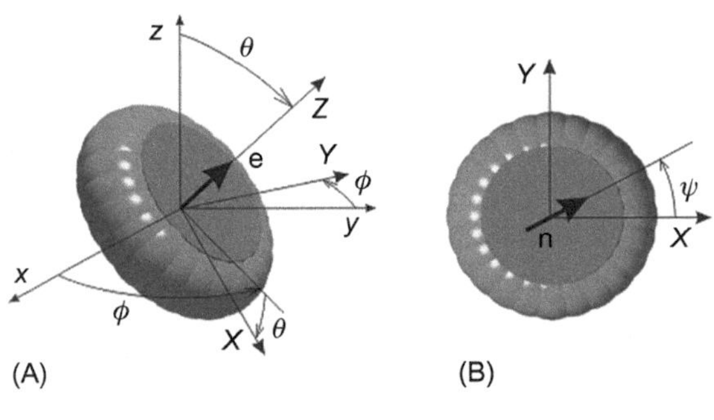

Figure 4.17 Particle-fixed coordinate system and absolute coordinate system.

This rotational matrix will allow us to express the relationship between an arbitrary position $\mathbf{a}^b = (a_x^b, a_y^b, a_z^b)$ in the XYZ-coordinate system and $\mathbf{a} = (a_x, a_y, a_z)$ in the xyz-coordinate system as

$$\mathbf{a}^b = \mathbf{R}\cdot\mathbf{a} \tag{4.44}$$

The inverse matrix $\mathbf{R}^{-1}$ of $\mathbf{R}$ is equal to the transpose matrix $\mathbf{R}^t$ of $\mathbf{R}$, so that $\mathbf{a}$ can be obtained from $\mathbf{a}^b$ as

$$\mathbf{a} = \mathbf{R}^{-1}\cdot\mathbf{a}^b \tag{4.45}$$

Thus, the particle direction $\mathbf{e}$ and the magnetic moment direction $\mathbf{n}$ of an arbitrary particle can be expressed as

$$\mathbf{e} = \mathbf{R}^{-1}\cdot\mathbf{e}^b, \quad \mathbf{n} = \mathbf{R}^{-1}\cdot\mathbf{n}^b \tag{4.46}$$

Since the XYZ-coordinate system is adopted so that the Z-axis is pointing in the particle direction, the unit vector $\mathbf{e}^b$ satisfies $\mathbf{e}^b = (0, 0, 1)$. This gives rise to the particle direction $\mathbf{e}$ in the xyz-coordinate system expressed as $\mathbf{e} = (e_x, e_y, e_z) = (\sin\theta\cos\phi,\ \sin\theta\sin\phi,\ \cos\theta)$. If necessary, known values of (e_x, e_y, e_z) yield the sine and cosine functions of θ and ϕ as $\cos\theta = e_z$, $\sin\theta = \sqrt{1-e_z^2}$, $\cos\phi = e_x/\sin\theta$, and $\sin\phi = e_y/\sin\theta$, in which it is noted that θ is defined in the range of $0 \le \theta \le \pi/2$. Several special features arising from this definition of range will be explained later.

We briefly explain the method for expressing the magnetic moment direction $\mathbf{n}$. As shown in Figure 4.17B, the direction of the magnetic moment can be specified by an angle ψ in the counterclockwise direction from the X-axis in the XYZ-coordinate system. That is, the magnetic moment direction $\mathbf{n}^b$ is expressed as $\mathbf{n}^b = (\cos\psi,\ \sin\psi,\ 0)$, so that the vector $\mathbf{n}$ in the xyz-coordinate system can be obtained from Eq. (4.46) as $\mathbf{n} = \mathbf{R}^{-1}\cdot\mathbf{n}^b$.

4.2.7 Attempt of Small Angular Changes in the Particle Axis and the Magnetic Moment

In MC simulations, an attempt is made to move each particle in translation and rotation with small displacements using uniform random numbers. Since the attempt of the translational movement is similar to that for a spherical particle system, we here show the method of rotating the particle direction and the magnetic moment direction.

We first consider the rotation of the particle direction. As shown previously, the particle direction (θ, ϕ) of an arbitrary particle is assumed to be made as $(\theta + \Delta\theta, \phi + \Delta\phi)$, with the small change $(\Delta\theta, \Delta\phi)$. Special treatment will be necessary if $(\theta + \Delta\theta)$ or $(\phi + \Delta\phi)$ is then larger than $\pi/2$ for θ or 2π for ϕ and also if smaller than zero for θ or ϕ, because the angles θ and ϕ are defined within the ranges of $0 \le \theta \le \pi/2$ and $0 \le \phi < 2\pi$.

1. For the case of $\theta + \Delta\theta < 0$:
   We make a modification such that $\theta' = -(\theta + \Delta\theta)$, $\phi' = \phi + \Delta\phi + \pi$, and $\psi' = \psi + \pi$, and use these values (θ', ϕ', ψ') for the rotational movement. Note that $(\phi' - 2\pi)$ needs to be adopted as ϕ' if $\phi' \geq 2\pi$ since ϕ' is defined in the range of $0 \leq \phi' < 2\pi$. Similar treatment is required for ψ'.
2. For the case of $\theta + \Delta\theta \geq \pi/2$:
   We make a modification such that $\theta' = \pi - (\theta + \Delta\theta)$, $\phi' = \phi + \Delta\phi + \pi$, and $\psi' = 2\pi - \psi$. If ϕ' or ψ' is outside the range of $0 \leq \phi', \psi' < 2\pi$, the above-mentioned treatment is applicable.
3. For the case of $0 \leq \theta + \Delta\theta < \pi/2$:
   In this case, a special modification is unnecessary and (θ', ϕ', ψ') are merely expressed as $\theta' = \theta + \Delta\theta$, $\phi' = \phi + \Delta\phi$, and $\psi' = \psi$, except that ϕ' is modified as in the previous case if ϕ' is outside the defined range.

For the above-modified θ', ϕ', and ψ', the rotational displacement is attempted and determined by the MC assessing procedure.

We next consider the rotation of the magnetic moment. The angle ψ specifying the direction is slightly displaced as $(\psi + \Delta\psi)$. Since ψ is defined in the range of $0 \leq \psi < 2\pi$, ψ' is modified such that $\psi' = \psi + \Delta\psi - 2\pi$ for $\psi + \Delta\psi \geq 2\pi$, $\psi' = \psi + \Delta\psi + 2\pi$ for $\psi + \Delta\psi < 0$ and $\psi' = \psi + \Delta\psi$ for the other cases. With this modified ψ', the magnetic moment direction $\mathbf{n}'^{b}$ is specified as $\mathbf{n}'^{b} = (\cos\psi', \sin\psi', 0)$ in the XYZ-coordinate system, and therefore the vector $\mathbf{n}'$ in the xyz-coordinate system can be obtained as $\mathbf{n}' = \mathbf{R}^{-1} \cdot \mathbf{n}'^{b}$ from Eq. (4.46). The magnetic interaction energies are calculated for the new magnetic moment direction, and the MC procedure determines whether this new state is accepted or rejected.

4.2.8 Parameters for Simulations

4.2.8.1 Initial Conditions

The assignment of an initial configuration of the circular disk-like particles explained in Section 2.1.2 is applied to the present system with different number of particles. As shown in Figure 2.5, four disk-like particles are located linearly along the x-axis, with the particles aligning in the y-direction. This stack of 4 particles is repeatedly placed in the y-direction, giving rise to 48 disk-like particles in the xy-plane at this stage. These particles are expanded in the z-direction to total 6 layers, giving a final sum of 288 particles placed in the simulation region. In this contact configuration, the size of the simulation region (L_x, L_y, L_z) is $(4r_p b_1, 12b_1, 6r_p b_1)$. The expansion of the distance between each pair of particles by α times each side length yields the desired volumetric fraction of particles ϕ_V. The relationship between α and ϕ_V can be expressed as

$$\alpha = \left[\frac{\pi}{24r_p^2\phi_V}\left\{6(r_p-1)^2 + 3\pi(r_p-1) + 4\right\}\right]^{\frac{1}{3}} \tag{4.47}$$

This configuration is perfectly regular, and therefore each particle is given a small translational displacement in order for the initial configuration to be able to transform to an equilibrium state straightforwardly. Then the direction of each particle is assigned as $\mathbf{e}_i = (0, 1, 0)$ $(i = 1, 2, \ldots, N)$.

Finally the direction of the magnetic moment is set to be arbitrary using random numbers. Thus, setting the ψ in the XY-plane gives rise to $\mathbf{n}^b = (n_x^b, n_y^b, 0) = (\cos\psi, \sin\psi, 0)$ and Eq. (4.46) finally yields the direction $\mathbf{n}$ in the xyz-coordinate system.

4.2.8.2 Assignment of Parameters

The simulations were conducted for the particle number $N = 288$ and the volumetric fraction ranging $\phi_V = 0.05 \sim 0.3$. An external magnetic field is applied in the z-direction as $\mathbf{h} = (0, 0, 1)$. We here employ the cutoff distance $r^*_{coff} = 5d^*_1$ for calculating magnetic interaction energies; an academic study may require a longer cutoff distance because magnetic energies are of long-range order. The nondimensional parameters ξ and λ representing the strengths of magnetic particle–field and particle–particle interactions are taken as $\xi = 0$, 1, 10, and 30 and $\lambda = 0$, 1, 10, 30, and 60. Note that the situation where $\xi \gg 1$ or $\lambda \gg 1$ means that the magnetic field or the magnetic particle–particle interaction is more dominant than the Brownian motion, respectively. The total number of MC steps $N_{mcsmplemx}$ is usually taken as $N_{mcsmplemx} = 100{,}000–1{,}000{,}000$, but the present exercise is only for the purpose of demonstration and therefore we employ a smaller value $N_{mcsmplemx} = 100{,}000$.

4.2.9 Results of Simulations

Figures 4.18–4.21 show the snapshots of the aggregate structures, which were obtained using the sample simulation program presented in the next subsection. Figure 4.18 is for no magnetic interactions between particles, that is, $\lambda = 0$, and

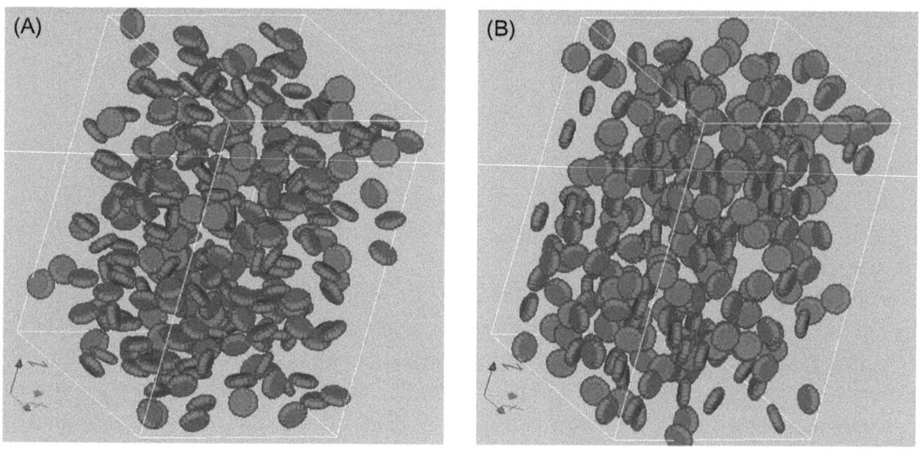

Figure 4.18 Aggregate structures for $\lambda = 0$: (A) $\xi = 0$ and (B) $\xi = 30$.

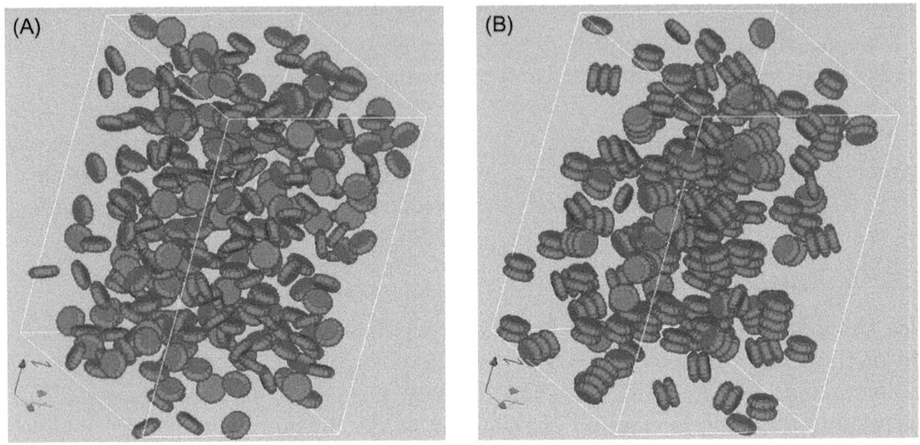

Figure 4.19 Aggregate structures for $\xi = 0$: (A) $\lambda = 10$ and (B) $\lambda = 30$.

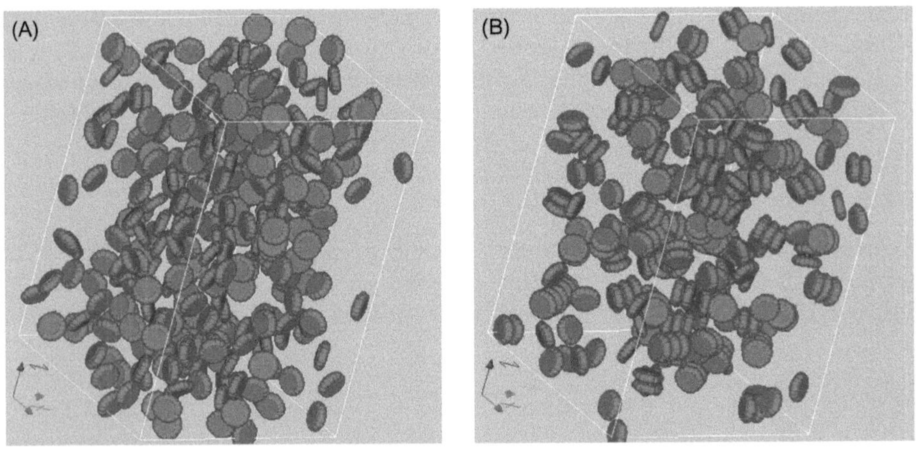

Figure 4.20 Aggregate structures for $\xi = 10$: (A) $\lambda = 10$ and (B) $\lambda = 30$.

Figures 4.19–4.21 are for the magnetic field strength $\xi = 0$ (i.e., no field), 10, and 30, respectively.

For the case of $\lambda = 0$ in Figure 4.18, no aggregates are formed because magnetic particle–particle interactions are absent. In addition, since an external magnetic field is also absent in Figure 4.18A, the particles do not show a specifically favored direction in their orientational characteristics. On the other hand, the application of a strong magnetic field, as shown in Figure 4.17B, makes the magnetic moment of each particle incline almost in the field direction (i.e., z-direction), resulting in the particle direction significantly fixed in the xy-plane.

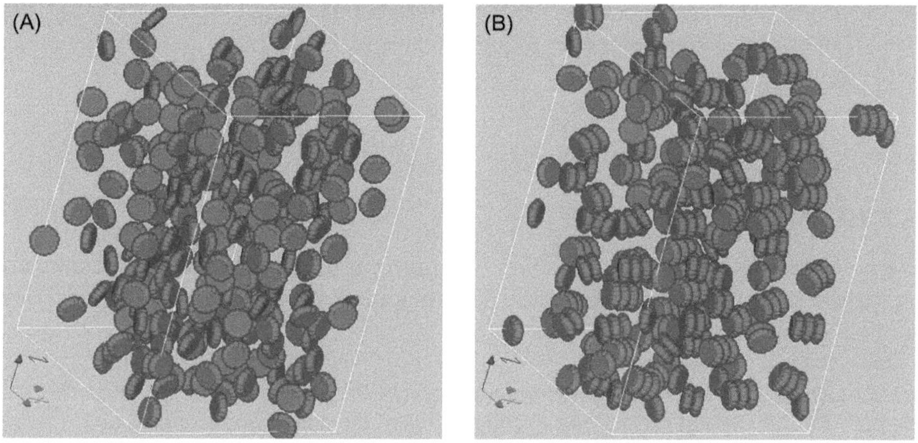

Figure 4.21 Aggregate structures for $\xi = 30$: (A) $\lambda = 30$ and (B) $\lambda = 60$.

Figure 4.19 is for no applied magnetic field $\xi = 0$, so that the particles have no tendency to incline in a specifically favored direction. In the case of $\lambda = 10$ shown in Figure 4.19A, short aggregates are found but are not significant. On the other hand, for the case of $\lambda = 30$ shown in Figure 4.19B, the disk-like particles aggregate to form column-like clusters in the particle direction (i.e., in the direction normal to the disk surface); each cluster inclines in its favored direction. This is because the magnetic particle–particle interaction is much more dominant than the Brownian motion. A careful observation of the column-like clusters indicates that the disk-like particles in the column-like cluster have their magnetic moments alternating in direction with the neighboring particles. This is because this type of internal structure gives rise to a minimum interaction energy for the magnetic particle–particle interaction. In the case of an external magnetic field of $\xi = 10$, shown in Figure 4.20B, the characteristic of the internal structure is the same as in Figure 4.19B because the magnetic interaction of $\lambda = 30$ is much more dominant than the applied magnetic field strength $\xi = 10$; that is, the magnetic particle–particle interaction tends to determine the internal structures of column-like clusters.

In contrast, for the strong applied magnetic field $\xi = 30$ shown in Figure 4.21A, column-like clusters obtained in Figure 4.20B are not formed, but the magnetic moment of each particle tends to incline toward the field direction and the particles move singly without forming clusters. The field strength $\xi = 30$ implies that an applied magnetic field significantly governs the aggregation process, so that the snapshot in Figure 4.21A is not essentially different from that in Figure 4.18B. A stronger interaction $\lambda = 60$ shown in Figure 4.21B recovers the formation of the column-like clusters that were seen in Figure 4.20B; in this case, the magnetic interactions significantly govern the aggregation process as compared with the external magnetic field. These discussions demonstrate that the internal structures of the aggregates are dependent on which factor is more dominant among the Brownian motion, the magnetic particle–particle interaction, and the magnetic field strength.

4.2.10 Simulation Program

We now present the sample simulation program, written in FORTRAN, for simulating the present physical phenomenon. The important variables used in the program are explained as follows:

`RX(I),RY(I),RZ(I)`	:	(x,y,z) components of the position vector $\mathbf{r}_i^*$ of particle i
`EX(I),EY(I),EZ(I)`	:	(x,y,z) components of the unit vector $\mathbf{e}_i$ of particle i denoting the particle direction
`NX(I),NY(I),NZ(I)`	:	(x,y,z) components of the unit vector $\mathbf{n}_i$ of particle i denoting the magnetic moment direction
`XL,YL,ZL`	:	Side lengths of the simulation box in the (x,y,z) directions
`N`	:	Number of particles
`D1`	:	Diameter of the circular disk-like particle d_1^*
`D`	:	Diameter of the cylinder part of the circular disk-like particle d^*
`RP`	:	Particle aspect ratio $d_1^*(=d_1/b_1)$
`VP`	:	Volume of the disk-like particle
`VDENS`	:	Volumetric fraction ϕ_V
`HX,HY,HZ`	:	(x,y,z) components of the unit vector $\mathbf{h}$ denoting the field direction
`RA`	:	Nondimensional parameter λ representing the strength of magnetic particle–particle interactions
`KU`	:	Nondimensional parameter ξ representing the strength of magnetic particle–field interactions
`RCOFF`	:	Cutoff distance for calculations of interaction energies
`DELR`	:	Maximum displacement in the translational movement
`DELT`	:	Maximum angle in the rotational movement
`RAN(J)`	:	Uniform random numbers ranging $0\sim1$ (`J` $=$ `1` $\sim$ `NRANMX`)
`NRAN`	:	Number of used random numbers
`E(I)`	:	Energy of particle i interacting with other particles
`MOMX(*),...,MOMZ(*)`	:	Mean value of the particle direction at each MC step
`MEANENE(*)`	:	Mean value of the system energy at each MC step

Brief comments have been added to the important features of the program in order to clarify the meaning for the reader. Note that the line numbers are merely for convenience and are unnecessary for the execution of the program.

The use of quasi-random numbers for saving the pseudo-random numbers `RAN(*)` has already been explained in Section 3.2.9.

```
C*********************************************************************
C*                                                                   *
C*       mcdisk3.f                                                   *
C*                                                                   *
C*          OPEN(9, FILE='@aaa1.dat', STATUS='UNKNOWN')              *
C*          OPEN(10,FILE='aaa11.dat', STATUS='UNKNOWN')              *
C*          OPEN(13,FILE='aaa41.mgf', STATUS='UNKNOWN')              *
C*          OPEN(21,FILE='aaa001.dat',STATUS='UNKNOWN')              *
C*          OPEN(22,FILE='aaa011.dat',STATUS='UNKNOWN')              *
C*          OPEN(23,FILE='aaa021.dat',STATUS='UNKNOWN')              *
C*          OPEN(24,FILE='aaa031.dat',STATUS='UNKNOWN')              *
C*          OPEN(25,FILE='aaa041.dat',STATUS='UNKNOWN')              *
C*          OPEN(26,FILE='aaa051.dat',STATUS='UNKNOWN')              *
C*          OPEN(27,FILE='aaa061.dat',STATUS='UNKNOWN')              *
C*          OPEN(28,FILE='aaa071.dat',STATUS='UNKNOWN')              *
C*          OPEN(29,FILE='aaa081.dat',STATUS='UNKNOWN')              *
C*          OPEN(30,FILE='aaa091.dat',STATUS='UNKNOWN')              *
C*                                                                   *
C*       --------      MONTE CARLO SIMULATIONS      --------         *
C*         THREE-DIMENSIONAL MONTE CARLO SIMULATION OF               *
C*         MAGNETIC COLLOIDAL DISPERSIONS COMPOSED OF                *
C*         MAGNETIC DISK-LIKE PARTICLES                              *
C*                                                                   *
C*       1. A PARTICLE IS MODELED AS A CIRCULAR DISK-LIKE PARTICLE.  *
C*       2. THE CLUSTER-MOVING METHOD IS NOT USED.                   *
C*       3. A STERIC LAYER IS NOT TAKEN INTO ACCOUNT.                *
C*                                                                   *
C*                                    VER.1  BY A.SATOH , '08 5/2    *
C*********************************************************************
C      N     : NUMBER OF PARTICLES (N=INIPX*INIPY*INIPZ)
C      D1    : DIAMETER OF OUTER CIRCLE OF A DISK-LIKE PARTICLE
C      D     : DIAMETER OF THE PART OF CYLINDER
C      B1    : THICKNESS OF PARTICLE (=1 FOR THIS CASE)
C      RP    : ASPECT RATIO (=D1/B1) (=D1 FOR THIS CASE)
C      VP    : VOLUME OF THE PARTICLE
C      NDENS : NUMBER DENSITY
C      VDENS : VOLUMETRIC FRACTION
C      IPTCLMDL : =1 FOR DIPOLE IN THE CENTER, =2 FOR TWO POINT CHARGES
C      RA    : NONDIMENSIONAL PARAMETER OF PARTICLE-PARTICLE INTERACT
C      RA0   : =RA/RP**3 FOR IPTCLMDL=1, =RA/RP FOR IPTCLMDL=2
C      KU    : NONDIMENSIONAL PARAMETER OF PARTICLE-FIELD INTERACTION
C      HX,HY,HZ : MAGNETIC FIELD DIRECTION (UNIT VECTOR)
C      RCOFF : CUTOFF RADIUS FOR CALCULATION OF INTERACTION ENERGIES
C      XL,YL,ZL : DIMENSIONS OF SIMULATION REGION
C                  (XL,YL,ZL)=(INIPX*RP, INIPY, INIPZ*RP) *ALPHA
C        (1) RP=3
C             INITREE=1 : (INIPX,INIPY,INIPZ)=( 3, 9,12), N= 324
C             INITREE=2 : (INIPX,INIPY,INIPZ)=( 4,12, 6), N= 288
C        (2) RP=4
C             INITREE=3 : (INIPX,INIPY,INIPZ)=( ?, ?, ?), N= ?
C             INITREE=4 : (INIPX,INIPY,INIPZ)=( ?, ?, ?), N= ?
C        (3) RP=5
C             INITREE=5 : (INIPX,INIPY,INIPZ)=( ?, ?, ?), N= ?
C             INITREE=6 : (INIPX,INIPY,INIPZ)=( ?, ?, ?), N= ?
C      RX(N),RY(N),RZ(N)      : PARTICLE POSITION
C      EX(N),EY(N),EZ(N)      : DIRECTION OF RODLIKE PARTICLE
C      NX(N),NY(N),NZ(N)      : DIRECTION OF MAGNETIC MOMENT
C      E(I)         : INTERACTION ENERGY OF PARTICLE I WITH THE OTHERS
C      MOMX(**),MOMY(**)      : MAG. MOMENT OF SYSTEM AT EACH TIME STEP
C      MOMZ(**)
C      MEANENE(**)          : MEAN ENERGY OF SYSTEM AT EACH MC STEP
C      ETHETA(N),EPHI(N)    : ANGLES DENOTING THE PARTICLE DIRECTION
C      NPSI(N)              : ANGLE DENOTING THE MAG.MOM. DIRECTION
C      RMAT(3,3,N)          : ROTATIONAL MATRIX
C      NXB(N), NYB(N)       : DIREC. OF MAG. MOM. IN THE BODY-FIXED AXIS
C                             SYSTEM
C
C      DELR   : MAXIMUM MOVEMENT DISTANCE
C      DELT   : MAXIMUM MOVEMENT IN ORIENTATION
C
C             0 < RX < XL ,  0 < RY < YL ,  0 < RZ < ZL
C--------------------------------------------------------------------
      IMPLICIT REAL*8 (A-H,O-Z), INTEGER (I-N)
C
      COMMON /BLOCK1/  RX   , RY    , RZ
```

```
      COMMON /BLOCK2/  NX    , NY     , NZ
      COMMON /BLOCK3/  N     , NDENS  , VDENS
      COMMON /BLOCK4/  D     , D1     , RP     ,  VP    , IPTCLMDL
      COMMON /BLOCK5/  XL  , YL  , ZL  , INIPX , INIPY , INIPZ , INITREE
      COMMON /BLOCK6/  RA    , RA0   , KU   , HX   , HY   , HZ
      COMMON /BLOCK7/  E     , ENEW  , EOLD
      COMMON /BLOCK8/  RCOFF, DELR  , DELT
      COMMON /BLOCK10/ MOMX  , MOMY  , MOMZ  , MEANENE
      COMMON /BLOCK11/ EX    , EY    , EZ
      COMMON /BLOCK12/ NXB   , NYB
      COMMON /BLOCK13/ ETHETA , EPHI , NPSI  , RMAT
      COMMON /BLOCK30/ NRAN    , RAN  , IX
C
      PARAMETER( NN=1360   , NNS=200000 )
      PARAMETER( NRANMX=1000000 , PI=3.141592653589793D0 )
C
      REAL*8    KU      , NDENS   , VDENS
      REAL*8    RX(NN) , RY(NN)  , RZ(NN)
      REAL*8    NX(NN) , NY(NN)  , NZ(NN) , E(NN)
      REAL*8    EX(NN) , EY(NN)  , EZ(NN)
      REAL*8    NXB(NN), NYB(NN)
      REAL*8    ETHETA(NN), EPHI(NN) , NPSI(NN)   , RMAT(3,3,NN)
      REAL      MOMX(NNS), MOMY(NNS) , MOMZ(NNS) , MEANENE(NNS)
C
      REAL      RAN(NRANMX)
      INTEGER   NRAN , IX , NRANCHK
C
      REAL*8    RXCAN , RYCAN , RZCAN
      REAL*8    NXCAN , NYCAN , NZCAN
      REAL*8    EXCAN , EYCAN , EZCAN
      REAL*8    RXI   , RYI   , RZI   , NXI   , NYI   , NZI
      REAL*8    EXI   , EYI   , EZI
      REAL*8    RXIJ  , RYIJ  , RZIJ  , RIJ   , RIJSQ , RCOFF2
      REAL*8    NXBI  , NYBI  ,  NXBC , NYBC  , NXC   , NYC   , NZC
      REAL*8    ETHETAI, EPHII, NPSII, ETHETAC, EPHIC, NPSIC
      REAL*8    RMATC(3,3)
      REAL*8    ECAN  , C1    , C2    , C3    , C4
      REAL*8    CX    , CY    , CZ
      INTEGER   MCSMPL , MCSMPLMX , MCSMPL1 , MCSMPL2 , NSMPL
      INTEGER   NGRAPH , NOPT     , DN      , DNSMPL
      INTEGER   ITHETA , IPHAI    , IT , IP
      INTEGER   NANIME , NANMCTR , NOPT1
      LOGICAL   OVRLAP
C
              OPEN(9,FILE='@baba1.dat'  , STATUS='UNKNOWN')
              OPEN(10,FILE='baba11.dat', STATUS='UNKNOWN')
              OPEN(13,FILE='baba41.mgf', STATUS='UNKNOWN')
              OPEN(21,FILE='baba001.dat',STATUS='UNKNOWN')
              OPEN(22,FILE='baba011.dat',STATUS='UNKNOWN')
              OPEN(23,FILE='baba021.dat',STATUS='UNKNOWN')
              OPEN(24,FILE='baba031.dat',STATUS='UNKNOWN')
              OPEN(25,FILE='baba041.dat',STATUS='UNKNOWN')
              OPEN(26,FILE='baba051.dat',STATUS='UNKNOWN')
              OPEN(27,FILE='baba061.dat',STATUS='UNKNOWN')
              OPEN(28,FILE='baba071.dat',STATUS='UNKNOWN')
              OPEN(29,FILE='baba081.dat',STATUS='UNKNOWN')
              OPEN(30,FILE='baba091.dat',STATUS='UNKNOWN')
                                                                NP=9
C
C                                                --- PARAMETER (1) ---
C                     -------------------------------------------------
C                     BE CAREFUL IN SETTING N, INIPX, ..., INITREE !!!
C                     -------------------------------------------------
      IPTCLMDL=  1
      VDENS   =  0.1D0
      KU      = 10.0D0
      RA      = 10.0D0
      INITREE =  2
      N       =  288
CCC   INITREE =  1
CCC   N       =  324
C                                                --- PARAMETER (2) ---
      HX      = 0.D0
      HY      = 0.D0
      HZ      = 1.D0
      RP      = 3.D0
```

• The given values are written out in @baba1, sampled magnetic moment directions are done in baba11, and data for MicroAVS are done in baba41. The particle positions and directions are written out in baba001 – baba091.

• The number of particles N=288, volumetric fraction ϕ_V= 0.1, λ=10, and ξ=10. The size of the simulation region is varied using INITREE.
• The aspect ratio r_p=3 and the field direction **h**=(0,0,1).

• The cutoff distance r^*_{coff}=5r_p, the volume of the particle VP, and the number density NDENS.

```
      D1       = RP
      D        = D1 - 1.D0
      RCOFF    = 5.D0*D1
      VP       = (PI/24.D0)*(6.D0*(RP-1.D0)**2+3.D0*PI*(RP-1.D0)+4.D0)
      NDENS    = VDENS/VP
      IF( IPTCLMDL .EQ. 1 )  RA0 = RA/RP**3
      IF( IPTCLMDL .EQ. 2 )  RA0 = RA/RP
C
      DELR     =  0.2D0
      DELT     =  (5.D0/180.D0 )*PI
C                                             --- PARAMETER (4) ---
CCC   MCSMPLMX= 100000
      MCSMPLMX= 10000
      NGRAPH   = MCSMPLMX/10
      NANIME   = MCSMPLMX/200
      DN       = 10
      DNSMPL   = 10
      NOPT     = 20
      RCOFF2   = RCOFF**2
C                                             --- PARAMETER (5) ---
      IX = 0
      CALL RANCAL( NRANMX, IX, RAN
      NRAN     = 1
      NRANCHK = NRANMX - 12*N
C
C     -----------------------------------------------------------------
C     -----------------    INITIAL CONFIGURATION    -----------------
C     -----------------------------------------------------------------
C
C                                          --- SET INITIAL CONFIG. ---
CCC   OPEN(19,FILE='aaba091.dat',STATUS='OLD')
CCC   READ(19,472)  N , XL , YL , ZL , D , D1 , RP
CCC   READ(19,473) (RX(I) ,I=1,N), (RY(I) ,I=1,N), (RZ(I) ,I=1,N)
CCC   READ(19,474) (NX(I) ,I=1,N), (NY(I) ,I=1,N), (NZ(I) ,I=1,N),
CCC  &             (EX(I) ,I=1,N), (EY(I) ,I=1,N), (EZ(I) ,I=1,N),
CCC  &             (NXB(I),I=1,N), (NYB(I),I=1,N)
CCC   READ(19,473) (ETHETA(I),I=1,N), (EPHI(I),I=1,N), (NPSI(I),I=1,N)
CCC   READ(19,474) (  ( (RMAT(II,JJ,I),II=1,3), JJ=1,3 ), I=1,N  )
CCC   CLOSE(19,STATUS='KEEP')
CCC   GOTO 7
C
      CALL INITIAL
C
    7 IF( XL .LE. YL ) THEN
        IF( RCOFF .GE. XL/2.D0 ) THEN
          RCOFF = XL/2.D0 - 0.00001D0
        END IF
      ELSE
        IF( RCOFF .GE. YL/2.D0 ) THEN
          RCOFF = YL/2.D0 - 0.00001D0
        END IF
      END IF
      RCOFF2 = RCOFF**2
      CRAD   = ( XL*YL*ZL/DBLE(N*N) ) / ( 4.D0*PI*DR )
C
C
C     ----------------------------------------------- PRINT OUT (1)---
      WRITE(NP,12) IPTCLMDL, N, VDENS, NDENS, RA, RA0, KU, RP,
     &             D, D1, XL, YL, ZL, RCOFF, DELR, DELT
      WRITE(NP,14) MCSMPLMX, NGRAPH, DN, DNSMPL
      WRITE(NP,15) HX, HY, HZ
C
C
      NANMCTR = 0
      NSMPL   = 0
C     -----------------------------------------------------------------
C     ---------------    START OF MONTE CARLO PROGRAM    -------------
C     -----------------------------------------------------------------
C
      DO 1000 MCSMPL = 1 , MCSMPLMX
C
        DO 400 I=1,N
C                                  +++++++++++++++++++ POSITION ++++++++
C                                                    --- OLD ENERGY ---
          RXI = RX(I)
```

• The maximum displacements in the MC method are δr^*_{max}=0.2 and $\delta\theta_{max}$=(5/180)π.

• The total number of MC steps is MCSMPLMX=10000 and sampling is carried out at every DNSMPL steps.
• The particle positions are written out at every NGRAPH steps. 200 sets of data are written out for making an animation.

• A sequence of uniform random numbers is prepared in advance. When necessary, random numbers are taken out from the variable RAN(*)

• These READ statements are for continuing the sequential simulation using the data saved previously.

• The initial positions and directions of particles are assigned.

• RCOFF has to be taken shorter than XL/2 and YL/2.

• The treatment concerning particle *i*.

```
             RYI = RY(I)
             RZI = RZ(I)
             NXI = NX(I)
             NYI = NY(I)
             NZI = NZ(I)
             EXI = EX(I)
             EYI = EY(I)
             EZI = EZ(I)
             ITREE = 0
```

• The interaction energies between particle *i* and its interacting particles.

```
             CALL ENECAL( I, RXI, RYI, RZI, EXI, EYI, EZI, NXI, NYI, NZI,
     &                            RCOFF2, EOLD, OVRLAP, ITREE, J )
C
C                                            ---------- (1) CANDIDATE
             NRAN  = NRAN + 1
             RXCAN = RX(I) + DELR*( 1.D0 - 2.D0*DBLE(RAN(NRAN)) )
             IF( RXCAN .GE. XL ) THEN
               RXCAN = RXCAN - XL
             ELSE IF( RXCAN .LT. 0.D0 ) THEN
               RXCAN = RXCAN + XL
             END IF
```

• Particle *i* is slightly moved according to Eq. (1.52).

```
             NRAN  = NRAN + 1
             RYCAN = RY(I) + DELR*( 1.D0 - 2.D0*DBLE(RAN(NRAN)) )
             IF( RYCAN .GE. YL ) THEN
               RYCAN = RYCAN - YL
             ELSE IF( RYCAN .LT. 0.D0 ) THEN
               RYCAN = RYCAN + YL
             END IF
```

• The treatment of the periodic BC.

```
             NRAN  = NRAN + 1
             RZCAN = RZ(I) + DELR*( 1.D0 - 2.D0*DBLE(RAN(NRAN)) )
             IF( RZCAN .GE. ZL ) THEN
               RZCAN = RZCAN - ZL
             ELSE IF( RZCAN .LT. 0.D0 ) THEN
               RZCAN = RZCAN + ZL
             END IF
```

• The interaction energies are calculated for this new state after the movement of particle *i*.

```
C                                            --- NEW ENERGY ---
             ITREE = 0
             CALL ENECAL( I , RXCAN, RYCAN, RZCAN, EXI, EYI, EZI,
     &                    NXI, NYI, NZI, RCOFF2, ECAN, OVRLAP, ITREE, J )
C
             IF( OVRLAP ) THEN
               GOTO 150
             END IF
```

• The adoption of the new state is determined according to the transition probability in Eq. (1.49).

```
C                                            -------- (2) ENERGY HANDAN
             C3 = ECAN - EOLD
             IF( C3 .GE. 0.D0 )THEN
               NRAN = NRAN + 1
               IF( DBLE(RAN(NRAN)) .GE. DEXP(-C3) )THEN
                 GOTO 150
               END IF
             END IF
C                                            ++++++++++++++++++++++
C                                            CANDIDATES ARE ACCEPTED
C                                            ++++++++++++++++++++++
             RX(I)  = RXCAN
             RY(I)  = RYCAN
             RZ(I)  = RZCAN
             EOLD   = ECAN
             E(I)   = ECAN
```

• The procedure after the acceptance of the new state.

```
C
C                          +++++++++++++++++++++ ROTATION ++++++++
  150        RXI = RX(I)
             RYI = RY(I)
             RZI = RZ(I)
             EXI = EX(I)
             EYI = EY(I)
             EZI = EZ(I)
             NXI = NX(I)
             NYI = NY(I)
             NZI = NZ(I)
             NXBI= NXB(I)
             NYBI= NYB(I)
             ETHETAI = ETHETA(I)
             EPHII   = EPHI(I)
             NPSII   = NPSI(I)
C
```

•The procedure for the rotation.

• The particle direction is described by the zenithal and azimuthal angles ETHETAI and EPHII. The magnetic moment direction is described by the angle NPSII taken counter clockwise from the *X*-axis about the *Z*-axis.

```
      NPSIC = NPSII
      NXBC  = NXBI
      NYBC  = NYBI
C                                            ---------- (3) CANDIDATE
      NRAN    = NRAN + 1
      C1      = DELT*DBLE(RAN(NRAN))
      NRAN    = NRAN + 1
      C1      = DSIGN( C1 , DBLE(RAN(NRAN)-0.5) )
      ETHETAC = ETHETAI + C1
      NRAN    = NRAN + 1
      C1      = DELT*DBLE(RAN(NRAN))
      NRAN    = NRAN + 1
      C1      = DSIGN( C1 , DBLE(RAN(NRAN)-0.5) )
      EPHIC = EPHII + C1
```

• The zenithal and azimuthal angles are slightly changed using random numbers to change the particle direction; the new angles are saved in ETHETAC and EPHIC.

```
C
      IF( ETHETAC .LT. 0.D0 ) THEN
        ETHETAC = DABS( ETHETAC )
        EPHIC   = EPHIC + PI
        IF( EPHIC .GE. 2.D0*PI ) EPHIC = EPHIC - 2.D0*PI
        NPSIC   = NPSII + PI
        IF( NPSIC .GE. 2.D0*PI ) NPSIC = NPSIC - 2.D0*PI
        NXBC = -NXBI
        NYBC = -NYBI
      ELSE IF ( ETHETAC .GT. PI/2.D0 ) THEN
        ETHETAC = PI - ETHETAC
        EPHIC   = EPHIC + PI
        IF( EPHIC .GE. 2.D0*PI ) EPHIC = EPHIC - 2.D0*PI
        NPSIC   = 2.D0*PI - NPSIC
      ELSE
        IF( EPHIC .GE. 2.D0*PI ) EPHIC = EPHIC - 2.D0*PI
        IF( EPHIC .LT. 0.D0 )    EPHIC = EPHIC + 2.D0*PI
      END IF
```

• The treatment shown in Section 4.2.7.

```
C                                                --- RMATC(3,3) ---
      C11 = DCOS( ETHETAC )
      C12 = DSIN( ETHETAC )
      C21 = DCOS( EPHIC )
      C22 = DSIN( EPHIC )
      RMATC(1,1) =  C11*C21
      RMATC(2,1) =  C11*C22
      RMATC(3,1) = -C12
      RMATC(1,2) = -C22
      RMATC(2,2) =  C21
      RMATC(3,2) =  0.D0
      RMATC(1,3) =  C12*C21
      RMATC(2,3) =  C12*C22
      RMATC(3,3) =  C11
```

• The rotational matrix $\mathbf{R}^{-1}$ (RMATC) is evaluated for the particle direction.

```
C
      EXC = RMATC(1,3)
      EYC = RMATC(2,3)
      EZC = RMATC(3,3)
      NXC = NXBC*RMATC(1,1) + NYBC*RMATC(1,2)
      NYC = NXBC*RMATC(2,1) + NYBC*RMATC(2,2)
      NZC = NXBC*RMATC(3,1) + NYBC*RMATC(3,2)
```

• The particle direction **e** and the magnetic moment direction **n** are calculated from Eq. (4.46).

```
C                                                --- NEW ENERGY ---
      ITREE = 0
      CALL ENECAL( I , RXI, RYI, RZI, EXC, EYC, EZC,
     &             NXC, NYC, NZC, RCOFF2, ECAN, OVRLAP, ITREE, J )
C
      IF( OVRLAP ) THEN
        GOTO 250
      END IF
```

• The interaction energies are calculated for the new direction of particle *i*.

```
C                                    -------- (4) ENERGY HANDAN --------
C
      C3 = ECAN - EOLD
      IF( C3 .GE. 0.D0 )THEN
        NRAN = NRAN + 1
        IF( DBLE(RAN(NRAN)) .GE. DEXP(-C3) )THEN
          GOTO 250
        END IF
      END IF
```

• The adoption of the new state is determined according to the transition probability in Eq. (1.49).

```
C                                          ++++++++++++++++++++++
C                                          CANDIDATES ARE ACCEPTED
C                                          ++++++++++++++++++++++
      EX(I)    = EXC
      EY(I)    = EYC
```

• The procedure after the acceptance of the new state.

```
          EZ(I)     = EZC
          NX(I)     = NXC
          NY(I)     = NYC
          NZ(I)     = NZC
          NXB(I)    = NXBC
          NYB(I)    = NYBC
          ETHETA(I) = ETHETAC
          EPHI(I)   = EPHIC
          NPSI(I)   = NPSIC
          DO 110 II=1,3
          DO 110 JJ=1,3
            RMAT(II,JJ,I) = RMATC(II,JJ)
  110     CONTINUE
          EOLD  = ECAN
          E(I)  = ECAN
C
C                          ++++++++++++++++++++ MOMENT ++++++++
```

• The particle direction, magnetic moment direction, and zenithal and azimuthal angles are renewed.
• The rotational matrix is renewed.
• The interaction energy of particle *i* is saved in E(I).

```
  250     RXI = RX(I)
          RYI = RY(I)
          RZI = RZ(I)
          NXI = NX(I)
          NYI = NY(I)
          NZI = NZ(I)
          EXI = EX(I)
          EYI = EY(I)
          EZI = EZ(I)
          NXBI= NXB(I)
          NYBI= NYB(I)
          ETHETAI = ETHETA(I)
          EPHII   = EPHI(I)
          NPSII   = NPSI(I)
C                                   ---------- (5) CANDIDATE
```

• The attempt for changing the magnetic moment direction.

• The particle direction is described by the zenithal and azimuthal angles ETHETAI and EPHII. The magnetic moment direction is described by the angle NPSII.

```
          NRAN  = NRAN + 1
          C1    = DELT*DBLE(RAN(NRAN))
          NRAN  = NRAN + 1
          C1    = DSIGN( C1 , DBLE(RAN(NRAN)-0.5) )
          NPSIC = NPSII + C1
C
          IF( NPSIC .GE. 2.D0*PI ) THEN
            NPSIC = NPSIC - 2.D0*PI
          ELSE IF ( NPSIC .LT. 0.D0 ) THEN
            NPSIC = NPSIC + 2.D0*PI
          END IF
C
          NXBC = DCOS( NPSIC )
          NYBC = DSIN( NPSIC )
          NXC  = RMAT(1,1,I)*NXBC + RMAT(1,2,I)*NYBC
          NYC  = RMAT(2,1,I)*NXBC + RMAT(2,2,I)*NYBC
          NZC  = RMAT(3,1,I)*NXBC + RMAT(3,2,I)*NYBC
C
C                                        --- NEW ENERGY ---
```

• The magnetic moment direction is slightly changed using random numbers; this new angle is saved in NPSIC.

• The new magnetic moment direction is evaluated from Eq. (4.46). The rotational matrix is unchanged, still valid.

• $\mathbf{n}^b$ = (NXBC,NYBC,0)and $\mathbf{n} = \mathbf{R}^{-1} \cdot \mathbf{n}^b$.

```
          ITREE = 1
          CALL ENECAL( I , RXI, RYI, RZI, EXI, EYI, EZI,
     &                 NXC, NYC, NZC, RCOFF2, ECAN, OVRLAP, ITREE, J )
C
CCC       IF( OVRLAP ) THEN
CCC         GOTO 400
CCC       END IF
C
C                          -------- (6) ENERGY HANDAN --------
```

• The interaction energies are calculated for this new state of particle *i*.

```
          C3 = ECAN - EOLD
          IF( C3 .GE. 0.D0 )THEN
            NRAN = NRAN + 1
            IF( DBLE(RAN(NRAN)) .GE. DEXP(-C3) )THEN
              GOTO 400
            END IF
          END IF
C                                  +++++++++++++++++++++++
C                                  CANDIDATES ARE ACCEPTED
C                                  +++++++++++++++++++++++
```

• The adoption of the new state is determined according to the transition probability in Eq. (1.49).

```
          NX(I)  = NXC
          NY(I)  = NYC
          NZ(I)  = NZC
          NXB(I) = NXBC
          NYB(I) = NYBC
```

• The procedure after the acceptance of the new state.

```
                NPSI(I)= NPSIC
                E(I)    = ECAN
C
C
  400      CONTINUE
C
C          -----------------------------------------------------------
C
C                                    ----- MOMENT AND ENERGY OF SYSTEM -----
           IF( MOD(MCSMPL,DNSMPL) .EQ. 0 ) THEN
             NSMPL    = NSMPL + 1
             C1 = 0.D0
             C2 = 0.D0
             C3 = 0.D0
             C4 = 0.D0
             DO 420 J=1,N
               C1 = C1 + NX(J)
               C2 = C2 + NY(J)
               C3 = C3 + NZ(J)
               C4 = C4 + E(J)
  420        CONTINUE
             MOMX(NSMPL)     = REAL(C1)/REAL(N)
             MOMY(NSMPL)     = REAL(C2)/REAL(N)
             MOMZ(NSMPL)     = REAL(C3)/REAL(N)
             MEANENE(NSMPL) = REAL( C4-KU*(C1*HX+C2*HY+C3*HZ) )/REAL(2*N)
           END IF
C
           IF( MOD(MCSMPL,NGRAPH) .EQ. 0 ) THEN
             NOPT = NOPT + 1
             WRITE(NOPT,472)  N , XL , YL , ZL , D , D1 , RP
             WRITE(NOPT,473) (RX(I),I=1,N),(RY(I),I=1,N),(RZ(I),I=1,N)
             WRITE(NOPT,474) (NX(I),I=1,N),(NY(I),I=1,N),(NZ(I),I=1,N),
     &                       (EX(I),I=1,N),(EY(I),I=1,N),(EZ(I),I=1,N),
     &                       (NXB(I),I=1,N),(NYB(I),I=1,N)
             WRITE(NOPT,473) (ETHETA(I),I=1,N), (EPHI(I),I=1,N),
     &                       (NPSI(I),I=1,N)
             WRITE(NOPT,474) (  ( (RMAT(II,JJ,I),II=1,3), JJ=1,3 ),
     &                                                   I=1,N  )
                                                 CLOSE(NOPT,STATUS='KEEP')
           END IF
C
C                                      --- DATA OUT
           IF( MOD(MCSMPL,NANIME) .EQ. 0 ) THEN
             NANMCTR = NANMCTR + 1
             NOPT1    = 13
             CALL ANIMEDAT( NOPT1, NANMCTR, MCSMPLMX, NANIME, N )
           END IF
C
C                                  --- CHECK OF THE SUM OF RANDOM NUMBERS ---
C
           IF( NRAN .GE. NRANCHK )THEN
             CALL RANCAL( NRANMX, IX, RAN )
             NRAN = 1
           END IF
C                                                --- NORMALIZATION ---
           IF( MOD(MCSMPL,DN) .EQ. 0 ) THEN
             DO 490 I=1,N
               C1 = DSQRT( NX(I)**2 + NY(I)**2 + NZ(I)**2 )
               NX(I)   = NX(I)/C1
               NY(I)   = NY(I)/C1
               NZ(I)   = NZ(I)/C1
               C1 = DSQRT( EX(I)**2 + EY(I)**2 + EZ(I)**2 )
               EX(I)   = EX(I)/C1
               EY(I)   = EY(I)/C1
               EZ(I)   = EZ(I)/C1
               C1 = DSQRT( NXB(I)**2 + NYB(I)**2 )
               NXB(I) = NXB(I)/C1
               NYB(I) = NYB(I)/C1
  490        CONTINUE
           END IF
C
C
 1000 CONTINUE
C
```

• To check the system convergence afterward, the average of the particle direction vector is calculated.

• The data of the particle positions and directions are written out at every NGRAPH MC steps for the postprocessing analysis.

• The data of the particle positions and directions are written out at every NANIME MC steps for making an animation.

• The number of the used random numbers is checked. If over NRANCHK, a uniform random number sequence is renewed.

• Each unit vector is modified at every DN steps so as to yield unit length.

```
C      ------------------------------------------------------------------
C      -------------------  END OF MONTE CARLO PROGRAM  ----------------
C      ------------------------------------------------------------------
C
C                                                 --- PRINT OUT (2) ---
       WRITE(NP,1002)
       MCSMPL1 = 1
CCC    MCSMPL2 = MCSMPLMX
       MCSMPL2 = NSMPL
       CALL PRNTDATA( MCSMPL1 , MCSMPL2 , NP )
       WRITE(NP,1004)  MCSMPL1 , MCSMPL2
```

• To check the system convergence afterward, the data of the particle directions and interaction energies are written out.

```
C
C                               --- DATA OUTPUT FOR GRAPHICS (3) ---
       WRITE(10,1012) IPTCLMDL, N, VDENS, NDENS, RA, RA0, KU
       WRITE(10,1014) RP, D, D1, XL, YL, ZL, RCOFF
       WRITE(10,1016) DELR, DELT
       WRITE(10,1017) HX, HY, HZ
       WRITE(10,1018) MCSMPLMX, NGRAPH, DN, DNSMPL
       WRITE(10,1022) MCSMPL1, MCSMPL2
       WRITE(10,1024)  ( MEANENE(I),I=MCSMPL1, MCSMPL2)
     &                ,( MOMX(I),    I=MCSMPL1, MCSMPL2)
     &                ,( MOMY(I),    I=MCSMPL1, MCSMPL2)
C
                                      CLOSE(9, STATUS='KEEP')
                                      CLOSE(10,STATUS='KEEP')
                                      CLOSE(13,STATUS='KEEP')
C      ------------------------- FORMAT ----------------------------
   12 FORMAT(/1H ,'-----------------------------------------------'
     &       /1H ,'-              MONTE CARLO METHOD              -'
     &       /1H ,'-----------------------------------------------'
     &      //1H ,'IPTCLMDL=',I4
     &       /1H ,'N=',I4, 2X, 'VDENS=',F4.2, 2X ,'NDENS=',F7.4
     &       /1H ,'RA=',F6.2, 2X, 'RA0=',F9.2, 2X ,'KU=',F6.2, 2X ,
     &            'RP=', F7.4
     &       /1H ,'D=',F5.2, 2X ,'D1=',F5.2, 2X,
     &            'XL=',F6.2, 2X,'YL=',F6.2, 2X, 'ZL=',F6.2, 2X,
     &            'RCOFF=',F6.2
     &       /1H ,'DELR=',F7.4, 2X ,'DELT=',F7.4)
   14 FORMAT( 1H ,'MCSMPLMX=',I8, 2X,'NGRAPH=',I8, 2X,'DN=',I4, 2X,
     &            'DNSMPL=',I4/)
   15 FORMAT( 1H ,'(HX,HY,HZ)=', 3F5.1 )
  472 FORMAT( I5 , 3F9.4 , 3F8.4 )
  473 FORMAT( (5F16.10) )
  474 FORMAT( (11F7.3) )
 1002 FORMAT(/1H ,'++++++++++++++++++++++++++++++'
     &       /1H ,'     WITHOUT CLUSTER MOVEMENT   '
     &       /1H ,'++++++++++++++++++++++++++++++'/)
 1004 FORMAT(///1H ,18X, 'START OF MC SAMPLING STEP=',I9
     &         /1H ,18X, 'END   OF MC SAMPLING STEP=',I9/)
 1012 FORMAT( I2 , I5 , 2F9.4 , 4F9.3 )
 1014 FORMAT( 3F7.2 , 3F9.3 , F9.3 )
 1016 FORMAT( 2F9.5 )
 1017 FORMAT( 3F7.2 )
 1018 FORMAT( 6I9 )
 1020 FORMAT( 2F7.3 , I4 , F7.3 , E12.4 )
 1022 FORMAT( 2I9 )
 1024 FORMAT( (7E11.4) )
 1367 FORMAT( 3I9, 2F9.4 )
 1368 FORMAT( I6 , F8.4 , 3F10.5 )
 1392 FORMAT( 2I9 )
 1394 FORMAT( (7E11.4) )
 1501 FORMAT( I8 )
 1502 FORMAT( (10F8.3) )
 1511 FORMAT( I8 )
 1513 FORMAT( (10I8) )
 1515 FORMAT( (10F8.3) )
 1521 FORMAT( I8 )
 1523 FORMAT( 2I8 )
 1525 FORMAT( (10F8.3) )
 1541 FORMAT( I8 )
 1543 FORMAT( (10I8) )
 1545 FORMAT( (10F8.3) )
                                                            STOP
                                                            END
```

```
C***********************************************************************
C**************************   SUBROUTINE   *************************
C***********************************************************************
C
C**** SUB PRNTDATA ****
      SUBROUTINE PRNTDATA( MCSST, MCSMX, NP )
C
      IMPLICIT REAL*8 (A-H,O-Z), INTEGER (I-N)
C
      COMMON /BLOCK10/ MOMX , MOMY , MOMZ , MEANENE
C
      PARAMETER( NN=1360   , NNS=200000 )
      PARAMETER( NRANMX=1000000 , PI=3.141592653589793D0 )
C
      INTEGER   MCSST     , MCSMX     , NP
      REAL      MOMX(NNS) , MOMY(NNS) , MOMZ(NNS) , MEANENE(NNS)
C
      REAL      AMOMX(10) , AMOMY(10) , AMOMZ(10) , AMEANENE(10) , C0
      INTEGER   IC , IMC(0:10) , JS , JE
C
C                                               ----- KEIKA INSATU -----
      IC = ( MCSMX-MCSST+1 )/50
      DO 20 I= MCSST-1+IC , MCSMX , IC
        WRITE(NP,10) I, MOMX(I), MOMY(I), MOMZ(I), MEANENE(I)
   20 CONTINUE
C                                     ----- MONTE CARLO STEP HEIKIN -----
      IC = ( MCSMX-MCSST+1 )/10
      DO 30 I=0,10
        IMC(I) = MCSST - 1 + IC*I
        IF( I .EQ. 10 ) IMC(I) =MCSMX
   30 CONTINUE
```

• The total MC steps are equally divided into 50 blocks, and the end value of each block is written out.

```
C
C
      DO 35 I=1,10
        AMOMX(I)    = 0.
        AMOMY(I)    = 0.
        AMOMZ(I)    = 0.
        AMEANENE(I) = 0.
   35 CONTINUE
```

• The total MC steps are equally divided into 10 blocks, and the subaverages are calculated for each block.

```
C
      DO 50 I=1,10
        JS = IMC(I-1) + 1
        JE = IMC(I)
        DO 40 J=JS,JE
          AMOMX(I)    = AMOMX(I)    + MOMX(J)
          AMOMY(I)    = AMOMY(I)    + MOMY(J)
          AMOMZ(I)    = AMOMZ(I)    + MOMZ(J)
          AMEANENE(I) = AMEANENE(I) + MEANENE(J)
   40   CONTINUE
   50 CONTINUE
C
      DO 70 I=1,10
        C0          = REAL( IMC(I)-IMC(I-1) )
        AMOMX(I)    = AMOMX(I)    /C0
        AMOMY(I)    = AMOMY(I)    /C0
        AMOMZ(I)    = AMOMZ(I)    /C0
        AMEANENE(I) = AMEANENE(I)/C0
   70 CONTINUE
C                                          ----- STEP HEIKIN INSATU -----
      WRITE(NP,75)
      DO 90 I=1,10
       WRITE(NP,80)I,IMC(I-1)+1,IMC(I),AMOMX(I),AMOMY(I),AMOMZ(I),
     &                                   AMEANENE(I)
   90 CONTINUE
C     -----------------------------------------------------------------
   10 FORMAT(1H ,'MCSMPL=',I8, 2X ,'MOMENT(X)=',F7.4, 2X ,
     &            'MOMENT(Y)=',F7.4, 2X ,'MOMENT(Z)=',F7.4
     &       /1H , 53X , 'MEAN ENERGY=',E12.5)
   75 FORMAT(//1H ,'-------------------------------------------'
     &         /1H ,'                MONTE CARLO HEIKIN          '
     &         /)
   80 FORMAT(1H ,'I=',I2, 2X ,'SMPLMN=',I8, 2X ,'SMPLMX=',I8
     &       /1H ,15X ,'MOMENT(X)=',F7.4, 2X ,
     &                 'MOMENT(Y)=',F7.4, 2X ,'MOMENT(Z)=',F7.4
     &       /1H ,53X, 'MEAN ENERGY=',E12.5/)
```

```
                                                                      RETURN
                                                                      END
C**** SUB ANIMEDAT ****
      SUBROUTINE ANIMEDAT( NOPT1, NANMCTR, MCSMPLMX, NANIME, N )
C
      IMPLICIT REAL*8 (A-H,O-Z), INTEGER (I-N)
C
      COMMON /BLOCK1/  RX    , RY    , RZ
      COMMON /BLOCK2/  NX    , NY    , NZ
      COMMON /BLOCK4/  D     , D1     , RP     ,  VP    , IPTCLMDL
      COMMON /BLOCK5/  XL , YL , ZL , INIPX , INIPY , INIPZ , INITREE
      COMMON /BLOCK11/ EX    , EY    , EZ
C
      PARAMETER( NN=1360 , PI=3.141592653589793D0 )
C
      REAL*8     RX(NN) , RY(NN) , RZ(NN)
      REAL*8     NX(NN) , NY(NN) , NZ(NN)
      REAL*8     EX(NN) , EY(NN) , EZ(NN)
      REAL*8     D02 , D102 , CX1 , CY1 , CZ1 , CX2 , CY2 , CZ2
      REAL*8     CNX(50) , CNY(50) , CNZ(50)
C
      D02      = D/2.D0
      D102     = D1/2.D0
C
      IF( NANMCTR .EQ. 1 ) THEN
        WRITE(NOPT1,181) (MCSMPLMX/NANIME)
      END IF
C
      IF( (NANMCTR.GE.1) .AND. (NANMCTR.LE.9) ) THEN
        WRITE(NOPT1,183) NANMCTR
      ELSE IF( (NANMCTR.GE.10) .AND. (NANMCTR.LE.99) ) THEN
        WRITE(NOPT1,184) NANMCTR
      ELSE IF( (NANMCTR.GE.100) .AND. (NANMCTR.LE.999) ) THEN
        WRITE(NOPT1,185) NANMCTR
      ELSE IF( (NANMCTR.GE.1000) .AND. (NANMCTR.LE.9999) ) THEN
        WRITE(NOPT1,186) NANMCTR
      END IF
C
C     -------------------------------------------------- CYLINDER (1) ---
      WRITE(NOPT1,211)  N
      DO 250 I=1,N
        CX1 = RX(I) - EX(I)*0.5D0
        CY1 = RY(I) - EY(I)*0.5D0
        CZ1 = RZ(I) - EZ(I)*0.5D0
        CX2 = RX(I) + EX(I)*0.5D0
        CY2 = RY(I) + EY(I)*0.5D0
        CZ2 = RZ(I) + EZ(I)*0.5D0
        WRITE(NOPT1,248) CX1,CY1,CZ1, CX2,CY2,CZ2, D02, 1.0, 0.0, 0.0
  250 CONTINUE
C
C     ---------------------------------------------------- SPHERE (1) ---
C                                         --- FOR MAKING OUTER SHAPE ---
      WRITE(NOPT1,311)  N*16
      DO 350 I=1,N
        CNX(1) = NX(I)
        CNY(1) = NY(I)
        CNZ(1) = NZ(I)
C
        C1X    = EY(I)*NZ(I) - EZ(I)*NY(I)
        C1Y    = EZ(I)*NX(I) - EX(I)*NZ(I)
        C1Z    = EX(I)*NY(I) - EY(I)*NX(I)
        C1     = DSQRT( C1X**2 + C1Y**2 + C1Z**2 )
        CNX(2) = C1X/C1
        CNY(2) = C1Y/C1
        CNZ(2) = C1Z/C1
        CNX(3) = - CNX(2)
        CNY(3) = - CNY(2)
        CNZ(3) = - CNZ(2)
C
        CNX(4) =  ( CNX(1) + CNX(2) )/1.4142D0
        CNY(4) =  ( CNY(1) + CNY(2) )/1.4142D0
        CNZ(4) =  ( CNZ(1) + CNZ(2) )/1.4142D0
        CNX(5) =  ( CNX(1) + CNX(3) )/1.4142D0
        CNY(5) =  ( CNY(1) + CNY(3) )/1.4142D0
        CNZ(5) =  ( CNZ(1) + CNZ(3) )/1.4142D0
```

• A subroutine for writing out the data, which can be directly used for making an animation based on MicroAVS.

• MicroAVS can make a visualization or animation by reading the data file baba41.mgf.

• Drawing of the cylindrical part.

• Drawing of the disk-like particle in Figure 4.12 by having the short cylinder surrounded by numerous spheres.

```
C
          CNX(6)  =  ( CNX(1)  + CNX(4)  )/1.8478D0
          CNY(6)  =  ( CNY(1)  + CNY(4)  )/1.8478D0
          CNZ(6)  =  ( CNZ(1)  + CNZ(4)  )/1.8478D0
          CNX(7)  =  ( CNX(2)  + CNX(4)  )/1.8478D0
          CNY(7)  =  ( CNY(2)  + CNY(4)  )/1.8478D0
          CNZ(7)  =  ( CNZ(2)  + CNZ(4)  )/1.8478D0
          CNX(8)  =  ( CNX(1)  + CNX(5)  )/1.8478D0
          CNY(8)  =  ( CNY(1)  + CNY(5)  )/1.8478D0
          CNZ(8)  =  ( CNZ(1)  + CNZ(5)  )/1.8478D0
          CNX(9)  =  ( CNX(3)  + CNX(5)  )/1.8478D0
          CNY(9)  =  ( CNY(3)  + CNY(5)  )/1.8478D0
          CNZ(9)  =  ( CNZ(3)  + CNZ(5)  )/1.8478D0
C
          CNX(10) = - CNX(1)
          CNY(10) = - CNY(1)
          CNZ(10) = - CNZ(1)
          CNX(11) = - CNX(4)
          CNY(11) = - CNY(4)
          CNZ(11) = - CNZ(4)
          CNX(12) = - CNX(5)
          CNY(12) = - CNY(5)
          CNZ(12) = - CNZ(5)
          CNX(13) = - CNX(6)
          CNY(13) = - CNY(6)
          CNZ(13) = - CNZ(6)
          CNX(14) = - CNX(7)
          CNY(14) = - CNY(7)
          CNZ(14) = - CNZ(7)
          CNX(15) = - CNX(8)
          CNY(15) = - CNY(8)
          CNZ(15) = - CNZ(8)
          CNX(16) = - CNX(9)
          CNY(16) = - CNY(9)
          CNZ(16) = - CNZ(9)
C
          DO 340 J=1,16
            CX1 = RX(I)  + CNX(J)*D02
            CY1 = RY(I)  + CNY(J)*D02
            CZ1 = RZ(I)  + CNZ(J)*D02
            WRITE(NOPT1,348) CX1, CY1, CZ1, 0.499 , 1.0, 0.2, 0.2
  340     CONTINUE
C
  350 CONTINUE
C
C     ----------------------------------------------- SPHERE (2) ---
C                                              --- FOR MAG MOMENT ---
      WRITE(NOPT1,311)  N
      DO 450 I=1,N
        CX1  = RX(I)  + NX(I)*D102
        CY1  = RY(I)  + NY(I)*D102
        CZ1  = RZ(I)  + NZ(I)*D102
        WRITE(NOPT1,348) CX1, CY1, CZ1, 0.12 , 0.0, 0.8, 1.0
  450 CONTINUE
C
```

• The magnetic moment direction is described by adding a small sphere to the surface of the torus part.

```
C     ------------------------------------- SIM.REGEON LINES  (3) ---
      WRITE(NOPT1,648)  17
      WRITE(NOPT1,649)  0. ,  0. ,  0.
      WRITE(NOPT1,649)  XL ,  0. ,  0.
      WRITE(NOPT1,649)  XL ,  YL ,  0.
      WRITE(NOPT1,649)  0. ,  YL ,  0.
      WRITE(NOPT1,649)  0. ,  0. ,  0.
      WRITE(NOPT1,649)  0. ,  0. ,  ZL
      WRITE(NOPT1,649)  XL ,  0. ,  ZL
      WRITE(NOPT1,649)  XL ,  YL ,  ZL
      WRITE(NOPT1,649)  0. ,  YL ,  ZL
      WRITE(NOPT1,649)  0. ,  0. ,  ZL
      WRITE(NOPT1,649)  0. ,  0. ,  0.
      WRITE(NOPT1,649)  0. ,  YL ,  0.
      WRITE(NOPT1,649)  0. ,  YL ,  ZL
      WRITE(NOPT1,649)  XL ,  YL ,  ZL
      WRITE(NOPT1,649)  XL ,  YL ,  0.
      WRITE(NOPT1,649)  XL ,  0. ,  0.
      WRITE(NOPT1,649)  XL ,  0. ,  ZL
C
```

• Drawing the frame of the simulation box.

```
C      ---------------------------- FORMAT ----------------------------
  181 FORMAT('# Micro AVS Geom:2.00'
     &      /'# Animation of MC simulation results'
     &      /I4)
  183 FORMAT('step',I1)
  184 FORMAT('step',I2)
  185 FORMAT('step',I3)
  186 FORMAT('step',I4)
  211 FORMAT( 'column'/'cylinder'/'dvertex_and_color'/'32'/I7 )
  248 FORMAT( 6F10.3 , F6.2 , 3F4.1)
  311 FORMAT( 'sphere'/'sphere_sample'/'color'/I7 )
  348 FORMAT( 3F10.3 , F6.2 , 3F5.2 )
  648 FORMAT( 'polyline'/'pline_sample'/'vertex'/I3 )
  649 FORMAT( 3F10.3 )
                                                                  RETURN
                                                                  END
C**** SUB INITIAL ****
      SUBROUTINE INITIAL
C
      IMPLICIT REAL*8 (A-H,O-Z), INTEGER (I-N)
C
      COMMON /BLOCK1/  RX    , RY    , RZ
      COMMON /BLOCK2/  NX    , NY    , NZ
      COMMON /BLOCK3/  N     , NDENS, VDENS
      COMMON /BLOCK4/  D     , D1     , RP     ,  VP    , IPTCLMDL
      COMMON /BLOCK5/  XL , YL , ZL , INIPX , INIPY , INIPZ , INITREE
      COMMON /BLOCK11/ EX    , EY    , EZ
      COMMON /BLOCK12/ NXB   , NYB
      COMMON /BLOCK13/ ETHETA , EPHI , NPSI , RMAT
C
      PARAMETER( NN=1360 , PI=3.141592653589793D0 )
C
      REAL*8    NDENS
      REAL*8    RX(NN) , RY(NN) , RZ(NN) , NX(NN) , NY(NN) , NZ(NN)
      REAL*8    EX(NN) , EY(NN) , EZ(NN)
      REAL*8    NXB(NN), NYB(NN)
      REAL*8    ETHETA(NN), EPHI(NN) , NPSI(NN) , RMAT(3,3,NN)
C
      INTEGER   PTCL , ICNTR
      REAL*8    XLUNT , YLUNT , ZLUNT,  RAN1 , RAN2 , RAN3
      REAL*8    VDENSMX , CRATIO , C0 , C1 , C2 , C3
C
      IF( INITREE .EQ. 1 ) THEN
        INIPX =  3
        INIPY =  9
        INIPZ = 12
        N     = 324
      ELSE IF( INITREE .EQ. 2 ) THEN
        INIPX =  4
        INIPY = 12
        INIPZ =  6
        N     = 288
      ELSE
        WRITE(6,*) '******************** SUB-INITIAL IS STOPPED *****'
        STOP
      END IF
C     ----------------------------------------------------------------
C
      VMN      = DBLE( INIPX*INIPY*INIPZ )*RP**2
      CRATIO   = (   ( DBLE(N)*VP )/( VMN*VDENS )   )**(1./3.)
      XLUNT    = RP
      YLUNT    = 1.D0
      ZLUNT    = RP
      XLUNT    = XLUNT*CRATIO
      YLUNT    = YLUNT*CRATIO
      ZLUNT    = ZLUNT*CRATIO
      XL       = XLUNT*DBLE(INIPX)
      YL       = YLUNT*DBLE(INIPY)
      ZL       = ZLUNT*DBLE(INIPZ)
C                                                  ----- POSITION -----
```

- A subroutine for setting the initial position and direction of each particle.

- (INIPX, INIPY, INIPZ) particles are placed in the *x*-, *y*-, and *z*-directions, respectively.

- The volumetric fraction ϕ_V satisfies $\phi_V=V_P\times N/(\alpha^3\times V_{mn})$, so that α can be obtained as $\alpha=(V_P\times N/(\phi_V\times V_{mn}))^{1/3}$, in which VMN=$V_{mn}$ and CRATIO=α.
- As shown in Figure 2.5, VMN is the minimum volume for a contact arrangement of the particles.

• (XLUNT, YLUNT, ZLUNT) are the distances between the neighboring particles in each axis direction.

• RAN1, RAN2, and RAN3 are quasi-random numbers.
• (INIPX, INIPY, INIPZ) particles are placed in each direction.
• Each particle is moved in parallel by the distance C0 to remove subtle situations at outer boundary surfaces. Also, to remove the regularity of the initial configuration, each particle is moved randomly by (C1*C0, C2*C0, C3*C0) in each direction.

```
      RAN1 = DSQRT( 2.D0 )
      RAN2 = DSQRT( 7.D0 )
      RAN3 = DSQRT( 11.D0 )
      C0   = 1.D-4
      PTCL = 0
      DO 10 K=0, INIPZ-1
      DO 10 J=0, INIPY-1
      DO 10 I=0, INIPX-1
        PTCL = PTCL + 1
        C1 = RAN1*DBLE(PTCL)
        C1 = C1 - DINT(C1)
        C1 = C1 - 0.5D0
        C2 = RAN2*DBLE(PTCL)
        C2 = C2 - DINT(C2)
        C2 = C2 - 0.5D0
        C3 = RAN3*DBLE(PTCL)
        C3 = C3 - DINT(C3)
        C3 = C3 - 0.5D0
        RX(PTCL) = DBLE(I)*XLUNT + C1*C0 + C0
        RY(PTCL) = DBLE(J)*YLUNT + C2*C0 + C0
        RZ(PTCL) = DBLE(K)*ZLUNT + C3*C0 + C0
   10 CONTINUE
      N = PTCL
C                                           ----- DIRECTION -----
```

• All the particles are set so as to point in the *y*-direction.

```
      RAN1 = DSQRT( 2.D0 )
      RAN2 = DSQRT( 3.D0 )
      DO 80 I=1,N
        C1 = PI/2.D0
        C2 = PI/2.D0
        EX(I) = DSIN(C1)*DCOS(C2)
        EY(I) = DSIN(C1)*DSIN(C2)
        EZ(I) = DCOS(C1)
C
        ETHETA(I)    = C1
        EPHI(I)      = C2
        RMAT(1,1,I) =  DCOS(C1)*DCOS(C2)
        RMAT(2,1,I) =  DCOS(C1)*DSIN(C2)
        RMAT(3,1,I) = -DSIN(C1)
        RMAT(1,2,I) = -DSIN(C2)
        RMAT(2,2,I) =  DCOS(C2)
        RMAT(3,2,I) =  0.D0
        RMAT(1,3,I) =  DSIN(C1)*DCOS(C2)
        RMAT(2,3,I) =  DSIN(C1)*DSIN(C2)
        RMAT(3,3,I) =  DCOS(C1)
   80 CONTINUE
C                                           ----- MOMENT -----
```

• The rotational matrix $\mathbf{R}^{-1}$ (=RMAT) can be evaluated as a transpose matrix in Eq. (4.43) using the particle direction data.

• The magnetic moment direction is randomly assigned using quasi-random numbers.

```
      RAN1 = DSQRT( 2.D0 )
      DO 90 I=1,N
        C1       = RAN1*DBLE(I)
        C1       = C1 - DINT(C1)
        NPSI(I)  = 2.D0*PI*C1
        NXB(I)   = DCOS(NPSI(I))
        NYB(I)   = DSIN(NPSI(I))
        NX(I)    = RMAT(1,1,I)*NXB(I) + RMAT(1,2,I)*NYB(I)
        NY(I)    = RMAT(2,1,I)*NXB(I) + RMAT(2,2,I)*NYB(I)
        NZ(I)    = RMAT(3,1,I)*NXB(I) + RMAT(3,2,I)*NYB(I)
   90 CONTINUE
                                                             RETURN
                                                             END
C**** SUB ENECAL *****
      SUBROUTINE ENECAL( I, RXI, RYI, RZI, EXI, EYI, EZI, NXI, NYI,
     &                   NZI, RCOFF2, ECAN, OVRLAP, ISTREET, JPTCL0 )
C
      IMPLICIT REAL*8 (A-H,O-Z), INTEGER (I-N)
C
      COMMON /BLOCK1/  RX    , RY    , RZ
      COMMON /BLOCK2/  NX    , NY    , NZ
      COMMON /BLOCK3/  N     , NDENS, VDENS
      COMMON /BLOCK4/  D     , D1    , RP    , VP    , IPTCLMDL
      COMMON /BLOCK5/  XL , YL , ZL , INIPX , INIPY , INIPZ , INITREE
      COMMON /BLOCK6/  RA    , RA0   , KU   ,  HX , HY , HZ
      COMMON /BLOCK7/  E     , ENEW  , EOLD
      COMMON /BLOCK8/  RCOFF, DELR  , DELT
      COMMON /BLOCK11/ EX    , EY    , EZ
      COMMON /BLOCK12/ NXB   , NYB
```

• A subroutine for calculating the interaction energies between particles.

```
      COMMON /BLOCK13/ ETHETA , EPHI , NPSI , RMAT
C
      PARAMETER( NN=1360 , PI=3.141592653589793D0 )
C
      REAL*8    NDENS   , KU
      REAL*8    RX(NN)  , RY(NN)  , RZ(NN)
      REAL*8    NX(NN)  , NY(NN)  , NZ(NN)  , E(NN)
      REAL*8    EX(NN)  , EY(NN)  , EZ(NN)
      REAL*8    NXB(NN), NYB(NN)
      REAL*8    ETHETA(NN), EPHI(NN) , NPSI(NN) , RMAT(3,3,NN)
      LOGICAL   OVRLAP
C
      REAL*8    RXI   , RYI   , RZI   , RXJ   , RYJ   , RZJ
      REAL*8    RXIP  , RYIP  , RZIP  , RXJP  , RYJP  , RZJP
      REAL*8    RXIJ  , RYIJ  , RZIJ  , RXJI  , RYJI  , RZJI
      REAL*8    RXIJQ, RYIJQ, RZIJQ
      REAL*8    NXI   , NYI   , NZI   , NXJ   , NYJ   , NZJ
      REAL*8    NXIJ  , NYIJ   , NZIJ
      REAL*8    NXIJ2, NYIJ2  , NZIJ2
      REAL*8    TXIJ  , TYIJ  , TZIJ  , TXIJS, TYIJS, TZIJS
      REAL*8    EXI   , EYI   , EZI   , EXJ   , EYJ   , EZJ
      REAL*8    EXIP  , EYIP  , EZIP  , EXJP  , EYJP  , EZJP
      REAL*8    EXIS  , EYIS  , EZIS  , EXJS  , EYJS  , EZJS
      REAL*8    KIS   , KJS   , KIJS  , KIJQ
      REAL*8    RRXI   , RRYI   , RRZI   , RRXJ   , RRYJ   , RRZJ
      REAL*8    RRXIJ  , RRYIJ  , RRZIJ  , RRXJI  , RRYJI  , RRZJI
      REAL*8    TTXIJ  , TTYIJ  , TTZIJ  , TTXIJS, TTYIJS, TTZIJS
      REAL*8    EEXI   , EEYI   , EEZI   , EEXJ   , EEYJ   , EEZJ
      REAL*8    EEXIS  , EEYIS  , EEZIS  , EEXJS  , EEYJS  , EEZJS
      REAL*8    KKIS   , KKJS   , KKIJS
      REAL*8    RIJ    , RIJSQ  , RIJ3   , R00 , R01 , R10 , R11
      REAL*8    RIJMN  , RIJMNFUN
      REAL*8    ECAN , RCOFF2  , RCHKSQ  , RCHKSQ2
      REAL*8    DSQ  , D1   , D1SQ , D02  , CHCK0  , CHCK1
      REAL*8    C0   , C1   , C2    , C00 , C01  , C02  , C03
      REAL*8    C11  , C21  , C12   , C22
      REAL*8    C1X  , C1Y  , C1Z   , C1SQ
      REAL*8    CEIEJ  , CEIRIJ , CEJEIX , CEJEIY   , CEJEIZ
      INTEGER   ITREE  , IPATH   , JPTCL
C
      OVRLAP   = .FALSE.
      ECAN     = - KU*( NXI*HX + NYI*HY + NZI*HZ )
      D1SQ     = D1**2
      DSQ      = D**2
      D02      = D/2.D0
```

• The treatment concerning particle *i*.

```
C
C     -------------------------------------------- MAIN LOOP START
C
      DO 1000 JPTCL=1,N
C
        J = JPTCL
        IF( J .EQ. I )                  GOTO 1000
```

• The treatment concerning particles *i* and *j*.

```
C
        RXJ = RX(J)
        RYJ = RY(J)
        RZJ = RZ(J)
c
        RXIJ = RXI - RXJ
        IF( RXIJ .GT. XL/2.D0 ) THEN
```

• The treatment of the periodic BC.

```
          RXIJ = RXIJ - XL
          RXJ  = RXJ  + XL
        ELSE IF( RXIJ .LT. -XL/2.D0 ) THEN
          RXIJ = RXIJ + XL
          RXJ  = RXJ  - XL
        END IF
        IF( DABS(RXIJ) .GE. RCOFF )  GOTO 1000
C
        RYIJ = RYI - RYJ
        IF( RYIJ .GT. YL/2.D0 ) THEN
          RYIJ = RYIJ - YL
          RYJ  = RYJ  + YL
        ELSE IF( RYIJ .LT. -YL/2.D0 ) THEN
          RYIJ = RYIJ + YL
          RYJ  = RYJ  - YL
        END IF
```

```
      IF( DABS(RYIJ) .GE. RCOFF )  GOTO 1000
C
      RZIJ = RZI - RZJ
      IF( RZIJ .GT. ZL/2.D0 ) THEN
        RZIJ = RZIJ - ZL
        RZJ  = RZJ  + ZL
      ELSE IF( RZIJ .LT. -ZL/2.D0 ) THEN
        RZIJ = RZIJ + ZL
        RZJ  = RZJ  - ZL
      END IF
      IF( DABS(RZIJ) .GE. RCOFF )  GOTO 1000
C
      RIJSQ= RXIJ**2 + RYIJ**2 + RZIJ**2
      IF( RIJSQ .GE. RCOFF2 )      GOTO 1000
      IF( RIJSQ .LT. 1.D0 ) THEN
        OVRLAP = .TRUE.
        RETURN
      END IF
C
      RIJ  = DSQRT(RIJSQ)
C
C     -------------------------------- START OF MAGNETIC ENERGY ---
      IF( IPTCLMDL .EQ. 1 ) THEN
C
        NXJ  = NX(J)
        NYJ  = NY(J)
        NZJ  = NZ(J)
        EXJ  = EX(J)
        EYJ  = EY(J)
        EZJ  = EZ(J)
        RXJI = -RXIJ
        RYJI = -RYIJ
        RZJI = -RZIJ
C
        C00  = NXI*NXJ  + NYI*NYJ  + NZI*NZJ
        C01  = NXI*RXIJ + NYI*RYIJ + NZI*RZIJ
        C02  = NXJ*RXIJ + NYJ*RYIJ + NZJ*RZIJ
        RIJ3 = RIJ*RIJSQ
C
        C1   = (RA/RIJ3)*( C00 - 3.D0*C01*C02/RIJSQ )
C
        ECAN = ECAN + C1
C
      ELSE IF( IPTCLMDL .EQ. 2 ) THEN
C
        NXJ   = NX(J)
        NYJ   = NY(J)
        NZJ   = NZ(J)
        NXIJ  = NXI - NXJ
        NYIJ  = NYI - NYJ
        NZIJ  = NZI - NZJ
        NXIJ2 = NXI + NXJ
        NYIJ2 = NYI + NYJ
        NZIJ2 = NZI + NZJ
        EXJ   = EX(J)
        EYJ   = EY(J)
        EZJ   = EZ(J)
        RXJI  = -RXIJ
        RYJI  = -RYIJ
        RZJI  = -RZIJ
C
        C11  = RXIJ*NXIJ  + RYIJ*NYIJ  + RZIJ*NZIJ
        C21  = RXIJ*NXIJ2 + RYIJ*NYIJ2 + RZIJ*NZIJ2
        C12  = 1.D0 - ( NXI*NXJ + NYI*NYJ + NZI*NZJ )
        C22  = 1.D0 + ( NXI*NXJ + NYI*NYJ + NZI*NZJ )
        C01  = D/RIJSQ
        C02  = D**2/(2.D0*RIJSQ)
C
        R00  = RIJ*( 1.D0 + C01*C11 + C02*C12 )**0.5
        R11  = RIJ*( 1.D0 - C01*C11 + C02*C12 )**0.5
        R01  = RIJ*( 1.D0 + C01*C21 + C02*C22 )**0.5
        R10  = RIJ*( 1.D0 - C01*C21 + C02*C22 )**0.5
        IF( (R00 .LT. 1.D0) .OR. (R11 .LT. 1.D0 ) .OR.
     &      (R01 .LT. 1.D0) .OR. (R10 .LT. 1.D0 )      ) THEN
          OVRLAP = .TRUE.
```

• If the two particles are separated over the cutoff distance r^*_{coff}, the calculation is unnecessary.

• The magnetic interaction energies are calculated.

• The treatment for the particle model with a magnetic dipole at the particle center.

• The treatment for the particle model with a plus and a minus magnetic charge at the torus part; this model is not used in the present exercise.

```
            RETURN
          END IF
C
          ECAN = ECAN
     &           + RA*( 1.D0/R00 + 1.D0/R11 - 1.D0/R01 - 1.D0/R10 )
C
        END IF
C       ---------------------------------- END OF MAGNETIC ENERGY ---
C
        IF( ISTREET .EQ. 1 )  GOTO 1000
C
        IF( RIJ .GE. D1 ) THEN
          OVRLAP = .FALSE.
          GOTO 1000
        END IF
C
C       ---------------------------------------------------------------
C       ---------    CHECK THE OVERLAP OF PARTICLES I AND J    -------
C       ---------------------------------------------------------------
C
        CEIEJ = EXI*EXJ +  EYI*EYJ +  EZI*EZJ
        TXIJ  = RXIJ/RIJ
        TYIJ  = RYIJ/RIJ
        TZIJ  = RZIJ/RIJ
        C11   = TXIJ*EXI + TYIJ*EYI + TZIJ*EZI
C
        IF( DABS(CEIEJ) .GT. 0.999D0) THEN
          IF(  DABS(C11) .LT. 0.001D0 )THEN
            ITREE = 2
          ELSE
            ITREE = 3
          END IF
        ELSE
          ITREE = 1
        END IF
C
C                                         ----------------------
C                                          ITREE=1: GENERAL
C                                          ITREE=2: ONE PLANE
C                                          ITREE=3: TWO PARALLEL
C                                                   PLANES
C                                         ----------------------
C
C       -------------------------------------------- (1) ITREE=2 ---
C
        IF( ITREE .EQ. 2 ) THEN
          IF( RIJ .GE. D1 ) THEN
            OVRLAP = .FALSE.
            GOTO 1000
          ELSE
            OVRLAP = .TRUE.
            RETURN
          END IF
        END IF
C
C       -------------------------------------------- (2) ITREE=3 ---
C
        IF( ITREE .EQ. 3 ) THEN
C
          CEIRIJ = EXI*RXIJ + EYI*RYIJ + EZI*RZIJ
          IF( DABS(CEIRIJ) .GE. 1.D0 ) THEN
            OVRLAP = .FALSE.
            GOTO 1000
          END IF
C
          RXIP = RXIJ - CEIRIJ*EXI
          RYIP = RYIJ - CEIRIJ*EYI
          RZIP = RZIJ - CEIRIJ*EZI
          C0   = DSQRT( RXIP**2 + RYIP**2 + RZIP**2 )
          IF( C0 .LE. D ) THEN
            OVRLAP = .TRUE.
            RETURN
          ELSE IF( C0 .GE. D1 ) THEN
            OVRLAP = .FALSE.
            GOTO 1000
          END IF
```

• The assessment of the overlap between particles *i* and *j*.

• The regime of particle overlap is assessed. There are three regimes: a general arrangement (ITREE=1), a one-plane arrangement (ITREE=2), and a parallel arrangement (ITREE=3).

• The treatment for a one-plane arrangement (ITREE=2).

• The occurrence of a particle overlap can be assessed by only the particle–particle distance.

• The treatment for a parallel arrangement (ITREE=3).

• No overlap if the condition (2.1) in Section 4.2.3 is satisfied.

• r_{ij}^{p} in Eq. (4.32) is evaluated.

• An overlap in the case of 2.2.1 in Section 4.2.3.

• No overlap in the case of 2.2.2 in Section 4.2.3.

```
            EXJP = RXIP/C0
            EYJP = RYIP/C0
            EZJP = RZIP/C0
            EXIP = -EXJP
            EYIP = -EYJP
            EZIP = -EZJP
            C1X  = RXI + D02*EXIP - ( RXJ + D02*EXJP )
            C1Y  = RYI + D02*EYIP - ( RYJ + D02*EYJP )
            C1Z  = RZI + D02*EZIP - ( RZJ + D02*EZJP )
            C1SQ = C1X**2 + C1Y**2 + C1Z**2
            IF( C1SQ .LT. 1.D0 ) THEN
              OVRLAP = .TRUE.
              RETURN
            ELSE
              OVRLAP = .FALSE.
              GOTO 1000
            END IF
C
          END IF
```

• An overlap in the case of 2.2.3 in Section 4.2.3.

• No overlap in the case of 2.2.4 in Section 4.2.3.

```
C
C         --------------------------------------------------- TIJS ---
          CEJEIX = EYJ*EZI - EZJ*EYI
          CEJEIY = EZJ*EXI - EXJ*EZI
          CEJEIZ = EXJ*EYI - EYJ*EXI
          C1     = DSQRT( CEJEIX**2 + CEJEIY**2 + CEJEIZ**2 )
          TXIJS  = CEJEIX / C1
          TYIJS  = CEJEIY / C1
          TZIJS  = CEJEIZ / C1
```

• $\mathbf{t}_{ij}^s$ (=(TXIJS,TYIJS,TZIJS)) in Eq. (4.23) is evaluated.

```
C
C         --------------------------------------------- EIS , EJS ---
          EXIS = -( EYI*TZIJS - EZI*TYIJS )
          EYIS = -( EZI*TXIJS - EXI*TZIJS )
          EZIS = -( EXI*TYIJS - EYI*TXIJS )
          EXJS =  ( EYJ*TZIJS - EZJ*TYIJS )
          EYJS =  ( EZJ*TXIJS - EXJ*TZIJS )
          EZJS =  ( EXJ*TYIJS - EYJ*TXIJS )
```

• $\mathbf{e}_i^s$ and $\mathbf{e}_j^s$ in Eq. (4.24) are evaluated.

• k_i^s, k_j^s, and k_{ij}^s in Eq. (4.26) are evaluated.

```
C
C         --------------------------------------------- KIS , KJS ---
          KIS  = -(EXJ*RXIJ + EYJ*RYIJ + EZJ*RZIJ)/
     &            (EXJ*EXIS + EYJ*EYIS + EZJ*EZIS)
          KJS  =  (EXI*RXIJ + EYI*RYIJ + EZI*RZIJ)/
     &            (EXI*EXJS + EYI*EYJS + EZI*EZJS)
          KIJS = RXIJ*TXIJS  + RYIJ*TYIJS + RZIJ*TZIJS
C
C         ------------------------ REPLACEMENT OF PARTICLES I AND J ---
          IF( DABS(KJS) .GE. DABS(KIS) ) THEN
            II     = I
            JJ     = J
            RRXI   = RXI
            RRYI   = RYI
            RRZI   = RZI
            RRXJ   = RXJ
            RRYJ   = RYJ
            RRZJ   = RZJ
            RRXIJ  = RXIJ
            RRYIJ  = RYIJ
            RRZIJ  = RZIJ
            RRXJI  = RXJI
            RRYJI  = RYJI
            RRZJI  = RZJI
            TTXIJ  = TXIJ
            TTYIJ  = TYIJ
            TTZIJ  = TZIJ
            TTXIJS= TXIJS
            TTYIJS= TYIJS
            TTZIJS= TZIJS
            EEXI   = EXI
            EEYI   = EYI
            EEZI   = EZI
            EEXJ   = EXJ
            EEYJ   = EYJ
            EEZJ   = EZJ
            EEXIS  = EXIS
            EEYIS  = EYIS
            EEZIS  = EZIS
```

• Treatment (1) shown in Section 4.2.5. The subscripts are exchanged between *i* and *j* so as to satisfy $|k_j^s|>|k_i^s|$.
• As a result, the particle names *i* and *j* are expressed as II and JJ in the program.

```
          EEXJS = EXJS
          EEYJS = EYJS
          EEZJS = EZJS
          KKIS  = KIS
          KKJS  = KJS
          KKIJS = KIJS
        ELSE
          II    = J
          JJ    = I
          RRXI  = RXJ
          RRYI  = RYJ
          RRZI  = RZJ
          RRXJ  = RXI
          RRYJ  = RYI
          RRZJ  = RZI
          RRXIJ = -RXIJ
          RRYIJ = -RYIJ
          RRZIJ = -RZIJ
          RRXJI = -RXJI
          RRYJI = -RYJI
          RRZJI = -RZJI
          TTXIJ = -TXIJ
          TTYIJ = -TYIJ
          TTZIJ = -TZIJ
          TTXIJS= -TXIJS
          TTYIJS= -TYIJS
          TTZIJS= -TZIJS
          EEXI  = EXJ
          EEYI  = EYJ
          EEZI  = EZJ
          EEXJ  = EXI
          EEYJ  = EYI
          EEZJ  = EZI
          EEXIS = EXJS
          EEYIS = EYJS
          EEZIS = EZJS
          EEXJS = EXIS
          EEYJS = EYIS
          EEZJS = EZIS
          KKIS  = KJS
          KKJS  = KIS
          KKIJS = KIJS
        END IF
C
C       ------------------- REPLACEMENT OF DIRECTIONS OF EI AND EJ ---
        CHCK0 = RRXJI*EEXI + RRYJI*EEYI + RRZJI*EEZI
        IF( CHCK0 .LT. 0.D0 ) THEN
          EEXI   = -EEXI
          EEYI   = -EEYI
          EEZI   = -EEZI
        END IF
C
        CEIEJ = EEXI*EEXJ + EEYI*EEYJ + EEZI*EEZJ
        IF( CEIEJ .LT. 0.D0 ) THEN
          EEXJ   = -EEXJ
          EEYJ   = -EEYJ
          EEZJ   = -EEZJ
          CEIEJ  = -CEIEJ
        END IF
C
C       ------------------------- REPLACEMENT OF DIRECTION OF TIJS ---
        CHCK0 = TTXIJS*RRXIJ + TTYIJS*RRYIJ + TTZIJS*RRZIJ
        IF( CHCK0 .LT. 0.D0 ) THEN
          TTXIJS = -TTXIJS
          TTYIJS = -TTYIJS
          TTZIJS = -TTZIJS
        END IF
C
C       -------- REPLACEMENT OF DIRECTIONS OF EIS,EJS,KIS,KJS,KIJS ---
        IF( KKIS .LT. 0.D0 ) THEN
          KKIS  = -KKIS
          EEXIS = -EEXIS
          EEYIS = -EEYIS
          EEZIS = -EEZIS
        END IF
```

• Treatment (2) shown in Section 4.2.5.

• Treatment (3) shown in Section 4.2.5.

• Treatment (4) shown in Section 4.2.5.

```
      IF( KKJS .LT. 0.D0 ) THEN
        KKJS  = -KKJS
        EEXJS = -EEXJS
        EEYJS = -EEYJS
        EEZJS = -EEZJS
      END IF
      IF( KKIJS .LT. 0.D0 ) THEN
        KKIJS = -KKIJS
      END IF
C
C
C     ---------------------------------------------- (3) ITREE=1 ---
C
      IF( ITREE .EQ. 1 ) THEN
C
C
        KIJQ = DABS( EEXJS*EEXI + EEYJS*EEYI + EEZJS*EEZI )
        IF( KKJS .GE. D02 ) THEN
          KIJQ  = ( KKJS - D02 )*KIJQ
          RXIJQ = RRXJ + D02*EEXJS - KIJQ*EEXI
          RYIJQ = RRYJ + D02*EEYJS - KIJQ*EEYI
          RZIJQ = RRZJ + D02*EEZJS - KIJQ*EEZI
          IPATH = 1
        ELSE
          KIJQ  = ( D02 - KKJS )*KIJQ
          RXIJQ = RRXJ + D02*EEXJS + KIJQ*EEXI
          RYIJQ = RRYJ + D02*EEYJS + KIJQ*EEYI
          RZIJQ = RRZJ + D02*EEZJS + KIJQ*EEZI
          IPATH = 2
        END IF
        CHCK1   = DSQRT( (RXIJQ-RRXI)**2 + (RYIJQ-RRYI)**2
     &                                   + (RZIJQ-RRZI)**2 )
        IF( CHCK1 .LE. D02 ) THEN
C                                          --- (3)-1 INNER CIRCLE ---
          IF( IPATH .EQ. 2 ) THEN
            OVRLAP = .TRUE.
            RETURN
          ELSE IF( IPATH .EQ. 1 ) THEN
            IF( KIJQ .LT. 1.D0 ) THEN
              OVRLAP = .TRUE.
              RETURN
            ELSE
              OVRLAP = .FALSE.
              GOTO 1000
            END IF
          END IF
        ELSE
C                                          --- (3)-2 OUTER CIRCLE ---
          IF( IPATH .EQ. 1 ) THEN
C                                             --- (3)-2-1 IPATH=1 ---
            IF( KIJQ .GE. 1.D0 ) THEN
              OVRLAP = .FALSE.
              GOTO 1000
            ELSE
              RIJMN = RIJMNFUN( EEXI, EEYI, EEZI, EEXJ, EEYJ, EEZJ,
     &                EEXIS, EEYIS, EEZIS, EEXJS, EEYJS, EEZJS,
     &                KKIS, KKJS, KKIJS, RRXIJ, RRYIJ, RRZIJ, D )
              IF( RIJMN .GE. 1.D0 ) THEN
                OVRLAP = .FALSE.
                GOTO 1000
              ELSE
                OVRLAP = .TRUE.
                RETURN
              END IF
            END IF
          ELSE IF( IPATH .EQ. 2 ) THEN
C                                                --- (3)-2-2 IPATH=2 --
            RIJMN = RIJMNFUN( EEXI, EEYI, EEZI, EEXJ, EEYJ, EEZJ,
     &                EEXIS, EEYIS, EEZIS, EEXJS, EEYJS, EEZJS,
     &                KKIS, KKJS, KKIJS, RRXIJ, RRYIJ, RRZIJ, D )
            IF( RIJMN .GE. 1.D0 ) THEN
              OVRLAP = .FALSE.
              GOTO 1000
            ELSE
```

- The treatment for a general arrangement (ITREE=1).
- $k^Q_{i(j)}$ in Eq. (4.27) and $\mathbf{r}^Q_{i(j)}$ in Eq. (4.28) are evaluated. IPATH=1 means $k^s_j \geq d/2$.
- $k^Q_{i(j)}$ in Eq. (4.29) and $\mathbf{r}^Q_{i(j)}$ in Eq. (4.30) are evaluated. IPATH=2 means $k^s_j < d/2$.
- An overlap in the case of 3.2.1 in Section 4.2.3.
- An overlap in the case of 3.1.2.a in Section 4.2.3.
- No overlap in the case of 3.1.1 in Section 4.2.3.
- No overlap in the case of 3.1.1 in Section 4.2.3.
- No overlap in the case of 3.1.2.b.1 in Section 4.2.3.
- An overlap in the case of 3.1.2.b.2 in Section 4.2.3.
- No overlap in the case of 3.2.2.a in Section 4.2.3.

```
                OVRLAP = .TRUE.
                RETURN
              END IF
C
            END IF
C
          END IF
C
C
        END IF
C
C
 1000 CONTINUE
                                                              RETURN
                                                              END
```

• An overlap in the case of 3.2.2.b in Section 4.2.3.

```
C#### FUN RIJMNFUN ####
      DOUBLE PRECISION FUNCTION RIJMNFUN( EXI, EYI, EZI, EXJ,EYJ,EZJ,
     &                           EXIS, EYIS, EZIS, EXJS, EYJS, EZJS,
     &                           KIS, KJS, KIJS, RXIJ, RYIJ, RZIJ, D )
C
      IMPLICIT REAL*8 (A-H,O-Z), INTEGER (I-N)
C
      PARAMETER( PI=3.141592653589793D0 )
C
      REAL*8    KIS, KJS, KIJS
C
      REAL*8    X0 , Y0 , Z0 , X1 , Y1 , Z1 , X2 , Y2 , Z2
      REAL*8    D02 , CS , SN , BETAN1 , BETAN2  , BETAN , DBETAN
      REAL*8    FBETAN , FPBETAN , GAB , GABMN
      REAL*8    DX1DB  , DY1DB   , DX2DB , DY2DB
      REAL*8    CX0X1  , CX0X1SQ , CY0CY1
      REAL*8    SNBETA , CSBETA  , CR2 , CRSQ  , CHCK0 , DDEG
      REAL*8    DELX , DELY , DELZ , C0 , C1 , C2
      INTEGER   ICTR
C
```

• A function subprogram for evaluating $r_{ij}^{(min)}$ by means of Newton's method.

```
      DDEG   = 10.D0 * (PI/180.D0)
      D02    = D/2.D0
      CS     = EXI*EXJ + EYI*EYJ + EZI*EZJ
      SN     = DSQRT( 1.D0 - CS**2 )
      X0     = KIJS
      CHCK0  = EXIS*EXJ + EYIS*EYJ + EZIS*EZJ
      IF( CHCK0 .LE. 0.D0 ) THEN
        DELX =  EXIS
        DELY =  EYIS
        DELZ =  EZIS
      ELSE
        DELX = -EXIS
        DELY = -EYIS
        DELZ = -EZIS
      END IF
      Y0 = -( RXIJ*DELX + RYIJ*DELY + RZIJ*DELZ )
      Z0 = KJS*DABS( EXJS*EXI + EYJS*EYI + EZJS*EZI )
C
```

• CS=cos(θ_0) and SN=sin(θ_0).
• x_0=(x_0, y_0, z_0) is evaluated.

```
C                                 --- FOR THE CASE OF COS(BETA)=0 ---
C                                 -   VALID ONLY FOR OUTER CIRCLE   -
      IF( DABS(X0) .LE. 0.05D0 ) THEN
        X2 =  X0
        Y2 =  Y0 - D02*CS
        Z2 =  Z0 - D02*SN
        X1 =  0.D0
        Z1 =  0.D0
        IF( Y2 .GE. 0.D0 ) THEN
          Y1 =  D02
        ELSE
          Y1 = -D02
        END IF
        GAB = (X2-X1)**2 + (Y2-Y1)**2 + (Z2-Z1)**2
        RIJMNFUN = DSQRT( GAB )
        RETURN
      END IF
C
```

• The case of $\mathbf{x}_0$=(0, y_0, z_0) and $|\mathbf{r}_{i(j)}^{Q}-\mathbf{r}_i|\geq d/2$ enables us to conduct simple treatment.

```
      X2      = X0 / 2.D0
      C1      = 1.0D0
      C2      =  -X0/D
      IF( DABS(C2) .GE. 1.D0 )  C2 = DSIGN( C1, C2 )
```

• A starting value of $\mathbf{x}_2$ is given. It is first assumed that X2=X0/2, yielding a starting value of β= BETAN.

```
      BETAN1 = DACOS( C2 )
      BETAN2 = 2.D0*PI - BETAN1
      C1     = DSIN(BETAN1)
      C2     = DSIN(BETAN2)
      IF( C1 .GE. C2 ) THEN
        BETAN = BETAN2
      ELSE
        BETAN = BETAN1
      END IF
C
C     ------------------------------- START OF NEWTON PROCEDURE -----
C
```

• The minimum value of $g(\alpha,\beta)$ is saved in GABMN.

• The start of the iteration procedure of Newton's method.

```
      GABMN  = 1.D5
      ICTR   = 0
   10 ICTR   = ICTR + 1
      SNBETA = DSIN( BETAN )
      CSBETA = DCOS( BETAN )
      X2     = D02*CSBETA    + X0
      Y2     = D02*SNBETA*CS + Y0
      Z2     = D02*SNBETA*SN + Z0
C
```

• $\mathbf{x}_2$=(X2,Y2,Z2) is calculated using BETAN which is an expected value of the solution β.

• $\mathbf{x}_1$=(X1,Y1,Z1) is calculated from procedure 3 of Newton's method in Section 4.2.3.

```
      CR2    = X2**2 + Y2**2
      CRSQ   = DSQRT( CR2 )
      X1     = (X2/CRSQ)*D02
      Y1     = (Y2/CRSQ)*D02
      Z1     = 0.D0
      C1     = (X2-X1)**2 + (Y2-Y1)**2 + (Z2-Z1)**2
      IF( C1 .LT. GABMN ) GABMN = C1
C
```

• $(x_0-x_1)^2 f(\beta_n)$=FBETAN is evaluated.

```
      CX0X1   = X0 - X1
      CX0X1SQ = CX0X1**2
      CY0Y1   = Y0 - Y1
      FBETAN  = CX0X1*( CX0X1*SNBETA/CSBETA - CS*CY0Y1 - SN*Z0 )
C
```

• $\partial x_2/\partial\beta$=DX2DB and $\partial y_2/\partial\beta$=DY2DB.
• $\partial x_1/\partial\beta$=DX1DB and $\partial y_1/\partial\beta$=DY1DB.

```
      DX2DB   = -D02*SNBETA
      DY2DB   =  D02*CSBETA*CS
      C0      =  X2*DX2DB + Y2*DY2DB
      C1      =  CRSQ/CR2
      C2      =  C0/(CRSQ*CR2)
      DX1DB   =  ( C1*DX2DB - C2*X2 )*D02
      DY1DB   =  ( C1*DY2DB - C2*Y2 )*D02
      CY0Y1   =  Y0 - Y1
      FPBETAN =  CX0X1SQ/CSBETA**2 - CS*( -DY1DB*CX0X1 + DX1DB*CY0Y1 )
     &                                 - Z0*SN*DX1DB
C
```

• $(x_0-x_1)^2 f'(\beta_n)$=FPBETAN is evaluated.

```
      BETAN1 = BETAN - FBETAN/FPBETAN
C                                               --- JUDGEMENT ---
      DBETAN = DABS(BETAN1-BETAN)
      IF( DBETAN .GT. 0.01D0 ) THEN
        IF( DBETAN .GT. DDEG ) THEN
          BETAN = DSIGN( DDEG, (BETAN1-BETAN) ) + BETAN
        ELSE
          BETAN = BETAN1
        ENDIF
        IF( ICTR .GT. 10 )  GOTO 900
        GOTO 10
      END IF
C
  900 GAB = (X2-X1)**2 + (Y2-Y1)**2 + (Z2-Z1)**2
      IF( GAB .GT. GABMN )  GAB = GABMN
      RIJMNFUN = DSQRT( GAB )
                                                                  RETURN
                                                                  END
C*********************************************************************
C   THIS SUBROUTINE IS FOR GENERATING UNIFORM RANDOM NUMBERS          *
C   (SINGLE PRECISION) FOR 32-BIT COMPUTER.                           *
C     N        : NUMBER OF RANDOM NUMBERS TO GENERATE                 *
C     IX       : INITIAL VALUE OF RANDOM NUMBERS (POSITIVE INTEGER)   *
C              : LAST GENERATED VALUE IS KEPT                         *
C     X(N)     : GENERATED RANDOM NUMBERS (0<X(N)<1)                  *
C*********************************************************************
C**** SUB RANCAL ****
      SUBROUTINE RANCAL( N, IX, X )
C
      IMPLICIT REAL*8 (A-H,O-Z), INTEGER (I-N)
C
```

• β_{x+1}=BETAN1 is evaluated from Eq. (4.41).

• A subroutine for generating a uniform random number sequence.

```
      REAL       X(N)
      INTEGER    INTEGMX, INTEGST, INTEG
C
      DATA INTEGMX/2147483647/
      DATA INTEGST,INTEG/584287,48828125/
C
      AINTEGMX = REAL( INTEGMX )
C
      IF ( IX.LT.0 ) STOP
      IF ( IX.EQ.0 ) IX = INTEGST
      DO 30 I=1,N
         IX = IX*INTEG
         IF (IX .LT. 0 )  IX    = (IX+INTEGMX)+1
         X(I) = REAL(IX)/AINTEGMX
   30 CONTINUE
      RETURN
      END
```

• This is for a 32-bit CPU based on the expression of two's complement.

5 Practice of Brownian Dynamics Simulations

In the previous chapters, we have shown how the MD method and MC method are applied in a practical simulation. In the present and successive chapters, we follow the same approach and demonstrate the microsimulation methods required for the application of the Brownian Dynamics (BD) method, the DPD method, and the lattice Boltzmann method. These further methods are very useful as simulation tools for a colloidal dispersion or a suspension composed of dispersed particles. These simulation methods have many applications in the pharmaceutical field, as well as in science and engineering.

The exercise in the present chapter is for a BD simulation to discuss how Lennard-Jones particles sediment in the gravitational field for cases when the Brownian motion is expected to be significant. This example of a physical phenomenon becomes attractive as a research subject when the particle aggregation is strongly related to the sedimentation. The sample simulation program is written in the C programming language.

5.1 Sedimentation Phenomena of Lennard-Jones Particles

We consider a thermodynamic equilibrium state of N Brownian particles with mass m dispersed in a base liquid contained in a rectangular parallelepiped box. For simplification, the Brownian particles are assumed to be the Lennard-Jones particle, where the particle–particle interactions can be expressed as a Lennard-Jones potential. The objective of the present practice is to discuss how the Brownian particles in thermodynamic equilibrium sediment after the gravitational field is switched on. The system temperature, gravitational force, and particle–particle interactions are expected to significantly influence the sedimentation phenomenon.

5.2 Specification of Problems in Equations

Since the particles sediment under the effect of the Brownian motion in a gravitational field, we are required to use the BD method, explained in Section 1.3, in order to simulate this phenomenon. In contrast to a magnetic particle system in which the particle rotation is restricted by an external magnetic field, the Lennard-Jones particles are only influenced by the isotropic force due to the Lennard-Jones potential. We

Introduction to Practice of Molecular Simulation. DOI: 10.1016/B978-0-12-385148-2.00005-7

therefore only need to treat the translational motion of the Brownian particles. The particles hydrodynamically interact through their ambient fluid, but it is difficult to take into account these multibody hydrodynamic interactions, even for the relatively simple spherical particle system. It is still more so for a nonspherical particle system, such as a rod-like or disk-like particle suspension. The difficulty of treating multibody hydrodynamic interactions forces us to take into account only the friction term as a first approximation, even in the case of a nondilute suspension. In the present exercise, we therefore take into account the nonhydrodynamic interaction but neglect the multibody hydrodynamic interaction among the particles.

If the position vector of an arbitrary particle i is denoted by $\mathbf{r}_i$, the velocity by $\mathbf{v}_i$, the nonhydrodynamic force by $\mathbf{f}_i$, and the random force by $\mathbf{f}_i^{\mathrm{B}}$, then the equation of motion of particle i is expressed as [1,4]

$$m\frac{\mathrm{d}^2\mathbf{r}_i}{\mathrm{d}t^2} = \mathbf{f}_i - \xi\mathbf{v}_i + \mathbf{f}_i^{\mathrm{B}} \tag{5.1}$$

in which ξ is the friction coefficient, expressed as $\xi = 3\pi\eta d$ (η is the liquid viscosity) under the assumption that the Lennard-Jones particles are spherical with diameter d. The random force $\mathbf{f}_i^{\mathrm{B}} = (f_{ix}^{\mathrm{B}}, f_{iy}^{\mathrm{B}}, f_{iz}^{\mathrm{B}})$ must satisfy the following stochastic properties:

$$\langle f_{ix}^{\mathrm{B}}(t)\rangle = \left\langle f_{iy}^{\mathrm{B}}(t)\right\rangle = \langle f_{iz}^{\mathrm{B}}(t)\rangle = 0 \tag{5.2}$$

$$\left\langle \{f_{ix}^{\mathrm{B}}(t)\}^2\right\rangle = \left\langle \left\{f_{iy}^{\mathrm{B}}(t)\right\}^2\right\rangle = \left\langle \{f_{iz}^{\mathrm{B}}(t)\}^2\right\rangle = 2\xi kT\delta(t-t') \tag{5.3}$$

The force $\mathbf{f}_{ij}$ acting on particle i by particle j due to the Lennard-Jones potential is expressed as

$$\mathbf{f}_{ij} = 24\varepsilon\left\{2\left(\frac{d}{r_{ij}}\right)^{12} - \left(\frac{d}{r_{ij}}\right)^{6}\right\}\frac{\mathbf{r}_{ij}}{r_{ij}^2} \tag{5.4}$$

in which $\mathbf{r}_{ij}$ is the position vector of particle i relative to particle j, expressed as $\mathbf{r}_{ij} = \mathbf{r}_i - \mathbf{r}_j$, and r_{ij} is the magnitude of $\mathbf{r}_{ij}$, that is, $r_{ij} = |\mathbf{r}_{ij}|$. The total force acting on particle i, $\mathbf{f}_i$, can be obtained by summing $\mathbf{f}_{ij}$ from the contributions of all the ambient particles.

The method of nondimensionalizing quantities is described in the next section.

5.3 Brownian Dynamics Algorithm

As explained in Section 1.3, the Ermak–McCammon method [24] enables us to transform the equation of motion in Eq. (5.1) into Eq. (1.59). We here show the nondimensional expressions in the following. It may be inappropriate to use the representative values usually employed for the Lennard-Jones system, because we consider a dispersion of fine particles—which are regarded as a Lennard-Jones particle performing Brownian motion—and not a pure molecular system. We therefore use the following representative values: the particle diameter d for distances; $mg/(3\pi\eta d)$ for the velocities, which is obtained by equating the friction force to the gravitational force; and the gravitational force mg for forces. With these representative values, the equation of an arbitrary particle i is written in nondimensional form as

$$\mathbf{r}_i^*(t^*+h^*)=\mathbf{r}_i^*(t^*)+h^*\mathbf{f}_i^*(t^*)+\Delta\mathbf{r}_i^{\mathrm{B}*} \tag{5.5}$$

in which the components $(\Delta x_i^{\mathrm{B}*}, \Delta y_i^{\mathrm{B}*}, \Delta z_i^{\mathrm{B}*})$ of the random displacement $\Delta\mathbf{r}_i^{\mathrm{B}*}$ must satisfy the following stochastic characteristics:

$$\left\langle \Delta x_i^{\mathrm{B}*}\right\rangle=\left\langle \Delta y_i^{\mathrm{B}*}\right\rangle=\left\langle \Delta z_i^{\mathrm{B}*}\right\rangle=0 \tag{5.6}$$

$$\left\langle \left(\Delta x_i^{\mathrm{B}*}\right)^2\right\rangle=\left\langle \left(\Delta y_i^{\mathrm{B}*}\right)^2\right\rangle=\left\langle \left(\Delta z_i^{\mathrm{B}*}\right)^2\right\rangle=2R_{\mathrm{B}}h^* \tag{5.7}$$

in which R_{B} is a nondimensional parameter representing the strength of the random force relative to the gravitational force, expressed as $R_{\mathrm{B}}=kT/(mgd)$. The gravitational force $\mathbf{f}_i^{(g)}$ acting on particle i and the force $\mathbf{f}_{ij}^{(\mathrm{LJ})}$ due to the Lennard-Jones interaction are expressed in nondimensional form as

$$\mathbf{f}_i^{(g)*}=\hat{\mathbf{g}} \tag{5.8}$$

$$\mathbf{f}_{ij}^{(\mathrm{LJ})*}=24R_{\mathrm{LJ}}\left\{2\left(\frac{1}{r_{ij}^*}\right)^{12}-\left(\frac{1}{r_{ij}^*}\right)^{6}\right\}\frac{\mathbf{r}_{ij}^*}{(r_{ij}^*)^2} \tag{5.9}$$

in which R_{LJ} is a nondimensional parameter presenting the strength of the force due to the Lennard-Jones potential relative to the gravitational force, expressed as $R_{\mathrm{LJ}}=\varepsilon/(mgd)$, and $\hat{\mathbf{g}}$ is the unit vector, denoting the gravitational direction. The consideration of these forces provides the nondimensional force $\mathbf{f}_i^*$ acting on particle i as

$$\mathbf{f}_i^*=\mathbf{f}_i^{(g)*}+\sum_{j(\neq i)}\mathbf{f}_{ij}^{(\mathrm{LJ})*}=\hat{\mathbf{g}}+24R_{\mathrm{LJ}}\sum_{j(\neq i)}\left\{2\left(\frac{1}{r_{ij}^*}\right)^{12}-\left(\frac{1}{r_{ij}^*}\right)^{6}\right\}\frac{\mathbf{r}_{ij}^*}{(r_{ij}^*)^2} \tag{5.10}$$

Since the particles sediment in the gravitational field direction, assumed to be the negative direction of the y-axis, the periodic boundary condition is not applicable at the sedimentation surface, but it is applicable to the boundary surfaces normal to the x- and z-directions. On the sedimentation surface, the elastic reflection condition is here employed for the boundary in order to ensure that a particle cannot cross the boundary surface. In the concrete treatment of a reflecting particle, the velocity component parallel to the boundary surface is unchanged, but the velocity component normal to the boundary surface is reversed in direction.

The main procedure of the BD simulation is summarized as follows. First, we set the number of particles N, the size of simulation region (L_x^*,L_y^*,L_z^*), and the volumetric fraction ϕ_V. Then, the assignment of the initial position of the particles enables us to begin the main loop in a simulation program. The particles are simulated according to the basic equations shown in Eq. (5.5) together with generating the random displacements of the particles based on the stochastic properties in Eqs. (5.6) and (5.7); these random displacements are sampled from the normal distribution specified by Eqs. (5.6) and (5.7). In order that we may discuss quantitatively the particle sedimentation phenomenon, the time variation in the local densities is evaluated for each thin-sliced volume along the y-direction. The pair correlation function is usually employed for an accurate quantitative discussion of the internal particle structure of a system, but we here focus only on the method of snapshots and employ the local number density.

5.4 Parameters for Simulations

In conducting the following BD simulations, the number of particles is taken as $N = 108$, and the volumetric fraction is taken as $\phi_V = 0.1$ to give a number density $n^* = 6\phi_V/\pi$. The face-centered cubic lattice system shown in Figure 2.2B is employed as an initial configuration of particles, yielding the lattice constant $a^* = (4/n^*)^{1/3}$ and $Q = (N/4)^{1/3}$; the replication of the unit cell $(Q - 1)$ times in each direction generates the particle configuration for the whole simulation region. The dimensions of the region are therefore $(L_x^*, L_y^*, L_z^*) = (Qa^*, Qa^*, Qa^*)$. An appropriate time interval h^* has to be chosen with sufficient consideration. Setting an unreasonably large time interval is likely to induce a serious particle overlap problem, which will result in the instability of the system. Choice of the appropriate time interval is strongly dependent on the nondimensional parameters R_{LJ} and R_B. The larger these quantities, the smaller the time interval (i.e., $h^* \ll 1$). In the present demonstration, $h^* = 0.00005$ was adopted for the case of $R_{LJ} = R_B = 1$. The simulations were carried out for various cases of R_{LJ} and R_B, where we have used $R_{LJ} = 1$ and 5 and also $R_B = 0.1$, 1, and 5.

5.5 Results of Simulations

Figures 5.1–5.3 show the snapshots of the Lennard-Jones particles in the sedimentation process under the influence of the gravitational field, which were obtained by conducting the sample simulation program explained in the next section. The snapshots in Figures 5.1 and 5.2 were obtained for different cases of R_B after the particle distribution attains to a steady state (in the macroscopical meaning). Those in Figure 5.3 are from the visualization of the sedimentation process with advancing time.

Figure 5.1A clearly shows that the particles have sedimented on the base surface area under the gravitational field. This is because the value of the nondimensional parameter $R_B = 0.1$ implies a significant influence of the gravitational force over the random Brownian force. On the other hand, in the case of $R_{LJ} = 5$ in

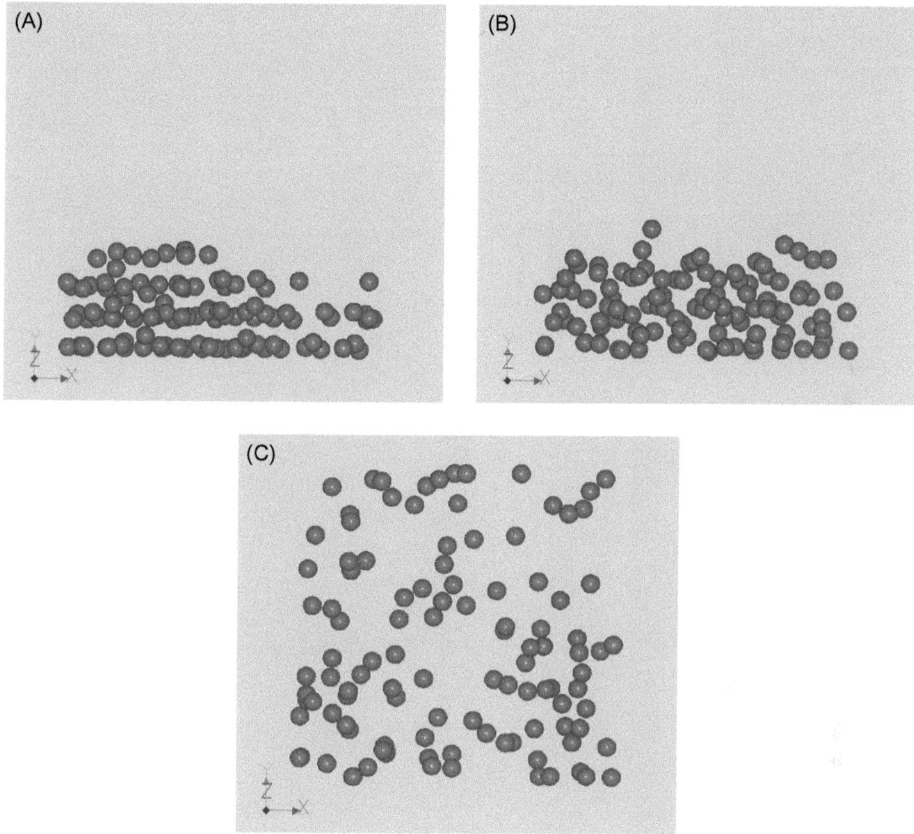

Figure 5.1 Snapshots in a steady state for $R_{LJ} = 1$: (A) $R_B = 0.1$, (B) $R_B = 1$, and (C) $R_B = 5$.

Figure 5.2A, the Lennard-Jones interactions significantly affect the sedimentation process, exhibiting characteristic aggregates formed differently from those in Figure 5.1A. For the case of the influence of the random force being equal to that of the gravitational force in Figure 5.2B, the particles have almost completely sedimented on the base area, but the internal structure seems to be considerably different from that found in Figure 5.1A. This is an example where the use of quantitative results from the pair correlation function would be required for a deeper discussion. In the case of $R_B = 5$ shown in Figures 5.1 and 5.2, the particles actively exhibit the Brownian motion without sedimenting on the base surface area; however, the particles tend to aggregate to form clusters with increasing values of R_{LJ} even in the case of $R_B = 5$. From these snapshots, we may conclude that the gravitational force mainly governs the sedimentation process, and the Lennard-Jones interactions between particles mainly determine the internal structures of the aggregates formed during the sedimentation process. As already pointed out, a higher-level academic study

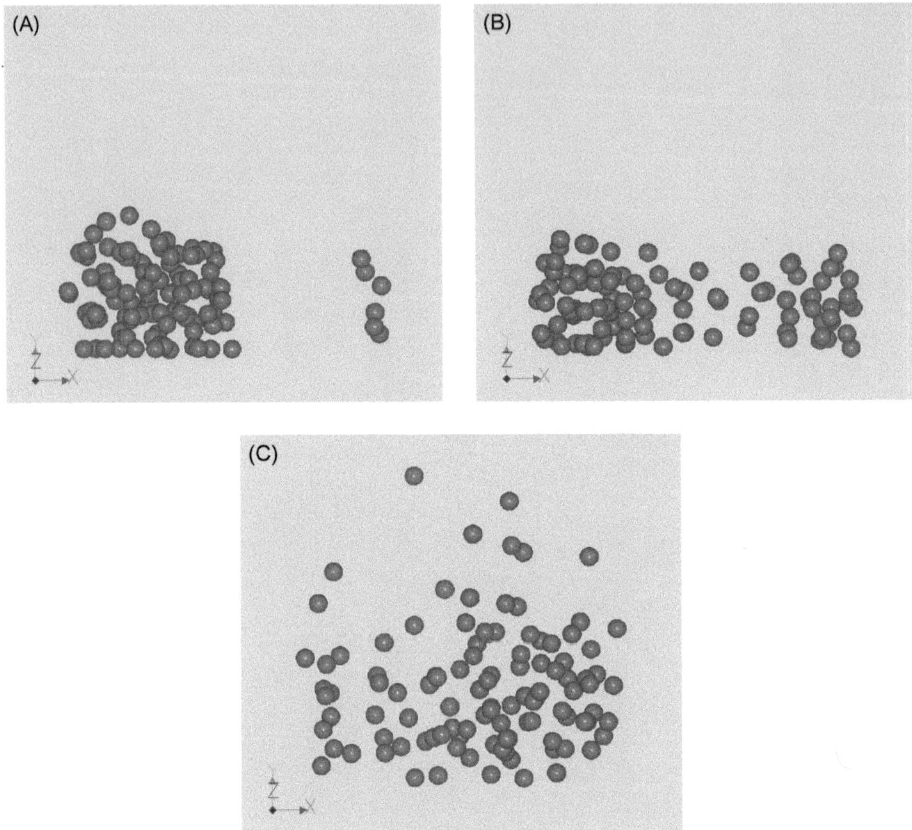

Figure 5.2 Snapshots in a steady state for $R_{LJ} = 5$: (A) $R_B = 0.1$, (B) $R_B = 1$, and (C) $R_B = 5$.

would require quantitative results, such as the pair correlation function, in addition to the qualitative results visualized here.

Figure 5.3 shows how the particles sediment with time, that is, the particle sedimentation process for the case of $R_{LJ} = 1$ and $R_B = 0.1$: Figures 5.3A–C are for nondimensional time $t^* = 1$, 4, and 8, respectively. In this case of $R_B = 0.1$, the gravitational force is much more dominant than the random force (i.e., the Brownian motion), so that the particles sediment, attain at the bottom surface, and form layer structures from the base with time.

Figure 5.4 shows the results of the local number density of particles n^* at the position y^* of each sliced layer taken from the base surface in the opposite direction to the gravitational field. Note that the nondimensional time is used, and the data or subaveraged values were calculated at every certain number of time steps. This figure demonstrates quantitative characteristics of the sedimentation process with time, which clearly suggests the layered structures of sedimented particles indicated previously.

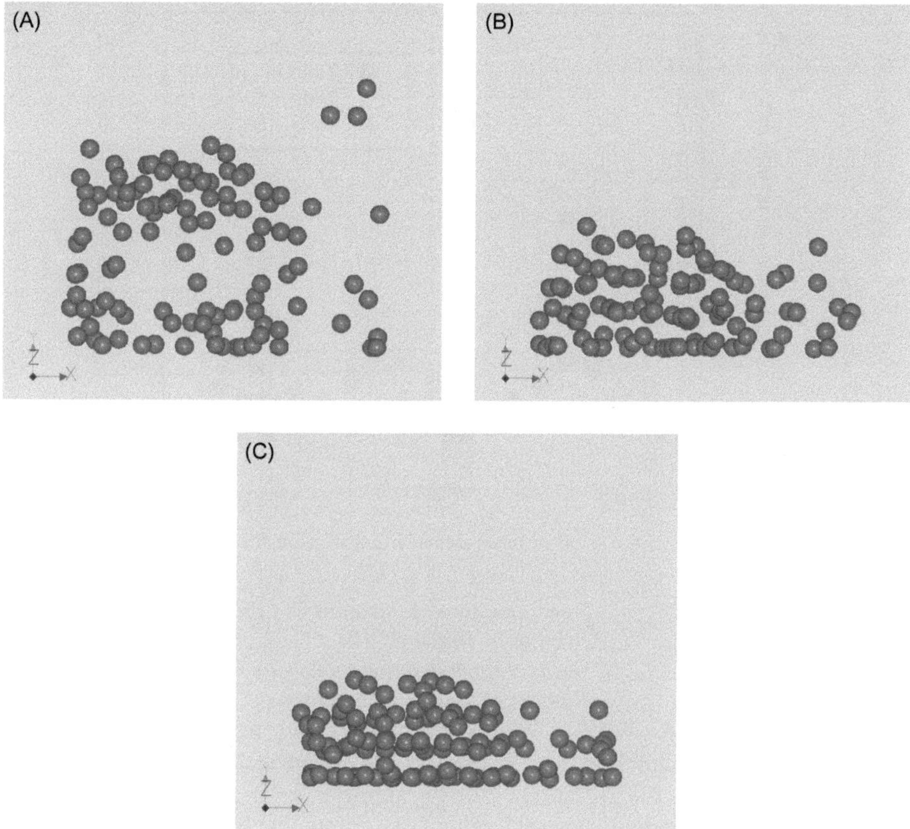

Figure 5.3 Time change of aggregate structures for $R_{LJ} = 1$ and $R_B = 0.1$: (A) $t^* = 1$, (B) $t^* = 4$, and (C) $t^* = 8$.

5.6 Simulation Program

We show a sample simulation program for the example of the present sedimentation phenomenon in the following. The program is written in the C language.

To aid the reader's understanding, the important variables used in the program are shown as follows:

`RX[i],RY[i],RZ[i]`	:	(x,y,z) components of the position vector $\mathbf{r}_i^*$ of particle i
`FX[i],FY[i],FZ[i]`	:	(x,y,z) components of the force $\mathbf{f}_i^*$ acting on particle i
`RXB[i],RYB[i], RZB[i]`	:	(x,y,z) components of the random displacements $\Delta\mathbf{r}_i^{B*}$ of particle i
`XL,YL,ZL`	:	Side lengths of the simulation box in the (x,y,z) directions
`h`	:	Time interval h^*
`ndens0`	:	Initial number density of particles
`phaiv0`	:	Initial volumetric fraction of particles

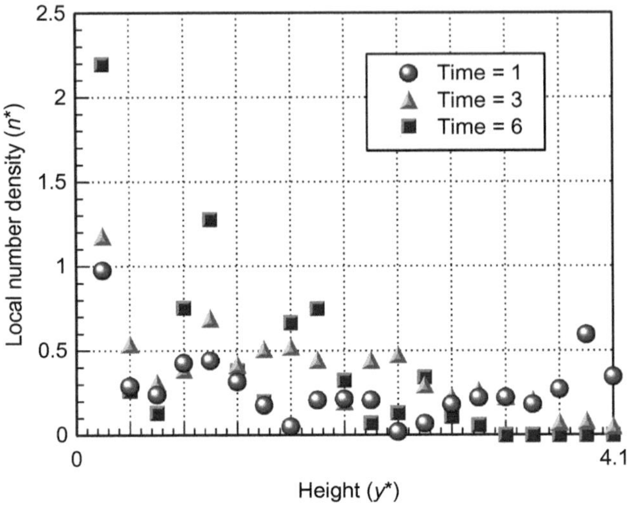

Figure 5.4 Time change in the local number density distribution for $R_{LJ} = 1$ and $R_B = 0.1$.

`RLJ,RB`	:	Nondimensional parameters R_{LJ} and R_B
`n`	:	Number of particles
`RAN[j]`	:	Uniform random numbers ranging 0∼1 (`j=1∼NRANMX`)
`NRAN`	:	Number of used random numbers

Note that the line numbers are added for convenience and are grammatically unnecessary.

In the following program, several explanatory comments have been added to the important features to assist the reader's understanding.

```
/*---------------------------------------------------------------------------*/
/*                             bdsedim1.c                                    */
/*                                                                           */
/*      ---------------------------------------------------------            */
/*      -  Brownian dynamics simulation of the sedimentation of  -           */
/*      -  spherical particles in gravity field.                 -           */
/*      ---------------------------------------------------------            */
/*                                                                           */
/*         np1    = fopen("@eaa1.data",   "w"); parameters                   */
/*         np2    = fopen("eaa11.data",   "w"); parameters                   */
/*         np[1]  = fopen("eaa001.data",  "w"); particle position            */
/*         np[2]  = fopen("eaa011.data",  "w"); particle position            */
/*         np[3]  = fopen("eaa021.data",  "w"); particle position            */
/*         np[4]  = fopen("eaa031.data",  "w"); particle position            */
/*         np[5]  = fopen("eaa041.data",  "w"); particle position            */
/*         np[6]  = fopen("eaa051.data",  "w"); particle position            */
/*         np[7]  = fopen("eaa061.data",  "w"); particle position            */
/*         np[8]  = fopen("eaa071.data",  "w"); particle position            */
/*         np[9]  = fopen("eaa081.data",  "w"); particle position            */
/*         np[10] = fopen("eaa091.data",  "w"); particle position            */
/*                                                                           */
/*         1. Lennard-Jones particle system.                                 */
/*                                                                           */
/*                                                                           */
/*                                      Ver.2 by A.Satoh , '04 3/10          */
/*---------------------------------------------------------------------------*/
```

```
/*                                                                          */
/*     RX[i],RY[i],RZ[i]     : particle position                            */
/*     RXB[i],RYB[i],RZB[i]  : random displace. due to Brownian motion      */
/*     FX[i],FY[i],FZ[i]     : forces acting on particle i                  */
/*     XL, YL, ZL  : size of simulation box along each axis                 */
/*     h           : time interval                                          */
/*     ndens0      : number density                                         */
/*     phaiv0      : volumetric fraction                                    */
/*     RB, RG, RLJ : nondimensional parameters                              */
/*     ychk[*]     : is used to calculate number density distribution       */
/*     dnsmpl,dtsmpl : data is sub-averaged using dnsmpl-data               */
/*                 : through dtsmpl-time                                    */
/*     ndens[*][+] : number density distribution                            */
/*     ntimemx     : maximum number of time step                            */
/*                                                                          */
/*               0<RX[i]<XL   ,   0<RY[i]<YL   ,   0<RZ[i]<ZL               */
/*--------------------------------------------------------------------------*/
#include <stdio.h>
#include <math.h>
#define  PI      3.141592653589793
#define  NN      501
#define  NS      2001
#define  NRANMX  2001
     double  RX[NN] , RY[NN] , RZ[NN]   ;
     double  RXB[NN], RYB[NN], RZB[NN]  ;
     double  FX[NN] , FY[NN] , FZ[NN]   ;
     double  XL, YL, ZL  ;
     double  RB, RG, RLJ ;
     float   RAN[NRANMX] ;
     int     NRAN, IX ;

/*------------------------------------------------------ main function ---*/
main()
  {
     int     n , nychk , dnsmpl ;
     double  h , ndens0, phaiv0 ;
     double  ndens[NN][NS], ychk[NN] ;
     double  cndns[NN] ;
     double  dtsmpl ;
     FILE    *fopen(), *np[11], *np1, *np2 ;

     double  rcoff , rcoff2 , rxi , ryi , rzi ;
     int     ntime , ntimemx , ntimemx1 , nsmpl , inp , ngraph ;
     int     i, j , nranchk ;

              np1     = fopen("@eaa1.data",   "w");
              np2     = fopen("eaa11.data",   "w");
              np[1]   = fopen("eaa001.data",  "w");
              np[2]   = fopen("eaa011.data",  "w");
              np[3]   = fopen("eaa021.data",  "w");
              np[4]   = fopen("eaa031.data",  "w");
              np[5]   = fopen("eaa041.data",  "w");
              np[6]   = fopen("eaa051.data",  "w");
              np[7]   = fopen("eaa061.data",  "w");
              np[8]   = fopen("eaa071.data",  "w");
              np[9]   = fopen("eaa081.data",  "w");
              np[10]  = fopen("eaa091.data",  "w");

                                                /*--- parameter (1) ---*/
               /*+++++++++++++++++++++++++++++++++++++++++++++++++++++++*/
               /*  n=32, 108, 256, 500, 864, 1372, 2048 must be chosen.  */
               /*+++++++++++++++++++++++++++++++++++++++++++++++++++++++*/
     n      = 108    ;
     rcoff  = 2.5    ;
     h      = 0.00005 ;
     rcoff2 = rcoff*rcoff ;
                                                /*--- parameter (2) ---*/
     RB     = 1.0    ;
     RLJ    = 1.0    ;  RG      = 1.0    ;
     phaiv0 = 0.1    ;
     ndens0 = phaiv0*6./PI ;
     nychk  = 40     ;

     ntimemx = 200000 ;
     dnsmpl  = 200    ;
     dtsmpl  = (double)dnsmpl*h ;
     ntimemx1= ntimemx/10 ;
     ngraph  = ntimemx/10 ;
     inp     = 0 ;

     IX      = 0 ;
```

- The given values and results are written out in @eaa1.data and eaa11.data.
- @eaa1 is for confirming the values set for starting a simulation, and eaa11 is for the postprocessing analysis.

- The particle position data are written out in eaa001–eaa091 for the postprocessing analysis.

- The particle number N=108, cutoff distance r^*_{coff}=2.5, and time interval h^*=0.00005.

- R_B=1, R_{LJ}=1, volumetric fraction ϕ_V=0.1, and number density $n^*=6\phi_V/\pi$. The simulation box is sliced into nychk equal pieces in the *y*-direction.

- The total number of time steps is 200,000 and data are sampled at every 200 time steps. The equilibration procedure is conducted until ntimemx1 steps. The particle positions are written out at every ngraph steps for the postprocessing analysis.

```
      rancal() ;
      NRAN    = 1 ;
      nranchk = NRANMX - 6*n ;
```

• A sequence of uniform random numbers is prepared in advance. When necessary, random numbers are taken out from the variable RAN[*].

```
      /*----------------------------
      /*-----------------    initial configuration    ------------------*/
      /*---------------------------------------------------------------*/
                                       /*--- set initial positions ---*/
      iniposit( n, ndens0 ) ; YL = XL ; ZL = XL ;
                                /*--- set grid for num.dens.dist. ---*/
      gridcal( nychk, ychk ) ;
                                          /*--- calculate energy ---*/
      forcecal( n, rcoff, rcoff2 ) ;
                                  /*--- cal random displacement ---*/
      randisp( n , h ) ;

                                     /*--- print out constants ---*/
      fprintf(np1,"----------------------------------------------\n") ;
      fprintf(np1,"              Brownian dynamics method         \n") ;
      fprintf(np1,"                                                \n") ;
      fprintf(np1,"       +++ Lennard-Jones particles system +++    \n") ;
      fprintf(np1,"n=%4d ndens=%8.3f phaiv0=%6.3f rcoff=%6.3f h=%10.8f\n",
                  n, ndens0, phaiv0, rcoff, h ) ;
      fprintf(np1,"XL=%6.3f YL=%6.3f ZL=%6.3f\n", XL, YL, ZL) ;
      fprintf(np1,"RB=%12.4e  RG=%12.4e  RLJ=%12.4e\n", RB, RG, RLJ) ;
      fprintf(np1,"ntimemx=%8d  nychk=%4d  dnsmple=%8d  dtsmpl=%12.4e\n",
                  ntimemx, nychk, dnsmpl, dtsmpl);
      fprintf(np1,"----------------------------------------------\n");

                                          /*--- initialization ---*/
      for( i=1 ; i <= nychk ; i++ ) {
          cndns[i] = 0. ;
      }
      nsmpl = 0 ;
```

• The variables are initialized for saving the local number densities afterward.

```
      /*-----------------------------------------------
      /*---------------------    equilibration    ------
      /*-----------------------------------------------
```

• The equilibration procedure is conducted below.

```
      for ( ntime = 1 ; ntime <= ntimemx1 ; ntime++ ) {

        for ( i=1 ; i<=n ; i++ ) {

          rxi = RX[i] + h*FX[i] + RXB[i] ;
          ryi = RY[i] + h*FY[i] + RYB[i] ;
          rzi = RZ[i] + h*FZ[i] + RZB[i] ;
          rxi += - rint( rxi/XL - 0.5 )*XL ;
          rzi += - rint( rzi/ZL - 0.5 )*ZL ;
          if( ryi < 0. ) ryi = - ryi ;
          if( ryi > YL ) ryi = YL - ( ryi - YL ) ;

          RX[i] = rxi ;
          RY[i] = ryi ;
          RZ[i] = rzi ;
        }

        forcecal( n, rcoff, rcoff2 ) ;
        randisp( n , h ) ;
                                 /*--- check of random numbers used ---*/
        if ( NRAN >= nranchk ) {
            rancal() ; NRAN = 1 ;
        }
      }
```

• The particle positions at the next time step are calculated from Eq. (5.5).
• The periodic BC is used for the x- and z-directions.

• The elastic collision model at the boundary surface is used for the y-direction.

• The forces acting on particles are calculated in the function forcecal. The random displacements are generated in the function randisp.

• The number of the used random numbers is checked. If over nranchk, a uniform random number sequence is renewed.

```
      /*---------------------------------------------------------------*/
      /*------------------    start of main loop    -------------------*/
      /*---------------------------------------------------------------*/

      for ( ntime = 1 ; ntime <= ntimemx ; ntime++ ) {

        for ( i=1 ; i<=n ; i++ ) {

          FY[i] = FY[i] - RG ;

          rxi = RX[i] + h*FX[i] + RXB[i] ;
          ryi = RY[i] + h*FY[i] + RYB[i] ;
          rzi = RZ[i] + h*FZ[i] + RZB[i] ;
          rxi += - rint( rxi/XL - 0.5 )*XL ;
          rzi += - rint( rzi/ZL - 0.5 )*ZL ;
          if( ryi < 0. ) ryi = - ryi ;
          if( ryi > YL ) ryi = YL - ( ryi - YL ) ;
```

• The particle positions at the next time step are evaluated according to Eq. (5.5).
• The periodic BC is used for the x- and z-directions.
• The elastic collision model at the boundary surface is used for the y-direction.

• The forces acting on particles are calculated in the function forcecal. The random displacements are generated in the function randisp.

```

        RX[i] = rxi ;
        RY[i] = ryi ;
        RZ[i] = rzi ;
      }
                                              /*--- cal force ---*/
      forcecal( n , rcoff, rcoff2 ) ;
                                  /*--- cal random displacement ---*/
      randisp( n , h ) ;

      /*-----------------------------------------------------------*/

                               /*--- ca
      ndnscal( n, nychk, ychk, cndns ) ;

      if( (ntime % dnsmpl) == 0 ) {
        nsmpl += 1 ;
        for ( j=1 ; j<=nychk ; j++ ) {
          cndns[j] /= (double)dnsmpl ;
          ndens[j][nsmpl] = cndns[j] / (XL*ZL*ychk[1] ) ;
          cndns[j] = 0. ;
        }
      }
```

• The value divided by the sampling number yields its average value, and then the average value divided by the volume of one sliced piece gives rise to the number density ndens[*]

```
                               /*--- data output for graphics (1) ---*/
      if( (ntime % ngraph) == 0 ) {
        inp += 1 ;
          fprintf(np[inp],"%6d%10.3f%10.3f%10.3f\n", n, XL, YL, ZL ) ;
        for (i=1 ; i<=n ; i++ ) {
          fprintf(np[inp],"%18.10e%18.10e%18.10e\n",

        }
        fclose(np[inp]) ;
      }
                               /*--- check of random numbers used ---*/
      if ( NRAN >= nranchk ) {
        rancal() ; NRAN = 1 ;
      }
    }
```

• The particle position data are written out at every ngraph time steps for the postprocessing analysis.

• The number of the used random numbers is checked. If over nranchk, a uniform random number sequence is renewed.

```

    /*-----------------------------------------------------------*/
    /*---------------------   end of main loop   -------------------*/
    /*-----------------------------------------------------------*/

                                                /*--- print out ---*/
    fprintf(np1,"nsmpl=%8d dnsmpl=%8d dtsmpl=%12.4e\n",
                                      nsmpl, dnsmpl, dtsmpl) ;
    for ( i= nsmpl/10 ; i<= nsmpl ; i += nsmpl/10 ) {
      fprintf(np1,"i=%8d time=%12.4e\n",
                                   i, dtsmpl*(double)i - dtsmpl/2. ) ;
      fprintf(np1,"ndens(1), ndens(2), ndens(3),...,ndens(nychk)\n") ;
      for ( j=1 ; j<=nychk ; j += 10 ) {
        fprintf(np1,
          "%8.4f%8.4f%8.4f%8.4f%8.4f%8.4f%8.4f%8.4f%8.4f%8.4f\n",
          ndens[j][i]  , ndens[j+1][i], ndens[j+2][i], ndens[j+3][i],
          ndens[j+4][i], ndens[j+5][i], ndens[j+6][i], ndens[j+7][i],
          ndens[j+8][i], ndens[j+9][i] ) ;
      }
    }
                                            /*--- data output (2)---*/
    fprintf(np2,"%4d%8.5f%8.5f%8.4f%14.6e%14.6e%14.6e%14.6e\n",
                            n, ndens0, phaiv0, rcoff, h, XL, YL, ZL ) ;
    fprintf(np2,"%14.6e%14.6e%14.6e\n", RB, RG, RLJ ) ;
    fprintf(np2,"%8d%8d\n", ntimemx, nychk) ;
                                            /*--- data output (3)---*/
    fprintf(np2,"%8d%8d%14.6e\n", nsmpl, dnsmpl, dtsmpl) ;
    for ( i= 1 ; i<=nsmpl ; i++ ) {
      fprintf(np2,"%8d%14.6e\n", i, dtsmpl*(double)i-dtsmpl/2) ;
      for ( j=1 ; j<=nychk ; j += 5 ) {
        fprintf(np2,"%12.4e%12.4e%12.4e%12.4e%12.4e\n",
            ndens[j][i], ndens[j+1][i], ndens[j+2][i], ndens[j+3][i],
            ndens[j+4][i] ) ;
      }
    }
              fclose (np1) ;
              fclose (np2) ;
  }
/*-------------------------------------------------------------------*/
/*------------------------- functions ------------------------------*/
/*-------------------------------------------------------------------*/
/*+++ fun iniposit +++*/
```

```
    iniposit( n , ndens )

    double  ndens ;
    int     n ;
    {
      double  rxi, ryi, rzi, rx0, ry0, rz0 , c0 ;
      int     q , k , ix , iy , iz , iface ;
                                                        /*--- start ---*/
      c0 =  pow( (4./ndens), (1./3.) ) ;
      q  =  rint(  pow( (double)(n/4), (1./3.) )  ) ;
      XL =  c0*(double)q ;
                                     /*--- set initial positions ---*/
      k  =  0 ;
      for (iface=1 ; iface<=4 ; iface++ ) {

        if( iface ==1 ) {
          rx0 = 0.0001 ; ry0 = 0.0001 ; rz0 = 0.0001 ;
        } else if( iface == 2 ) {
          rx0 = c0/2.   ; ry0 = c0/2.   ; rz0 = 0.0001 ;
        } else if( iface == 3 ) {
          rx0 = c0/2.   ; ry0 = 0.0001 ; rz0 = c0/2.   ;
        } else {
          rx0 = 0.0001 ; ry0 = c0/2.   ; rz0 = c0/2.   ;
        }

        for ( iz=0 ; iz <= q-1 ; iz++ ) {
          rzi = (double)iz*c0 + rz0 ;
          if( rzi >= XL )  break ;
          for ( iy=0 ; iy <= q-1 ; iy++ ) {
            ryi = (double)iy*c0 + ry0 ;
            if( ryi >= XL )  break ;
            for ( ix=0 ; ix <= q-1 ; ix++ ) {
              rxi = (double)ix*c0 + rx0 ;
              if( rxi >= XL )  break ;

              k += 1 ;
              RX[k] = rxi ;  RY[k] = ryi ; RZ[k] = rzi ;
            }
          }
        }
      }
    }
/*+++ fun gridcal +++*/
    gridcal( nychk, ychk )

    int     nychk ;
    double  ychk[NN] ;
    {
      double  c1 ;
      int     i ;

      c1 = YL/(double)nychk ;
      for ( i=1 ; i<= nychk ; i++ ) {
        ychk[i] = c1 * (double)i ;
      }
    }
/*+++ ndnscal +++*/
    ndnscal( n, nychk, ychk, cndns )

    int     n , nychk ;
    double  ychk[NN], cndns[NN] ;
    {
      int   i, j ;

      for ( i=1 ; i<=n      ; i++ ) {
        for ( j=1 ; j<=nychk ; j++ ) {
          if( ychk[j] >= RY[i] ) {
            cndns[j] += 1. ;
            goto L2 ;
          }
        }
        cndns[nychk] += 1. ;
 L2:    continue ;
      }
    }
/*+++ forcecal +++*/
    forcecal( n, rcoff, rcoff2 )

    double  rcoff, rcoff2 ;
    int     n ;
    {
```

• A function for setting the initial particle positions.

• $n^*=4/a^{*3}$, $a^*=(4/n^*)^{1/3}$, and $Q=(N/4)^{1/3}$. a^* and Q are saved in the variables c0 and q, respectively.

• The particles are placed in the face-centered cubic lattice formation shown in Figure 2.2(B).
• The four ways of setting provides this initial formation of particles.
• Each particle is moved in parallel by a small distance 0.0001 to remove subtle situations at outer boundary surfaces.

• In order to evaluate the local number densities, the simulation box is divided into equal volumes sliced in the y-direction.
• The y-axis side length of each volume is YL/nychk, in which nychk is the number of the sliced volumes.

• The number of the particles belonging to each volume is calculated in order to evaluate the local number density.
• The later procedure of dividing cndns[*] by the volume, leading to the number density of particles.

• A function for calculating the forces acting on particles.

```
      double  rxi   , ryi   , rzi   , rxij , ryij , rzij , rijsq ;
      double  fxi   , fyi   , fzi   , fxij , fyij , fzij , fij   ;
      double  sr2   , sr6   , sr12 ;
      int     i , j ;

      for ( i=1 ; i<=n ; i++ ) {
        FX[i] = 0. ; FY[i] = 0. ; FZ[i] = 0. ;
      }
```

- The variables for saving forces are initialized.

```
      for ( i=1 ; i<=n-1 ; i++ ) {

        rxi = RX[i] ; ryi = RY[i] ; rzi = RZ[i] ;
        fxi = FX[i] ; fyi = FY[i] ; fzi = FZ[i] ;

        for ( j=i+1 ; j<=n ; j++ ) {
```

- The consideration of the action–reaction law enables us to calculate only the pairs of particles satisfying *i*<*j*.

```
          rxij  = rxi  - RX[j] ;
          rxij += - rint(rxij/XL)*XL ;
          if( fabs(rxij) >= rcoff )  goto L10 ;
          ryij  = ryi  - RY[j] ;
/*        ryij += - rint(ryij/YL)*YL ;     */
          if( fabs(ryij) >= rcoff )  goto L10 ;
          rzij  = rzi  - RZ[j] ;
          rzij += - rint(rzij/ZL)*ZL ;
          if( fabs(rzij) >= rcoff )  goto L10 ;
```

- The treatment of the periodic BC.
- If the two particles are separated over the cutoff distance r^*_{coff}, the calculation is unnecessary.

```
          rijsq= rxij*rxij + ryij*ryij + rzij*rzij ;
          if( rijsq >= rcoff2 )      goto L10 ;

          sr2  = 1./rijsq ; sr6 = sr2*sr2*sr2 ; sr12 = sr6*sr6 ;
          fij  = ( 2.*sr12 - sr6 )/rijsq ;
          fxij = fij*rxij ;
          fyij = fij*ryij ;
          fzij = fij*rzij ;
          fxi += fxij ;
          fyi += fyij ;
          fzi += fzij ;

          FX[j] += - fxij ;
          FY[j] += - fyij ;
          FZ[j] += - fzij ;

 L10:     continue ;
        }

        FX[i] = fxi ;
        FY[i] = fyi ;
        FZ[i] = fzi ;

      }

      for( i=1 ; i<= n ; i++ ) {
        FX[i] *= RLJ*24. ;
        FY[i] *= RLJ*24. ;
        FZ[i] *= RLJ*24. ;
      }
    }
```

- The forces acting on particles are calculated according to Eq. (5.9); the constant 24 is multiplied in the later procedure.
- The action–reaction law can provide the force acting on particle *j* as $(-f_{xij})$, $(-f_{yij})$, and $(-f_{zij})$.

```
/*+++ randisp +++*/
    randisp( n , h )

    int     n ;
    double  h ;
    {
      double  ran1, ran2 ;
      int     i , j ;

      for ( i=1 ; i<= n ; i++ ) {
                                                  /*--- random disp x ---*/
        NRAN += 1 ;
        ran1 = (double)( RAN[NRAN] ) ;
        NRAN += 1 ;
        ran2 = (double)( RAN[NRAN] ) ;
        RXB[i] = pow( -2.*(2.*h*RB)*log(ran1) , 0.5 ) * cos(2.*PI*ran2);
                                                  /*--- random disp y ---*/
        NRAN += 1 ;
        ran1 = (double)( RAN[NRAN] ) ;
```

- The random displacements can be generated from Eq. (A2.3) with the variance of the right-hand side term in Eq. (5.7).

```
        NRAN += 1 ;
        ran2 = (double)( RAN[NRAN] ) ;
        RYB[i] = pow( -2.*(2.*h*RB)*log(ran1) , 0.5 ) * cos(2.*PI*ran2);
                                               /*--- random disp z ---*/
        NRAN += 1 ;
        ran1 = (double)( RAN[NRAN] ) ;
        NRAN += 1 ;
        ran2 = (double)( RAN[NRAN] ) ;
        RZB[i] = pow( -2.*(2.*h*RB)*log(ran1) , 0.5 ) * cos(2.*PI*ran2);
      }
    }
/*--- rancal ---*/
    rancal()

    {
      float   aintegmx ;
      int     integmx, integst, integ ;
      int     i ;

      integmx = 2147483647 ;
      integst = 584287     ;
      integ   = 48828125   ;

      aintegmx = (float)integmx ;

      if ( IX == 0 ) IX = integst ;
      for (i=1 ; i<NRANMX ; i++ ) {
        IX *= integ ;
        if (IX < 0 )  IX = (IX+integmx)+1 ;
        RAN[i] = (float)IX/aintegmx ;
      }
    }
```

• A function for generating a uniform random number sequence.

• This is for a 32-bit CPU based on the expression of two's complement.

6 Practice of Dissipative Particle Dynamics Simulations

In this chapter we consider an alternative microsimulation method called the dissipative particle dynamics (DPD) method," which is also available for simulating a particle suspension system. In the DPD method [4–8], the fluid is assumed to be composed of virtual fluid particles called "dissipative particles," and therefore the solution of a flow field can be obtained from the motion of the dissipative particles in a way similar to the MD method. A significant advantage of this method is that when it is applied to the simulation of a particle suspension, the multibody hydrodynamic interaction is taken into account without introducing a special technique. This characteristic of the DPD method provides it with a great potential as a simulation tool for particle suspensions; the present method is thus available for various fields of scientific research, including the pharmaceutical sciences and specialized engineering fields. The sample simulation program is written in the FORTRAN programming language.

6.1 Aggregation Phenomena of Magnetic Particles

For our example, a system composed of N magnetic particles with mass m dispersed in a base liquid is assumed to be in thermodynamic equilibrium. The main objective of the present exercise is to discuss the feasibility of the DPD method for successfully capturing the aggregate formations of the magnetic particles, which are dependent on the strength of magnetic particle–particle interactions. It is important to note that in the present demonstration we assume the applied magnetic field to be very strong, so that we only need to consider the translational motion of magnetic particles. The rotational motion may be neglected.

6.2 Specification of Problems in Equations

6.2.1 *Kinetic Equation of Dissipative Particles*

A ferromagnetic colloidal suspension is composed of ferromagnetic particles and the molecules of a base liquid. If a base liquid is regarded as being composed of dissipative particles, the motion of magnetic particles is governed by the interaction with both the other magnetic particles and the ambient dissipative particles. In the following, we show the kinetic equation for the dissipative particles.

Introduction to Practice of Molecular Simulation. DOI: 10.1016/B978-0-12-385148-2.00006-9

Three kinds of forces act on dissipative particle i: a repulsive conservative force $\mathbf{F}_{ij}^{C}$, exerted by the other particles; a dissipative force $\mathbf{F}_{ij}^{D}$, providing a viscous drag to the system; and a random or stochastic force $\mathbf{F}_{ij}^{R}$, inducing the thermal motion of particles. The force acting on the dissipative particles by magnetic particles is not taken into account in this subsection, since that force will be addressed later. The equation of motion of particle i is therefore written as

$$m_{\mathrm{d}}\frac{\mathrm{d}\mathbf{v}_i}{\mathrm{d}t} = \sum_{j(\neq i)}\mathbf{F}_{ij}^{\mathrm{C}} + \sum_{j(\neq i)}\mathbf{F}_{ij}^{\mathrm{D}} + \sum_{j(\neq i)}\mathbf{F}_{ij}^{\mathrm{R}} \tag{6.1}$$

in which

$$\mathbf{F}_{ij}^{\mathrm{D}} = -\gamma w_{\mathrm{D}}(r_{ij})(\mathbf{e}_{ij}\cdot\mathbf{v}_{ij})\mathbf{e}_{ij}, \quad \mathbf{F}_{ij}^{\mathrm{R}} = \sigma w_{\mathrm{R}}(r_{ij})\mathbf{e}_{ij}\zeta_{ij}, \quad \mathbf{F}_{ij}^{\mathrm{C}} = \alpha w_{\mathrm{R}}(r_{ij})\mathbf{e}_{ij} \tag{6.2}$$

In these equations, m_{d} is the mass of particle i, and $\mathbf{v}_i$ is the velocity. Regarding the use of subscripts, as an example, $\mathbf{F}_{ij}^{C}$ is the force acting on particle i by particle j. Moreover, α, γ, and σ are constants representing the strengths of the repulsive, the dissipative, and the random forces, respectively. The weight functions $w_{\mathrm{D}}(r_{ij})$ and $w_{\mathrm{R}}(r_{ij})$ are introduced such that the interparticle force decreases with increasing particle–particle separation. The expression for $w_{\mathrm{R}}(r_{ij})$ is written as

$$w_{\mathrm{R}}(r_{ij}) = \begin{cases} 1 - \dfrac{r_{ij}}{d_{\mathrm{c}}} & \text{for } r_{ij} \le d_{\mathrm{c}} \\ 0 & \text{for } r_{ij} > d_c \end{cases} \tag{6.3}$$

The weight functions $w_{\mathrm{D}}(r_{ij})$ and $w_{\mathrm{R}}(r_{ij})$, as well as γ and σ, must satisfy the following relationships, respectively:

$$w_{\mathrm{D}}(r_{ij}) = w_{\mathrm{R}}^2(r_{ij}), \quad \sigma^2 = 2\gamma kT \tag{6.4}$$

In the above equations, d_{c} is the apparent diameter of dissipative particles, $\mathbf{r}_{ij}$ is the relative position ($r_{ij} = |\mathbf{r}_{ij}|$), given by $\mathbf{r}_{ij} = \mathbf{r}_i - \mathbf{r}_j$; $\mathbf{e}_{ij}$ is the unit vector denoting the direction of particle i relative to particle j, expressed as $\mathbf{e}_{ij} = \mathbf{r}_{ij}/r_{ij}$; $\mathbf{v}_{ij}$ is the relative velocity, expressed as $\mathbf{v}_{ij} = \mathbf{v}_i - \mathbf{v}_j$; k is Boltzmann's constant; and T is the liquid temperature. Also, ζ_{ij} is a random variable inducing the random motion of the particles.

If Eq. (6.1) is integrated with respect to time over a small time interval Δt from t to $t + \Delta t$, then the finite difference equations governing the particle motion in simulations can be obtained as

$$\Delta\mathbf{r}_i = \mathbf{v}_i\Delta t \tag{6.5}$$

$$\begin{aligned}\Delta\mathbf{v}_i = {} & \frac{\alpha}{m_{\mathrm{d}}}\sum_{j(\neq i)} w_{\mathrm{R}}(r_{ij})\mathbf{e}_{ij}\Delta t - \frac{\gamma}{m_{\mathrm{d}}}\sum_{j(\neq i)} w_{\mathrm{R}}^2(r_{ij})(\mathbf{e}_{ij}\cdot\mathbf{v}_{ij})\mathbf{e}_{ij}\Delta t \\ & + \frac{(2\gamma kT)^{1/2}}{m_{\mathrm{d}}}\sum_{j(\neq i)} w_{\mathrm{R}}(r_{ij})\mathbf{e}_{ij}\theta_{ij}\sqrt{\Delta t}\end{aligned} \tag{6.6}$$

in which θ_{ij} is the stochastic variable that must satisfy the following stochastic properties:

$$\langle\theta_{ij}\rangle = 0, \quad \langle\theta_{ij}\theta_{i'j'}\rangle = (\delta_{ii'}\delta_{jj'} + \delta_{ij'}\delta_{ji'}) \tag{6.7}$$

in which δ_{ij} is the Kronecker delta. During the simulation, the stochastic variable θ_{ij} is sampled from a uniform or normal distribution with zero average value and unit variance.

6.2.2 Model of Particles

A magnetic particle is idealized as a spherical particle with a central point dipole and is coated with a uniform steric layer (or surfactant layer). Using the notation d_s for the diameter of the particle, δ for the thickness of the steric layer, and d $(=d_s + 2\delta)$ for the diameter, including the steric layer, then the magnetic interaction energy between particles i and j, $u_{ij}^{(m)}$, and the particle–field interaction energy, $u_i^{(H)}$, and the interaction energy arising due to the overlap of the steric layers, $u_{ij}^{(V)}$, are expressed, respectively, as [31]

$$u_{ij}^{(m)} = \frac{\mu_0}{4\pi r_{ij}^3}\{\mathbf{m}_i\cdot\mathbf{m}_j - 3(\mathbf{m}_i\cdot\mathbf{t}_{ij})(\mathbf{m}_j\cdot\mathbf{t}_{ij})\} \tag{6.8}$$

$$u_i^{(H)} = -\mu_0\mathbf{m}_i\cdot\mathbf{H} \tag{6.9}$$

$$u_{ij}^{(V)} = kT\lambda_V\left\{2 - \frac{2r_{ij}/d_s}{t_\delta}\ln\left(\frac{d}{r_{ij}}\right) - 2\frac{r_{ij}/d_s - 1}{t_\delta}\right\} \tag{6.10}$$

in which μ_0 is the permeability of free space, $\mathbf{m}_i$ is the magnetic moment $(m_0 = |\mathbf{m}_i|)$, $\mathbf{t}_{ij}$ is the unit vector given by $\mathbf{r}_{ij}/r_{ij}$, $\mathbf{r}_{ij} = \mathbf{r}_i - \mathbf{r}_j$, $r_{ij} = |\mathbf{r}_{ij}|$, $\mathbf{H}$ is the applied magnetic field $(H = |\mathbf{H}|)$, and t_δ is the ratio of the thickness of the steric layer δ to the radius of the solid part of the particle, equal to $2\delta/d_s$. The nondimensional parameter λ_V, appearing in Eq. (6.10), represents the strength of the steric particle–particle interaction relative to the thermal energy, expressed as $\lambda_V = \pi d_s^2 n_s/2$, in which n_s is the number of surfactant molecules per unit area on the particle surface.

From Eqs. (6.8) and (6.10), the forces acting on particle i are derived as

$$\mathbf{F}_{ij}^{(m)} = -\frac{3\mu_0}{4\pi r_{ij}^4}\left[-(\mathbf{m}_i\cdot\mathbf{m}_j)\mathbf{t}_{ij} + 5(\mathbf{m}_i\cdot\mathbf{t}_{ij})(\mathbf{m}_j\cdot\mathbf{t}_{ij})\mathbf{t}_{ij} - \{(\mathbf{m}_j\cdot\mathbf{t}_{ij})\mathbf{m}_i + (\mathbf{m}_i\cdot\mathbf{t}_{ij})\mathbf{m}_j\}\right] \tag{6.11}$$

$$\mathbf{F}_{ij}^{(V)} = \frac{kT\lambda_V}{\delta}\cdot\frac{\mathbf{r}_{ij}}{r_{ij}}\ln\left(\frac{d}{r_{ij}}\right) \quad (d_s \le d_{ij} \le d) \tag{6.12}$$

In addition to these forces, the forces due to dissipative particles have to be taken into account, but are not treated here, since they will be addressed in the following subsection.

The motion of magnetic particles is specified by Newton's equations and are discretized in time to obtain the finite difference equations governing the particle motion in simulations:

$$\Delta \mathbf{r}_i = \mathbf{v}_i \Delta t \tag{6.13}$$

$$\Delta \mathbf{v}_i = \sum_{j(\neq i)} \mathbf{F}_{ij} \Delta t / m_{\mathrm{m}} \tag{6.14}$$

in which m_{m} is the mass of magnetic particles and $\mathbf{F}_{ij} = \mathbf{F}_{ij}^{(\mathrm{m})} + \mathbf{F}_{ij}^{(\mathrm{V})}$.

6.2.3 Model Potential for Interactions Between Dissipative and Magnetic Particles

Each colloidal particle is modeled as a group of dissipative particles. In the ordinary application of the method, the interaction of a magnetic particle with the ambient dissipative particles is treated as the interaction between the ambient dissipative particles and the constituent dissipative particles of the magnetic particle. However, in a real dispersion, the interaction between colloidal particles and the solvent molecules should depend on the characteristics of the dispersion of interest. Such interactions are strongly dependent on the ratio of the mass and the diameter of the colloidal particles to that of solvent molecules together with the properties of the interaction potential.

Therefore, instead of regarding a colloidal particle as a group of dissipative particles, it may be possible to use a model potential to describe the interaction between the magnetic and the ambient dissipative particles.

The simplest potential model may be the hard sphere potential, in which magnetic particles are regarded as a hard sphere and dissipative particles are elastically reflected on the contact with a magnetic particle. Another simple potential model may be the Lennard-Jones potential. Although the present exercise adopts the latter model potential and attempts to discuss its validity, the simple form of the Lennard-Jones potential based on each particle center may cause a nonphysical overlap. Hence, as shown in Figure 6.1, we consider an inscribed sphere with the same diameter as the dissipative particles, which is located on the line connecting each center of dissipative and magnetic particle. The Lennard-Jones potential is then employed using the inscribed particle and dissipative particles such that the interaction energy u_{ip} for dissipative particle p and magnetic particle i is expressed as

$$u_{ip} = 4\varepsilon \left\{ \left(\frac{d_{\mathrm{c}}}{r_{ip}'} \right)^m - \left(\frac{d_{\mathrm{c}}}{r_{ip}'} \right)^n \right\} \tag{6.15}$$

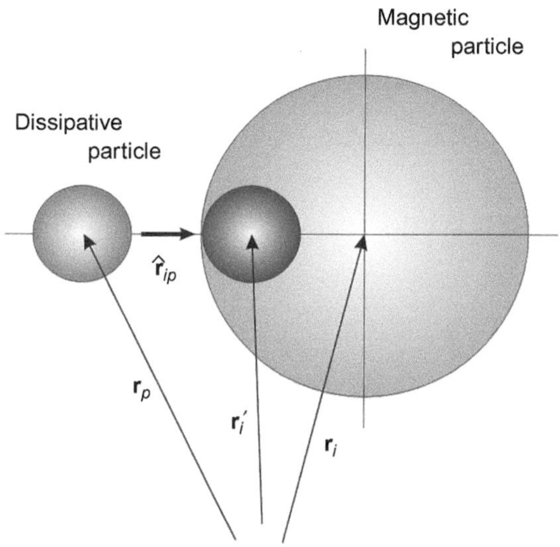

Figure 6.1 Model of the interaction between magnetic and dissipative particles.

in which ε is a constant representing the strength of such an interaction, $\mathbf{r}_{ip}' = \mathbf{r}_i' - \mathbf{r}_p$, $r_{ip}' = |\mathbf{r}_{ip}'|$, $\mathbf{r}_i$ is the position vector of the center of magnetic particle i, $\mathbf{r}_p$ is similarly the position vector of dissipative particle p, and $\mathbf{r}_i'$ is the position vector of the inscribed sphere. The expression for $\mathbf{r}_i'$ is written as

$$\mathbf{r}_i' = \mathbf{r}_i - (d - d_c/2)\hat{\mathbf{r}}_{ip} \tag{6.16}$$

in which $\hat{\mathbf{r}}_{ip} = \mathbf{r}_{ip}/r_{ip}$, $\mathbf{r}_{ip} = \mathbf{r}_i - \mathbf{r}_p$, and $r_{ip} = |\mathbf{r}_{ip}|$. If we set $m = 12$ and $n = 6$ in Eq. (6.15), the model potential leads to the well-known Lennard-Jones 12–6 potential, and this potential is employed in the present simulation.

From the expression of the interaction energy in Eq. (6.15), the force acting on dissipative particle p by magnetic particle i, $\mathbf{F}_{ip}^{(\mathrm{int})}$ is derived as

$$\mathbf{F}_{ip}^{(\mathrm{int})} = 4n\varepsilon \left\{ \frac{m}{n} \left(\frac{d_c}{r_{ip}'} \right)^m - \left(\frac{d_c}{r_{ip}'} \right)^n \right\} \frac{\hat{\mathbf{r}}_{ip}}{r_{ip}'} \tag{6.17}$$

6.2.4 Nondimensionalization of the Equation of Motion and Related Quantities

For the nondimensionalization of each quantity, the following representative values are used: d for distances, m_m for masses, kT for energies, $(kT/m_m)^{1/2}$ for velocities, $d(m_m/kT)^{1/2}$ for time, kT/d for forces, and so forth. With these representative values, Eqs. (6.5) and (6.6) are nondimensionalized as

$$\Delta \mathbf{r}_i^* = \mathbf{v}_i^* \Delta t^* \tag{6.18}$$

$$\Delta \mathbf{v}_i^* = \frac{1}{m_d^* d_c^*}\alpha^* \sum_{j(\neq i)} w_R(r_{ij}^*)\mathbf{e}_{ij}\Delta t^* - \frac{1}{(m_d^*)^{1/2} d_c^*}\gamma^* \sum_{j(\neq i)} w_R^2(r_{ij}^*)(\mathbf{e}_{ij}\cdot \mathbf{v}_{ij}^*)\mathbf{e}_{ij}\Delta t^*$$
$$- \frac{1}{(m_d^*)^{3/4} d_c^{*1/2}}(2\gamma^*)^{1/2} \sum_{j(\neq i)} w_R(r_{ij}^*)\mathbf{e}_{ij}\theta_{ij}\sqrt{\Delta t^*} - \frac{1}{m_d{}^*}\sum_k \mathbf{F}_{ki}^{(\mathrm{int})*}\Delta t^* \tag{6.19}$$

in which

$$w_R(r_{ij}^*) = \begin{cases} 1 - r_{ij}^*/d_c^* & \text{for } r_{ij}^*/d_c^* \leq 1 \\ 0 & \text{for } r_{ij}^*/d_c^* > 1 \end{cases} \tag{6.20}$$

$$\alpha^* = \alpha \frac{d_c}{kT}, \quad \gamma^* = \gamma \frac{d_c}{(m_d kT)^{1/2}} \tag{6.21}$$

In the above equations, the superscript $*$ indicates the nondimensionalized quantities. Note that Eq. (6.19) includes the forces due to the interaction with magnetic particles, described in Section. 6.2.3.

Similarly, the nondimensional form of Eqs. (6.13), (6.14), (6.11), and (6.12) are expressed as

$$\Delta \mathbf{r}_i{}^* = \mathbf{v}_i{}^* \Delta t^* \tag{6.22}$$

$$\Delta \mathbf{v}_i^* = \sum_{j(\neq i)} \mathbf{F}_{ij}^* \Delta t^* + \sum_p \mathbf{F}_{ip}^{(\mathrm{int})*}\Delta t^* \tag{6.23}$$

$$\mathbf{F}_{ij}^{(\mathrm{m})*} = -3\lambda \frac{1}{r_{ij}^{4*}}\left[-(\mathbf{n}_i\cdot\mathbf{n}_j)\mathbf{t}_{ij} + 5(\mathbf{n}_i\cdot\mathbf{t}_{ij})(\mathbf{n}_j\cdot\mathbf{t}_{ij})\mathbf{t}_{ij} - \left\{(\mathbf{n}_j\cdot\mathbf{t}_{ij})\mathbf{n}_i + (\mathbf{n}_i\cdot t_{ij})\mathbf{n}_j\right\}\right] \tag{6.24}$$

$$\mathbf{F}_{ij}^{(\mathrm{V})*} = \lambda_V \frac{1}{t_\delta^*}\cdot \mathbf{t}_{ij}\ln\left(\frac{1}{r_{ij}{}^*}\right) \quad (d_s{}^* \leq r_{ij}{}^* \leq 1) \tag{6.25}$$

in which $\mathbf{F}_{ij}^* = \mathbf{F}_{ij}{}^{(\mathrm{m})*} + \mathbf{F}_{ij}{}^{(\mathrm{V})*}$, $\mathbf{n}_i$ is the unit vector denoting the direction of the magnetic moment $\mathbf{m}_i$, expressed as $\mathbf{n}_i = \mathbf{m}_i/m_0$ ($m_0 = |\mathbf{m}_i|$). The nondimensional parameter λ in Eq. (6.24) is the strength of magnetic particle interactions relative to the thermal energy, expressed as $\lambda = \mu_0 m_0{}^2/4\pi d^3 kT$. A slightly different parameter $\lambda_s = (d/d_s)^3\lambda$ ($=\mu_0 m_0{}^2/4\pi d_s{}^3 kT$), which is defined based on the diameter of the solid part, will be useful in order to compare the present results with the previous MC and BD simulations.

The expression of the force between a dissipative and a magnetic particle is written in nondimensional form as

$$\mathbf{F}_{ip}^{*} = \lambda_{\varepsilon}\left\{\frac{m}{n}\left(\frac{d_{c}^{*}}{r_{ip}^{\prime *}}\right)^{m} - \left(\frac{d_{c}^{*}}{r_{ip}^{\prime *}}\right)^{n}\right\}\frac{\hat{\mathbf{r}}_{ip}}{r_{ip}^{\prime *}/d_{c}^{*}} \tag{6.26}$$

in which λ_{ε} is a nondimensional parameter representing the strength of the interaction, expressed as $\lambda_{\varepsilon} = 4n\varepsilon/(kTd_{c}^{*})$.

In the present simulation we consider a two-dimensional system in thermodynamic equilibrium, and therefore the relationship between the system temperature and the mean kinetic energy of one dissipative particle is expressed from the equipartition law of energies as

$$\overline{\frac{1}{2}m_{d}v_{d}^{2}} = 2\frac{kT}{2} \tag{6.27}$$

From this equation, the mean square velocity of dissipative particles $\overline{v_{d}^{*2}}$ is written as

$$\overline{v_{d}^{*2}} = 2/m_{d}^{*} \tag{6.28}$$

Similarly, the mean square velocity of magnetic particles $\overline{v_{m}^{*2}}$ is expressed as

$$\overline{v_{m}^{*2}} = 2 \tag{6.29}$$

The number density of dissipative particles is nondimensionalized as

$$n_{d}^{*} = n_{d}d^{2} = n_{d}d_{c}^{2}(d/d_{c})^{2} = \hat{n}_{d}^{*}/d_{c}^{*2} \tag{6.30}$$

In addition to n_{d}^{*}, the nondimensional density $\hat{n}_{d}^{*}$ based on the diameter of dissipative particles may be useful for quantifying the packing characteristics of the dissipative particles. The nondimensional number density of magnetic particles is expressed as $n_{m}^{*} = n_{m}d^{2}$.

6.3 Parameters for Simulations

In this chapter, we are considering a two-dimensional dispersion composed of ferromagnetic particles in order to investigate the validity of using the method for this type of problem. The equations of motion of a dissipative particle include many indefinite factors, so we have chosen to focus on a simplified case in which the external magnetic field is strong enough that we may neglect the rotational motion of magnetic particles. In this situation, each magnetic moment will point along the magnetic field direction. Also, we will only focus on the one specific model potential of $(m,n) = (12,6)$. Representative parameters used for the present simulations are $\gamma^{*} = 10$, $\alpha^{*} = \gamma^{*}/10$, $m_{d}^{*} = 0.01$, $d_{c}^{*} = 0.4$, $\lambda_{\varepsilon} = 10$, $\hat{n}_{d}^{*} = 1$, and $\Delta t^{*} = 0.0001$.

Eq. (6.19) shows that the displacement distance of a dissipative particle per unit time step becomes greater with decreasing values of m_d^* and d_c^*, and for this reason the time interval Δt^* will be adjusted in proportion to the product of m_d^* and d_c^*. In this way, a smaller value of the time interval is employed as the value of $m_d^* d_c^*$ decreases. The total number of simulation steps, N_{timemx}, is expected to be sufficient when the condition of $\Delta t^* N_{timemx} = 100$ is satisfied.

6.4 Results of Simulations

We treat a multiparticle system with the number density of $n_m^* \simeq 0.4$, composed of 81 magnetic particles, to investigate the influence of the mass of dissipative particles on the aggregate structures. Figure 6.2 illustrates the results for aggregate structures in thermodynamic equilibrium for two cases of magnetic particle–particle interactions, $\lambda_s = 10$ and 3. Unless specifically noted, all simulation results were obtained for the case of $d_c^* = 0.4$ using the other representative values of the parameters given in Section 6.3. Figures 6.2A and B are for a value of the mass of dissipative particles, $m_d^* = 0.05$. Figures 6.2C and D are for $m_d^* = 0.01$. Figures 6.2E and F are for $m_d^* = 0.005$. Figures 6.2A, C, and E were obtained for $\lambda_\varepsilon = 10$. Figures 6.2B, D, and F are for $\lambda_s = 3$. In the figures, small and large circles indicate the dissipative and magnetic particles, respectively.

Since the magnetic particle–particle interaction is much more dominant than the thermal energy for $\lambda_s = 10$, magnetic particles tend to aggregate to form chain-like clusters along the magnetic field direction, which was clearly shown in the

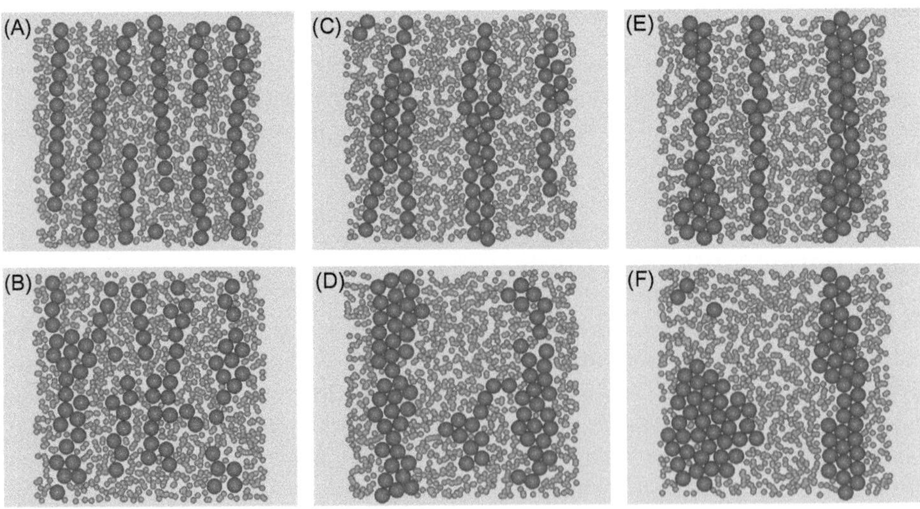

Figure 6.2 Influence of the particle mass m_d^* on the aggregate structures for $d_c^* = 0.4$: (A) for $m_d^* = 0.05$ and $\lambda_s = 10$, (B) for $m_d^* = 0.05$ and $\lambda_s = 3$, (C) for $m_d^* = 0.01$ and $\lambda_s = 10$, (D) for $m_d^* = 0.01$ and $\lambda_s = 3$, (E) for $m_d^* = 0.005$ and $\lambda_s = 10$, and (F) for $m_d^* = 0.005$ and $\lambda_s = 3$.

previous MC simulations. As shown in Figures 6.2A, C, and E, the present DPD simulation results also reproduce this type of cluster formation well. However, the aggregate structures seem to be strongly dependent on the mass of the dissipative particles. That is, although only thin chain-like clusters are formed for the case of a relatively large mass, such as $m_d^* = 0.05$, magnetic particles form thicker chain-like clusters with decreasing values of the particle mass.

Now, we consider why much thicker chain-like clusters tend to form with decreasing mass of the dissipative particles. If the mass of dissipative particles is small, the magnetic particles should move easily by separating the ambient dissipative particles so they can force a path and approach each other. The thin chain-like clusters shown in Figure 6.2A, therefore, have a sufficient probability to aggregate to form the thicker chain-like clusters shown in Figure 6.2E. On the other hand, Eq. (6.28) shows that dissipative particles with smaller mass move with larger average velocity for a given system temperature. Hence, although a chain-like cluster can thicken to a certain degree, after that further growth is limited by the Brownian motion of the magnetic particles due to the influence of the active motion of dissipative particles. Since the magnetic particle–particle interaction is of a slightly larger order than the thermal energy for the case of $\lambda_s = 3$, significant aggregates should not be formed. However, the present DPD simulations exhibit significant cluster formation with decreasing mass of dissipative particles; such unexpected aggregate formation is significant for $m_d^* = 0.005$, and we also find that relatively long chain-like clusters are formed even for the case of $m_d^* = 0.05$. In order to explain these results, the first consideration must be that we do not use an equation of motion which can simulate the rotational motion of the magnetic particles, although the transnational motion is taken into account in the present exercise. Another consideration must be the model potential we have employed for the interaction between the magnetic and the dissipative particles.

For reference, the aggregate structures for $d_c^* = 0.2$ are shown in Figure 6.3 under the same conditions as in Figure 6.2 except for the particle diameter. We here focus on the differences between the aggregate structures in Figures 6.2 and 6.3 without addressing the features of each aggregate structure in detail. The aggregates in Figure 6.3 have a more compact or denser internal structure, and it appears that large clusters are formed to a certain degree but do not grow any further. It seems as if the Brownian motion of the magnetic particles due to the interaction with the dissipative particles is not significant. The snapshot in Figure 6.3F also shows aggregates with a dense internal structure, and the effect of the particle Brownian motion does not appear significantly in the formation of these internal structures.

Finally, we consider what the appropriate mass of a dissipative particle should be for obtaining physically reasonable results. As pointed out previously, dissipative particles are virtual and regarded as groups or clusters of the real solvent molecules, so that it seems to be reasonable for the mass density of dissipative particles to be taken as roughly equal to the mass density of the base liquid of the dispersion system, which one must consider for evaluating physical quantities experimentally. In the present demonstration, for example, we consider a ferromagnetic colloidal

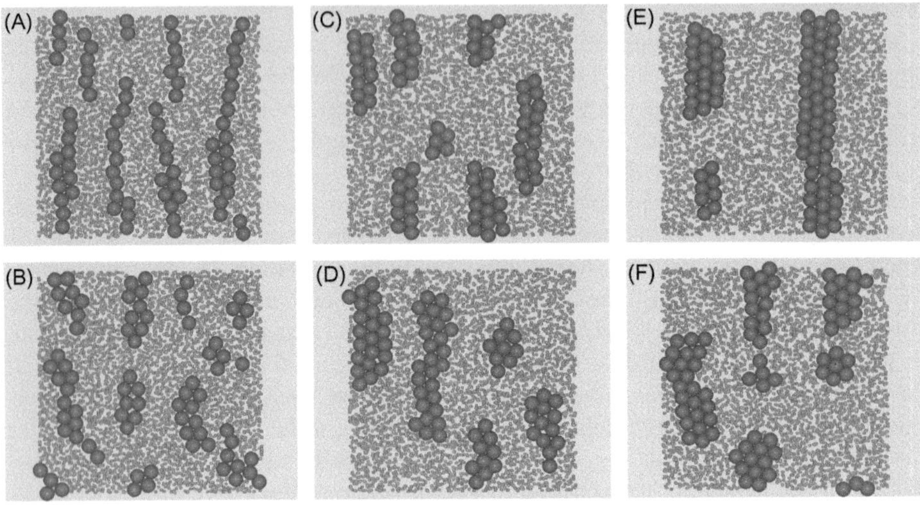

Figure 6.3 Influence of the particle mass m_d^* on aggregate structures for $d_c^* = 0.2$: (A) for $m_d^* = 0.05$ and $\lambda_s = 10$, (B) for $m_d^* = 0.05$ and $\lambda_s = 3$, (C) for $m_d^* = 0.01$ and $\lambda_s = 10$, (D) for $m_d^* = 0.01$ and $\lambda_s = 3$, (E) for $m_d^* = 0.005$ and $\lambda_s = 10$, and (F) for $m_d^* = 0.005$ and $\lambda_s = 3$.

dispersion in which metallic ferromagnetic fine particles are assumed to be dispersed into a base liquid, such as kerosene or water. In this case, if the ratio of the mass density of magnetic particles to dissipative ones is regarded as 5–8, then the ratio of mass is 0.013–0.008 for $d_c^* = 0.4$, and 0.0016–0.001 for $d_c^* = 0.2$. Hence, it is for the case of $d_c^* = 0.4$ and $m_d^* = 0.01$ that physically reasonable aggregate structures can be regarded as being reproduced. This consideration is verified by comparing it with the results obtained by MC and BD simulations.

In addition to the previous discussion, it may be necessary to verify that the aggregate formation is truly induced by the magnetic interaction between magnetic particles in a physically reasonable manner and not by certain false mechanisms arising from the improper interaction between dissipative and magnetic particles. Figure 6.4A and B show the results that were obtained for the strength of magnetic interaction $\lambda_s = 0$ by using the aggregate structures in Figures 6.2C and 6.3C as an initial configuration. Since the snapshot in Figure 6.4B from an initial configuration in Figure 6.3C for $d_c^* = 0.2$ and $m_d^* = 0.01$ exhibits the formation of large aggregates, we may conclude that this case does not give rise to physically reasonable results. In contrast, for the case of an initial configuration in Figure 6.2C for $d_c^* = 0.4$ and $m_d^* = 0.01$, Figure 6.4A shows that the thick chain-like clusters, formed in the field direction, are dissociated sufficiently. However, a large aggregate (i.e., not chain-like) still remains, although the internal structure of this aggregate is considerably looser. The dissociation of the chain-like clusters indicates that the Brownian motion has been sufficiently effective. On the other hand, this type of loose aggregate structure of magnetic particles may be the result of employing a kinetic equation without including the rotational motion, as adopted here, or from

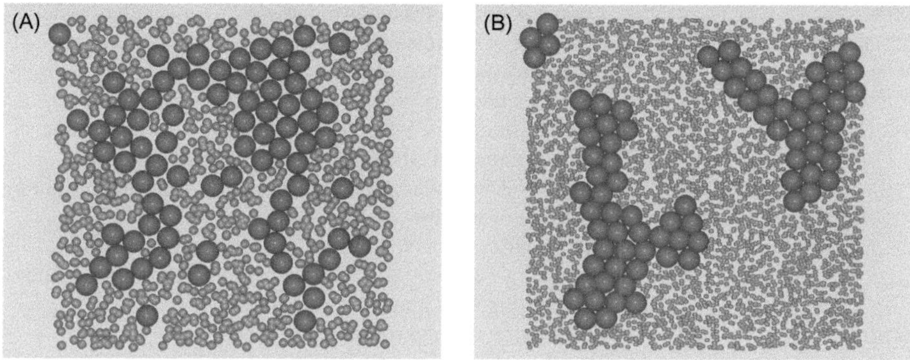

Figure 6.4 Snapshots for $\lambda_s = 0$ for the two initial configurations: Figures 6.2C and 6.3C were used as an initial configuration for (A) and (B), respectively.

employing the model potential for the interaction between dissipative and magnetic particles.

6.5 Simulation Program

A sample simulation program is shown below for conducting the simulation of the present exercise: the program is written in FORTRAN.

The important variables used in the program are explained as follow:

`RX(I),RY(I)`	:	(x,y) coordinates of the position vector $\mathbf{r}_i^*$ of magnetic particle i
`NX(I),NY(I)`	:	(x,y) coordinates of the magnetic moment direction $\mathbf{n}_i^*$
`VX(I),VY(I)`	:	(x,y) coordinates of the velocity $\mathbf{v}_i^*$ of magnetic particle i
`FX(I),FY(I)`	:	(x,y) coordinates of the magnetic force $\mathbf{F}_i^*$ acting on magnetic particle i
`FXMD(I),FYMD(I)`	:	(x,y) coordinates of the force acting on magnetic particle i by dissipative particles
`N,NDENS,VDENS`	:	Number of particles N, number density n^*, volumetric fraction ϕ_V^* concerning magnetic particles
`D,DS,DEL`	:	Diameter, the diameter of solid part, the thickness of the steric layer of magnetic particles
`(XL,YL)`	:	Side lengths of the simulation box in the (x,y) directions
`RAS,RA,RV,RE`	:	Nondimensional parameters λ_s, λ, λ_V, and λ_ε
`OVRLAP(I)`	:	`OVERLAP(I)` = `.TRUE.` in the case of an extraordinary overlap of magnetic particles
`RXD(I),RYD(I)`	:	Position vector $\mathbf{r}_i^*$ of dissipative particle i
`VXD(I),VYD(I)`	:	Velocity vector $\mathbf{v}_i^*$ of dissipative particle i
`FDXD(I),FDYD(I)`	:	Dissipative force $\mathbf{F}_i^{D*}$ acting on dissipative particle i
`FCXD(I),FCYD(I)`	:	Conservative force $\mathbf{F}_i^{C*}$ acting on dissipative particle i
`FRXD(I),FRYD(I)`	:	Random force $\mathbf{F}_i^{R*}$ acting on dissipative particle i

`FXDM(I),FYDM(I)`	:	Force acting on dissipative particle i by magnetic particles
`ND,MD,DC,VDENSD`	:	Number of particles, mass m_d^*, diameter d_c^*, volumetric fraction concerning dissipative particles
`NDENSD,NDENSDH`	:	Number densities of dissipative particles n_d^*, $\hat{n}_d^*$
`TMX(I),TABLE(*,I)`	:	Names of cells to which dissipative particles interacting with the magnetic particle of interest belong
`VTMX(I), VTABLE(#)`	:	Names of magnetic particles interacting with the magnetic particle of interest
`VPLACE(I)`	:	Information starts to appear from the position `VPLACE(I)` in the variable `VTABLE(I)` concerning magnetic particles interacting with magnetic particle i
`TMXD(GRP), TABLED(*,GRP))`	:	Cell index method for dissipative particles
`GRPX(I),GRPY(I)`	:	Name of cell to which dissipative particle i belongs is saved
`ALP,GAM`	:	Parameters α^* and γ^* representing the strengths of repulsive and dissipative forces acting between dissipative particles, respectively

As an aid to understanding the program, explanatory comments have been added to important features. The line numbers are only for the reader's convenience, and unnecessary for executing a FORTRAN program.

```
C***********************************************************************
C*                            dpdmag3.f                                *
C*                                                                     *
C*       OPEN(9, FILE='@daa1.data', STATUS='UNKNOWN'); parameters      *
C*       OPEN(10,FILE='daa11.data', STATUS='UNKNOWN'); para. & data    *
C*       OPEN(11,FILE='daa21.mgf' , STATUS='UNKNOWN'); anime data      *
C*       OPEN(21,FILE='daa001.data',STATUS='UNKNOWN'); particle pos.   *
C*       OPEN(22,FILE='daa011.data',STATUS='UNKNOWN'); particle pos.   *
C*       OPEN(23,FILE='daa021.data',STATUS='UNKNOWN'); particle pos.   *
C*       OPEN(24,FILE='daa031.data',STATUS='UNKNOWN'); particle pos.   *
C*       OPEN(25,FILE='daa041.data',STATUS='UNKNOWN'); particle pos.   *
C*       OPEN(26,FILE='daa051.data',STATUS='UNKNOWN'); particle pos.   *
C*       OPEN(27,FILE='daa061.data',STATUS='UNKNOWN'); particle pos.   *
C*       OPEN(28,FILE='daa071.data',STATUS='UNKNOWN'); particle pos.   *
C*       OPEN(29,FILE='daa081.data',STATUS='UNKNOWN'); particle pos.   *
C*       OPEN(30,FILE='daa091.data',STATUS='UNKNOWN'); particle pos.   *
C*                                                                     *
C*    ----------  DPD SIMULATION OF MAGNETIC PARTICLES  ----------     *
C*        TWO-DIMENSIONAL DPD SIMULATION OF MAGNETIC SPHERICAL         *
C*        PARTICLES IN DISSIPATIVE PARTICLES                           *
C*                                                                     *
C*          1. FOR A STRONG MAGNETIC FIELD CASE (Y-DIRECTION).         *
C*          2. FERROMAGNETIC SPHERICAL PARTICLES WITH STERIC LAYER.    *
C*          3. LENNARD-JONES MODEL FOR INTERACTIONS BETWEEN            *
C*             MAGNETIC AND DISSIPATIVE PARTICLES.                     *
C*          4. NNN SHOULD BE SUFFICIENTLY LARGE (NNN=10000)            *
C*          5. OVRLAP(*) IS INTRODUCED.                                *
C*                                                                     *
C*                                       VER.1 BY A.SATOH, '09  4/5    *
C***********************************************************************
C
C     N     : NUMBER OF MAGNETIC PARTICLES (M. PTCL.)
C     D     : DIAMETER OF PARTICLE INCLUDING SURFACTANT LAYER
C              ( =1 FOR THIS CASE )
C     DS    : DIAMETER OF SOLID PARTICLE WITHOUT STERIC LAYER
C     DEL   : THICKNESS OF STERIC LAYER
C     TD    : DIMENSIONLESS THICKNESS OF STERIC LAYER BASED ON RADIUS
C     NDENS : NUMBER DENSITY OF M. PTCL
```

```
C     VDENS : VOLUMETRIC FRACTION OF PARTICLES
C     RA    : NONDIMENSIONAL PARAMETER OF PARTICLE-PARTICLE INTERACT
C     RAS   : NONDIMENSIONAL PARAMETER OF PARTICLE-PARTICLE INTERACT
C             BASED ON THE DIAMETER OF THE SOLID PART
C     KU    : NONDIMENSIONAL PARAMETER OF PARTICLE-FIELD INTERACTION
C     RV    : NONDIMENSIONAL PARAMETER OF STERIC REPULSION (=120)
C     RVS   : NONDIMENSIONAL PARAMETER OF STERIC REPULSION
C             BASED ON THE DIAMETER OF THE SOLID PART (=150)
C     RE    : NONDIMENSIONAL PARAMETER OF M.PTCL.-D.PTCL INTERACTION
C     RCOFF    :CUTOFF RADIUS FOR CALCULATION OF MAG. FORCES
C     RCOFFMD  :CUTOFF RADIUS FOR FORCES BETWEEN M.PTCL. AND D.PTCL.
C     RCOFFDDM :CUTOFF RADIUS FOR FORCES BETWEEN P.PTCL. AND VIRTUAL
C               PTCL. INSIDE M.PTCL.
C     XL,YL : DIMENSIONS OF SIMULATION REGION
C     H     : TIME INTERVAL FOR DPD SIMULATIONS
C     (HX,HY,HZ)  : APPLIED MAGNETIC FIELD (UNIT VECTOR)
C     VELTHRY     : AVERAGE OF (VX**2+VY**2) (DESIRED) FOR M-PTCL
C     VELTHRYD    : AVERAGE OF (VX**2+VY**2) (DESIRED) FOR D-PTCL
C     NVELSC      : VELOCITIES OF M-PTCL ARE SCALED EVERY NVELSC
C                   TIME STEP TO SATISFY THE DESIRED VELOCITY
C     NVELSCD     : VELOCITIES OF D-PTCL ARE SCALED EVERY NVELSCD
C                   TIME STEP TO SATISFY THE DESIRED VELOCITY
C
C     RX(N),RY(N) : PARTICLE POSITION
C     NX(N),NY(N) : DIRECTION OF MAGNETIC MOMENT
C     VX(N),VY(N) : PARTICLE VELOCITY
C     FX(N),FY(N) : PARTICLE FORCE DUE TO MAGNETIC FORCES
C     FXMD(N),FYMD(N) : PARTICLE FORCE BY D. PTCL. ON M. PTCL.
C     TMX(I)      : TOTAL NUMBER OF INDEX CELLS OF D. PTCL. WHICH MAY
C                   INTERACT WITH M. PTCL. I
C     TABLE(*,I)  : NAME OF INDEX CELLS WHICH MAY INTERACT WITH M. PTCL.
C     VTMX(I)     : TOTAL NUMBER OF NEIGHBORING M.PTCL. WHICH MAY
C                   INTERACT WITH M.PTCL. WITHIN THE CUTOFF RANGE
C     VTABLE(NNN): NAME OF M.PTCL. IS SAVED IN ORDER (VERLET METHOD)
C     VPLACE(I)   : THE FIRST PTCL., WHICH INTERACTS WITH PTCL. I,
C                   APPEARS AT VPLACE(I) IN THE TABLE OF VTABLE(**)
C     VRADIUS     : CUTOFF RADIUS FOR VERLET METHOD
C     NVTABLE     : VERLET TABLE IS RENEWED EVERY NVTABLE TIME STEP
C
C     OVRLAP(*)   : OVRLAP(I)=.TRUE. FOR OVERLAPING
C
C     ND      : NUMBER OF DISSIPATIVE PARTICLES (D.PTCL.)
C     MD      : MASS OF D.PTCL.
C     DC      : DIAMETER OF D.PTCL.
C     RCOFFD  : CUTOFF DISTANCE FOR INTERACTIONS BETWEEN D. PTCL.
C     ALP     : COEFFICIENT REPRESENTING REPULSIVE FORCE OF D.PTCL.
C     GAM     : COEFFICIENT REPRESENTING DISSIPATIVE FORCE OF D.PTCL.
C
C     RXD(ND),RYD(ND)    : POSITIONS OF D.PTCL.
C     VXD(ND),VYD(ND)    : VELOCITIES OF D.PTCL.
C     FCXD(ND),FCYD(ND)  : CONSERVATIVE FORCES ACTING ON A PARTICLE
C     FDXD(ND),FDYD(ND)  : DISSIPATIVE FORCES  ACTING ON A PARTICLE
C     FRXD(ND),FRYD(ND)  : RANDOM FORCES ACTING ON A PARTICLE
C     FXDM(ND),FYDM(ND)  : PARTICLE FORCE BY M. PTCL. ON D. PTCL.
C     NDENSDH : NUMBER DENSITY WITH HAT
C     NDENSD  : NUMBER DENSITY OF D.PTCL.
C     VDENSD  : VOLUMETRIC FRACTION OF D.PTCL.
C
C     GRPX(ND),GRPY(ND)   : GROUP TO WHICH D.PTCL. I BELONGS
C     PXD             : NUMBER OF CUT-OFF CELLS IN EACH DIRECTION
C     TMXD(GRP)       : TOTAL NUMBER OF PTCL. BELONGING TO GROUP(GRP)
C     TABLED(*,GRP): NAME OF PTCL. BELONGING TO GROUP(GRP)
C     GRPLXD(PXD)     : IS USED FOR DETERMINE THE CELL TO WHICH A
C                       PARTICLE IS BELONG
C
C     RAN(NRANMX)  : RANDOM NUMBERS BETWEEN  0  AND  1
C
C                 -XL/2 <RX(I) <XL/2 , -YL/2 <RY(I)< YL/2
C-----------------------------------------------------------------
      IMPLICIT REAL*8 (A-H,O-Z), INTEGER (I-N)
C
      COMMON /BLOCK1/  RX , RY
```

```
      COMMON /BLOCK2/  VX , VY
      COMMON /BLOCK3/  NX , NY
      COMMON /BLOCK5/  FX , FY
      COMMON /BLOCK7/  N  , NDENS , VDENS , D , DS , DEL , TD
      COMMON /BLOCK8/  RA , RV , RE
      COMMON /BLOCK9/  TMX  , TABLE
      COMMON /BLOCK10/ VTMX , VTABLE , VPLACE , NVTABLE , VRADIUS
      COMMON /BLOCK11/ FXMD , FYMD , RCOFFMD , RCOFFDDM
      COMMON /BLOCK13/ OVRLAP
      COMMON /BLOCK15/ H , XL , YL , RCOFF
      COMMON /BLOCK16/ VELTHRY, VELTHRYD, NVELSC, NVELSCD
C
      COMMON /BLOCK21/ RXD  , RYD
      COMMON /BLOCK22/ VXD  , VYD
      COMMON /BLOCK23/ FCXD , FCYD
      COMMON /BLOCK24/ FDXD , FDYD
      COMMON /BLOCK25/ FRXD , FRYD
      COMMON /BLOCK26/ ND   , NDENSDH , NDENSD , VDENSD , MD
      COMMON /BLOCK27/ DC   , ALP , GAM , RCOFFD
      COMMON /BLOCK28/ GRPX , GRPY
      COMMON /BLOCK29/ TMXD , TABLED
      COMMON /BLOCK30/ PXD  , GRPLXD , PXYD
      COMMON /BLOCK31/ FXDM , FYDM
C
      COMMON /BLOCK35/ NRAN , RAN , IX
C
      INTEGER     TT, PPXD, PPXYD, TTD
      PARAMETER( NN=100 , NNN=10000 , TT=500  )
      PARAMETER( NRANMX=100000000 )
      PARAMETER( PI=3.141592653589793D0 )
      PARAMETER( NND=50000 , PPXD=500 , PPXYD=250000 , TTD=20 )
C
      REAL*8      RX(NN)    , RY(NN)   , VX(NN)    , VY(NN)
      REAL*8      FX(NN)    , FY(NN)   , NX(NN)    , NY(NN)
      REAL*8      NDENS
      REAL*8      FXMD(NN) , FYMD(NN)
      INTEGER     TMX(NN)  , TABLE(TT,NN)
      INTEGER     VTMX(NN) , VTABLE(NNN) , VPLACE(NN)
      LOGICAL     OVRLAP(NN)
C
      REAL*8      RXD(NND) , RYD(NND)  , VXD(NND)  , VYD(NND)
      REAL*8      FCXD(NND), FCYD(NND) , FDXD(NND) , FDYD(NND)
      REAL*8      FRXD(NND), FRYD(NND) , FXDM(NND) , FYDM(NND)
      REAL*8      NDENSDH  , NDENSD    , MD
      REAL*8      GRPLXD(PPXD)
      INTEGER     GRPX(NND), GRPY(NND)
      INTEGER     TMXD(PPXYD), TABLED(TTD,PPXYD) , PXD , PXYD
C
      REAL*8      VELTHRY, VELTHRYD
      INTEGER     NVELSC, NVELSCD
C
      REAL        RAN(NRANMX)
      INTEGER     NRAN  , IX  ,  NRANCHK
C
      REAL*8      RXI   , RYI  , RXID  , RYID  , RCOFF2, HSQ2  , H2
      REAL*8      VXI   , VYI  , VXID  , VYID  , VELAV , VELAVD
      REAL*8      VELMX , VELDMX
      REAL*8      EVELX, EVELY, EVELSQ, EVELXD, EVELYD, EVELSQD
      INTEGER     NTIME , NTIMEMX , NGRAPH , NANIME , NANMCTR
      INTEGER     NVELAV, NVELAVD , NP , NOPT
      INTEGER     TMX00 , TMXD00  , VTABLE00
C
            OPEN(9, FILE='@acka1.data'  ,STATUS='UNKNOWN')
            OPEN(10,FILE='acka11.data'  ,STATUS='UNKNOWN')
            OPEN(11,FILE='acka21.mgf'   ,STATUS='UNKNOWN')
            OPEN(21,FILE='acka001.data',STATUS='UNKNOWN')
            OPEN(22,FILE='acka011.data',STATUS='UNKNOWN')
            OPEN(23,FILE='acka021.data',STATUS='UNKNOWN')
            OPEN(24,FILE='acka031.data',STATUS='UNKNOWN')
            OPEN(25,FILE='acka041.data',STATUS='UNKNOWN')
            OPEN(26,FILE='acka051.data',STATUS='UNKNOWN')
            OPEN(27,FILE='acka061.data',STATUS='UNKNOWN')
```

• The given values are written out in @acka1 and acka11.

```
                  OPEN(28,FILE='acka071.data',STATUS='UNKNOWN')
                  OPEN(29,FILE='acka081.data',STATUS='UNKNOWN')
                  OPEN(30,FILE='acka091.data',STATUS='UNKNOWN')
                                                                   NP=9
C                                  ++++++++++++++++++++++++++++++++++++++
C                                   N=25, 36, 49, 64, 81, 100, 121, ...
C                                   H=0.001 FOR RAS=10
C                                  ++++++++++++++++++++++++++++++++++++++
C
         N         = 81
         VDENS     = 0.3D0
         NDENS     = VDENS*(4.D0/PI)
         RAS       = 10.D0
         RV        = 120.D0
         RE        = 10.D0
         D         = 1.D0
         TD        = 0.3D0
         DEL       = TD/2.D0
         DS        = 1.D0 - TD
         RCOFF     = 8.D0
         VRADIUS   = RCOFF*1.3D0
         RCOFF2    = RCOFF**2
         VELMX     = 2.D0*5.5D0**2
         RA        = RAS*DS**3
C
         NDENSDH   = 1.0D0
         DC        = 0.4D0
         NDENSD    = NDENSDH/DC**2
         VDENSD    = NDENSDH*PI/4.D0
         GAM       = 10.D0
         ALP       = GAM/10.D0
         MD        = 0.05D0
         RCOFFD    = DC
         RCOFFDDM= 3.D0*DC
         RCOFFMD   = 0.5D0 + RCOFFDDM - DC/2.D0
         VELDMX    = (2.D0/MD)*5.5D0**2
C
         H         = 0.001D0
         NTIMEMX   = 100000
         NGRAPH    = NTIMEMX/10
         NANIME    = NTIMEMX/200
         NVTABLE = INT( 0.0001D0/H )
         IF( NVTABLE .LE. 0 ) NVTABLE = 1
C
         NVELSC    = INT( 0.1D0 /H + 0.001D0)
         NVELSCD = INT( 0.01D0/H + 0.001D0)
         VELTHRY = 2.D0**0.5
         VELTHRYD= (2.D0/MD)**0.5
         IF( NVELSC   .LE. 0 )   NVELSC    = 1
         IF( NVELSCD .LE. 0 )   NVELSCD = 1
C
         IX        = 0
         CALL RANCAL( NRANMX, IX, RAN )
         NRAN      = 1
C
C        ----------------------------------------------------------------
C        -----------------    INITIAL CONFIGURATION    -----------------
C        ----------------------------------------------------------------
C                                         --- SET INITIAL POSIT. AND VEL. ---
CCC      OPEN(19,FILE='acka091.data',STATUS='OLD')
CCC        READ(19,592)  N , XL, YL
CCC        READ(19,594)  (RX(I),I=1,N) , (RY(I),I=1,N) ,
CCC    &                 (VX(I),I=1,N) , (VY(I),I=1,N) ,
CCC    &                 (NX(I),I=1,N) , (NY(I),I=1,N)
CCC        READ(19,596)  ND
CCC        READ(19,598)  (RXD(I),I=1,ND) , (RYD(I),I=1,ND) ,
CCC    &                 (VXD(I),I=1,ND) , (VYD(I),I=1,ND)
CCC      CLOSE(19,STATUS='KEEP')
CCC      GOTO 7
C
         CALL INIPOSIT( N , VDENS , NDENS, PI , VRADIUS )
```

• The positions and velocities of particles are written out in acka001 to acka091, and the data for MicroAVS are written out in acka21.

Concerning magnetic particles:
• The number of particles N=81, volumetric fraction ϕ_V^*=0.3, λ_s=10, λ_V=120 and λ_ε=10. The particle diameter d^*=1, the surfactant layer thickness δ^*=0.15 and t_δ=0.3.
• The cutoff distance r^*_{coff}=8, r_l^* is used for the Verlet neighbor list method (see Figure 2.12).

Concerning dissipative particles:
• The number density $\hat{n}_d^*$=1, diameter d_c^*=0.4 and mass m_d^*=0.05. γ^*=10, $\alpha^*=\gamma^*/10$, and cutoff distance d_c^*.
• The cutoff distance between magnetic and dissipative particles is denoted by RCOFFMD.
• The maximum velocity is assumed to be VELDMX.

• Time interval h^*=0.001.

•The number of the total time steps is 100,000. The particle positions are written out at every NGRAPH time steps, and 200 sets of data are written out for making an animation.

• The data table in the Verlet neighbor list method is renewed at every NVTABLE time steps. The velocity scaling of magnetic and dissipative particles is carried out at every NVELSC and NVELSCD, respectively. The theoretical averaged velocities of magnetic and dissipative particles are denoted by VELTHRY and VELTHRYD, respectively.

• Pseudo-random numbers are saved in the variable RAN(*).

• The initial positions of magnetic and dissipative particles are assigned in the subroutines INIPOSIT and INIPOSID. Similarly, the initial velocities are set in INIVEL and INIVELD.

```
      CALL INIPOSID( DC , RCOFFD , N )
      CALL INIVEL( N , PI , VELMX )
      CALL INIVELD( ND , MD , PI , VELDMX )
C                                --- (A1
    7 CALL GRIDGENE( XL , RCOFFD )
C                                --- (A
      CALL GROUP( ND )
C
      CALL TABLECAL( ND , PXD )
C                                --- (B
C                                -        FOR M. PTCL.            -
      CALL VTABLEDP( N , RCOFFD , RCOFFMD , XL , YL , DC )
C                                --- (B2) SET UP VERLET TABLE OF M.PTCL ---
C                                -        FOR M. PTCL.            -
      CALL VTABLEMA( N , XL , YL )
C
      CALL FORCEMAG( RCOFF2 , NTIME )
      CALL FORCEDPD( PI )
      CALL FORCEINT( N , ND , RE , DC )
C
C                                           --- PRINT OUT CONSTANTS ---
      WRITE(NP,10) N, VDENS, NDENS, RAS, RA, RV, RE, D, TD, DEL, DS,
     &             RCOFF, VRADIUS, RCOFFMD, RCOFFDDM, XL, YL, H
      WRITE(NP,12) ND, NDENSDH, DC, NDENSD, VDENSD, MD, ALP, GAM,
     &             RCOFFD
      WRITE(NP,14) H, NTIMEMX, NGRAPH, NVTABLE
C
C
      NVELAV   = 0
      VELAV    = 0.D0
      NVELAVD  = 0
      VELAVD   = 0.D0
      NOPT     = 20
      NRANCHK  = NRANMX - ND*ND
      NANMCTR  = 0
C
      EVELX    = 0.D0
      EVELY    = 0.D0
      EVELSQ   = 0.D0
      EVELXD   = 0.D0
      EVELYD   = 0.D0
      EVELSQD  = 0.D0
C
C     -----------------------------------------------------------------
C     -------------------   START OF MAIN LOOP   ----------------------
C     -----------------------------------------------------------------
C
      DO 1000 NTIME = 1,NTIMEMX
C
C                                    --------- (1) D. PTCL. CASE ---
        DO 100 I = 1,ND
C
          RXID    = RXD(I) + VXD(I)*H
          RYID    = RYD(I) + VYD(I)*H
          RXID    = RXID   - DNINT( RXID/XL )*XL
          RYID    = RYID   - DNINT( RYID/YL )*YL
          RXD(I)  = RXID
          RYD(I)  = RYID
C                                               --- VELOCITIES ---
          VXID    = VXD(I) + FCXD(I) + FDXD(I) + FRXD(I) + FXDM(I)*H/MD
          VYID    = VYD(I) + FCYD(I) + FDYD(I) + FRYD(I) + FYDM(I)*H/MD
          VXD(I)  = VXID
          VYD(I)  = VYID
          C1      = VXID**2 + VYID**2
          IF( C1 .GT. VELDMX ) THEN
            C1 = DSQRT( VELDMX/C1 )
            VXD(I)  = VXID*C1
            VYD(I)  = VYID*C1
          END IF
C
          IF( NTIME .GT. NTIMEMX/2 ) VELAVD=VELAVD+VXD(I)**2+VYD(I)**2
```

• Cells are set for using the cell index method.

• The name of the cell to which each dissipative particle belongs is grasped. Also, the name of dissipative particles belonging to each cell is grasped.

• The names of the cells interacting with each magnetic particle are grasped.

• The names of magnetic particles interacting with each magnetic particle are grasped in VTABLEMA.

• The forces acting on magnetic and dissipative particles are calculated in the subroutines FORCEMAG and FORCEDPD, respectively. The forces acting between magnetic and dissipative particles are calculated in FORCEINT.

• The positions of dissipative particles at the next time step are evaluated according to Eq. (6.18).

• The treatment of the periodic BC.

• The velocities of dissipative particles at the next time step are evaluated according to Eq. (6.19).

• The velocity of each particle is modified so as to be smaller than the maximum value.

```
C
  100   CONTINUE
C
        IF( NTIME .GT. NTIMEMX/2 )   NVELAVD = NVELAVD + 1
C                                    --------- (2) M. PTCL. CASE ---
        HSQ2 = H*H/2.D0
        H2   = H/2.D0
        DO 200 I = 1,N
ccc                  RXI    = RX(I) + VX(I)*H + ( FX(I)+FXMD(I) )*HSQ2
ccc                  RYI    = RY(I) + VY(I)*H + ( FY(I)+FYMD(I) )*HSQ2
          RXI    = RX(I) + VX(I)*H
          RYI    = RY(I) + VY(I)*H
          C1     = VX(I)**2 + VY(I)**2
```

• The positions of magnetic particles at the next time step are evaluated according to Eq. (6.22).

```
          IF( C1 .GT. VELMX ) THEN
            C1   = DSQRT( VELMX/C1 )
            VXI  = VX(I)*C1
            VYI  = VY(I)*C1
            RXI  = RX(I) + VXI*H
            RYI  = RY(I) + VYI*H
          END IF
          RXI    = RXI - DNINT( RXI/XL )*XL
          RYI    = RYI - DNINT( RYI/YL )*YL
          RX(I)  = RXI
          RY(I)  = RYI
```

• The treatment of the periodic BC.

```
C                                       --- PART (1) OF VEL ---
         IF( OVRLAP(I) ) THEN
           VXI   = VX(I) +    FX(I) * H
           VYI   = VY(I) +    FY(I) * H
         ELSE
           VXI   = VX(I) + ( FX(I)+FXMD(I) )*H
           VYI   = VY(I) + ( FY(I)+FYMD(I) )*H
         END IF
```

• The treatment in the case of the solid parts of the two magnetic particles overlapping.

```
         VX(I) = VXI
         VY(I) = VYI
         C1    = VXI**2 + VYI**2
```

• The velocities of magnetic particles at the next time step are evaluated according to Eq. (6.23).

```
         IF( C1 .GT. VELMX ) THEN
           C1 = DSQRT( VELMX/C1 )
           VX(I) = VXI*C1
           VY(I) = VYI*C1
         END IF
  200   CONTINUE
C
```

• The velocity of each particle is modified so as to be smaller than the maximum value.

```
        CALL GROUP( ND )
        CALL TABLECAL( ND , PXD )
        CALL VTABLEDP( N , RCOFFD , RCOFFMD , XL , YL , DC )
        IF( MOD(NTIME,NVTABLE) .EQ. 0 ) THEN
          CALL VTABLEMA( N , XL , YL )
        END IF
C
```

• The information in the cell index table and in the Verlet neighbor list table is renewed.

```
        CALL FORCEMAG( RCOFF2 , NTIME )
        CALL FORCEDPD( PI )
        CALL FORCEINT( N , ND , RE , DC )
```

• The forces acting between magnetic particles, between dissipative particles, and between magnetic and dissipative particles are calculated.

```
C                                              --- SAMPLING ---
        DO 220 I = 1,N
          IF( NTIME .GT. NTIMEMX/2 )  VELAV = VELAV+VX(I)**2+VY(I)**2
  220   CONTINUE
C
        IF( NTIME .GT. NTIMEMX/2 )    NVELAV = NVELAV + 1
C
C       ------------------------------------ FOR VELOCITY SCALING ---
        DO 255 I = 1, N
          EVELX   = EVELX  + VX(I)
          EVELY   = EVELY  + VY(I)
          EVELSQ  = EVELSQ + VX(I)**2 + VY(I)**2
  255   CONTINUE
        DO 260 I = 1, ND
          EVELXD  = EVELXD  + VXD(I)
          EVELYD  = EVELYD  + VYD(I)
          EVELSQD = EVELSQD + VXD(I)**2 + VYD(I)**2
  260   CONTINUE
C                                   --- MAG VELOCITY SCALING ---
```

• The velocities are sampled for scaling the particle velocities afterward.

```
      IF( MOD(NTIME,NVELSC) .EQ. 0 ) THEN
        EVELX  = EVELX /DBLE(N*NVELSC)
        EVELY  = EVELY /DBLE(N*NVELSC)
        EVELSQ = EVELSQ/DBLE(N*NVELSC)
        CALL SCALEVEL( N, VX, VY, VELTHRY, EVELX, EVELY, EVELSQ )
        EVELX  = 0.D0
        EVELY  = 0.D0
        EVELSQ = 0.D0
      END IF
```

• The velocities of magnetic particles are scaled so as to yield the desired system temperature.

```
C                                 --- DPD VELOCITY SCALING ---
      IF( MOD(NTIME,NVELSCD) .EQ. 0 ) THEN
        EVELXD  = EVELXD /DBLE(ND*NVELSCD)
        EVELYD  = EVELYD /DBLE(ND*NVELSCD)
        EVELSQD = EVELSQD/DBLE(ND*NVELSCD)
        CALL SCALEVEL( ND,VXD,VYD,VELTHRYD,EVELXD,EVELYD,EVELSQD )
        EVELXD  = 0.D0
        EVELYD  = 0.D0
        EVELSQD = 0.D0
      END IF
```

• The velocities of dissipative particles are scaled so as to yield the desired system temperature.

```
C
C     --------------------------------------------------------------
C
C                          --- DATA OUTPUT FOR GRAPHICS (1) ---
      IF( MOD(NTIME,NGRAPH) .EQ. 0 ) THEN
        NOPT = NOPT + 1
        WRITE(NOPT,592)  N , XL, YL
        WRITE(NOPT,594) (RX(I),I=1,N) , (RY(I),I=1,N) ,
     &                  (VX(I),I=1,N) , (VY(I),I=1,N) ,
     &                  (NX(I),I=1,N) , (NY(I),I=1,N)
        WRITE(NOPT,596)  ND
        WRITE(NOPT,598) (RXD(I),I=1,ND) , (RYD(I),I=1,ND) ,
     &                  (VXD(I),I=1,ND) , (VYD(I),I=1,ND)
        CLOSE(NOPT,STATUS='KEEP')
      END IF
C                          --- DATA OUTPUT (2) FOR ANIMATION ---
      IF( MOD(NTIME,NANIME) .EQ. 0 ) THEN
        NANMCTR = NANMCTR + 1
C
        IF( NANMCTR .EQ. 1 ) THEN
          WRITE(11,381) (NTIMEMX/NANIME)
        END IF
C
        IF( (NANMCTR.GE.1) .AND. (NANMCTR.LE.9) ) THEN
          WRITE(11,383) NANMCTR
        ELSE IF( (NANMCTR.GE.10) .AND. (NANMCTR.LE.99) ) THEN
          WRITE(11,384) NANMCTR
        ELSE IF( (NANMCTR.GE.100) .AND. (NANMCTR.LE.999) ) THEN
          WRITE(11,385) NANMCTR
        ELSE IF( (NANMCTR.GE.1000) .AND. (NANMCTR.LE.9999) ) THEN
          WRITE(11,386) NANMCTR
        END IF
C
        WRITE(11,388) ( N+ND )
C
        DO 400 I=1,N
          WRITE(11,398) RX(I) ,RY(I) ,0.0, D/2.D0,  1.0, 0.0, 0.0
  400   CONTINUE
        DO 410 I=1,ND
          WRITE(11,398) RXD(I),RYD(I),0.0, DC/2.D0, 0.0, 0.8, 1.0
  410   CONTINUE
      END IF
```

• The data are written out for making an animation based on the commercial software MicroAVS.

```
C
C                              --- CHECK RANDOM NUMBERS USED ---
      IF( NRAN .GT. NRANCHK ) THEN
        CALL RANCAL( NRANMX, IX, RAN )
        NRAN    = 1
      END IF
C
C
 1000 CONTINUE
C
C    ----------------------------------------------------------------
```

• The number of the used random numbers is checked. If over NRANCHK, a uniform random number sequence is renewed.

```
C     ---------------------- END OF MAIN LOOP --------------------
C     -----------------------------------------------------------
C
      VELAV  =  VELAV /DBLE(NVELAV*N)
      VELAVD =  VELAVD/DBLE(NVELAVD*ND)
C
      TMX00  = 0
      TMXD00 = 0
      DO 1006 I=1,N
        IF( TMX(I)  .GT. TMX00  ) TMX00  = TMX(I)
 1006 CONTINUE
      DO 1007 I=1,PXYD
        IF( TMXD(I) .GT. TMXD00 ) TMXD00 = TMXD(I)
 1007 CONTINUE
      VTABLE00 = VPLACE(N) + VTMX(N) - 1
C
C                                            ---- PRINT OUT (1) ----
      WRITE(NP,1011) TMX00 , TMXD00 , VTABLE00 ,
     &               REAL(TMX00)/REAL(TT) , REAL(TMXD00)/REAL(TTD) ,
     &               REAL(VTABLE00)/REAL(NNN) , REAL(PXYD)/REAL(PPXYD)
      WRITE(NP,1013) PXD   , PXYD
      WRITE(NP,1014) DSQRT(VELAV)       , DSQRT(VELAVD) ,
     &               DSQRT(VELAV/2.D0) , DSQRT(VELAVD*MD/2.D0)
C
C                               ---- DATA OUTPUT FOR GRAPHICS (1) ----
      WRITE(10,1210) N, VDENS, NDENS, RAS, RA, RV, RE, D, TD, DEL, DS
      WRITE(10,1211) RCOFF, VRADIUS, RCOFFMD, RCOFFDDM, XL, YL, H
      WRITE(10,1213) ND, NDENSDH, DC, NDENSD, VDENSD, MD, ALP, GAM,
     &               RCOFFD
      WRITE(10,1214) H, NTIMEMX, NGRAPH, NVTABLE
C
                                CLOSE( 9,STATUS='KEEP')
                                CLOSE(10,STATUS='KEEP')
                                CLOSE(11,STATUS='KEEP')
C
C     ------------------------ FORMAT -----------------------------
   10 FORMAT(/1H ,'-----------------------------------------------'
     &       /1H ,'      DPD SIMULATION OF MAGNETIC PARTICLES      '
     &       /1H ,'      IN DISSIPATIVE PARTICLES IN EQUILIBRIUM   '
     &       /1H ,'     +++ TWO-DIMENSIONAL EQUILIBRIUM CASE +++   '
     &       /1H ,'-----------------------------------------------'
     &      //1H ,'N=',I3 ,2X, 'VDENS=', F6.3, 2X, 'NDENS=',F6.3 ,2X,
     &            'RAS=', F7.3, 2X, 'RA=', F7.3, 2X, 'RV=', F8.2
     &       /1H ,'RE=', F7.3, 2X, 'D=', F3.1, 2X, 'TD=', F4.2, 2X,
     &            'DEL=', F5.3, 2X, 'DS=', F5.2
     &       /1H ,'RCOFF=',F6.2, 2X, 'VRADIUS=',F6.2, 2X, 'RCOFFMD=',
     &             F5.2, 2X, 'RCOFFDDM=', F5.2
     &       /1H ,'XL=', F6.2, 2X, 'YL=', F6.2, 2X, 'H=', E9.2)
   12 FORMAT(/1H ,'ND=',I4, 2X, 'NDENSDH=', F6.3, 2X, 'DC=', F6.2, 2X,
     &            'NDENSD=', F6.2, 2X, 'VDENSD=', F6.2, 2X, 'MD=',F5.3
     &       /1H ,'ALP=', F6.2, 2X, 'GAM=', F6.2, 2X, 'RCOFFD=', F6.2)
   14 FORMAT(/1H ,'H=', E9.2, 2X, 'NTIMEMX=', I8, 2X, 'NGRAPH=',I7,
     &             2X, 'NVTABLE=', I4/)
  381 FORMAT('# Micro AVS Geom:2.00'
     &      /'# Animation of DPD simulation results'/I4)
  383 FORMAT('step',I1)
  384 FORMAT('step',I2)
  385 FORMAT('step',I3)
  386 FORMAT('step',I4)
  388 FORMAT( 'sphere'/'sphere_sample'/'color'/I7 )
  398 FORMAT( 3F10.4 , F6.2 , 3F5.2 )
  592 FORMAT( I8, 2F12.6 )
  594 FORMAT( (5E16.8) )
  596 FORMAT( I8 )
  598 FORMAT( (5E16.8) )
 1011 FORMAT(/1H ,'TMX00=',I5, 2X, 'TMXD00=',I5, 2X, 'VTABLE00=',I5
     &       /1H ,'REAL(TMX00)/REAL(TT)=',F5.3, 2X,
     &            'REAL(TMXD00)/REAL(TTD)=',F5.3
     &       /1H ,'REAL(VTABLE00)/REAL(NNN)=',F5.3, 2X,
     &            'REAL(PXYD)/REAL(PPXYD)=',F5.3)
 1013 FORMAT(1H ,'PXD=', I5, 2X, 'PXYD=', I6/)
 1014 FORMAT(1H ,'VELAV=', F9.4, 2X, 'VELAVD=', F9.4
```

```
     &        /1H ,'VELAV/THEORY=', F9.4, 2X, 'VELAVD/THEORY=', F9.4)
 1210 FORMAT( I4 , 2F6.3 , 3F8.3 , 5F7.3 )
 1211 FORMAT( 6F8.3 , E11.3 )
 1213 FORMAT( I4 , F6.3 , 7F8.3 )
 1214 FORMAT( E11.3 , 3I8 )
                                                                  STOP
                                                                  END
C*********************************************************************
C***************************   SUBROUTINE   **************************
C*********************************************************************
C
C**** SUB INIPOSIT ****
      SUBROUTINE INIPOSIT( N , VDENS , NDENS , PI , VRADIUS )
C
      IMPLICIT REAL*8 (A-H,O-Z), INTEGER (I-N)
C
      COMMON /BLOCK1/  RX , RY
      COMMON /BLOCK3/  NX , NY
      COMMON /BLOCK15/ H  , XL , YL , RCOFF
C
      PARAMETER( NN=100 )
C
      REAL*8    RX(NN) , RY(NN) , NX(NN) , NY(NN) , NDENS
      REAL*8    A , RAN , C1
      INTEGER   Q , PTCL
C
      A  = DSQRT( PI/(4.D0*VDENS) )
      Q  = NINT( SQRT(REAL(N+1)) )
      XL = A*DBLE(Q)
      YL = XL
C                                               ----- POSITION -----
      RAN1 = DSQRT(2.D0)
      RAN2 = DSQRT(3.D0)
      PTCL = 0
      DO 10 J=0,Q-1
      DO 10 I=0,Q-1
        PTCL = PTCL + 1
        C1 = RAN1*DBLE(PTCL)
        C1 = C1 - DINT(C1)
        C1 = C1 - 0.5D0
        C2 = RAN2*DBLE(PTCL)
        C2 = C2 - DINT(C2)
        C2 = C2 - 0.5D0
        RX(PTCL) = DBLE(I)*A - XL/2.D0 + 0.1D0 + C1*0.091D0
        RY(PTCL) = DBLE(J)*A - YL/2.D0 + 0.1D0 + C2*0.091D0
   10 CONTINUE
      N = PTCL
C
      DO 20 I=1,N
        NX(I) = 0.D0
        NY(I) = 1.D0
   20 CONTINUE
C
      IF( VRADIUS .GT. XL/2.D0 ) THEN
        VRADIUS = XL/2.D0
        RCOFF   = XL/2.D0
      END IF
                                                               RETURN
                                                               END
C**** SUB INIPOSID *****
      SUBROUTINE INIPOSID( DC , RCOFFD , N )
C
      IMPLICIT REAL*8 (A-H,O-Z), INTEGER (I-N)
C
      COMMON /BLOCK1/  RX , RY
      COMMON /BLOCK15/ H  , XL , YL , RCOFF
      COMMON /BLOCK21/ RXD , RYD
      COMMON /BLOCK26/ ND  , NDENSDH , NDENSD , VDENSD , MD
C
      PARAMETER( NN=100 , NND=50000 )
C
      REAL*8    RX(NN) , RY(NN) , RXD(NND), RYD(NND)
```

• A subroutine for setting the initial positions of magnetic particles.

• $\phi_V=(\pi/4)/a^{*2}$, $a^*=(\pi/(4\phi_V))^{1/2}$ and $Q=N^{1/2}$. The values of a^* and Q are saved in A and Q, respectively.

• RAN1 and RAN2 are quasi-random numbers.
• Each particle is moved in parallel by the distance (0.1, 0.1) to remove subtle situations at the outer boundary surfaces. Also, to remove the regularity of the initial configuration, each particle is moved randomly by the maximum displacement (1/2)×(0.091, 0.091) using quasi-random numbers.

• Additionally each particle is moved in parallel by (1/2)×(–XL, –YL) so that the center of the simulation box is the origin of the coordinate system.

• The direction of each magnetic moment is set in the *y*-direction.

• A subroutine for setting the initial positions of dissipative particles.

```
      REAL*8    NDENSDH, NDENSD , MD , B , RSQCHK , RXID , RYID
      REAL*8    RXI , RYI , RXIJ , RYIJ , RIJSQ, RCOFFMN, RCOFFMN2
      INTEGER   P   , PTCL
C
      B  = DSQRT( 1.D0/NDENSD )
      P  = INT( XL/B )
      RSQCHK   = (0.5D0 + DC/2.D0)**2
      RCOFFMN  = 0.5D0 + ( DC/2.D0 )*0.3D0
      RCOFFMN2 = RCOFFMN**2
```

• $n_d^*=1/b^{*2}$ and $b^*=(1/n_d^*)^{1/2}$. Particles are placed in each axis direction.

```
C                                            ------- POSITION (1) ---
      PTCL=0
      DO 120 IY=0,P-1
        RYID = DBLE(IY)*B - YL/2.D0 + 0.0001D0
        IF( RYID .GE. YL/2.D0 )   GOTO 120
      DO 100 IX=0,P-1
        RXID = DBLE(IX)*B - XL/2.D0 + 0.0001D0
        IF( RXID .GE. XL/2.D0 )   GOTO 100
```

• Each particle is moved in parallel by (1/2)×(–XL,–YL), so that the center of the simulation box is the origin of the coordinate system.

```
C                              --- REMOVE OVERLAP WITH MAG.PTCL. ---
        DO 50 I=1,N
          RXI = RX(I)
          RYI = RY(I)
          RXIJ = RXID - RXI
          RXIJ = RXIJ - DNINT(RXIJ/XL)*XL
          IF( DABS(RXIJ) .GT. RCOFFMN )    GOTO 50
          RYIJ = RYID - RYI
          RYIJ = RYIJ - DNINT(RYIJ/YL)*YL
          IF( DABS(RYIJ) .GT. RCOFFMN )    GOTO 50
          RIJSQ= RXIJ**2 + RYIJ**2
          IF( RIJSQ .LT. RCOFFMN2 )        GOTO 100
   50   CONTINUE
C
        PTCL = PTCL + 1
        RXD(PTCL) =  RXID
        RYD(PTCL) =  RYID
  100 CONTINUE
  120 CONTINUE
      ND = PTCL
                                                     RETURN
                                                     END
C**** SUB INIVEL *****
      SUBROUTINE INIVEL( N , PI , VELMX )
C
      IMPLICIT REAL*8 (A-H,O-Z), INTEGER (I-N)
C
      COMMON /BLOCK2/  VX   , VY
      COMMON /BLOCK35/ NRAN , RAN , IX
C
      PARAMETER( NN=100 , NRANMX=100000000 )
C
      REAL*8  VX(NN) , VY(NN) , MOMX , MOMY , CC1 , CC2
      REAL    RAN(NRANMX)
C
      DO 10 I=1,N
        NRAN = NRAN + 1
        CC1 = DSQRT(  -2.D0*(1.D0)*DLOG( DBLE(RAN(NRAN)) )  )
        NRAN = NRAN + 1
        CC2 = 2.D0*PI*DBLE(RAN(NRAN))
        VX(I) = CC1*DCOS(CC2)
C
        NRAN = NRAN + 1
        CC1 = DSQRT(  -2.D0*(1.D0)*DLOG( DBLE(RAN(NRAN)) )  )
        NRAN = NRAN + 1
        CC2 = 2.D0*PI*DBLE(RAN(NRAN))
        VY(I) = CC1*DSIN(CC2)
C
        C1   = VX(I)**2 + VY(I)**2
        IF( C1 .GT. VELMX ) THEN
          C1 = DSQRT( VELMX/C1 )
          VX(I) = VX(I)*C1
          VY(I) = VY(I)*C1
        END IF
   10 CONTINUE
```

• The dissipative particles are not placed if the separation between magnetic and dissipative particles is shorter than RCOFFMN.

• A subroutine for setting the initial velocities of magnetic particles.

• The initial velocities are assigned according to Eq. (A2.3).

• The initial velocities are modified so as to be smaller than the maximum velocity.

```
C                                        --- SET TOTAL MOMENTUM ZERO ---
      MOMX = 0.D0
      MOMY = 0.D0
      DO 20 I=1,N
        MOMX = MOMX + VX(I)
        MOMY = MOMY + VY(I)
   20 CONTINUE
      MOMX = MOMX/DBLE(N)
      MOMY = MOMY/DBLE(N)
C                              --- CORRECT VELOCITIES TO SATISFY  ---
C                              ---     ZERO TOTAL MOMENTUM        ---
      DO 30 I=1,N
        VX(I) = VX(I) - MOMX
        VY(I) = VY(I) - MOMY
   30 CONTINUE
                                                                RETURN
                                                                END
```

• The velocities are modified so as to yield zero total system momentum.

```
C**** SUB INIVELD *****
      SUBROUTINE INIVELD( ND , MD , PI , VELDMX )
C
      IMPLICIT REAL*8 (A-H,O-Z), INTEGER (I-N)
C
      COMMON /BLOCK22/ VXD  , VYD
      COMMON /BLOCK35/ NRAN , RAN  , IX
C
      PARAMETER( NND=50000 , NRANMX=100000000 )
C
      REAL*8  VXD(NND), VYD(NND) , MD , MOMX , MOMY , CC1 , CC2
      REAL    RAN(NRANMX)
C
      DO 10 I=1,ND
        NRAN = NRAN + 1
        CC1  = DSQRT(  -2.D0*(1.D0/MD)*DLOG( DBLE(RAN(NRAN)) )  )
        NRAN = NRAN + 1
        CC2  = 2.D0*PI*DBLE(RAN(NRAN))
        VXD(I) = CC1*DCOS(CC2)
C
        NRAN = NRAN + 1
        CC1  = DSQRT(  -2.D0*(1.D0/MD)*DLOG( DBLE(RAN(NRAN)) )  )
        NRAN = NRAN + 1
        CC2  = 2.D0*PI*DBLE(RAN(NRAN))
        VYD(I) = CC1*DSIN(CC2)
C
        C1     = VXD(I)**2 + VYD(I)**2
        IF( C1 .GT. VELDMX ) THEN
          C1 = DSQRT( VELDMX/C1 )
          VXD(I) = VXD(I)*C1
          VYD(I) = VYD(I)*C1
        END IF
   10 CONTINUE
```

• A subroutine for setting the initial velocities of dissipative particles.

• The initial velocities are assigned according to Eq. (A2.3).

• The initial velocities are modified so as to be smaller than the maximum velocity.

```
C                                        --- SET TOTAL MOMENTUM ZERO ---
      MOMX = 0.D0
      MOMY = 0.D0
      DO 20 I=1,ND
        MOMX = MOMX + VXD(I)
        MOMY = MOMY + VYD(I)
   20 CONTINUE
      MOMX = MOMX/DBLE(ND)
      MOMY = MOMY/DBLE(ND)
C                              --- CORRECT VELOCITIES TO SATISFY  ---
C                              ---     ZERO TOTAL MOMENTUM        ---
      DO 30 I=1,ND
        VXD(I) = VXD(I) - MOMX
        VYD(I) = VYD(I) - MOMY
   30 CONTINUE
                                                                RETURN
                                                                END
C**** SUB SCALEVEL ****
      SUBROUTINE SCALEVEL( N, VX, VY, VELTHRY, VELX, VELY, VELSQ )
```

• The velocities are modified so as to yield zero total system momentum.

• A subroutine for scaling the velocities (common for both magnetic and dissipative particles).

```
C
      IMPLICIT REAL*8 (A-H,O-Z), INTEGER (I-N)
C
      REAL*8   VX(N), VY(N)
C                                 --- ZERO TOTAL MOMENTUM FOR EACH AXIS ---
      DO 10 I = 1,N
        VX(I) = VX(I) - VELX
        VY(I) = VY(I) - VELY
   10 CONTINUE
```

• The velocities are modified so as to yield zero total momentum.

```
C                                     --- CORRECT VELOCITIES TO SATISFY ---
C                                     -   SPECIFIED TEMPERATURE           -
      C1 = VELTHRY/DSQRT( VELSQ - VELX**2 - VELY**2 )
      DO 50 I = 1,N
          VXI   = VX(I)
          VYI   = VY(I)
          VX(I) = VXI*C1
          VY(I) = VYI*C1
   50 CONTINUE
                                                            RETURN
                                                            END
```

• The velocities are modified so as to yield the desired system temperature.

```
C**** SUB GRIDGENE ****
      SUBROUTINE GRIDGENE( XL , RCOFFD )
C
      IMPLICIT REAL*8 (A-H,O-Z), INTEGER (I-N)
C
      COMMON /BLOCK30/ PXD  , GRPLXD , PXYD
C
      INTEGER    PPXD
      PARAMETER( PPXD=500 )
C
      REAL*8   GRPLXD(PPXD) , C0
      INTEGER  PXD , PXYD
C
      PXD  = INT( XL/RCOFFD )
      PXYD = PXD**2
      C0   = XL/DBLE(PXD)
      DO 10 I=1,PXD
        GRPLXD(I) = C0*DBLE(I) - XL/2.D0
   10 CONTINUE
                                                            RETURN
                                                            END
```

• A subroutine for generating cells for the cell index method in the case of dissipative particles.

• The cells are made by dividing the simulation box into PXD equal cells in each axis-direction. The position of the *x*-coordinate (equal to y-coordinate) is saved in GRPLXD.

```
C**** SUB GROUP *****
      SUBROUTINE GROUP( ND )
C
      IMPLICIT REAL*8 (A-H,O-Z), INTEGER (I-N)
C
      COMMON /BLOCK21/ RXD  , RYD
      COMMON /BLOCK28/ GRPX , GRPY
      COMMON /BLOCK30/ PXD  , GRPLXD , PXYD
C
      INTEGER    PPXD, PPXYD, TTD
      PARAMETER( NND=50000 , PPXD=500 , PPXYD=250000 , TTD=20 )
C
      REAL*8   RXD(NND) , RYD(NND)
      REAL*8   GRPLXD(PPXD)
      INTEGER  GRPX(NND), GRPY(NND) , PXD , PXYD
C
      DO 100 I=1,ND
C                                                    --- X AXIS ---
        DO 10 J=1,PXD
          IF( GRPLXD(J) .GT. RXD(I) ) THEN
            GRPX(I) = J
            GOTO 15
          END IF
   10   CONTINUE
        GRPX(I) = PXD
C                                                    --- Y AXIS ---
   15   DO 20 J=1,PXD
          IF( GRPLXD(J) .GT. RYD(I) ) THEN
            GRPY(I) = J
            GOTO 100
          END IF
```

• A subroutine for grasping the name of the cell to which each dissipative particle belongs.

• If particle *i* belongs to the cell which is assumed to be the (GRPX(I)-th, GRPY(I)-th) cell in *x*- and *y*-directions, the name of the cell is GP=GRPX(I)+(GRPY(I)−1)*PXD.

```
   20   CONTINUE
          GRPY(I) = PXD
C
  100 CONTINUE
                                                            RETURN
                                                            END
C**** SUB TABLECAL *****
      SUBROUTINE TABLECAL( ND , PXD )
C
      IMPLICIT REAL*8 (A-H,O-Z), INTEGER (I-N)
C
      COMMON /BLOCK28/ GRPX , GRPY
      COMMON /BLOCK29/ TMXD , TABLED
C
      INTEGER    PPXD, PPXYD, TTD
      PARAMETER( NND=50000 , PPXD=500 , PPXYD=250000 , TTD=20 )
C
      INTEGER  GRPX(NND), GRPY(NND)
      INTEGER  TMXD(PPXYD), TABLED(TTD,PPXYD) , PXD , GX , GY , GP
C
      DO 10 GY=1,PXD
      DO 10 GX=1,PXD
        GP = GX + (GY-1)*PXD
        TMXD(GP)      = 0
        TABLED(1,GP) = 0
   10 CONTINUE
C
      DO 20 I=1,ND
        GX = GRPX(I)
        GY = GRPY(I)
        GP = GX + (GY-1)*PXD
        TMXD(GP) = TMXD(GP) + 1
        TABLED( TMXD(GP),GP ) = I
   20 CONTINUE
                                                            RETURN
                                                            END
```

• A subroutine for grasping the names of dissipative particles belonging to each cell.

• If particle *i* belongs to the cell which is assumed to be the (GX-th, GY-th) cell in the *x*- and *y*-directions, the name of the cell is GP=GX+(GY−1)*PXD.
•The name of particle *i* is therefore saved in the variable in TABLED(*,GP) concerning cell GP.

```
C**** SUB VTABLEDP *****
      SUBROUTINE VTABLEDP( N , RCOFFD , RCOFFMD , XL , YL , DC )
C
      IMPLICIT REAL*8 (A-H,O-Z), INTEGER (I-N)
C
      COMMON /BLOCK1/  RX   , RY
      COMMON /BLOCK9/  TMX  , TABLE
      COMMON /BLOCK30/ PXD  , GRPLXD  , PXYD
C
      INTEGER    TT, PPXD
      PARAMETER( NN=100 , NNN=10000 , TT=500 , PPXD=500 )
C
      INTEGER  TMX(NN), TABLE(TT,NN) , PXD , PXYD
      REAL*8   RX(NN) , RY(NN) , GRPLXD(PPXD)
      REAL*8   RXI, RYI, RX1, RY1, RX2, RY2, XI, YI, CL, MODX, MODY
      REAL*8   RSQCHK  , RSQCHK2 , RRISQ , RCHK
      INTEGER  GPX1 , GPX2 , GPY1 , GPY2 , GP
C
      CL      = GRPLXD(2) - GRPLXD(1)
      RCHK    = RCOFFMD + (CL/2.D0)*1.415D0
      RSQCHK  = RCHK**2
      RSQCHK2 = ( 0.5D0-DC/2.D0-(CL/2.D0)*1.415D0 )**2
      DO 10 I=1,N
        TMX(I)      = 0
        TABLE(1,I) = 0
   10 CONTINUE
C
      DO 200 I=1,N
        RXI = RX(I)
        RYI = RY(I)
        RX1 = RXI - RCHK
        RY1 = RYI - RCHK
        RX2 = RXI + RCHK
        RY2 = RYI + RCHK
```

• A subroutine for grasping the cells in which dissipative particles possibly interact with magnetic particles.

• The dissipative particles only in the neighboring cells possibly interact with magnetic particle *i*.

```
          GPX1 = INT( (RX1+XL/2.D0)/CL ) - 1
          GPX2 = INT( (RX2+XL/2.D0)/CL ) + 2
          GPY1 = INT( (RY1+YL/2.D0)/CL ) - 1
          GPY2 = INT( (RY2+YL/2.D0)/CL ) + 2
C
          DO 150 IY0 = GPY1, GPY2
            IY   = IY0
            MODY = 0.D0
            IF( IY0 .LE. 0 ) THEN
              IY   =  IY0 + PXD
              MODY = -YL
            END IF
            IF( IY0 .GT. PXD ) THEN
              IY   =  IY0 - PXD
              MODY =  YL
            END IF
            YI = GRPLXD(IY) - CL/2.D0 + MODY
C
          DO 140 IX0 = GPX1, GPX2
            IX   = IX0
            MODX = 0.D0
            IF( IX0 .LE. 0 ) THEN
              IX   =  IX0 + PXD
              MODX = -XL
            END IF
            IF( IX0 .GT. PXD ) THEN
              IX   =  IX0 - PXD
              MODX =  XL
            END IF
            XI = GRPLXD(IX) - CL/2.D0 + MODX
C
            GP    = IX + PXD*(IY-1)
            RRISQ= (XI-RXI)**2 + (YI-RYI)**2
            IF( RRISQ .GE. RSQCHK )  GOTO 140
            IF( RRISQ .LE. RSQCHK2 ) GOTO 140
C
            TMX(I) = TMX(I) + 1
            TABLE( TMX(I),I) = GP
  140     CONTINUE
  150     CONTINUE
  200 CONTINUE
                                                      RETURN
                                                      END
```

• The treatment of the periodic BC.

• If the distance between magnetic particle *i* and a cell is shorter than RSQCHK, the cell is regarded as a possible interacting cell.

• A subroutine for grasping the names of magnetic particles interacting with magnetic particle themselves according to the Verlet neighbor list method.

```
C**** SUB VTABLEMA *****
      SUBROUTINE VTABLEMA( N , XL , YL )
C
      IMPLICIT REAL*8 (A-H,O-Z), INTEGER (I-N)
C
      COMMON /BLOCK1/  RX , RY
      COMMON /BLOCK10/ VTMX , VTABLE , VPLACE , NVTABLE , VRADIUS
C
      PARAMETER( NN=100 , NNN=10000 )
C
      REAL*8   RX(NN)    , RY(NN)
      INTEGER  VTMX(NN)  , VTABLE(NNN)  , VPLACE(NN)  , N2
      REAL*8   RXI  , RYI  , RXIJ  , RYIJ  , RIJ2  , VRADIUS2
C
      VRADIUS2 = VRADIUS**2
      N2       = N**2
      IF( N2 .GT. NNN )  N2 = NNN
      DO 10 I=1,N
        VTMX(I)   = 0
        VPLACE(I) = 0
   10 CONTINUE
      DO 15 I=1,N2
        VTABLE(I) = 0
   15 CONTINUE
C
C
      DO 200 I=1,N
C
        RXI    = RX(I)
```

• The number of the magnetic particles interacting with particle *i* is saved in VTMX(I), and the names of the interacting particles are saved in VTABLE(*). The name of the particle interacting with particle *i* first appears in the VPLACE(I)-th position of the variable VTABLE(*).

```
        RYI    = RY(I)
        IF( I .EQ. 1 ) THEN
          VPLACE(I) = 1
        ELSE
          VPLACE(I) = VPLACE(I-1) + VTMX(I-1)
        END IF
C
      DO 150 J=1,N
C
        IF( J.EQ.I )                         GOTO 150
        RXIJ = RXI  - RX(J)
        RXIJ = RXIJ - DNINT(RXIJ/XL)*XL
        IF( DABS(RXIJ) .GE. VRADIUS )        GOTO 150
        RYIJ = RYI  - RY(J)
        RYIJ = RYIJ - DNINT(RYIJ/YL)*YL
        IF( DABS(RYIJ) .GE. VRADIUS )        GOTO 150
C
        RIJ2 = RXIJ*RXIJ + RYIJ*RYIJ
        IF( RIJ2 .GE. VRADIUS2 )             GOTO 150
C
        VTMX(I) = VTMX(I) + 1
        VTABLE( VPLACE(I) + VTMX(I) - 1 ) = J
C
  150 CONTINUE
  200 CONTINUE
                                                      RETURN
                                                      END
```

• The treatment for the periodic BC.

• If the distance between magnetic particles is within VRADIUS, the names of the magnetic particles are saved in VTABLE(*).

```
C**** SUB FORCEMAG *****
      SUBROUTINE FORCEMAG( RCOFF2 , NTIME )
C
      IMPLICIT REAL*8 (A-H,O-Z), INTEGER (I-N)
C
      COMMON /BLOCK1/  RX , RY
      COMMON /BLOCK2/  VX , VY
      COMMON /BLOCK3/  NX , NY
      COMMON /BLOCK5/  FX , FY
      COMMON /BLOCK7/  N  , NDENS , VDENS , D , DS , DEL , TD
      COMMON /BLOCK8/  RA , RV , RE
      COMMON /BLOCK10/ VTMX , VTABLE , VPLACE , NVTABLE , VRADIUS
      COMMON /BLOCK13/ OVRLAP
      COMMON /BLOCK15/ H , XL , YL , RCOFF
C
      INTEGER    TT
      PARAMETER( NN=100 , NNN=10000 , TT=500 )
C
      REAL*8    RX(NN)   , RY(NN)    , VX(NN)    , VY(NN)
      REAL*8    FX(NN)   , FY(NN)    , NX(NN)    , NY(NN)
      REAL*8    NDENS
      LOGICAL   OVRLAP(NN)
      INTEGER   VTMX(NN) , VTABLE(NNN) , VPLACE(NN)
C
      REAL*8    RXI , RYI , RXIJ , RYIJ
      REAL*8    NXI , NYI , NXJ  , NYJ
      REAL*8    FXI , FYI , FXIJ , FYIJ
      REAL*8    TXIJ , TYIJ , RIJ , RIJ2 , RIJ4 , RIJORGN
      REAL*8    RA3, RMN, RMN2
      REAL*8    C0 , C1 , C2 , C3
      INTEGER   IVPLACE
C
      RA3   = 3.D0*RA
      RMN   = DS
      RMN2  = RMN**2
      DO 10 I=1,N
        FX(I) = 0.D0
        FY(I) = 0.D0
        OVRLAP(I) = .FALSE.
   10 CONTINUE
C
C
      DO 100 I=1,N
C
```

• A subroutine for calculating the magnetic forces acting on magnetic particles.

• Whether or not an overlap of the solid parts of the two magnetic particles appears is described by the logical variable OVRLAP(*).

• The name of the particles interacting with particle *i* first appears in the VPLACE(I)-th position of the variable VTABLE(*). The number of the magnetic particles interacting with particle *i* is VTMX(I).

```
          RXI = RX(I)
          RYI = RY(I)
          NXI = NX(I)
          NYI = NY(I)
          FXI = FX(I)
          FYI = FY(I)
C
          IF( VTMX(I) .EQ. 0 )               GOTO 100
C
          IVPLACE = VPLACE(I)
          DO 50 JJ=1, VTMX(I)
C
            J    = VTABLE( IVPLACE + JJ - 1 )
            IF( J.EQ.I )                     GOTO 50
            RXIJ = RXI  - RX(J)
            RXIJ = RXIJ - DNINT(RXIJ/XL)*XL
            IF( DABS(RXIJ) .GE. RCOFF )      GOTO 50
            RYIJ = RYI  - RY(J)
            RYIJ = RYIJ - DNINT(RYIJ/YL)*YL
            IF( DABS(RYIJ) .GE. RCOFF )      GOTO 50
C
            RIJ2   = RXIJ*RXIJ + RYIJ*RYIJ
            IF( RIJ2 .GE. RCOFF2 )           GOTO 50
            RIJ       = DSQRT(RIJ2)
            RIJORGN   = RIJ
C
```

• The treatment for the periodic BC.

• If the solid particles overlap, OVRLAP(I)=OVRLAP(J)=.TRUE. are set.

```
            IF( RIJ2 .LT. RMN2 ) THEN
              RXIJ = RMN*RXIJ/RIJ
              RYIJ = RMN*RYIJ/RIJ
              RIJ  = RMN
              RIJ2 = RMN2
              OVRLAP(I) = .TRUE.
              OVRLAP(J) = .TRUE.
            END IF
            RIJ4 = RIJ2**2
            TXIJ = RXIJ/RIJ
            TYIJ = RYIJ/RIJ
            NXJ = NX(J)
            NYJ = NY(J)
C
            C1   = NXI*NXJ   + NYI*NYJ
            C2   = NXI*TXIJ  + NYI*TYIJ
            C3   = NXJ*TXIJ  + NYJ*TYIJ
C                                             --- MAGNETIC FORCE ---
            FXIJ = - ( RA3/RIJ4) * (  ( - C1 + 5.D0*C2*C3 )*TXIJ
     &                                  - ( C3*NXI + C2*NXJ )  )
            FYIJ = - ( RA3/RIJ4) * (  ( - C1 + 5.D0*C2*C3 )*TYIJ
     &                                  - ( C3*NYI + C2*NYJ )  )
C                                           --- STERIC REPULSION ---
            IF( RIJORGN .LT. 1.D0 ) THEN
               C0   = DLOG( 1.D0 / RIJORGN )
               FXIJ = FXIJ + RV*TXIJ*C0/DEL
               FYIJ = FYIJ + RV*TYIJ*C0/DEL
            END IF
C
            FXI   = FXI   + FXIJ
            FYI   = FYI   + FYIJ
C
   50     CONTINUE
C
          FX(I) = FXI
          FY(I) = FYI
C
  100 CONTINUE
                                                        RETURN
                                                        END
C**** SUB FORCEDPD *****
      SUBROUTINE FORCEDPD( PI )
C
      IMPLICIT REAL*8 (A-H,O-Z), INTEGER (I-N)
C
```

• The magnetic forces acting on particles are calculated according to Eq. (6.24).

•The repulsive force arising from the overlap of the surfactant layers is calculated according to Eq. (6.25).

• A subroutine for calculating the forces acting between dissipative particles.

```
      COMMON /BLOCK15/ H , XL , YL , RCOFF
      COMMON /BLOCK21/ RXD  , RYD
      COMMON /BLOCK22/ VXD  , VYD
      COMMON /BLOCK23/ FCXD , FCYD
      COMMON /BLOCK24/ FDXD , FDYD
      COMMON /BLOCK25/ FRXD , FRYD
      COMMON /BLOCK26/ ND   , NDENSDH , NDENSD , VDENSD , MD
      COMMON /BLOCK27/ DC   , ALP , GAM , RCOFFD
      COMMON /BLOCK28/ GRPX , GRPY
      COMMON /BLOCK29/ TMXD , TABLED
      COMMON /BLOCK30/ PXD  , GRPLXD , PXYD
      COMMON /BLOCK35/ NRAN , RAN    , IX
C
      INTEGER    PPXD, PPXYD, TTD
      PARAMETER( NND=50000 , PPXD=500 , PPXYD=250000 , TTD=20 )
      PARAMETER( NRANMX=100000000 )
C
      REAL*8   RXD(NND) , RYD(NND)  , VXD(NND) , VYD(NND)
      REAL*8   FCXD(NND), FCYD(NND) , FDXD(NND), FDYD(NND)
      REAL*8   FRXD(NND), FRYD(NND)
      REAL*8   NDENSDH  , NDENSD    , MD
      REAL*8   GRPLXD(PPXD)
      INTEGER  GRPX(NND), GRPY(NND)
      INTEGER  TMXD(PPXYD), TABLED(TTD,PPXYD) , PXD , PXYD
C
      REAL     RAN(NRANMX)
      INTEGER  NRAN  , IX  ,  NRANCHK
C
      REAL*8   RXI  , RYI  , RXIJ , RYIJ , RIJSQ , RIJ
      REAL*8   VXI  , VYI  , VXIJ , VYIJ
      REAL*8   FCXI , FCYI , FCXIJ, FCYIJ
      REAL*8   FDXI , FDYI , FDXIJ, FDYIJ
      REAL*8   FRXI , FRYI , FRXIJ, FRYIJ
      REAL*8   FXIJ , FYIJ
      REAL*8   EXIJ , EYIJ
      REAL*8   WR   , WR2  , TTAIJ , RAN1 , RAN2 , RCOFFD2
      REAL*8   MODX , MODY , C1
      INTEGER  GX   , GY   , GRP
C
      RCOFFD2 = RCOFFD**2
      DO 10 I=1,ND
        FCXD(I) = 0.D0
        FCYD(I) = 0.D0
        FDXD(I) = 0.D0
        FDYD(I) = 0.D0
        FRXD(I) = 0.D0
        FRYD(I) = 0.D0
   10 CONTINUE
C
      DO 500 I=1,ND
C
        RXI  = RXD(I)
        RYI  = RYD(I)
        VXI  = VXD(I)
        VYI  = VYD(I)
        FCXI = FCXD(I)
        FCYI = FCYD(I)
        FDXI = FDXD(I)
        FDYI = FDYD(I)
        FRXI = FRXD(I)
        FRYI = FRYD(I)
C                                               +++ NEIGHBORING GROUP +++
        DO 300 JJ=-1,1
          GY = GRPY(I) + JJ
          IF( GY .EQ. 0 ) THEN
            GY   = PXD
            MODY = -YL
            GOTO 150
          END IF
          IF( GY .EQ. PXD+1 ) THEN
            GY   = 1
            MODY = YL
```

• The conservative force, i.e., the first term on the right-hand side of Eq. (6.19), is saved in FCXD(*) and FCYD(*). Similarly, the dissipative term, i.e., the second term, is saved in FDXD(*) and FDYD(*). The random term, i.e., the third term, is saved in FRXD(*) and FRYD(*).

• The name of the cell in which the particles possibly interact with particle *i* of interest is GRP=GX+(GY−1)*PXD.
• (MODX, MODY) are used in treating the periodic BC.

```
              GOTO 150
            END IF
            MODY =0.D0
C
   150   DO 300 II=-1,1
            GX = GRPX(I) + II
            IF( GX .EQ. 0 ) THEN
              GX   = PXD
              MODX =-XL
              GOTO 160
            END IF
            IF( GX .EQ. PXD+1 ) THEN
              GX   = 1
              MODX = XL
              GOTO 160
            END IF
            MODX =0.D0
C
   160      GRP = GX + (GY-1)*PXD
            IF( TMXD(GRP) .EQ. 0 )              GOTO 300
C                                                                  +++ ENERGY +++
            DO 200 JJJ=1,TMXD(GRP)
C
              J = TABLED(JJJ,GRP)
              IF( J .LE. I )                    GOTO 200
C
              RXIJ  = RXI - (RXD(J) + MODX)
              IF( DABS(RXIJ) .GE. RCOFFD )      GOTO 200
              RYIJ  = RYI - (RYD(J) + MODY)
              IF( DABS(RYIJ) .GE. RCOFFD )      GOTO 200
              RIJSQ = RXIJ**2 + RYIJ**2
              IF( RIJSQ .GE. RCOFFD2 )          GOTO 200
              RIJ   = DSQRT(RIJSQ)
              VXIJ  = VXI - VXD(J)
              VYIJ  = VYI - VYD(J)
C
              EXIJ  = RXIJ/RIJ
              EYIJ  = RYIJ/RIJ
              IF(RIJ .LE. DC ) THEN
                WR  = 1.D0 - RIJ/DC
                WR2 = WR*WR
              ELSE
                WR  = 0.D0
                WR2 = 0.D0
              END IF
C                                                                  --- FC ---
              FCXIJ   = WR*EXIJ
              FCYIJ   = WR*EYIJ
              FCXI    = FCXI    + FCXIJ
              FCYI    = FCYI    + FCYIJ
              FCXD(J)= FCXD(J)  - FCXIJ
              FCYD(J)= FCYD(J)  - FCYIJ
C                                                                  --- FD ---
              C1      = EXIJ*VXIJ + EYIJ*VYIJ
              FDXIJ   = - WR2*C1*EXIJ
              FDYIJ   = - WR2*C1*EYIJ
              FDXI    = FDXI    + FDXIJ
              FDYI    = FDYI    + FDYIJ
              FDXD(J)= FDXD(J)  - FDXIJ
              FDYD(J)= FDYD(J)  - FDYIJ
C                                                                  --- FR ---
              NRAN  = NRAN + 1
              RAN1  = DBLE(RAN(NRAN))
              IF( RAN1 .LE. 0.D0 ) RAN1 = 0.99999D0
              NRAN  = NRAN + 1
              RAN2  = DBLE(RAN(NRAN))
              TTAIJ = DSQRT(-2.D0*DLOG(RAN1))*DCOS(2.D0*PI*RAN2)
CCC           IF( DABS(TTAIJ) .GT. 6.D0 ) TTAIJ = DSIGN( 6.D0, TTAIJ )
C
```

• The treatment of the periodic BC.
• If the two particles are separated over the cutoff distance RCOFFD, the calculation is unnecessary.

• The action–reaction law can provide the force acting on particle *j* as (−FCXIJ) and (−FCYIJ).

• The calculation of the first conservative force in Eq. (6.19).

• The calculation of the second dissipative force in Eq. (6.19).

• The calculation of the third random force in Eq. (6.19).

• TTAIJ means θ_{ij}.

```
               FRXIJ   = WR*EXIJ*TTAIJ
               FRYIJ   = WR*EYIJ*TTAIJ
               FRXI    = FRXI     + FRXIJ
               FRYI    = FRYI     + FRYIJ
               FRXD(J)= FRXD(J)  - FRXIJ
               FRYD(J)= FRYD(J)  - FRYIJ
C
   200       CONTINUE
C
   300     CONTINUE
C
           FCXD(I)  = FCXI
           FCYD(I)  = FCYI
           FDXD(I)  = FDXI
           FDYD(I)  = FDYI
           FRXD(I)  = FRXI
           FRYD(I)  = FRYI
C
   500  CONTINUE
C
        DO 520 I=1,ND
          FCXD(I)  = FCXD(I)*H*ALP/(MD*DC)
          FCYD(I)  = FCYD(I)*H*ALP/(MD*DC)
          FDXD(I)  = FDXD(I)*H*GAM/(DC*MD**0.5)
          FDYD(I)  = FDYD(I)*H*GAM/(DC*MD**0.5)
          FRXD(I)  = FRXD(I)*(H*2.D0*GAM)**0.5/(MD**0.75*DC*0.5)
          FRYD(I)  = FRYD(I)*(H*2.D0*GAM)**0.5/(MD**0.75*DC*0.5)
   520  CONTINUE
                                                              RETURN
                                                              END
C**** SUB FORCEINT *****
        SUBROUTINE FORCEINT( N , ND , RE , DC )
C
        IMPLICIT REAL*8 (A-H,O-Z), INTEGER (I-N)
C
        COMMON /BLOCK1/  RX , RY
        COMMON /BLOCK9/  TMX  , TABLE
        COMMON /BLOCK11/ FXMD , FYMD , RCOFFMD , RCOFFDDM
        COMMON /BLOCK15/ H , XL , YL , RCOFF
        COMMON /BLOCK21/ RXD  , RYD
        COMMON /BLOCK29/ TMXD , TABLED
        COMMON /BLOCK31/ FXDM , FYDM
C
        INTEGER    TT, PPXD, PPXYD, TTD
        PARAMETER( NN=100 , NNN=10000 , TT=500  )
        PARAMETER( NND=50000 , PPXD=500 , PPXYD=250000 , TTD=20 )
C
        REAL*8   RX(NN)    , RY(NN)    , FXMD(NN) , FYMD(NN)
        REAL*8   RXD(NND) , RYD(NND) , FXDM(NND), FYDM(NND)
        INTEGER  TMX(NN)   , TABLE(TT,NN)
        INTEGER  TMXD(PPXYD), TABLED(TTD,PPXYD)
C
        REAL*8   RCOFFMD2     , FCOFFDDM     , RCOFFMN     , RCOFFMN2
        REAL*8   RXI  , RYI   , RXIJ , RYIJ , RZIJ , RIJ , RIJ2
        REAL*8   RXID , RYID , RRIJ , TXIJ , TYIJ
        REAL*8   FIJ  , FXIJ , FYIJ , SR2  , SR4
        INTEGER  GP
C
        RCOFFMD2 = RCOFFMD**2
        FCOFFDDM = 2.D0*(DC/RCOFFDDM)**12 - (DC/RCOFFDDM)**6
        RCOFFMN  = 0.5D0 + ( DC/2.D0 )*0.3D0
        RCOFFMN2 = RCOFFMN**2
        DO 10 I=1,N
          FXMD(I) = 0.D0
          FYMD(I) = 0.D0
    10  CONTINUE
        DO 12 I=1,ND
          FXDM(I) = 0.D0
          FYDM(I) = 0.D0
    12  CONTINUE
C
```

• A subroutine for calculating the forces between magnetic and dissipative particles.

• The force acting on magnetic particle *i* by dissipative particles is saved in FXMD(I) and FYMD(I). The force acting on dissipative particle *i* by magnetic particles is saved in FXDM(I) and FYDM(I).

```
C
      DO 200 I=1,N
        RXI = RX(I)
        RYI = RY(I)
        IF( TMX(I) .EQ. 0 )                           GOTO 200
C
        DO 150 J=1, TMX(I)
          GP = TABLE(J,I)
          IF( TMXD(GP) .EQ. 0 )                       GOTO 150
C
          DO 120 K=1, TMXD(GP)
            II   = TABLED(K,GP)
            RXID = RXD(II)
            RYID = RYD(II)
C
```

• The name of the cell in which the dissipative particles possibly interact with magnetic particle *i* is GP. The names of such dissipative particles are read from the variable TABLED (*,GP).

```
            RXIJ = RXI - RXID
            RXIJ = RXIJ - DNINT(RXIJ/XL)*XL
            IF( DABS(RXIJ) .GE. RCOFFMD )             GOTO 120
            RYIJ = RYI - RYID
            RYIJ = RYIJ - DNINT(RYIJ/YL)*YL
            IF( DABS(RYIJ) .GE. RCOFFMD )             GOTO 120
            RIJ2 = RXIJ**2 + RYIJ**2
            IF( RIJ2 .GT. RCOFFMD2 )                  GOTO 120
            IF( RIJ2 .LT. RCOFFMN2 )  RIJ2 = RCOFFMN2
C
            RIJ  = DSQRT( RIJ2 )
            TXIJ = RXIJ/RIJ
            TYIJ = RYIJ/RIJ
```

• If the magnetic particle and the dissipative particle are separated over RCOFFMD, the force is regarded to be zero. If the two particles significantly overlap, the separation is regarded as RCOFFMN in order to prevent the system from diverging.

```
            RRIJ = RIJ - 0.5D0 + DC/2.D0
            SR1  = (DC/RRIJ)
            SR2  = (DC/RRIJ)**2
            SR4  = SR2*SR2
            SR6  = SR2*SR4
            SR12 = SR6*SR6
            FIJ  = (RE*DC/RRIJ)*( 2.D0*SR12 - SR6 - FCOFFDDM )
            FXIJ = FIJ*TXIJ
            FYIJ = FIJ*TYIJ
C
            FXMD(I)   = FXMD(I)   + FXIJ
            FYMD(I)   = FYMD(I)   + FYIJ
            FXDM(II)  = FXDM(II)  - FXIJ
            FYDM(II)  = FYDM(II)  - FYIJ
C
  120     CONTINUE
  150   CONTINUE
  200 CONTINUE
                                                                RETURN
                                                                END
C*********************************************************************
C   THIS SUBROUTINE IS FOR GENERATING UNIFORM RANDOM NUMBERS         *
C   (SINGLE PRECISION) FOR 32-BIT COMPUTER.                          *
C      N       : NUMBER OF RANDOM NUMBERS TO GENERATE                *
C      IX      : INITIAL VALUE OF RANDOM NUMBERS (POSITIVE INTEGER)  *
C              : LAST GENERATED VALUE IS KEPT                        *
C      X(N)    : GENERATED RANDOM NUMBERS (0<X(N)<1)                 *
C*********************************************************************
C**** SUB RANCAL ****
      SUBROUTINE RANCAL( N, IX, X )
C
      IMPLICIT REAL*8 (A-H,O-Z), INTEGER (I-N)
C
      REAL      X(N)
      INTEGER   INTEGMX, INTEGST, INTEG
C
      DATA INTEGMX/2147483647/
      DATA INTEGST,INTEG/584287,48828125/
C
      AINTEGMX = REAL( INTEGMX )
C
      IF ( IX.LT.0 ) STOP
```

• The forces are calculated according to Eq. (6.26).

• A subroutine for generating a uniform random number sequence.

• This is for a 32-bit CPU based on the expression of two's complement.

```
      IF ( IX.EQ.0 ) IX = INTEGST
      DO 30 I=1,N
         IX = IX*INTEG
         IF (IX .LT. 0 )  IX    = (IX+INTEGMX)+1
         X(I) = REAL(IX)/AINTEGMX
   30 CONTINUE
      RETURN
      END
```

7 Practice of Lattice Boltzmann Simulations

In this chapter, we consider the lattice Boltzmann method, which is generally used as a simulation technique for a pure liquid system but has a different approach to the molecular simulation and microsimulation methods. The lattice Boltzmann method is also a potential simulation technique for taking into account multibody hydrodynamic interactions among particles in a particle suspension or polymers in a polymeric liquid. Therefore, the lattice Boltzmann method may be a promising simulation tool in various fields in science and engineering.

In treating fluid properties, such as the flow field, the lattice Boltzmann method employs an abstract approach that makes use of the particle distribution function, whereas the usual fluid simulation method deals with quantities that are intuitively understandable, such as velocities and pressures. The reader may therefore find that the basic principle behind the lattice Boltzmann method is slightly more difficult to understand. However, once mastered, the concept of the particle distribution function and the theoretical background of this simulation method will enable a research scientist to apply the lattice Boltzmann method to various types of flow problems in a relatively straightforward manner.

The present exercise addresses a uniform flow around a circular cylinder, which will be a foundation for applying the lattice Boltzmann method to flow problems in a particle dispersion or a polymeric liquid. The validity of the solution obtained by this method can be evaluated by comparing it with that obtained by a fully developed simulation method, such as the finite difference method. The sample simulation program has been developed from the viewpoint of applying it to a particle suspension; it may thus be very valuable in a practical context.

7.1 Uniform Flow Around a Two-Dimensional Circular Cylinder

We here consider solving the problem of uniform flow past a circular cylinder by means of the lattice Boltzmann method. In a certain limited range of the Reynolds number, a pair of vortices appears behind the cylinder. The formation of these vortices is very sensitive to the type of boundary model used for the interaction between the cylinder and the neighboring virtual fluid particles.

Introduction to Practice of Molecular Simulation. DOI: 10.1016/B978-0-12-385148-2.00007-0

7.2 Specification of Problems in Equations

The important task in the formalization of the present problem is the treatment of the boundary condition between the cylinder and the virtual fluid particles in the neighboring lattice sites in addition to the outer boundary conditions.

We consider a uniform flow past a two-dimensional circular cylinder in the x-direction, as shown in Figure 7.1. The present flow problem is treated as a two-dimensional flow, so we use the D2Q9 lattice model, as explained in Section 1.5. The simulation region is divided into the lattice system shown in Figure 7.2. The two-dimensional circular cylinder with diameter D is fixed at the origin of the coordinate system. Numbering the velocity direction α in the unit cell is as shown in Figure 1.5B, and α is taken as $\alpha = 0,1,2,\ldots,8$. If $\mathbf{r}$ is the position vector of an arbitrary lattice point and $f_\alpha(\mathbf{r},t)$ is the particle distribution function at time t, the function after the time interval Δt, $f_\alpha(\mathbf{r}+\mathbf{c}_\alpha\Delta t,\ t+\Delta t)$, can be evaluated from Eq. (1.91) as

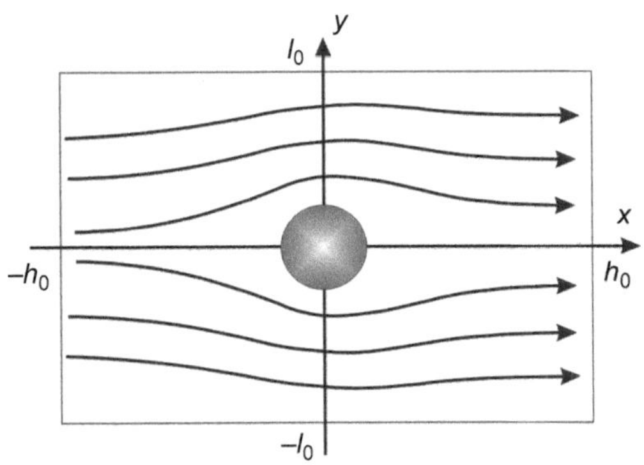

Figure 7.1 Uniform flow past a circular cylinder.

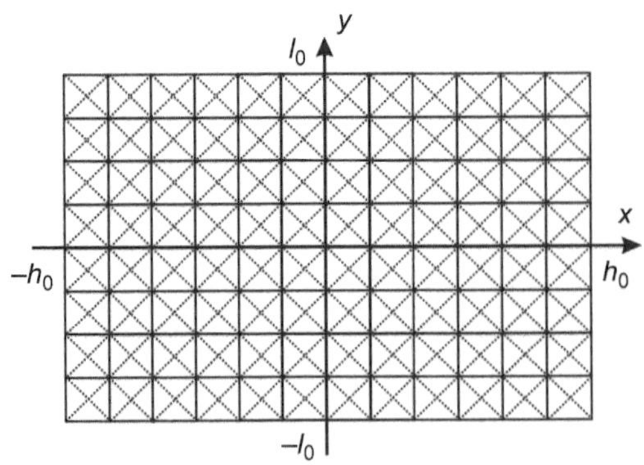

Figure 7.2 Simulation region made up of square lattices.

$$\left.\begin{aligned} f_\alpha(\mathbf{r}+\mathbf{c}_\alpha\Delta t, t+\Delta t) &= \tilde{f}_\alpha(\mathbf{r}, t) \\ \tilde{f}_\alpha(\mathbf{r}, t) &= f_\alpha(\mathbf{r}, t) + \frac{1}{\tau}\{f_\alpha^{(0)}(\mathbf{r}, t) - f_\alpha(\mathbf{r}, t)\} \end{aligned}\right\} \tag{7.1}$$

in which τ is the relaxation time, $f_\alpha^{(0)}$ is the thermodynamic equilibrium distribution function, and $\mathbf{c}_\alpha$ is the lattice velocity in the α-direction. With the notation $\mathbf{u}$ for the macroscopic velocity and ρ for the density, the equilibrium distribution function is written as

$$f_\alpha^{(0)} = \rho w_\alpha \left\{ 1 + 3\frac{\mathbf{c}_\alpha \cdot \mathbf{u}}{c^2} - \frac{3u^2}{2c^2} + \frac{9}{2} \cdot \frac{(\mathbf{c}_\alpha \cdot \mathbf{u})^2}{c^4} \right\} \tag{7.2}$$

in which w_α is a weighting constant. For the case of the D2Q9 model, these terms are written as

$$w_\alpha = \begin{cases} 4/9 & \text{for } \alpha = 0 \\ 1/9 & \text{for } \alpha = 1, 2, 3, 4 \\ 1/36 & \text{for } \alpha = 5, 6, 7, 8 \end{cases} \quad |\mathbf{c}_\alpha| = \begin{cases} 0 & \text{for } \alpha = 0 \\ c & \text{for } \alpha = 1, 2, 3, 4 \\ \sqrt{2}c & \text{for } \alpha = 5, 6, 7, 8 \end{cases} \tag{7.3}$$

In these expressions, c is the velocity of the movement for the shortest lattice distance, expressed as $c = \Delta x/\Delta t$, in which Δx is the shortest distance between two neighboring sites. The lattice velocities given in Eq. (7.3) guarantee that the fluid particles can move from site to site during the time interval Δt. If the particle distributions f_α ($\alpha = 0,1,2,\ldots,8$) are known for all the directions, the macroscopic density and momentum can be evaluated from Eqs. (1.88) and (1.89). That is,

$$\rho(\mathbf{r}, t) = \sum_{\alpha=0}^{8} f_\alpha(\mathbf{r}, t), \quad \rho(\mathbf{r}, t)\mathbf{u}(\mathbf{r}, t) = \sum_{\alpha=0}^{8} f_\alpha(\mathbf{r}, t)\mathbf{c}_\alpha \tag{7.4}$$

In the present case, a uniform flow is generated by employing a thermodynamic equilibrium distribution with a given uniform velocity at the upstream boundary surface at $x = -h_0$. In order to ensure that we obtain reasonable solutions for the present flow problem, we must give careful attention to the interaction between the cylinder and the neighboring lattice sites, and to the outer boundary condition. In the next section we consider the treatment of the boundary conditions.

7.3 Boundary Conditions

We are now ready to formalize the boundary conditions that complement the basic equations explained previously. The boundary surfaces to be treated are the upstream and downstream boundaries, both outer side boundaries, and the cylinder surface boundary. Among these boundary surfaces, the boundary between the cylinder

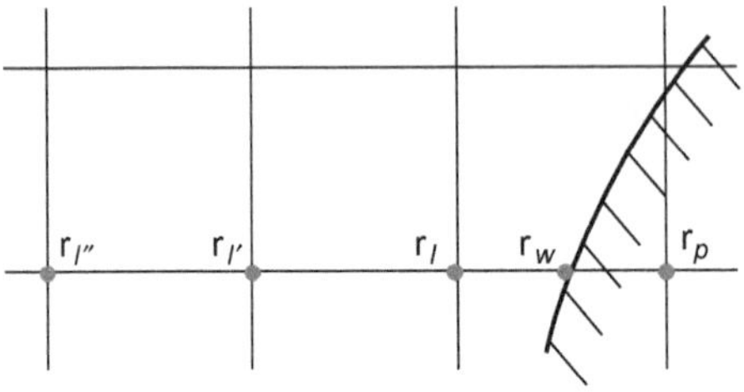

Figure 7.3 Boundary condition on the material surface.

and its neighboring lattice points is the most important and complex. In the following paragraphs, the boundary conditions relating to the cylinder surface are treated first.

We first explain the Yu–Mei–Luo–Shyy (YMLS) model [34] using Figure 7.3. The particle distribution function in the α-direction is considered ($\alpha = 2$ in Figure 7.3). In Figure 7.3, $\mathbf{r}_w$ is the point at the cylinder surface, $\mathbf{r}_p$ is the neighboring point inside the cylinder, $\mathbf{r}_l$ is the neighboring site in the liquid area, and $\mathbf{r}_{l'}$ is the next neighboring point. Since the next point of $\mathbf{r}_l$ in the direction of $\alpha = 1$ is inside the cylinder, $f_2(\mathbf{r}_l, t + \Delta t)$ cannot be obtained from Eq. (7.1). That is, $f_2(\mathbf{r}_l, t + \Delta t)$ is dependent on the particle distribution function at the cylinder surface $\mathbf{r}_w$, and not on that at $\mathbf{r}_p$. If the particle distribution function at $\mathbf{r}_w$, $f_2(\mathbf{r}_w, t + \Delta t)$ is known, $f_2(\mathbf{r}_l, t + \Delta t)$ at $\mathbf{r}_l$ can be evaluated from the linear interpolation method using those at $\mathbf{r}_{l'}$ and $\mathbf{r}_w$ as

$$f_2(\mathbf{r}_l, t + \Delta t) = \frac{\Delta_w}{1 + \Delta_w} f_2(\mathbf{r}_{l'}, t + \Delta t) + \frac{1}{1 + \Delta_w} f_2(\mathbf{r}_w, t + \Delta t) \tag{7.5}$$

in which $\Delta_w = |\mathbf{r}_l - \mathbf{r}_w| / |\mathbf{r}_l - \mathbf{r}_p|$. Figure 7.3 shows the treatment for the direction of $\overline{\alpha} = 2$ (in the opposite direction to $\alpha = 1$), and Eq. (7.5) is simply applied to the direction $\overline{\alpha} = 2$, in which the connecting line in the opposite direction ($\alpha = 1$) crosses the cylinder surface. In order to evaluate $f_2(\mathbf{r}_l, t + \Delta t)$ from Eq. (7.5), $f_2(\mathbf{r}_w, t + \Delta t)$ at the surface is necessary, and this method uses the following equation:

$$f_2(\mathbf{r}_w, t + \Delta t) = (1 - \Delta_w)\tilde{f}_1(\mathbf{r}_{l'}, t) + \Delta_w \tilde{f}_1(\mathbf{r}_l, t) \tag{7.6}$$

This expression means that the particle distribution function on the right-hand side, which is obtained from the linear interpolation method, becomes that in the opposite direction at the next time step. The linear YMLS method [34] uses the linear interpolation procedure with Eqs. (7.5) and (7.6) to obtain $f_2(\mathbf{r}_l, t + \Delta t)$. In this method, only two lattice points are used for the interpolation procedure, so it is suitable for many particle dispersions in which a near-contact situation of particles frequently arises.

In addition to the present YMLS boundary model, for the purpose of study, we will employ three other methods explained in Chapter 8: the historical bounce-back rule [35,36] in Eq. (8.106); the quadratic YMLS method, based on the quadratic curve with the additional point $\mathbf{r}_{l''}$ (Eq. (8.121)); and the Bouzidi–Firdaouss–Lallemand (BFL) model [37] in Eqs. (8.113) and (8.116), or in Eqs. (8.117) and (8.118), which uses the slightly different interpolation scheme. The two different procedures are adopted for $\Delta_w \leq 1/2$ and $\Delta_w > 1/2$ in order not to lose the accuracy of the interpolation.

Next, we specify the treatment at the upstream and downstream surfaces. At the upstream surface, the equilibrium distribution with a given uniform velocity U is specified. On the other hand, the extrapolation condition, which is widely used in numerical analysis methods, may be employed at the downstream boundary surface. As will be shown in Chapter 8, the extrapolation method regards the last three values at $\mathbf{r}_{N-2}$, $\mathbf{r}_{N-1}$, and $\mathbf{r}_N$ as having a linear relationship, expressed as

$$f_{\overline{\alpha}}(\mathbf{r}_N, t+\Delta t) = 2f_{\overline{\alpha}}(\mathbf{r}_{N-1}, t+\Delta t) - f_{\overline{\alpha}}(\mathbf{r}_{N-2}, t+\Delta t) \tag{7.7}$$

in which $\overline{\alpha}$ is the direction leaving the outer boundary toward the inside of the simulation region.

Similarly, the zero-gradient condition may be applicable, and in this condition the differential away from the boundary is regarded as zero:

$$f_{\overline{\alpha}}(\mathbf{r}_N, t+\Delta t) = f_{\overline{\alpha}}(\mathbf{r}_{N-1}, t+\Delta t) \tag{7.8}$$

This condition is inferior to the previous extrapolation in accuracy but superior on the point of divergence. In addition, the uniform flow condition is employed, in which a uniform flow is assumed outside the simulation region.

Finally, the outer side boundary surfaces of the simulation region are specified. If the simulation region is sufficiently large compared with the cylinder diameter, the periodic boundary condition, which is generally used in molecular simulations, is applicable. With this condition, the particle distribution function at the upper surface in Figure 7.1, $f_\alpha(x,y,t)\,|_{\text{upper}}$ ($\alpha = 0,1,\ldots,8$), is regarded as equal to $f_\alpha(x,y,t)\,|_{\text{lower}}$ at the lower surface. Also, the equilibrium distribution in Eq. (7.2) and the bounce-back rule may be applied at both side boundaries. However, these boundary models may cause significant distortion of the flow field, unless a sufficiently large simulation region is employed. The most effective method for removing the influences of the outer boundary surfaces is expected to be the extrapolation condition. Hence, we next discuss the relative accuracy of the uniform flow condition (i.e., the equilibrium distribution condition), the extrapolation condition, and the zero-gradient condition.

7.4 Various Treatments in the Simulation Program

7.4.1 Definition and Evaluation of the Drag Coefficient

The cylinder located in the fluid acts as a resistance to the smooth fluid flow. The drag coefficient C_{D} for a uniform flow past a two-dimensional circular cylinder can be evaluated using the force F per unit length in the flow direction exerted by the ambient fluid, defined as

$$C_{\mathrm{D}} = \frac{F}{\rho U^2 D/2} \tag{7.9}$$

in which ρ is the density of the fluid, U is the uniform flow velocity, and D is the cylinder diameter.

We now show the method of evaluating F. It is assumed that the point $\mathbf{r}_l^{\mathrm{cyl}}$ is the nearest neighbor site in the liquid to the cylinder surface, and the neighbor lattice point from the site in the α-direction is inside the cylinder. The momentum toward the cylinder surface from $\mathbf{r}_l^{\mathrm{cyl}}$ at time t is $\mathbf{c}_{\alpha_l^{\mathrm{cyl}}}\tilde{f}_{\alpha_l^{\mathrm{cyl}}}(\mathbf{r}_l^{\mathrm{cyl}}, t)\Delta x\Delta y$, and that after the collision with the cylinder surface at $(t+\Delta t)$ is $-\mathbf{c}_{\alpha_l^{\mathrm{cyl}}}f_{\overline{\alpha}_l^{\mathrm{cyl}}}(\mathbf{r}_l^{\mathrm{cyl}}, t+\Delta t)\Delta x\Delta y$. The change in the momentum during the time interval Δt is equal to the impulse $\mathbf{F}_{\alpha_l^{\mathrm{cyl}}}\Delta t$. Hence, $\mathbf{F}_{\alpha_l^{\mathrm{cyl}}}\Delta t$ can be obtained as

$$\mathbf{F}_{\alpha_l^{\mathrm{cyl}}} = \left\{\mathbf{c}_{\alpha_l^{\mathrm{cyl}}}\tilde{f}_{\alpha_l^{\mathrm{cyl}}}(\mathbf{r}_l^{\mathrm{cyl}}, t)\Delta x\Delta y + \mathbf{c}_{\alpha_l^{\mathrm{cyl}}}f_{\overline{\alpha}_l^{\mathrm{cyl}}}(\mathbf{r}_l^{\mathrm{cyl}}, t+\Delta t)\Delta x\Delta y\right\}/\Delta t \tag{7.10}$$

The force acting on the cylinder by the fluid $\mathbf{F}$ can be evaluated by summing the contributions from the neighbor lattice sites interacting with the cylinder as

$$\mathbf{F} = \sum_l \sum_{\alpha_l^{\mathrm{cyl}}} \mathbf{F}_{\alpha_l^{\mathrm{cyl}}} \tag{7.11}$$

In the present flow, the absolute value of $F = |\mathbf{F}|$ is used to calculate the drag coefficient in Eq. (7.9).

The flow field and the drag coefficient have already been obtained theoretically and numerically as a function of the Reynolds number Re for a uniform flow past a cylinder, so the accuracy of the present results can be evaluated by comparison with such theoretical and numerical solutions. The Reynolds number Re is defined as $Re = DU/\nu$, in which the kinematic viscosity ν is expressed in Eq. (8.94) for the D2Q9 model. That is,

$$\nu = \frac{\Delta t c^2}{3}(\tau - 1/2) \tag{7.12}$$

7.4.2 Choice of the Procedures by Coloring Lattice Sites

All the lattice points can be classified into one of several groups. That is, the group is composed of (1) lattice points at the upstream and downstream boundary surfaces, (2) lattice points at the outer side boundary surfaces, (3) lattice points interacting with the cylinder, (4) lattice points inside the cylinder, and (5) all other usual lattice points. In the simulation program, this discrimination is expressed using the function "*color*." The following values are given to *color*(i) in the sample program:

color(i) = 0 : all the lattice points in the simulation region not included below
color(i) = 1 : lattice points at the upstream boundary (both end points are included)

$color(i) = 2$	:	lattice points at the downstream boundary (both end points are included)
$color(i) = 3$	:	lattice points at the outer upper boundary surfaces (neither end point is included)
$color(i) = 4$	:	lattice points at the outer lower boundary surfaces (neither end point is included)
$color(i) = 5$	:	lattice points interacting with the cylinder
$color(i) = 6$	:	lattice points inside the cylinder, interacting with the neighboring outside points
$color(i) = 7$	:	lattice points inside the cylinder, not interacting with the neighboring outside points

In the present study, since the cylinder is fixed and does not move, the above checking procedure is only required once before starting the main loop in the program. The introduction of the *color* variable is useful to make the logical flow clear in the program, which is important in developing a simulation program. Moreover, this approach is directly applicable when the dispersed particles move with time, so that the checking procedure must be regularly undertaken until the end of the simulation.

7.4.3 Treatment of Interactions on the Cylinder Surface

In order to use the above-mentioned boundary conditions at the cylinder surface, the quantity $\Delta_w = |\mathbf{r}_l - \mathbf{r}_w|/|\mathbf{r}_l - \mathbf{r}_p|$ must be evaluated. Since the point $\mathbf{r}_w$ is at the cylinder surface, the following equation has to be satisfied:

$$|(1 - \Delta_w)(\mathbf{r}_l - \mathbf{r}_p) + \mathbf{r}_p - \mathbf{r}_{\text{cyl}}| = R_{\text{cyl}} \tag{7.13}$$

in which R_{cyl} is the cylinder radius ($R_{\text{cyl}} = D/2$), and $\mathbf{r}_{\text{cyl}}$ is the cylinder position vector ($\mathbf{r}_{\text{cyl}} = 0$ in the present exercise). Equation (7.13) reduces to an easily solved quadratic equation:

$$\Delta_w = \frac{(\hat{\mathbf{r}}_l^2 - \hat{\mathbf{r}}_p \cdot \hat{\mathbf{r}}_l) - \sqrt{(\hat{\mathbf{r}}_l^2 - \hat{\mathbf{r}}_p \cdot \hat{\mathbf{r}}_l)^2 - (\hat{\mathbf{r}}_l - \hat{\mathbf{r}}_p)^2(\hat{\mathbf{r}}_l^2 - R_{\text{cyl}}^2)}}{(\hat{\mathbf{r}}_l - \hat{\mathbf{r}}_p)^2} \tag{7.14}$$

in which the notation of $\hat{\mathbf{r}}_l = \mathbf{r}_l - \mathbf{r}_{\text{cyl}}$ and $\hat{\mathbf{r}}_p = \mathbf{r}_p - \mathbf{r}_{\text{cyl}}$ is used for simplification. In simulations, the value of Δ_w for all pairs of the two interacting points on either side of the cylinder surface is calculated and saved.

7.4.4 Evaluation of the Velocity and Density

In order to employ the equilibrium distribution function, the macroscopic velocity $\mathbf{u}$ and density ρ at an arbitrary lattice point must be evaluated. The definition of the lattice velocities and the coordinate system are shown in Figure 1.4 and Figure 7.1, respectively. First, the density $\rho(\mathbf{r},t)$ at an arbitrary point $\mathbf{r}$ is evaluated from

Eq. (7.4), and then the velocity $\mathbf{u}=(u_x,u_y)$ is calculated from the following equations:

$$\begin{aligned}\rho(\mathbf{r},t)u_x(\mathbf{r},t) &= c(f_1(\mathbf{r},t)-f_2(\mathbf{r},t))+\sqrt{2}c\left(\frac{\sqrt{2}}{2}f_5(\mathbf{r},t)-\frac{\sqrt{2}}{2}f_6(\mathbf{r},t)\right)\\ &\quad+\sqrt{2}c\left(\frac{\sqrt{2}}{2}f_7(\mathbf{r},t)-\frac{\sqrt{2}}{2}f_8(\mathbf{r},t)\right)\\ &= c(f_1(\mathbf{r},t)-f_2(\mathbf{r},t)+f_5(\mathbf{r},t)-f_6(\mathbf{r},t)+f_7(\mathbf{r},t)-f_8(\mathbf{r},t))\end{aligned} \tag{7.15}$$

$$\rho(\mathbf{r},t)u_y(\mathbf{r},t)=c(f_3(\mathbf{r},t)-f_4(\mathbf{r},t)+f_5(\mathbf{r},t)-f_6(\mathbf{r},t)-f_7(\mathbf{r},t)+f_8(\mathbf{r},t)) \tag{7.16}$$

7.5 Nondimensionalization of the Basic Equations

In simulations, it is usual practice for each quantity to be nondimensionalized and for the nondimensionalized equations to be treated. Since this has been explained in Section 8.6, we briefly show the nondimensionalized results. Here time is nondimensionalized by Δt, velocities by c ($=\Delta x/\Delta t$), and the particle distribution function by ρ_0, so that the basic equation (7.1) is expressed in nondimensional form as

$$\left.\begin{aligned}&f_\alpha^*(\mathbf{r}^*+\mathbf{c}_\alpha^*,t^*+1)=\tilde{f}_\alpha^*(\mathbf{r}^*,t^*)\\ &\tilde{f}_\alpha^*(\mathbf{r}^*,t^*)=f_\alpha^*(\mathbf{r}^*,t^*)+\frac{1}{\tau}\{f_\alpha^{(0)*}(\mathbf{r}^*,t^*)-f_\alpha^*(\mathbf{r}^*,t^*)\}\end{aligned}\right\} \tag{7.17}$$

in which

$$f_\alpha^{(0)*}=w_\alpha^*\rho\left\{1+3\mathbf{c}_\alpha^*\cdot\mathbf{u}^*+\frac{9}{2}(\mathbf{c}_\alpha^*\cdot\mathbf{u}^*)^2-\frac{3}{2}u^{*2}\right\} \tag{7.18}$$

$$|\mathbf{c}_\alpha^*|=\begin{cases}0 & \text{for } \alpha=0\\ 1 & \text{for } \alpha=1,2,3,4\\ \sqrt{2} & \text{for } \alpha=5,6,7,8\end{cases} \tag{7.19}$$

In these equations, w_α has already been shown in Eq. (7.3), and τ is originally a nondimensional quantity. Note that the relationship $c^*=1$ has been taken into account in the above derivations. The nondimensional expressions of Eq. (7.4) are:

$$\rho^*(\mathbf{r}^*,t^*)=\sum_{\alpha=0}^{8}f_\alpha^*(\mathbf{r}^*,t^*),\quad \rho^*(\mathbf{r}^*,t^*)\mathbf{u}^*(\mathbf{r}^*,t^*)=\sum_{\alpha=0}^{8}f_\alpha^*(\mathbf{r}^*,t^*)\mathbf{c}_\alpha^* \tag{7.20}$$

Since the velocities of fluid particles are nondimensionalized by the lattice speed c, the nondimensional speed of sound c_s^* is expressed as $c_s^* = 1/\sqrt{3}$ in Eq. (8.46). Hence, it should be noted that one needs to treat flow problems for a macroscopic velocity u^* of $u^* \ll 1$, unless the density significantly varies in the simulation region. The nondimensional kinematic viscosity, which is necessary for evaluating the Reynolds number, is expressed as $\nu^* = (2\tau - 1)/6$.

7.6 Conditions for Simulations

7.6.1 Initial Distribution

As an initial distribution, the equilibrium distribution with a uniform velocity U and density ρ_0 is used here for the inner simulation region, as well as for the entrance boundary surface. It is possible to use an equilibrium distribution with zero velocity, but this may induce a divergence of the system with time. It is important to discuss the validity of the various initial conditions adopted in order to clarify the characteristics of the simulation program.

7.6.2 Parameters for Simulations

The solution of the flow field and the drag coefficient for the case of a uniform flow past a two-dimensional circular cylinder has already been solved theoretically for $Re \lesssim 1$ and numerically for $Re \gtrsim 1$. Since a pair of stable vortices appears behind the cylinder in the range of $7 \lesssim Re \lesssim 40$, it is quite reasonable to focus on a pair of vortices for $7 \lesssim Re \lesssim 40$; these vortices are very sensitive to the type of surface model employed. Hence, the present simulations have been conducted within the range of $1 \leq Re \leq 20$. The Reynolds number can be expressed as $Re = U^* D^* /((2\tau - 1)/6)$, so that in order to take a large Reynolds number, the relaxation time τ is chosen as $\tau \simeq 1/2$. The uniform velocity U^* cannot be large due to the restriction of the use of a slow uniform velocity compared with the speed of sound. From these considerations, the uniform flow velocity is taken as $U^* = 0.005-0.01$ and the relaxation time as $\tau = 0.515-0.8$. The cylinder diameter D^* is $D^* = 3-20$, and the size of the simulation region is taken as $2h_0^* = 4D^* - 14D^*$ and $2l_0^* = 3D^* - 11D^*$. The influence of the boundary model will appear to be more significant for a smaller simulation region.

7.7 Results of Simulations

It is known that the flow field for outer flow problems is significantly distorted unless a sufficiently large simulation region is used. The results for a relatively small simulation region $(2h_0^*, 2l_0^*) = (7D^*, 6D^*)$ are shown in Figure 7.4 for $Re = 20$. Figures 7.4A and B depict the uniform flow condition and the zero-gradient

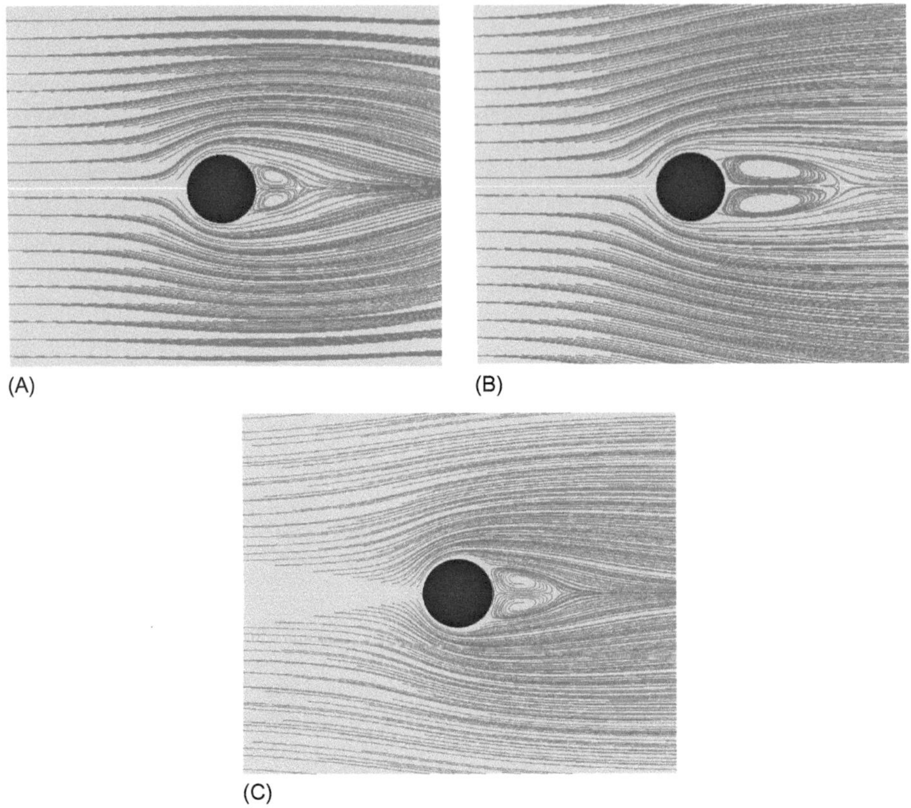

Figure 7.4 Dependence of the flow field on the outer boundary conditions for $Re = 20$; the bounce-back rule is used for the cylinder surface: (A) uniform flow condition, (B) zero-gradient condition, and (C) numerical solution of Navier–Stokes equation.

condition, respectively, and Figure 7.4C shows the Navier–Stokes solution. The historical bounce-back rule has been used for the treatment of the interactions with the cylinder. In the case of $Re = 20$, the length of the pair of vortices is approximately the same as the cylinder diameter, and the formation of these vortices is quite sensitive to the outer boundary condition that has been adopted. The result in Figure 7.4C is the numerical solution obtained by the ordinary finite difference method, and it can be regarded as an exact solution. As shown in Figure 7.4A, for the uniform flow condition (the equilibrium distribution case), the pair of vortices behind the cylinder is significantly distorted and shortened, and the fluid flows along and does not tend to cross the outer side boundary surfaces. This is quite understandable in this case, because a uniform flow is assumed just outside the boundary surfaces; therefore, the flow crossing the boundaries does not tend to arise. The pair of distorted vortices is due to a similar reason—the flow crossing the downstream boundary surface is

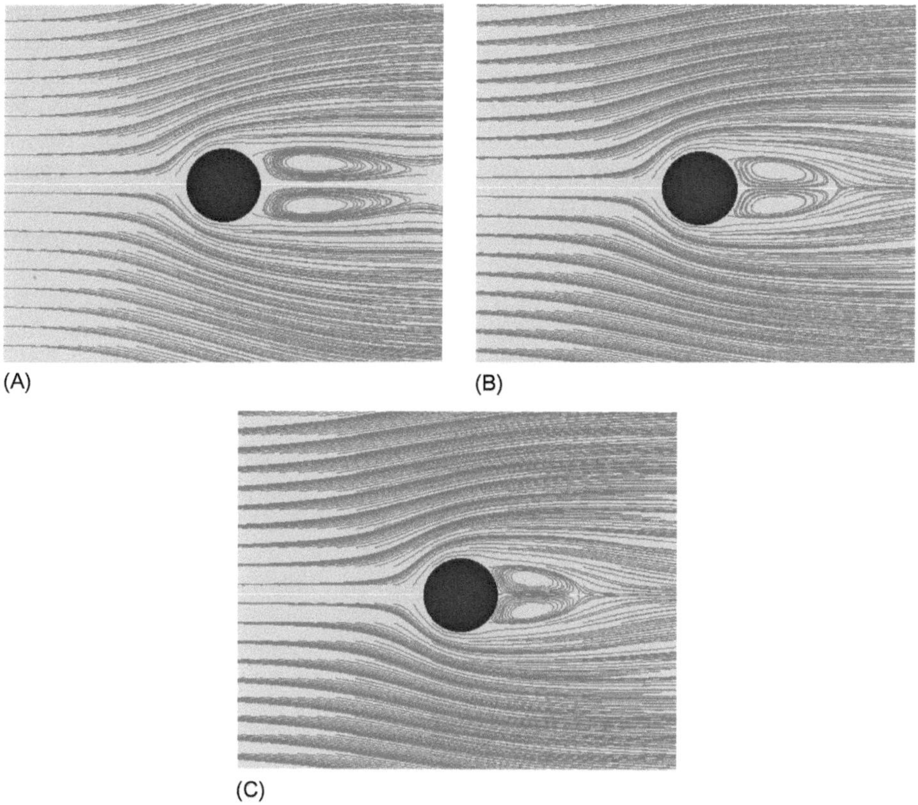

Figure 7.5 Dependence of the flow field on the size of the simulation region ($Re = 20$, the bounce-back method): (A) $(2h_0^*, 2l_0^*) = (6D^*, 5D^*)$; (B) $(2h_0^*, 2l_0^*) = (9D^*, 7D^*)$; and (C) $(2h_0^*, 2l_0^*) = (14D^*, 11D^*)$.

significantly distorted. These results clearly show that a uniform flow condition has the tendency to distort the flow field significantly unless a sufficiently large simulation region is used, although this condition is found to exhibit less divergence in the calculation procedures during a simulation run. In contrast, the result for the zero-gradient condition shown in Figure 7.4B is in agreement with the Navier–Stokes solution, but the pair of vortices is significantly distorted. As discussed in the following, this is again due to the use of a small simulation region. For the extrapolation boundary condition, it was found that stable solutions could not be obtained because the flow field diverged during the advance of the time steps.

Figure 7.5 shows the influence of the size of the simulation region on the formation of a pair of vortices for the three cases of $(2h_0^*, 2l_0^*) = (6D^*, 5D^*)$, $(9D^*, 7D^*)$, and $(14D^*, 11D^*)$, which correspond to Figure 7.5A–C, respectively. The bounce-back rule has been used for the collision with the cylinder, and the zero-gradient condition has been used for the boundaries of the simulation box. The Reynolds number Re is 20, as in the previous case. For the case of our smallest simulation region, shown in Figure 7.5A, the pair of vortices unreasonably lengthens in the downstream area due to the significantly small region used. The results obtained by the lattice Boltzmann method tend to approach the Navier–Stokes solution shown in Figure 7.4C with the size of the simulation region, and the flow field is in agreement with the exact solution. This clearly demonstrates the importance of grasping the influence of this effect by investigating several cases with different size simulation regions.

Figure 7.6 shows the influence of the boundary model employed at the cylinder surface on the formation of the pair of vortices. Figure 7.6A–D illustrate the bounce-back rule, the linear YMLS method, the liner BFL method, and the Navier–Stokes solution. These results were obtained for $Re = 20$, the simulation region $(2h_0^*, 2l_0^*) = (14D^*, 11D^*)$, and the zero-gradient condition for the outer boundary surfaces. The quadratic YMLS and BFL methods give rise to a divergence of the flow field. As clearly seen in Figure 7.6, no significant difference can be observed among these flow fields, and these three boundary models show agreement concerning the formation of the pair of vortices behind the cylinder. Qualitative and quantitative agreement with the exact solution was also confirmed

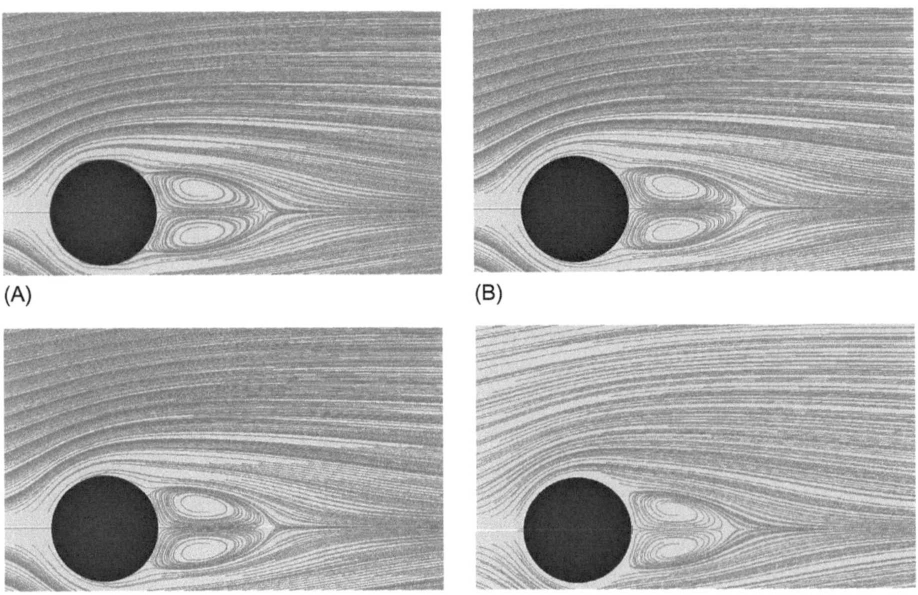

(A) (B) (C) (D)

Figure 7.6 Dependence of the flow field on the surface models on the cylinder surface ($Re = 20$, the zero-gradient condition): (A) bounce-back rule, (B) linear YMLS method, (C) linear BFL method, and (D) numerical solution of Navier–Stokes equation.

concerning the drag coefficient and the velocity distributions, although not shown here. In particular, agreement for the linear YMLS method is good, which may indicate there is some advantage to be found in the application of this boundary method for particle dispersions. As previously discussed, this is because the method that uses the fewer lattice points in the interpolation scheme is the most desirable.

7.8 Simulation Program

The following list is an example simulation program written in FORTRAN for the case discussed in this chapter, and it explains the significance of the important variables used in the program:

`RX(I,J),RY(I,J)`	:	(x,y) components of the position $\mathbf{r}^*_{i,j}$ of lattice site (i,j) (`I = 0,1,...,PX; J = 0,1,...,PY`)
`VX(I,J),VY(I,J)`	:	Macroscopic velocity $\mathbf{u}^*_{i,j}$ at lattice site (i,j)
`RHO(I,J)`	:	Macroscopic density at lattice site (i,j)
`F(I,J,K)`	:	Particle distribution function (`K = 0,1,...,8`) at lattice site (i,j)
`FTILD(I,J,K)`	:	Particle distribution function after the collision at lattice (i,j)
`W(K)`	:	Weighting constant w_α
`CVEL(2,K)`	:	Lattice velocity $\mathbf{c}_\alpha$ (`CVEL(1,K)` is x-component, and `CVEL(2,K)` is y-component)
`XL,YL`	:	Dimensions of the simulation region in the (x,y) directions
`DNS0`	:	Density of an inflow fluid
`DCYL`	:	Diameter of the cylinder
`UVELX`	:	Uniform flow velocity U^*
`RE`	:	Reynolds number Re
`TAU`	:	Relaxation time τ
`RXCYL,RYCYL`	:	Center of the cylinder (equal to the origin in this practice)
`ICYL,JCYL`	:	Lattice site (in the (x,y) direction) representing the cylinder center
`COLOR(ITH)`	:	*Color* function representing the type of lattice site (i,j) (`ITH = (1 + PX)*J + I + 1`)
`TBLNAM(II)`	:	Save the name of lattice sites interacting with the cylinder
`POSINTBL(ITH)`	:	Save the order in which each lattice site appears in `TBLNAM`
`TBLPOS(II)`	:	Save the order in which quantities relate to lattice site `TBLNAM(II)` appear in the variable `TBLDW`
`TBLNUM(II)`	:	Save the number of velocities interacting with the cylinder concerning lattice site `TBLNAM(II)`
`TBLDW(III)`	:	Save the value of Δ_w
`TBLAL(III)`	:	Save the name of the lattice directions α interacting with the cylinder

In order to assist the reader in understanding the program, explanatory statements have been added to the important features.

```
C***********************************************************************
C*                           LBcyl5.f                                  *
C*                                                                     *
C*    OPEN(9, FILE='@bbba1.dat' ,STATUS='UNKNOWN'); para, results     *
C*    OPEN(11,FILE='bbba11.dat' ,STATUS='UNKNOWN'); parameters        *
C*    OPEN(12,FILE='bbba21.dat' ,STATUS='UNKNOWN'); VEL data          *
C*    OPEN(21,FILE='bbba001.dat',STATUS='UNKNOWN'); VEL field         *
C*    OPEN(22,FILE='bbba011.dat',STATUS='UNKNOWN'); VEL field         *
C*    OPEN(23,FILE='bbba021.dat',STATUS='UNKNOWN'); VEL field         *
C*    OPEN(24,FILE='bbba031.dat',STATUS='UNKNOWN'); VEL field         *
C*    OPEN(25,FILE='bbba041.dat',STATUS='UNKNOWN'); VEL field         *
C*    OPEN(26,FILE='bbba051.dat',STATUS='UNKNOWN'); VEL field         *
C*    OPEN(27,FILE='bbba061.dat',STATUS='UNKNOWN'); VEL field         *
C*    OPEN(28,FILE='bbba071.dat',STATUS='UNKNOWN'); VEL field         *
C*    OPEN(29,FILE='bbba081.dat',STATUS='UNKNOWN'); VEL field         *
C*    OPEN(30,FILE='bbba091.dat',STATUS='UNKNOWN'); VEL field         *
C*    OPEN(41,FILE='avsvel1.fld',STATUS='UNKNOWN'); MicroAVS fld      *
C*    OPEN(42,FILE='avsvel1.dat',STATUS='UNKNOWN'); MicroAVS data     *
C*                                                                     *
C*    -----  LATTICE BOLTZMANN SIMULATION OF A FLOW PAST    -----      *
C*           A CIRCULAR CYLINDER IN A TWO-DIMENSIONAL SYSTEM           *
C*                                                                     *
C*    VER.1:                                                           *
C*      1. D2Q9 MODEL IS USED                                          *
C*      2. EQUILIBRIUM BC WITH GIVEN UNIFORM VEL. IS USED FOR          *
C*         UPSTREAM BC                                                 *
C*      3. THREE FOLLOWING BC'S ARE USED FOR BOTH SIDES BC OF CYL      *
C*         (1) EXTRAPOLATION BC     (ITREESID=1)                       *
C*         (2) DEF=0                (ITREESID=2)                       *
C*         (3A) UIFORM FLOW (Const)(ITREESID=3)                        *
C*         (3B) UIFORM FLOW (DEF=0)(ITREESID=4)                        *
C*         (3C) UIFORM FLOW (Extra)(ITREESID=5)                        *
C*      4. THREE FOLLOWING BC'S ARE USED FOR DOWNSTREAM BC             *
C*         (1)  EXTRAPOLATION BC    (ITREEDWN=1)                       *
C*         (2)  DEF=0               (ITREEDWN=2)                       *
C*         (3A) UIFORM FLOW (Const)(ITREEDWN=3)                        *
C*         (3B) UIFORM FLOW (DEF=0)(ITREEDWN=4)                        *
C*         (3C) UIFORM FLOW (Extra)(ITREEDWN=5)                        *
C*      5. THREE FOLLOWING BC'S ARE USED FOR COLLISION BETWEEN         *
C*         SITES AND CYLINDER                                          *
C*         (1)  BOUNCE-BACK                (ITREECYL=1)                *
C*         (2A) YMLS METHOD(Quadratic)     (ITREECYL=2)                *
C*         (2B) YMLS METHOD(Liner)         (ITREECYL=3)                *
C*         (3A) BFL METHOD(Quadratic)      (ITREECYL=4)                *
C*         (3B) BFL METHOD(Linear)         (ITREECYL=5)                *
C*                                                                     *
C*                                         VER.1 BY A.SATOH, '08  7/4  *
C***********************************************************************
C     ---- THE FOLLOWING NOTATIONS ARE USED FOR LATTICE BOLTZMANN ---
C     F(I,J,K)       : DENSITY DISTRIBUTION FUNCTION
C                      I=0,1,2,...,PX : J=0,1,2,...PY : K=0,1,...,8
C     FTILD(I,J,K)   : DENSITY DISTRIBUTION FUNCTION BEFORE TRAVEL
C     CVEL(2,K)      : C_ALPHA
C                      C_0=(0,0)
C                      C_1=(1,0), C_2=(-1, 0), C_3=(0, 1), C_4=( 0,-1)
C                      C_5=(1,1), C_6=(-1,-1), C_7=(1,-1), C_8=(-1, 1)
C     W(K)           : WEIGHT CONSTANTS
C                      W(0)=4/9, W(ALPHA)=1/9  (ALPHA=1,2,3,4),
C                                W(ALPHA)=1/36 (ALPHA=5,6,7,8)
C     ALPHAMX        : =8 FOR D2Q9
C     IINC(2,K)      : INCREMENT IN EACH DIRECTION FOR TRANSFER
C                      FOR ALPHA DIRECTION
C                      (E.X., IINC(1,1)=1, IINC(2,1)=0)
C     ANTIALPH(K)    : NAME OF THE OPPOSITE DIRECTION SITE FOR ALPHA
C                      (E.X., ANTIALPH(1)=2)
C     RHO(I,J)       : DENSITY AT (I,J)
C     RX(I,J),RY(I,J) : LATTICE POSITION
C     VX(I,J),VY(I,J) : VELOCITY COMPONENTS IN X- AND Y-DIRECTIONS
C     DNS0       : MEAN DENSITY (CONSTANT FOR NON-COMPRESSIVE FLOW)
C     PX,PY      : NUMBER OF CELLS IN EACH DIRECTION (EVEN VALUES)
C     PXY        : = (PX+1)*(PY+1)
C     XL,YL      : LENGTHS OF SIMULATION REGION IN EACH DIRECTION
C     TAU        : NON-DIMENSIONAL RELAXATION TIME
C     DX         : UNIT LENGTH (=1)
C     DT         : TIME INTERVAL(=1)
C     CLAT       : LATTICE VELOCITY (=1)
C     (UVELX,UVELY) : UNIFORM VELOCITY COMPONENTS
C
C     ------  THE FOLLOWING NOTATIONS ARE USED FOR THE CYLINDER ------
```

```
C      DCYL            : DIAMETER OF CYLINDER
C      RXCYL,RYCYL     : POSITION OF CYLINDER(=(0,0) FOR THE PRESENT CASE)
C      ICYL , JCYL     : SITE POSITION OF CYLINDER
C      COLOR(PXY)      : COLOR FOR DISTINGUISHING PROCEDURES FOR EACH SITE
C      POSINTBL(PXY)   : POSITION OF THE SITE IN TBLNAM(*) FOR EACH SITE
C      TBLNAM(NTBL)    : NAMES OF INTERACTING SITES WITH CYLINDER
C      TBLNUM(NTBL)    : NUMBERS OF INTERACTING ALPHA-VELS FOR EACH SITE
C      TBLPOS(NTBL)    : POSITION OF THE SITE APPEARING IN TBLDW(*) AND
C                        TBLAL(*)
C      TBLDW(NTBLDW)   : VALUES OF DW ARE SAVED IN TBLDW(*) FOR EACH SITE
C      TBLAL(NTBLDW)   : VALUES OF ALPHA-VEL ARE SAVED FOR EACH SITE
C      TBLNAMIN(NTBLNAMI) : THE NAMES OF SITES INSIDE CYLINDER
C
C      ------------------ CONCERNING DRAG COEFFICIENT ----------------
C      CD             : DRAG COEFFICIENT
C      CDFORCE(NSMPLCD) : FORCE IS SAVED FOR EACH TIME STEP
C      CDFORCE0       : COEFFICIENT (U**2)D/2 IS USED FOR CAL. CD
C      NSMPLCD        : TOTAL SAMPLING NUMBER
C      RE             : REYNOLDS NUMBER
C                         U=0.005 D=20 TAU=0.80  Re=  1
C                         U=0.005 D=20 TAU=0.60  Re=  3
C                         U=0.005 D=20 TAU=0.55  Re=  6
C                         U=0.005 D=20 TAU=0.53  Re= 10
C                         U=0.005 D=20 TAU=0.52  Re= 15
C                         U=0.005 D=20 TAU=0.515 Re= 20
C                         U=0.005 D=20 TAU=0.51  Re= 30
C
C        +++  -XL1<RX(I)<XL2 , -YL1<RY(I)<YL2 , -ZL1<RZ(I)<ZL2  +++
C-------------------------------------------------------------------
       IMPLICIT REAL*8 (A-H,O-Z), INTEGER (I-N)
C
       COMMON /BLOCK1/  F     , FTILD
       COMMON /BLOCK2/  CVEL  , W    , IINC  , ANTIALPH, ALPHAMX
       COMMON /BLOCK3/  RHO   , RX   , RY    , VX    , VY
       COMMON /BLOCK4/  DNS0  , TAU  , DX    , DT    , CLAT
       COMMON /BLOCK5/  XL , YL , XL1 , YL1 , XL2 , YL2 , PX , PY , PXY
       COMMON /BLOCK6/  UVELX , UVELY
C
       COMMON /BLOCK14/ RXCYL , RYCYL , ICYL , JCYL , DCYL
       COMMON /BLOCK15/ COLOR     , POSINTBL
       COMMON /BLOCK16/ TBLNAM    , TBLNUM    , TBLPOS    , NTBL
       COMMON /BLOCK17/ TBLDW     , TBLAL     , NTBLDW
       COMMON /BLOCK18/ TBLNAMIN , NTBLNAMI
C
       COMMON /BLOCK21/ CD , CDFORCE0 , CDFORCE , RE , NSMPLCD
C
C      ----------------------------
       INTEGER     PP , QQ , KK
       PARAMETER( PP=300 , QQ=400 , KK=8 , PI=3.141592653589793D0 )
C
       REAL*8     F(0:PP,0:QQ,0:KK), FTILD(0:PP,0:QQ,0:KK)
       REAL*8     CVEL(2,0:KK)      , W(0:KK)
       REAL*8     RHO(0:PP,0:QQ)
       REAL*8     RX( 0:PP,0:QQ)    , RY(0:PP,0:QQ)
       REAL*8     VX( 0:PP,0:QQ)    , VY(0:PP,0:QQ)
       INTEGER    ALPHAMX , IINC(2,0:KK) , ANTIALPH(0:KK)
       INTEGER    PX , PY ,  PXY
C      ----------------------------
       INTEGER     PPXY
       PARAMETER( PPXY=150000 , NNTBL=2200 , NNTBL2=4400 , NNTBL3=4400 )
C
       REAL*8     TBLDW(NNTBL2)
       INTEGER    COLOR(PPXY)     , POSINTBL(PPXY)
       INTEGER    TBLNAM(NNTBL) , TBLNUM(NNTBL) , TBLPOS(NNTBL) , NTBL
       INTEGER    TBLAL(NNTBL2) , NTBLDW
       INTEGER    TBLNAMIN(NNTBL3) , NTBLNAMI
C      ----------------------------
       INTEGER     NNCD
       PARAMETER( NNCD=1000000 )
C
       REAL*8     CDFORCE(NNCD)
C      ----------------------------
       REAL*8     VXSUM(0:PP,0:QQ), VYSUM(0:PP,0:QQ), RHOSUM(0:PP,0:QQ)
       REAL*8     H , DCYL2SQ , CD99 , C1
       INTEGER    NTIMEMX , NGRAPH  , NANIME , NOPT, NSMPLCD, NDUM
       INTEGER    NTHROW  , NSMPLVEL, NANMCTR, NSMPL1
       INTEGER    ITREECYL, ITREESID, ITREEDWN
```

• The given values are written out in @bbbd1 and bbbd11, and the velocities are written out in bbbd21.

```
C
      OPEN(9,FILE='@bbbd1.dat'   ,STATUS='UNKNOWN')
      OPEN(11,FILE='bbbd11.dat'  ,STATUS='UNKNOWN')
      OPEN(12,FILE='bbbd21.dat'  ,STATUS='UNKNOWN')
      OPEN(21,FILE='bbbd001.dat',STATUS='UNKNOWN')
      OPEN(22,FILE='bbbd011.dat',STATUS='UNKNOWN')
      OPEN(23,FILE='bbbd021.dat',STATUS='UNKNOWN')
      OPEN(24,FILE='bbbd031.dat',STATUS='UNKNOWN')
      OPEN(25,FILE='bbbd041.dat',STATUS='UNKNOWN')
      OPEN(26,FILE='bbbd051.dat',STATUS='UNKNOWN')
      OPEN(27,FILE='bbbd061.dat',STATUS='UNKNOWN')
      OPEN(28,FILE='bbbd071.dat',STATUS='UNKNOWN')
      OPEN(29,FILE='bbbd081.dat',STATUS='UNKNOWN')
      OPEN(30,FILE='bbbd091.dat',STATUS='UNKNOWN')
      OPEN(41,FILE='avsvel1.fld',STATUS='UNKNOWN')
      OPEN(42,FILE='avsvel1.dat',STATUS='UNKNOWN')
                                                          NP=9
C                                          --- PARAMETER (1) ---
C                              ++ PX=140, PY=120 ; PX=280, PY=220 ++
C                              ++ PX=180, PY=140 ; PX=320, PY=260 ++
C                              ++ PX=220, PY=180 ; PX=340, PY=280 ++
      AU     = 0.515D0
      UVELX  = 0.005D0
      UVELY  = 0.0D0
      PX     = 140
      PY     = 120
C                                          --- PARAMETER (2) ---
        ALPHAMX =  8
        XL      = DBLE(PX)
        YL      = DBLE(PY)
        XL1     = XL/2.D0
        YL1     = YL/2.D0
        XL2     = XL - XL1
        YL2     = YL - YL1
        DNS0    =  1.D0
        DX      =  1.D0
        DT      =  1.D0
        CLAT    =  1.D0
        PXY     = (PX+1)*(PY+1)
C                                          --- PARAMETER (2) ---
        DCYL    = 20.D0 - 0.0001D0
        DCYL2SQ = DCYL**2 / 4.D0
        RXCYL   = 0.D0
        RYCYL   = 0.D0
C                                          --- PARAMETER (4) ---
C                    ++ (1)  BOUNCE-BACK                (ITREECYL=1) ++
C                    ++ (2A) YMLS METHOD(Quadratic)     (ITREECYL=2) ++
C                    ++ (2B) YMLS METHOD(Liner)         (ITREECYL=3) ++
C                    ++ (3A) BFL METHOD(Quadratic)      (ITREECYL=4) ++
C                    ++ (3B) BFL METHOD(Linear)         (ITREECYL=5) ++
C                         ++ (1) EXTRAPOLATION BC       (ITREESID=1) ++
C                         ++ (2) DEF=0                  (ITREESID=2) ++
C                         ++ (3A) UIFORM FLOW(Const)(ITREESID=3) ++
C                         ++ (3B) UIFORM FLOW(DEF=0)(ITREESID=4) ++
C                         ++ (3C) UIFORM FLOW(Extra)(ITREESID=5) ++
C                        ++ (1) EXTRAPOLATION BC        (ITREEDWN=1) ++
C                        ++ (2) DEF=0                   (ITREEDWN=2) ++
C                        ++ (3A) UIFORM FLOW(Const)     (ITREEDWN=3) ++
C                        ++ (3B) UIFORM FLOW(DEF=0)     (ITREEDWN=4) ++
C                        ++ (3C) UIFORM FLOW(Extra)     (ITREEDWN=5) ++
        ITREECYL= 1
        ITREESID= 2
        ITREEDWN= 2
C
        NTIMEMX = 200000
        NGRAPH  = NTIMEMX/10
        NANIME  = NTIMEMX/10
        NOPT    = 20
C                                          --- PARAMETER (5) ---
C                                          -  NSMPLCD FOR CD   -
C
        NSMPLCD =  NTIMEMX
        NSMPL1  =  5
        NTHROW  =  NTIMEMX/10
C                                          --- PARAMETER (6) ---
        CDFORCE0= (DNS0*(UVELX)**2)*DCYL /2.D0
        RE      = UVELX*DCYL/( (2.D0*TAU - 1.D0 )/6.D0 )
C
C       ---------------------------------------------------------
C       --------------------    INITIAL SETTING    -------------
```

• The velocities and densities are written out in bbbd001 to bbbd091, and the data are written out in avsvel1 for Micro-AVS.

• τ=0.515 and U^*=0.005. The numbers of the lattice points in the x- and y-directions are (PX, PY), respectively. α=0,...,8. The size of the simulation box is (XL,YL) in each direction.

• The name of the first lattice point is 0, so that the total number of the lattice points is PXY=(PX+1)×(PY+1).
• The cylinder diameter centered at the origin is DCYL= 20, and its center is therefore (RXCYL,RYCYL)=(0,0).

• The boundary condition is adopted according to the values of ITREECYL, ITREESID, and ITREEDWN.
• The total number of time steps is NTIMEMX= 200000. The velocity field data are written out at every NGRAPH time steps.

• 10 sets of data are written out for making an animation based on MicroAVS.

• The velocities and positions of the lattice points are assigned.

```
C     -----------------------------------------------------------
C                       --- SET C_VEL(2,8),W(8),IINC(2,8),ANTIALPH(8) ---
      CALL INICVEL
C                            --- SET LATTICE POSITION RX(*,*),RY(*,*) ---
      CALL INILAT
C                                     --- SET INITIAL POSIT. AND VEL. ---
C
CCC   OPEN(19,FILE='bbbd091.dat',STATUS='OLD')
CCC     READ(19,201) PX, PY, ALPHAMX
CCC     READ(19,202) (  ( ( F(I,J,K),K=0,ALPHAMX ),J=0,PY ), I=0,PX  )
CCC     READ(19,204) ( (RX( I,J),J=0,PY),I=0,PX )
CCC     READ(19,204) ( (RY( I,J),J=0,PY),I=0,PX )
CCC     READ(19,206) ( (VX( I,J),J=0,PY),I=0,PX )
CCC     READ(19,206) ( (VY( I,J),J=0,PY),I=0,PX )
CCC     READ(19,208) ( (RHO(I,J),J=0,PY),I=0,PX )
CCC   CLOSE(19,STATUS='KEEP')
CCC   GOTO 7
C                                    --- S
      CALL INIDIST( DNSO , ALPHAMX )
C                                        -
C                                        -
    7 CALL INICOLOR( PX , PY , DCYL2SQ )
C                                --- MAKE T
C                                -   WITH C
      CALL MAKETBLE( DCYL2SQ , NTBL , NTBLDW )
C
C                                       --- SET ZERO
      DO 9 J=0, PY
      DO 8 I=0, PX
        ITH = (PX+1)*J + (I+1)
        IF( (COLOR(ITH).EQ.6) .OR. (COLOR(ITH).EQ.7) ) THEN
          VX( I,J) = 0.D0
          VY( I,J) = 0.D0
          RHO(I,J) = DNS0
        END IF
    8 CONTINUE
    9 CONTINUE
C     --------------------------------------- PRINT OUT CONSTANTS ---
      WRITE(NP,10) DNS0, TAU, DX, DT, CLAT, ALPHAMX
      WRITE(NP,11) PX, PY, PXY, XL, YL, XL1, YL1, XL2, YL2,
     &             UVELX, UVELY
      WRITE(NP,13) DCYL, ITREECYL, ITREESID, ITREEDWN
      WRITE(NP,14) NTIMEMX, NGRAPH, NANIME, NSMPLCD, NTHROW, NSMPL1
      WRITE(NP,15) CDFORCE0, RE
C     -------------------------------------------- INITIALIZATION ---
C
C                                           --- INITIALIZE(1) ---
      NSMPLCD  = 0
      DO 20 I=1, NTIMEMX
        CDFORCE(I) = 0.D0
   20 CONTINUE
C                                           --- INITIALIZE(2) ---
      DO 30 J=0, PY
      DO 25 I=0, PX
        VXSUM( I,J) = 0.D0
        VYSUM( I,J) = 0.D0
        RHOSUM(I,J) = 0.D0
   25 CONTINUE
   30 CONTINUE
      NSMPLVEL = 0
C
      NANMCTR  = 0
C
C     -----------------------------------------------------------
C     -------------------    START OF MAIN LOOP    -----------------
C     -----------------------------------------------------------
C
      DO 1000 NTIME = 1,NTIMEMX
C
C                                --- CAL. VEL AT EACH LAT. POS. ---
C                                -   VX(*,*),VY(*,*),RHO(*,*)      -
        CALL VELCAL( COLOR , ITREESID , ITREEDWN, NTIME )
C                            --- COLLISION PROCEDURE FTILD(*,*,8)  ---
        CALL COLLPROC( COLOR , ALPHAMX )
C                      --- PROPAGATION PROCEDURE,FORCE EVALUATION ---
C                      -                      F(*,*,8) WITHOUT BC   -
        NSMPLCD = NSMPLCD + 1
        CALL MOVEPROC( PX , PY , ANTIALPH , RHO , DNSO , ITREECYL )
C                                --- BOUNDARY CONDITION PROC. ---
```

• The initial values of the distribution function are assigned, and the values of the variable *color* in Section 7.4.2 are evaluated. This procedure is conducted only once because of the cylinder being fixed.

• The lattice points interacting with the cylinder are checked.

• The velocities at the lattice points inside the cylinder are set to be zero.

• The following procedure is conducted in the main loop: (1) the velocities at each lattice point are evaluated in VELCAL, (2) the collision treatment is carried out in COLLPROC, (3) the transfer of the distribution function is conducted in MOVEPROC, and (4) the BC treatment is conducted in BCPROC.

```
C                                  -          FX(*,*,8) FOR BC     -
      CALL BCPROC( PX , PY , DNS0 , ALPHAMX , ITREESID ,
     &                                                ITREEDWN )
C
C                              --- DATA OUTPUT (1) FOR GRAPHICS ---
C
      IF( MOD(NTIME,NGRAPH) .EQ. 0 ) THEN
C
        CALL VELCAL( COLOR , ITREESID , ITREEDWN , NTIME )
C
        NOPT = NOPT + 1
        WRITE(NOPT,201) PX, PY, ALPHAMX
        WRITE(NOPT,202) (   (   ( F(I,J,K),K=0,ALPHAMX ),J=0,PY   ),
     &                                                  I=0,PX    )
        WRITE(NOPT,204) ( (RX( I,J),J=0,PY),I=0,PX )
        WRITE(NOPT,204) ( (RY( I,J),J=0,PY),I=0,PX )
        WRITE(NOPT,206) ( (VX( I,J),J=0,PY),I=0,PX )
        WRITE(NOPT,206) ( (VY( I,J),J=0,PY),I=0,PX )
        WRITE(NOPT,208) ( (RHO(I,J),J=0,PY),I=0,PX )
C
        CLOSE(NOPT,STATUS='KEEP')
      END IF
C                             --- DATA OUTPUT (2) FOR ANIMATION ---
C
      IF( MOD(NTIME,NANIME) .EQ. 0 ) THEN
C
        CALL VELCAL( COLOR , ITREESID , ITREEDWN , NTIME )
C
        NANMCTR = NANMCTR + 1
        CALL GRAPHVEL( NANMCTR )
C
      END IF
C
C                                  --- DATA BETWEEN NTIME=0 AND ---
C                                  --- =NTHROW ARE THROWN AWAY. ---
      IF( NTIME .LT. NTHROW ) GOTO 1000
C
C     ------------------------------------------------------------
      IF( NTIME .EQ. NTHROW ) THEN
C                                                 +++ INITIALIZE +++
        NSMPLCD = 0
        DO 302 I=1, NTIMEMX
          CDFORCE(I) = 0.D0
  302   CONTINUE
C
        DO 310 J=0, PY
        DO 305 I=0, PX
          VXSUM( I,J) = 0.D0
          VYSUM( I,J) = 0.D0
          RHOSUM(I,J) = 0..D0
  305   CONTINUE
  310   CONTINUE
        NSMPLVEL     = 0
C
        GOTO 1000
      END IF
C                                    --- CAL. SUM OF VELOCITIES ---
C
      IF( MOD(NTIME,NSMPL1) .EQ. 0 )THEN
        NSMPLVEL = NSMPLVEL + 1
        CALL VELCAL( COLOR , ITREESID , ITREEDWN , NTIME )
C
        DO 500 J=0, PY
        DO 490 I=0, PX
          VXSUM( I,J) = VXSUM( I,J) + VX( I,J)
          VYSUM( I,J) = VYSUM( I,J) + VY( I,J)
          RHOSUM(I,J) = RHOSUM(I,J) + RHO(I,J)
  490   CONTINUE
  500   CONTINUE
      END IF
C
C
 1000 CONTINUE
C
C     ------------------------------------------------------------
C     ---------------------- END OF MAIN LOOP --------------------
C     ------------------------------------------------------------
C
C                                                 --- CAL. CD ---
      C1 = 0.D0
      DO 1100 I=1, NSMPLCD
```

• The velocity data, etc., are written out at every NGRAPH time steps for the post processing analysis.

• The velocity data, etc., are written out at every NANIME time steps for making an animation.

• In order to evaluate average values, the velocity data, etc., are sampled at every NSMPL1 time steps.

• The drag coefficient is calculated.

```
          C1 = C1 + CDFORCE(I)
 1100 CONTINUE
      CD   = ( C1/DBLE(NSMPLCD) ) / CDFORCE0
      CD99 =    CDFORCE( NSMPLCD ) / CDFORCE0
C
C                       --- CAL AVE. AND PRINT OUT CONSTANTS (1) ---
      CALL AVECAL( NP, NSMPLVEL, VXSUM, VYSUM, RHOSUM,
     &                                        PX, PY, ALPHAMX )
C
C                                             --- DATA OUTPUT (3) ---
      WRITE(11,1101) DNS0, TAU, DX, DT, CLAT, ALPHAMX
      WRITE(11,1103) PX, PY, PXY, XL, YL, XL1, XL2, YL1, YL2
      WRITE(11,1105) UVELX, UVELY
      WRITE(11,1107) DCYL, ITREECYL, ITREESID, ITREEDWN
      WRITE(11,1109) NTIMEMX, NGRAPH, NANIME, NSMPLCD, NTHROW, NSMPL1
      WRITE(11,1111) CD, RE
C                                             --- DATA OUTPUT (4) ---
      WRITE(12,1121) PX, PY
      WRITE(12,1123) (  ( VXSUM( I,J),J=0,PY ), I=0,PX    )
      WRITE(12,1123) (  ( VYSUM( I,J),J=0,PY ), I=0,PX    )
      WRITE(12,1125) (  ( RHOSUM(I,J),J=0,PY ), I=0,PX    )
C
C                                               --- PRINT OUT (2) ---
      WRITE(NP,1131) CD99 , CD , RE
C                                             --- DATA OUTPUT (5) ---
      WRITE(12,1133) CD , RE , NSMPLCD
      WRITE(12,1135) ( CDFORCE(I), I=1, NSMPLCD )
C
                              CLOSE( 9,STATUS='KEEP')
                              CLOSE(11,STATUS='KEEP')
                              CLOSE(12,STATUS='KEEP')
                              CLOSE(41,STATUS='KEEP')
                              CLOSE(42,STATUS='KEEP')
C
C     ------------------------- FORMAT -----------------------------
C
   10 FORMAT(/1H ,'-----------------------------------------------'
     &       /1H ,1X,'LATTICE BOLTZMANN SIMULATION OF',
     &                ' A FLOW AROUND A CYLINDER'
     &       /1H ,10X,' +++ TWO-DIMENSIONAL FLOW +++'
     &       /1H ,'-----------------------------------------------'
     &      //1H ,'DNS0=', F6.3, 2X, 'TAU=',F6.4,
     &              2X, 'DX=', F6.2, 2X, 'DT=', F6.2, 2X, 'CLAT=',
     &              F6.3
     &       /1H , 'ALPHAMX=', I3)
   11 FORMAT(1H ,'PX=', I3, 1X, 'PY=', I3, 1X, 'PXY=', I6, 1X,
     &            'XL=', F6.2, 1X, 'YL=', F6.2, 1X, 'XL1=', F6.2, 1X,
     &            'YL1=',F6.2
     &      /1H ,'XL2=', F6.2, 1X, 'YL2=', F6.2, 2X,
     &            'UVELX=', F6.2, 2X, 'UVELY=',F6.2)
   13 FORMAT(1H ,'DCYL=', F7.3, 2X,'ITREECYL=',I3, 2X,'ITREESID=',I3,
     &             2X,'ITREEDWN=',I3)
   14 FORMAT(1H ,'NTIMEMX=', I8, 2X, 'NGRAPH=', I8, 2X, 'NANIME=', I8
     &      /1H ,'NSMPLCD=',I8, 2X, 'NTHROW=',I8, 2X, 'NSMPL1=',I8)
   15 FORMAT(1H ,'CDFORCE0=', F9.4, 2X, 'RE=', F9.3)
  201 FORMAT( 3I9 )
  202 FORMAT( (6E13.6) )
  204 FORMAT( (6E13.6) )
  206 FORMAT( (6E13.6) )
  208 FORMAT( (6E13.6) )
 1101 FORMAT( 5F9.4, I8 )
 1103 FORMAT( 3I8, 6F9.3 )
 1105 FORMAT( 2F11.5 )
 1107 FORMAT( F6.2 , 3I3 )
 1109 FORMAT( 6I10 )
 1111 FORMAT( 2F12.6 )
 1121 FORMAT( 2I10  )
 1123 FORMAT( ( 8E10.3 ) )
 1125 FORMAT( ( 8E10.3 ) )
 1131 FORMAT(/1H ,'CD99=', F10.5, 3X, 'CD=', F10.5, 3X, 'RE=', F10.5)
 1133 FORMAT( 2F10.4 , I9 )
 1135 FORMAT( ( 7E11.4 ) )
                                                                  STOP
                                                                  END
C*********************************************************************
C***********************  SUBROUTINE  ******************************
C*********************************************************************
C
C**** SUB AVECAL *****
      SUBROUTINE AVECAL( NP, NSMPLVEL, VXSUM, VYSUM, RHOSUM,
```

```
     &                                     PX, PY, ALPHAMX )
C
      IMPLICIT REAL*8 (A-H,O-Z), INTEGER (I-N)
```

• The velocity field is calculated by an averaging procedure.

```
C
      INTEGER   PP , QQ , KK
      PARAMETER( PP=300 , QQ=400 , KK=8 , PI=3.141592653589793D0 )
C
      INTEGER   PX , PY , ALPHAMX
      REAL*8    VXSUM(0:PP,0:QQ), VYSUM(0:PP,0:QQ), RHOSUM(0:PP,0:QQ)
C
C                                      --- CAL VELOCITY FIELD ---
      DO 1010 J=0, PY
      DO 1008 I=0, PX
        VXSUM( I,J) = VXSUM( I,J) / DBLE(NSMPLVEL)
        VYSUM( I,J) = VYSUM( I,J) / DBLE(NSMPLVEL)
        RHOSUM(I,J) = RHOSUM(I,J) / DBLE(NSMPLVEL)
 1008 CONTINUE
 1010 CONTINUE
C                             --- PRINT OUT (2) VELOCITY FIELD ---
C                                                      +++ VX +++
      WRITE(NP,1021)
      DO 1030 I=0, PX
      DO 1029 J=0, PY, 17
        WRITE(NP,1026) VXSUM(I,J   ),VXSUM(I,J+ 1),VXSUM(I,J+ 2),
     &                 VXSUM(I,J+ 3),VXSUM(I,J+ 4),VXSUM(I,J+ 5),
     &                 VXSUM(I,J+ 6),VXSUM(I,J+ 7),VXSUM(I,J+ 8),
     &                 VXSUM(I,J+ 9),VXSUM(I,J+10),VXSUM(I,J+11),
     &                 VXSUM(I,J+12),VXSUM(I,J+13),VXSUM(I,J+14),
     &                 VXSUM(I,J+15),VXSUM(I,J+16)
 1029 CONTINUE
 1030 CONTINUE
C                                                      +++ VY +++
      WRITE(NP,1041)
      DO 1050 I=0, PX
      DO 1049 J=0, PY, 17
        WRITE(NP,1026) VYSUM(I,J   ),VYSUM(I,J+ 1),VYSUM(I,J+ 2),
     &                 VYSUM(I,J+ 3),VYSUM(I,J+ 4),VYSUM(I,J+ 5),
     &                 VYSUM(I,J+ 6),VYSUM(I,J+ 7),VYSUM(I,J+ 8),
     &                 VYSUM(I,J+ 9),VYSUM(I,J+10),VYSUM(I,J+11),
     &                 VYSUM(I,J+12),VYSUM(I,J+13),VYSUM(I,J+14),
     &                 VYSUM(I,J+15),VYSUM(I,J+16)
 1049 CONTINUE
 1050 CONTINUE
C                                                      +++ RHO +++
      WRITE(NP,1061)
      DO 1070 I=0, PX
      DO 1069 J=0, PY, 17
        WRITE(NP,1062) RHOSUM(I,J   ),RHOSUM(I,J+ 1),RHOSUM(I,J+ 2),
     &                 RHOSUM(I,J+ 3),RHOSUM(I,J+ 4),RHOSUM(I,J+ 5),
     &                 RHOSUM(I,J+ 6),RHOSUM(I,J+ 7),RHOSUM(I,J+ 8),
     &                 RHOSUM(I,J+ 9),RHOSUM(I,J+10),RHOSUM(I,J+11),
     &                 RHOSUM(I,J+12),RHOSUM(I,J+13),RHOSUM(I,J+14),
     &                 RHOSUM(I,J+15),RHOSUM(I,J+16)
 1069 CONTINUE
 1070 CONTINUE
C
 1021 FORMAT(/1H ,' VX1, VX2, VX3, VX4, VX5, VX6,...')
 1026 FORMAT( (7E11.4) )
 1041 FORMAT(/1H ,' VY1, VY2, VY3, VY4, VY5, VY6,...')
 1061 FORMAT(/1H ,' RHO1, RHO2, RHO3, RHO4, RHO5, RHO6,...')
 1062 FORMAT( (6E13.6) )
                                                            RETURN
                                                            END
C**** SUB INICVEL *****
      SUBROUTINE INICVEL
```

• A subroutine for setting the lattice velocities, etc.

```
C
      IMPLICIT REAL*8 (A-H,O-Z), INTEGER (I-N)
C
      COMMON /BLOCK2/  CVEL , W  , IINC , ANTIALPH, ALPHAMX
C
      INTEGER    PP , QQ , KK
      PARAMETER( PP=300 , QQ=400 , KK=8 , PI=3.141592653589793D0 )
C
      REAL*8    CVEL(2,0:KK) , W(0:KK)
      INTEGER   ALPHAMX , IINC(2,0:KK) , ANTIALPH(0:KK)
C
```

• The lattice velocity $\boldsymbol{c}_\alpha$ is set.

```
      CVEL(1,0) =  0.D0
      CVEL(2,0) =  0.D0
      CVEL(1,1) =  1.D0
      CVEL(2,1) =  0.D0
      CVEL(1,2) = -1.D0
```

```
      CVEL(2,2) =  0.D0
      CVEL(1,3) =  0.D0
      CVEL(2,3) =  1.D0
      CVEL(1,4) =  0.D0
      CVEL(2,4) = -1.D0
      CVEL(1,5) =  1.D0
      CVEL(2,5) =  1.D0
      CVEL(1,6) = -1.D0
      CVEL(2,6) = -1.D0
      CVEL(1,7) =  1.D0
      CVEL(2,7) = -1.D0
      CVEL(1,8) = -1.D0
      CVEL(2,8) =  1.D0
C
      W(0) = 4.D0/9.D0
      W(1) = 1.D0/9.D0
      W(2) = W(1)
      W(3) = W(1)
      W(4) = W(1)
      W(5) = 1.D0/36.D0
      W(6) = W(5)
      W(7) = W(5)
      W(8) = W(5)
```

• The weighting coefficient w_α is set.

```
C
      IINC(1,1) =  1
      IINC(2,1) =  0
      IINC(1,2) = -IINC(1,1)
      IINC(2,2) = -IINC(2,1)
      IINC(1,3) =  0
      IINC(2,3) =  1
      IINC(1,4) = -IINC(1,3)
      IINC(2,4) = -IINC(2,3)
      IINC(1,5) =  1
      IINC(2,5) =  1
      IINC(1,6) = -IINC(1,5)
      IINC(2,6) = -IINC(2,5)
      IINC(1,7) =  1
      IINC(2,7) = -1
      IINC(1,8) = -IINC(1,7)
      IINC(2,8) = -IINC(2,7)
```

• IINC is used for describing the relationship between the lattice point and the α-direction. For example, the neighboring site in the α-direction of α =1 is arrived at by moving (+1,0) in the *x*- and *y*-direction from the site of interest. In this case, the movement is described as IINC(1,1)=1 and IINC(2,1)=0.

```
C
      ANTIALPH(1) = 2
      ANTIALPH(2) = 1
      ANTIALPH(3) = 4
      ANTIALPH(4) = 3
      ANTIALPH(5) = 6
      ANTIALPH(6) = 5
      ANTIALPH(7) = 8
      ANTIALPH(8) = 7
                                                  RETURN
                                                  END
C**** SUB INILAT *****
      SUBROUTINE INILAT
```

• The opposite direction of the α-direction is saved in ATIALPH(*).

• A subroutine for setting the lattice positions.

```
C
      IMPLICIT REAL*8 (A-H,O-Z), INTEGER (I-N)
C
      COMMON /BLOCK3/  RHO   , RX , RY   , VX   , VY
      COMMON /BLOCK5/  XL , YL , XL1 , YL1 , XL2 , YL2 , PX , PY , PXY
C
      INTEGER     PP , QQ , KK
      PARAMETER( PP=300 , QQ=400 , KK=8 , PI=3.141592653589793D0 )
C
      REAL*8    RHO(0:PP,0:QQ)
      REAL*8    RX( 0:PP,0:QQ)    , RY(0:PP,0:QQ)
      REAL*8    VX( 0:PP,0:QQ)    , VY(0:PP,0:QQ)
      INTEGER   PX , PY , PXY
C
      C1 = XL/DBLE(PX)
      C2 = YL/DBLE(PY)
      DO 100 J=0, PY
      DO  90 I=0, PX
        RX(I,J) = DBLE(I)*C1 - XL1
        RY(I,J) = DBLE(J)*C1 - YL1
   90 CONTINUE
  100 CONTINUE
                                                  RETURN
                                                  END
C**** SUB INIDIST *****
```

• (PX+1,PY+1) lattice points are set in the *x*- and *y*-direction.

• A subroutine for setting the initial value of the distribution functions.

```
      SUBROUTINE INIDIST( DNS0 , ALPHAMX )
C
      IMPLICIT REAL*8 (A-H,O-Z), INTEGER (I-N)
C
      COMMON /BLOCK1/  F   , FTILD
      COMMON /BLOCK5/  XL  , YL  , XL1  , YL1  , XL2  , YL2  , PX  , PY  , PXY
      COMMON /BLOCK6/  UVELX  , UVELY
C
      INTEGER    PP , QQ , KK
      PARAMETER( PP=300  , QQ=400  , KK=8  , PI=3.141592653589793D0 )
C
      REAL*8    F(0:PP,0:QQ,0:KK), FTILD(0:PP,0:QQ,0:KK)
      INTEGER   PX , PY  , PXY  , ALPHAMX
C
      REAL*8    FEQ, CDNS0
C
      CDNS0 = DNS0
C
      DO 110 J=0, PY
      DO 100 I=0, PX
        DO 10  K=0, ALPHAMX
          IF( I.EQ.0 ) THEN
            F(I,J,K)= FEQ( UVELX, UVELY, K, CDNS0 )
          ELSE
CCC         F(I,J,K)= FEQ( 0.D0  , 0.D0  , K, CDNS0 )
            F(I,J,K)= FEQ( UVELX, UVELY, K, CDNS0 )
          END IF
   10   CONTINUE
  100 CONTINUE
  110 CONTINUE
                                                       RETURN
                                                       END
C**** SUB INICOLOR ****
      SUBROUTINE INICOLOR( PX , PY , DCYL2SQ )
C                                   --------------------------------
C                                    0 : USUAL TREATMENT
C                                    1 : TREATMENT AT Bupstream
C                                    2 : TREATMENT AT Bdownstream
C                                    3 : TREATMENT AT Bupper_side
C                                    4 : TREATMENT AT Blower_side
C                                    5 : TREATMENT AT Bcyl_surface
C                                    6 : NO BC TREAT. INSIDE PTCL,
C                                        BUT INTERACTING OUTER SITES
C                                    7 : NO BC TREAT. INSIDE PTCL,
C                                        NOT INTERACTING OUTER SITES
C                                   --------------------------------
      IMPLICIT REAL*8 (A-H,O-Z), INTEGER (I-N)
C
      COMMON /BLOCK3/  RHO   , RX , RY   , VX   , VY
C
      COMMON /BLOCK14/ RXCYL , RYCYL , ICYL , JCYL , DCYL
      COMMON /BLOCK15/ COLOR     , POSINTBL
      COMMON /BLOCK18/ TBLNAMIN , NTBLNAMI
C
C     ----------------------------
      INTEGER    PP , QQ , KK
      PARAMETER( PP=300  , QQ=400  , KK=8  , PI=3.141592653589793D0 )
C
      REAL*8    RHO(0:PP,0:QQ)
      REAL*8    RX( 0:PP,0:QQ)    , RY(0:PP,0:QQ)
      REAL*8    VX( 0:PP,0:QQ)    , VY(0:PP,0:QQ)
      INTEGER   PX , PY , PXY
C     ----------------------------
      INTEGER    PPXY
      PARAMETER( PPXY=150000 , NNTBL=2200 , NNTBL2=4400 , NNTBL3=4400 )
C
      INTEGER   COLOR(PPXY )      , POSINTBL(PPXY)
      INTEGER   TBLNAMIN(NNTBL3) , NTBLNAMI
C     ----------------------------
      REAL*8    RJDG1 , RJDG2 , RJDG2SQ , RXI , RYI , C1
      REAL*8    RXIJ  , RYIJ  , RIJSQ
      INTEGER   ISITE, IC1, IS, IE, JS, JE
C
      DO 120 J=0, PY
      DO 100 I=0, PX
        ISITE  = (PX+1)*J + (I+1)
C
        POSINTBL( ISITE ) = 0
C
```

• An equilibrium distribution with the uniform velocity U is used as an initial distribution.

• A subroutine for evaluating the values of the variable *color* explained in Section 7.4.2.

```
          IF(      I.EQ.0   ) THEN
                                COLOR(ISITE) = 1
          ELSE IF( I.EQ.PX  ) THEN
                                COLOR(ISITE) = 2
          ELSE IF( J.EQ.PY  ) THEN
                                COLOR(ISITE) = 3
          ELSE IF( J.EQ.0   ) THEN
                                COLOR(ISITE) = 4
          ELSE
                                COLOR(ISITE) = 0
          END IF
  100 CONTINUE
  120 CONTINUE
C              --- FOR SPECIAL TREATMENT OF SITES INSIDE CYLINDER ---
```

• The values shown in Section 7.4.2 are assigned to the lattice sites next to each boundary surface.

```
      DO 150 I=1, PX
        IF( RX(I,0) .GE. RXCYL ) THEN
          ICYL = I
          GOTO 170
        END IF
  150 CONTINUE
      ICYL = PX
C
  170 DO 160 J=1, PY
        IF( RY(0,J) .GE. RYCYL ) THEN
          JCYL = J
          GOTO 180
        END IF
  160 CONTINUE
      JCYL = PY
C
```

• The treatment concerning the sites related to the cylinder.

```
  180 C1     = (DCYL/2.D0+0.01D0) / ( RX(2,0)-RX(1,0) )
CCC   IC1    = IDINT(C1)
      IC1    = IDINT(C1) + 2
      IS     = ICYL - IC1 - 1
      IE     = ICYL + IC1
      JS     = JCYL - IC1 - 1
      JE     = JCYL + IC1
      RJDG1   = (DCYL/2.D0) + 3.D0*( RX(2,0)-RX(1,0) )
      RJDG2   = RJDG1
      RJDG2SQ = RJDG2**2
C
      NTBLNAMI = 0
      DO 220 J=JS, JE
      DO 200 I=IS, IE
C
        RXI = RX(I,J)
        RYI = RY(I,J)
C
        ISITE  = (PX+1)*J + (I+1)
        RXIJ   = RXI - RXCYL
        IF( DABS(RXIJ) .GE. RJDG1 )  GOTO 200
        RYIJ   = RYI - RYCYL
        IF( DABS(RYIJ) .GE. RJDG1 )  GOTO 200
        RIJSQ  = RXIJ**2 + RYIJ**2
        IF( RIJSQ .GE. RJDG2SQ )     GOTO 200
C
```

• The sites to be checked are limited to the neighboring sites around the cylinder to a certain degree.

```
        IF( RIJSQ .LE. DCYL2SQ ) THEN
          COLOR(ISITE) = 7
          NTBLNAMI = NTBLNAMI +1
          TBLNAMIN(NTBLNAMI) = ISITE
        END IF
C
  200 CONTINUE
  220 CONTINUE
                                                     RETURN
                                                     END
```

• *color*(*)=7 is set for the site inside the cylinder.

```
C**** SUB MAKETBLE ****
      SUBROUTINE MAKETBLE( DCYL2SQ , NTBL , NTBLDW )
C
      IMPLICIT REAL*8 (A-H,O-Z), INTEGER (I-N)
C
      COMMON /BLOCK3/  RHO   , RX , RY   , VX   , VY
      COMMON /BLOCK5/  XL  , YL , XL1 , YL1 , XL2 , YL2 , PX , PY , PXY
C
      COMMON /BLOCK14/ RXCYL , RYCYL , ICYL , JCYL , DCYL
      COMMON /BLOCK15/ COLOR , POSINTBL
C
C     ----------------------------
      INTEGER    PP , QQ , KK
      PARAMETER( PP=300 , QQ=400 , KK=8 , PI=3.141592653589793D0 )
C
```

• A subroutine for making a list of the lattice sites interacting with the cylinder.

```
      REAL*8    RHO(0:PP,0:QQ)
      REAL*8    RX( 0:PP,0:QQ)    , RY(0:PP,0:QQ)
      REAL*8    VX( 0:PP,0:QQ)    , VY(0:PP,0:QQ)
      INTEGER   PX , PY ,  PXY
C     ----------------------------
      INTEGER    PPXY
      PARAMETER( PPXY=150000 , NNTBL=2200 , NNTBL2=4400 , NNTBL3=4400 )
C
      INTEGER   COLOR(PPXY)    , POSINTBL(PPXY)
C     ----------------------------
      REAL*8    C1 , RXI , RYI
      REAL*8    RJDG1 , RJDG2 , RJDG2SQ
      INTEGER   IS , IE , JS , JE , IC1 , ISITE
CCC   INTEGER   JJ , JJ2 , II , II2 , IALPHA(0:8) , NIALPHA
      INTEGER   IALPHA(0:8) , NIALPHA
C
C                 --- CYLINDER POSITION IS (ICYL,JCYL) IN LATTICE ---
      NTBL   = 0
      NTBLDW= 0
C                   --- CHECK WHETHER OR NOT SITES ARE INSIDE CYL ---
C
   40 C1     = (DCYL/2.D0+0.01D0) / ( RX(2,0)-RX(1,0) )
CCC   IC1    = IDINT(C1)
      IC1    = IDINT(C1) + 2
      IS     = ICYL - IC1 - 1
      IE     = ICYL + IC1
      JS     = JCYL - IC1 - 1
      JE     = JCYL + IC1
C
      RJDG1   = (DCYL/2.D0) + 3.D0*( RX(2,0)-RX(1,0) )
      RJDG2   = RJDG1
      RJDG2SQ = RJDG2**2
C
C
```

• The sites to be checked are limited to the neighboring sites around the cylinder to a certain degree.

```
      DO 220 J=JS, JE
      DO 200 I=IS, IE
C
        RXI = RX(I,J)
        RYI = RY(I,J)
C       --------------------------------- FOR THE MOST OUTER SITES ---
        IF(      (I.EQ.IS) .AND. (J.EQ.JS) ) THEN
C                                            +++AT LEFT-DOWN CORNER+++
          IALPHA(1) = 5
          NIALPHA   = 1
          CALL INTERACT( I, J, RXI, RYI, RXCYL, RYCYL, NIALPHA, IALPHA,
     &                   RJDG2SQ , DCYL2SQ )
        ELSE IF( (I.EQ.IE) .AND. (J.EQ.JS) ) THEN
C                                           +++AT RIGHT-DOWN CORNER+++
          IALPHA(1) = 8
          NIALPHA   = 1
          CALL INTERACT( I, J, RXI, RYI, RXCYL, RYCYL, NIALPHA, IALPHA,
     &                   RJDG2SQ , DCYL2SQ )
        ELSE IF( (I.EQ.IS) .AND. (J.EQ.JE) ) THEN
C                                              +++AT LEFT-UP CORNER+++
          IALPHA(1) = 7
          NIALPHA   = 1
          CALL INTERACT( I, J, RXI, RYI, RXCYL, RYCYL, NIALPHA, IALPHA,
     &                   RJDG2SQ , DCYL2SQ )
        ELSE IF( (I.EQ.IE) .AND. (J.EQ.JE) ) THEN
C                                             +++AT RIGHT-UP CORNER+++
          IALPHA(1) = 6
          NIALPHA   = 1
          CALL INTERACT( I, J, RXI, RYI, RXCYL, RYCYL, NIALPHA, IALPHA,
     &                   RJDG2SQ , DCYL2SQ )
C
C      ------- FOR OUTER CIRCUMFERENCE SIT
C
```

• The treatment for the four corner sites of the outermost rectangle.

• For the left-down site.

• For the right-down site.

• For the left-up site.

• For the right-up site.

• The treatment for the sites on the outermost rectangle, except the four corner sites.

```
        ELSE IF ( J.EQ.JS ) THEN
C                                           +++ALONG X-AXIS (DOWN)+++
          IALPHA(1) = 3
          IALPHA(2) = 5
          IALPHA(3) = 8
          NIALPHA   = 3
C
          CALL INTERACT( I, J, RXI, RYI, RXCYL, RYCYL, NIALPHA, IALPHA,
     &                   RJDG2SQ , DCYL2SQ )
        ELSE IF ( I.EQ.IS ) THEN
C                                      +++ ALONG Y-AXIS (LEFT)  +++
```

• For the sites on the bottom line along the *x*-axis.

• For the sites on the left line along the *y*-axis.

```
            IALPHA(1) = 1
            IALPHA(2) = 5
            IALPHA(3) = 7
            NIALPHA   = 3
C
            CALL INTERACT( I, J, RXI, RYI, RXCYL, RYCYL, NIALPHA, IALPHA,
     &                    RJDG2SQ , DCYL2SQ )
          ELSE IF ( I.EQ.IE ) THEN
C                                             +++ ALONG Y-AXIS (RIGHT) +++
            IALPHA(1) = 2
            IALPHA(2) = 6
            IALPHA(3) = 8
            NIALPHA   = 3
C
            CALL INTERACT( I, J, RXI, RYI, RXCYL, RYCYL, NIALPHA, IALPHA,
     &                    RJDG2SQ , DCYL2SQ )
          ELSE IF ( J.EQ.JE ) THEN
C                                             +++ ALONG X-AXIS (UP)    +++
            IALPHA(1) = 4
            IALPHA(2) = 6
            IALPHA(3) = 7
            NIALPHA   = 3
C
            CALL INTERACT( I, J, RXI, RYI, RXCYL, RYCYL, NIALPHA, IALPHA,
     &                    RJDG2SQ , DCYL2SQ )
C
C         ------------------ FOR INNER SITES  OF CHECKING RECTANGLE ---
          ELSE
C
            IALPHA(1) = 1
            IALPHA(2) = 2
            IALPHA(3) = 3
            IALPHA(4) = 4
            IALPHA(5) = 5
            IALPHA(6) = 6
            IALPHA(7) = 7
            IALPHA(8) = 8
            NIALPHA   = 8
C
            CALL INTERACT( I,J, RXI,RYI, RXCYL,RYCYL, NIALPHA,IALPHA,
     &                    RJDG2SQ , DCYL2SQ )
C
          END IF
C
  200 CONTINUE
  220 CONTINUE
                                                        RETURN
                                                        END
C**** SUB INTERACT *****
      SUBROUTINE INTERACT( I, J, RXI, RYI, RXCYL , RYCYL ,
     &                     NIALPHA, IALPHA, RJDG2SQ, DCYL2SQ )
C
      IMPLICIT REAL*8 (A-H,O-Z), INTEGER (I-N)
C
      COMMON /BLOCK2/  CVEL , W   , IINC , ANTIALPH, ALPHAMX
      COMMON /BLOCK3/  RHO  , RX , RY  , VX  , VY
      COMMON /BLOCK5/  XL , YL , XL1 , YL1 , XL2 , YL2 , PX , PY , PXY
C
      COMMON /BLOCK15/ COLOR    , POSINTBL
      COMMON /BLOCK16/ TBLNAM   , TBLNUM   , TBLPOS   , NTBL
      COMMON /BLOCK17/ TBLDW    , TBLAL    , NTBLDW
C
C     ----------------------------
      INTEGER    PP , QQ , KK
      PARAMETER( PP=300 , QQ=400 , KK=8 , PI=3.141592653589793D0 )
C
      REAL*8     CVEL(2,0:KK)      , W(0:KK)
      REAL*8     RHO(0:PP,0:QQ)
      REAL*8     RX( 0:PP,0:QQ)    , RY(0:PP,0:QQ)
      REAL*8     VX( 0:PP,0:QQ)    , VY(0:PP,0:QQ)
      INTEGER    ALPHAMX , IINC(2,0:KK) , ANTIALPH(0:KK)
      INTEGER    PX , PY ,  PXY    , NIALPHA , IALPHA(0:8)
C     ----------------------------
      INTEGER    PPXY
      PARAMETER( PPXY=150000 , NNTBL=2200 , NNTBL2=4400 , NNTBL3=4400 )
C
      REAL*8     TBLDW(NNTBL2)
      INTEGER    COLOR(PPXY)    , POSINTBL(PPXY)
      INTEGER    TBLNAM(NNTBL)  , TBLNUM(NNTBL)  , TBLPOS(NNTBL)  , NTBL
      INTEGER    TBLAL(NNTBL2)  , NTBLDW
C     ----------------------------
```

• For the sites on the right line along the *y*-axis.

• For the sites on the top line along the *x*-axis.

• For the sites inside the outermost rectangle.

• A subroutine for assessing whether or not the neighboring site is inside the cylinder.

```
C
      INTEGER   ISITE, ISITE1, IPATH, IALPHA0
      INTEGER   IX1 , IY1 , JJJ
      REAL*8    RXI , RYI , RXIJ , RYIJ , RIJSQ
      REAL*8    RXI1, RYI1, RXIJ1, RYIJ1, RIJSQ1 , RJDG2SQ
      REAL*8    RXCYL, RYCYL , DCYL2SQ
      REAL*8    C01 , C1 , C2 , C3 , CDW
C
C
      ISITE  = (PX+1)*J + (I+1)
      RXIJ   = RXI - RXCYL
      RYIJ   = RYI - RYCYL
      RIJSQ  = RXIJ**2 + RYIJ**2
      IF( RIJSQ .GE. RJDG2SQ ) THEN
        COLOR(ISITE) = 0
        RETURN
      END IF
      IF( RIJSQ .LE. DCYL2SQ ) RETURN
```

• The sites being far over the RJDG2SQ distance (note the square) have no interaction with the cylinder.

```
C
      IPATH = 0
      DO 200 JJJ = 1, NIALPHA
C                                      --- (I,J)       : ORIGINAL  ---
C                                      --- ISITE       : ORIGINAL  ---
C                                      --- (RXI,RYI)   : ORIGINAL  ---
C                                      --- (IX1,IY1)   : CANDIDATE ---
C                                      --- ISITE1      : CANDITATE ---
C                                      --- (RXI1,RYI1) : CANDIDATE ---
C                                      --- (RXCYL,RYCYL) : CYLINDER---
        IALPHA0= IALPHA(JJJ)
        IX1    = I + IINC(1,IALPHA0)
        IY1    = J + IINC(2,IALPHA0)
        RXI1   = RX(IX1,IY1)
        RYI1   = RY(IX1,IY1)
        RXIJ1  = RXI1 - RXCYL
        RYIJ1  = RYI1 - RYCYL
        RIJSQ1 = RXIJ1**2 + RYIJ1**2
C
        IF( RIJSQ1 .LE. DCYL2SQ ) THEN
```

• If the neighboring site is inside the cylinder, then the variable *color* is set to be 5 for this site, its site name is saved in TBLNAM, and the order of the site appearing in TBLNAM is saved in POSINTBL.

```
          IPATH = IPATH + 1
          IF( IPATH .EQ. 1 ) THEN
            NTBL = NTBL + 1
            TBLNAM(NTBL) = ISITE
            COLOR(ISITE) = 5
            POSINTBL(ISITE) = NTBL
          END IF
```

• Δ_w=CDW is calculated from Eq. (7.14). The direction of the neighboring site inside the cylinder is saved in TBLAL, and the value of Δ_w is saved in TBLDW.

```
C                               --- FOR OUTSIDE SITES OF CYLINDER ---
          C01 = RXIJ*RXIJ1 + RYIJ*RYIJ1
          C1  = RIJSQ + RIJSQ1 - 2.D0*C01
          C2  = -RIJSQ + C01
          C3  = RIJSQ - DCYL2SQ
          CDW = ( - C2 - DSQRT( C2**2 - C1*C3 ) ) / C1
C
          NTBLDW = NTBLDW + 1
          TBLDW( NTBLDW ) = CDW
          TBLAL( NTBLDW ) = IALPHA0
          IF( IPATH .EQ. 1 )  TBLPOS(NTBL) = NTBLDW
C
C                                --- FOR INSIDE SITES OF CYLINDER ---
          ISITE1 = (PX+1)*IY1 + (IX1+1)
          COLOR(ISITE1) = 6
        END IF
C
  200 CONTINUE
C
      IF( IPATH .GE. 1 ) THEN
        TBLNUM(NTBL) = IPATH
      ELSE
        COLOR(ISITE) = 0
      END IF
                                                       RETURN
                                                       END
C**** SUB VELCAL *****
      SUBROUTINE VELCAL( COLOR , ITREESID , ITREEDWN , NTIME )
C
      IMPLICIT REAL*8 (A-H,O-Z), INTEGER (I-N)
C
      COMMON /BLOCK1/  F    , FTILD
      COMMON /BLOCK3/  RHO  , RX , RY  , VX  , VY
      COMMON /BLOCK4/  DNS0 , TAU , DX , DT , CLAT
      COMMON /BLOCK5/  XL , YL , XL1 , YL1 , XL2 , YL2 , PX , PY , PXY
      COMMON /BLOCK6/  UVELX , UVELY
```

• The order of the quantities, related to the site of interest, first appearing in TBLAL and TBLDW, is saved in TBLPOS.

• The number of the sites, inside the cylinder, interacting with the site of interest is saved in TBLNUM.

• A subroutine for calculating the velocities and densities at each lattice site.

```
C
C      ----------------------------
       INTEGER    PP , QQ , KK
       PARAMETER( PP=300 , QQ=400 , KK=8 , PI=3.141592653589793D0 )
C
       REAL*8     F(0:PP,0:QQ,0:KK), FTILD(0:PP,0:QQ,0:KK)
       REAL*8     RHO(0:PP,0:QQ)
       REAL*8     RX( 0:PP,0:QQ)   , RY(0:PP,0:QQ)
       REAL*8     VX( 0:PP,0:QQ)   , VY(0:PP,0:QQ)
       INTEGER    PX , PY ,  PXY
C      ----------------------------
       INTEGER    PPXY
       PARAMETER( PPXY=150000 , NNTBL=2200 , NNTBL2=4400 , NNTBL3=4400 )
C
       INTEGER    COLOR(PPXY)
C      ----------------------------
       REAL*8     VX0, VY0, RHO0
       INTEGER    ITH, ICLR
C                                                  --- Bupstream ---
       DO 50 J=0, PY
         VX(  0, J) = UVELX
         VY(  0, J) = UVELY
         RHO( 0, J) = DNS0
    50 CONTINUE
C                                                --- INSIDE AREA ---
       DO 100 I=1, PX
       DO  90 J=0, PY
         ITH  = ( PX+1 )*J + ( I+1 )
         ICLR = COLOR(ITH)
         IF( (ICLR .EQ. 6) .OR. (ICLR .EQ. 7) ) GOTO 90
C
         VX0  = F(I,J,1) - F(I,J,2) + F(I,J,5) - F(I,J,6)
     &                              + F(I,J,7) - F(I,J,8)
         VY0  = F(I,J,3) - F(I,J,4) + F(I,J,5) - F(I,J,6)
     &                              + F(I,J,8) - F(I,J,7)
         RHO0 = F(I,J,0) + F(I,J,1) + F(I,J,2) + F(I,J,3) + F(I,J,4)
     &                   + F(I,J,5) + F(I,J,6) + F(I,J,7) + F(I,J,8)
         VX( I,J) = VX0 /RHO0
         VY( I,J) = VY0 /RHO0
         RHO(I,J) = RHO0
         IF(   (ICLR.EQ.1) .OR. (ICLR.EQ.2) .OR. (ICLR.EQ.3) .OR.
     &         (ICLR.EQ.4)   ) THEN
           IF( RHO(I,J) .LT. DNS0 )  RHO(I,J) = DNS0
         END IF
    90 CONTINUE
   100 CONTINUE
C
       IF( (ITREESID.EQ.3).OR.(ITREESID.EQ.4).OR.(ITREESID.EQ.5) ) THEN
         DO 120 I=1, PX-1
           IF( ITREESID.EQ.3 ) THEN
             VX( I,PY) = UVELX
             VY( I,PY) = UVELY
             RHO(I,PY) = DNS0
             VX( I, 0) = UVELX
             VY( I, 0) = UVELY
             RHO(I, 0) = DNS0
           ELSE IF( ITREESID.EQ.4 ) THEN
             VX( I,PY) = VX( I,PY-1)
             VY( I,PY) = VY( I,PY-1)
             RHO(I,PY) = RHO(I,PY-1)
             IF( RHO(I,PY) .LT. DNS0 )  RHO(I,PY) = DNS0
             VX( I, 0) = VX( I,1)
             VY( I, 0) = VY( I,1)
             RHO(I, 0) = RHO(I,1)
             IF( RHO(I, 0) .LT. DNS0 )  RHO(I, 0) = DNS0
           ELSE IF( ITREESID.EQ.5 ) THEN
             VX( I,PY) = 2.D0*VX( I,PY-1) - VX( I,PY-2)
             VY( I,PY) = 2.D0*VY( I,PY-1) - VY( I,PY-2)
             RHO(I,PY) = 2.D0*RHO(I,PY-1) - RHO(I,PY-2)
             IF( RHO(I,PY) .LT. DNS0 )  RHO(I,PY) = DNS0
             VX( I, 0) = 2.D0*VX( I,1) - VX( I,2)
             VY( I, 0) = 2.D0*VY( I,1) - VY( I,2)
             RHO(I, 0) = 2.D0*RHO(I,1) - RHO(I,2)
             IF( RHO(I, 0) .LT. DNS0 )  RHO(I, 0) = DNS0
           END IF
   120   CONTINUE
       END IF
C                                                --- Bdownstream ---
       IF( (ITREEDWN.EQ.3).OR.(ITREEDWN.EQ.4).OR.(ITREEDWN.EQ.5) ) THEN
```

• A uniform flow is set at the upstream boundary surface.

• The local velocities and densities are calculated inside the cylinder from Eq. (7.20).

• The densities are assumed to be not smaller than the given density at the outer boundary surfaces.

• The treatment at the side boundary surfaces.

• (1) The equilibrium distribution.

• (2) The zero-gradient condition (Eq. (7.8)).

• (3) The extrapolation condition (Eq. (7.7)).

• The treatment at the downstream boundary surface.

• (1) The equilibrium distribution.

• (2) The zero-gradient condition (Eq. (7.8)).

• (3) The extrapolation condition (Eq. (7.7)).

```
            DO 140 J=1, PY-1
              IF( ITREEDWN.EQ.3 ) THEN
                VX( PX,J) = UVELX
                VY( PX,J) = UVELY
                RHO(PX,J) = DNS0
              ELSE IF( ITREEDWN.EQ.4 ) THEN
                VX( PX,J) = VX( PX-1,J)
                VY( PX,J) = VY( PX-1,J)
                RHO(PX,J) = RHO(PX-1,J)
                IF( RHO(PX,J) .LT. DNS0 )  RHO(PX,J) = DNS0
              ELSE IF( ITREEDWN.EQ.5 ) THEN
                VX( PX,J) = 2.D0*VX( PX-1,J) - VX( PX-2,J)
                VY( PX,J) = 2.D0*VY( PX-1,J) - VY( PX-2,J)
                RHO(PX,J) = 2.D0*RHO(PX-1,J) - RHO(PX-2,J)
                IF( RHO(PX,J) .LT. DNS0 )  RHO(PX,J) = DNS0
              END IF
    140     CONTINUE
C                                                 ++ Corners ++
            IF( ITREEDWN.EQ.3 ) THEN
              VX( PX,PY) = UVELX
              VY( PX,PY) = UVELY
              RHO(PX,PY) = DNS0
              VX( PX, 0) = UVELX
              VY( PX, 0) = UVELY
              RHO(PX, 0) = DNS0
            ELSE IF( ITREEDWN.EQ.4 ) THEN
              VX( PX,PY) = VX( PX-1,PY-1)
              VY( PX,PY) = VY( PX-1,PY-1)
              RHO(PX,PY) = RHO(PX-1,PY-1)
              IF( RHO(PX,PY) .LT. DNS0 )  RHO(PX,PY) = DNS0
              VX( PX, 0) = VX( PX-1,1)
              VY( PX, 0) = VY( PX-1,1)
              RHO(PX, 0) = RHO(PX-1,1)
              IF( RHO(PX, 0) .LT. DNS0 )  RHO(PX, 0) = DNS0
            ELSE IF( ITREEDWN.EQ.5 ) THEN
              VX( PX,PY) = 2.D0*VX( PX-1,PY-1) - VX( PX-2,PY-2)
              VY( PX,PY) = 2.D0*VY( PX-1,PY-1) - VY( PX-2,PY-2)
              RHO(PX,PY) = 2.D0*RHO(PX-1,PY-1) - RHO(PX-2,PY-2)
              IF( RHO(PX,PY) .LT. DNS0 )  RHO(PX,PY) = DNS0
              VX( PX, 0) = 2.D0*VX( PX-1,1) - VX( PX-2,2)
              VY( PX, 0) = 2.D0*VY( PX-1,1) - VY( PX-2,2)
              RHO(PX, 0) = 2.D0*RHO(PX-1,1) - RHO(PX-2,2)
              IF( RHO(PX, 0) .LT. DNS0 )  RHO(PX, 0) = DNS0
            END IF
          END IF
                                                              RETURN
                                                              END
C**** SUB COLLPROC *****
          SUBROUTINE COLLPROC( COLOR , ALPHAMX )
C                                  --------- COLLISION PROCEDURE ---
          IMPLICIT REAL*8 (A-H,O-Z), INTEGER (I-N)
C
          COMMON /BLOCK1/  F    , FTILD
          COMMON /BLOCK3/  RHO  , RX , RY  , VX  , VY
          COMMON /BLOCK4/  DNS0 , TAU , DX , DT , CLAT
          COMMON /BLOCK5/  XL , YL , XL1 , YL1 , XL2 , YL2 , PX , PY , PXY
C
C         -----------------------------
          INTEGER    PP , QQ , KK
          PARAMETER( PP=300 , QQ=400 , KK=8 , PI=3.141592653589793D0 )
C
          REAL*8    F(0:PP,0:QQ,0:KK), FTILD(0:PP,0:QQ,0:KK)
          REAL*8    RHO(0:PP,0:QQ)
          REAL*8    RX( 0:PP,0:QQ)   , RY(0:PP,0:QQ)
          REAL*8    VX( 0:PP,0:QQ)   , VY(0:PP,0:QQ)
          INTEGER   ALPHAMX , PX , PY ,  PXY
C         -----------------------------
          INTEGER    PPXY
          PARAMETER( PPXY=150000 , NNTBL=2200 , NNTBL2=4400 , NNTBL3=4400 )
C
          INTEGER   COLOR(PPXY)
C         -----------------------------
          REAL*8    FEQ, CDNS0, UVELX0, UVELY0
          INTEGER   ITH, ICLR
C
CCC       CDNS0 = DNS0
C
          DO 210 I=0, PX
          DO 200 J=0, PY
C
```

• (1) The equilibrium distribution.

• (2) The zero-gradient condition (Eq. (7.8)).

• (3) The extrapolation condition (Eq. (7.7)).

• A subroutine for treating the collision at each site.

• The treatment for the sites at the downstream boundary surface and inside the simulation region, and also for the sites interacting with the cylinder according to Eq. (7.17).

```
          ITH = (PX+1)*J + (I+1)
          ICLR = COLOR(ITH)
          IF( (ICLR.EQ.6) .OR. (ICLR.EQ.7) )  GOTO 200
C                              --- FOR Busual,Bdownstream,Bcyl_surface ---
          IF( (ICLR .EQ. 0) .OR. (ICLR .EQ. 2) .OR. (ICLR .EQ. 5) ) THEN
            UVELX0 = VX( I,J)
            UVELY0 = VY( I,J)
            CDNS0  = RHO(I,J)
            DO 100 K=0, ALPHAMX
              FTILD(I,J,K) = F(I,J,K) * (TAU-1.D0)/TAU
     &                   +  FEQ( UVELX0, UVELY0, K, CDNS0 ) / TAU
  100       CONTINUE
C                                                 --- FOR  Bupstream ---
          ELSE IF( ICLR .EQ. 1 ) THEN
            UVELX0 = VX( 0,J)
            UVELY0 = VY( 0,J)
            CDNS0  = RHO(0,J)
            DO 120 K=0, ALPHAMX
              FTILD(0,J,K) = FEQ( UVELX0, UVELY0, K, CDNS0 )
  120       CONTINUE
```

• The treatment for the sites at the upstream boundary surface. The equilibrium distribution is used.

```
C                                               --- FOR  Bupper_side ---
          ELSE IF( ICLR .EQ. 3 ) THEN
            UVELX0 = VX( I,PY)
            UVELY0 = VY( I,PY)
            CDNS0  = RHO(I,PY)
            DO 140 K=0, ALPHAMX
              FTILD(I,PY,K) = F(I,PY,K) * (TAU-1.D0)/TAU
     &                    +  FEQ( UVELX0, UVELY0, K, CDNS0 ) / TAU
  140       CONTINUE
```

• The treatment for the sites at the upper side boundary surface. Eq. (7.17) is treated.

```
C                                                 --- FOR Blower_side ---
          ELSE IF( ICLR .EQ. 4 ) THEN
            UVELX0 = VX( I,0)
            UVELY0 = VY( I,0)
            CDNS0  = RHO(I,0)
            DO 160 K=0, ALPHAMX
              FTILD(I,0,K) = F(I,0,K) * (TAU-1.D0)/TAU
     &                   +  FEQ( UVELX0, UVELY0, K, CDNS0 ) / TAU
  160       CONTINUE
          END IF
C
  200 CONTINUE
  210 CONTINUE
                                                                RETURN
                                                                END
C**** SUB MOVEPROC *****
      SUBROUTINE MOVEPROC( PX , PY , ANTIALPH , RHO , DNS0 , ITREECYL )
C                                     --------- MOVEMENT PROCEDURE ---
      IMPLICIT REAL*8 (A-H,O-Z), INTEGER (I-N)
C
      COMMON /BLOCK1/  F     , FTILD
C
      COMMON /BLOCK14/ RXCYL , RYCYL , ICYL , JCYL , DCYL
      COMMON /BLOCK15/ COLOR     , POSINTBL
      COMMON /BLOCK16/ TBLNAM    , TBLNUM   , TBLPOS    , NTBL
      COMMON /BLOCK17/ TBLDW     , TBLAL    , NTBLDW
C
      COMMON /BLOCK21/ CD , CDFORCE0 , CDFORCE , RE , NSMPLCD
C
C     ----------------------------
      INTEGER    PP , QQ , KK
      PARAMETER( PP=300 , QQ=400 , KK=8 , PI=3.141592653589793D0 )
C
      REAL*8     F(0:PP,0:QQ,0:KK), FTILD(0:PP,0:QQ,0:KK)
      REAL*8     RHO(0:PP,0:QQ)
      INTEGER    PX , PY ,  ANTIALPH(0:KK)
C     ----------------------------
      INTEGER    PPXY
      PARAMETER( PPXY=150000 , NNTBL=2200 , NNTBL2=4400 , NNTBL3=4400 )
C
      REAL*8     TBLDW(NNTBL2)
      INTEGER    COLOR(PPXY)     , POSINTBL(PPXY)
      INTEGER    TBLNAM(NNTBL) , TBLNUM(NNTBL) , TBLPOS(NNTBL) , NTBL
      INTEGER    TBLAL(NNTBL2) , NTBLDW
C     ----------------------------
      INTEGER    NNCD
      PARAMETER( NNCD=1000000 )
C
      REAL*8     CDFORCE(NNCD)
C     ----------------------------
      INTEGER    ITH , ICLR , ITBL , INUM , IPOS , IALPHA , K , KANTI
      INTEGER    I1 , I2 , ID , J1 , J2 , JD , I00 , J00
```

• The treatment for the sites at the lower side boundary surface. Eq. (7.17) is treated.

• A subroutine for the transfer process of the particle distribution function.

```
      INTEGER   I11, J11, I21, I22, J21, J22
      REAL*8    FWALL, CDW  , C1   ,  C2
      REAL*8    CA11 , CA12 , CA21 , CA22 , CA23
      REAL*8    CB11 , CB12 , CB21 , CB22 , CB23
      REAL*8    CD11 , CD12 , CD21 , CD22 , CD23
C
C     ------------------------------------------- 0-DIRECTION ---
      DO 3 I=0, PX
      DO 1 J=0, PY
        F(I,J,0)=FTILD(I,J,0)
    1 CONTINUE
    3 CONTINUE
C
C     -------------------------------- 1,2,3,4,5,6,7,8-DIRECTION ---
C
      DO 100 K=1,8
```

• K means the α-direction.

```
C
        IF( K.EQ.1 ) THEN
          I1 =  PX-1
          I2 =  1
          ID = -1
          J1 =  PY-1
          J2 =  1
          JD = -1
```

• The sites to be treated begin from I1 to I2 at interval ID for the *x*-direction.

• The sites to be treated begin from J1 to J2 at interval JD for the *y*-direction.

```
        ELSE IF( K.EQ.2 ) THEN
          I1 =  1
          I2 =  PX-1
          ID =  1
          J1 =  PY-1
          J2 =  1
          JD = -1
        ELSE IF( K.EQ.3 ) THEN
          I1 =  1
          I2 =  PX-1
          ID =  1
          J1 =  PY-1
          J2 =  1
          JD = -1
        ELSE IF( K.EQ.4 ) THEN
          I1 =  1
          I2 =  PX-1
          ID =  1
          J1 =  1
          J2 =  PY-1
          JD =  1
        ELSE IF( K.EQ.5 ) THEN
          I1 =  PX-1
          I2 =  1
          ID = -1
          J1 =  PY-1
          J2 =  1
          JD = -1
        ELSE IF( K.EQ.6 ) THEN
          I1 =  1
          I2 =  PX-1
          ID =  1
          J1 =  1
          J2 =  PY-1
          JD =  1
        ELSE IF( K.EQ.7 ) THEN
          I1 =  PX-1
          I2 =  1
          ID = -1
          J1 =  1
          J2 =  PY-1
          JD =  1
        ELSE IF( K.EQ.8 ) THEN
          I1 =  1
          I2 =  PX-1
          ID =  1
          J1 =  PY-1
          J2 =  1
          JD = -1
        END IF
C
C
        DO 40 I= I1, I2, ID
        DO 20 J= J1, J2, JD
C
          ITH  = (PX+1)*J + (I+1)
          ICLR = COLOR(ITH)
```

• The ITH-th site is treated in the following.

```
            IF( (ICLR.EQ.6) .OR. (ICLR.EQ.7) )  GOTO 20
C
            IF( K.EQ.1) THEN
              I00 = I-1
              J00 = J
            ELSE IF( K.EQ.2) THEN
              I00 = I+1
              J00 = J
            ELSE IF( K.EQ.3) THEN
              I00 = I
              J00 = J-1
            ELSE IF( K.EQ.4) THEN
              I00 = I
              J00 = J+1
            ELSE IF( K.EQ.5) THEN
              I00 = I-1
              J00 = J-1
            ELSE IF( K.EQ.6) THEN
              I00 = I+1
              J00 = J+1
            ELSE IF( K.EQ.7) THEN
              I00 = I-1
              J00 = J+1
            ELSE IF( K.EQ.8) THEN
              I00 = I+1
              J00 = J-1
            END IF
C                                        ----- FOR CYL_surface -----
            IF( ICLR .EQ. 5 ) THEN
C
              ITBL = POSINTBL(ITH)
              INUM = TBLNUM(ITBL)
              IPOS = TBLPOS(ITBL)
C
              DO 10 JJ=0,INUM-1
C
                IALPHA = TBLAL( IPOS+JJ )
                KANTI  = ANTIALPH(K)
C
                IF( IALPHA .EQ. KANTI ) THEN
C
                  IF(  (K.EQ.1) .OR. (K.EQ.5) .OR. (K.EQ.7)  ) THEN
                    CDFORCE(NSMPLCD)=CDFORCE(NSMPLCD) - FTILD(I,J,KANTI)
                  END IF
C
                  IF( K.EQ.1 ) THEN
                    I11 = I+1
                    J11 = J
                    I21 = I+1
                    I22 = I+2
                    J21 = J
                    J22 = J
                  ELSE IF( K.EQ.2 ) THEN
                    I11 = I-1
                    J11 = J
                    I21 = I-1
                    I22 = I-2
                    J21 = J
                    J22 = J
                  ELSE IF( K.EQ.3 ) THEN
                    I11 = I
                    J11 = J+1
                    I21 = I
                    I22 = I
                    J21 = J+1
                    J22 = J+2
                  ELSE IF( K.EQ.4 ) THEN
                    I11 = I
                    J11 = J-1
                    I21 = I
                    I22 = I
                    J21 = J-1
                    J22 = J-2
                  ELSE IF( K.EQ.5 ) THEN
                    I11 = I+1
                    J11 = J+1
                    I21 = I+1
                    I22 = I+2
                    J21 = J+1
                    J22 = J+2
```

• The position (name) of the site in the opposite direction to the α-direction (K) is described as (I00,J00). For example, if α=2 (K=2), such a site is I00=I+1 and J00=J, where (I,J) is the position (name) of the site of interest.

• The treatment of the site interacting with the cylinder.

• The order of the ITH-site, in which its information is saved in TBLNUM and TBLPOS, is extracted from POSINTBL. The result is saved in ITBL.

• INUM is the number of the interacting sites inside the cylinder. IPOS is the first position of such sites appearing in the corresponding variables.

• (I) For IALPHA=KANTI.

• IALPHA is the direction of the ITH-th site toward the neighboring site inside the cylinder, and the opposite direction to K is KANTI.

• The variables (I11,J11) are used in the linear interpolation procedure of the BFL and YMLS methods.
• The variables (I21,J21) and (I22,J22) are used in the quadratic interpolation procedure for the BFL and YMLS methods.

```
      ELSE IF( K.EQ.6 ) THEN
        I11 = I-1
        J11 = J-1
        I21 = I-1
        I22 = I-2
        J21 = J-1
        J22 = J-2
      ELSE IF( K.EQ.7 ) THEN
        I11 = I+1
        J11 = J-1
        I21 = I+1
        I22 = I+2
        J21 = J-1
        J22 = J-2
      ELSE IF( K.EQ.8 ) THEN
        I11 = I-1
        J11 = J+1
        I21 = I-1
        I22 = I-2
        J21 = J+1
        J22 = J+2
      END IF
C
C
      IF( (ITREECYL.EQ.2) .OR. (ITREECYL.EQ.3) ) THEN
        CDW    = TBLDW(IPOS+JJ)
        FWALL =   (1.D0-CDW) * FTILD(I11,J11,KANTI)
     &                + CDW * FTILD(I   ,J   ,KANTI)
C
        C1     =   1.D0+CDW
        C2     =   2.D0+CDW
        CA11   =   CDW /C1
        CA12   =   1.D0/C1
        CA21   =   2.D0*CA12/C2
        CA22   =   2.D0*CA12*CDW
        CA23   =  -CDW/C2
      END IF
```

• CA11 and CA12 are used in the linear interpolation procedure of YMLS expressed in Eq. (7.5), and CA21, CA22, and CA23 are used in the quadratic interpolation procedure of YMLS in Eq. (8.121); in advance, the coefficients are calculated and saved in these variables for the successive procedures.

```
C
      IF( (ITREECYL.EQ.4) .OR. (ITREECYL.EQ.5) ) THEN
        CDW    =   TBLDW(IPOS+JJ)
        C1     =   1.D0+2.D0*CDW
        C2     =   1.D0-2.D0*CDW
        CB11   =   C2
        CB12   =   2.D0*CDW
        CB21   =   CDW*C1
        CB22   =   C1*C2
        CB23   =  -CDW*C2
        CD11   =   (-C2)/(2.D0*CDW)
        CD12   =   1.D0/(2.D0*CDW)
        CD21   =   1.D0/CB21
        CD22   =   (-C2)/CDW
        CD23   =   C2/C1
      END IF
```

• CB11 and CB12 are used in the linear interpolation procedure of BFL in Eq. (8.117), and CB21, CB22, and CB23 are used in the quadratic interpolation procedure of BFL in Eq. (8.112); in advance, the coefficients are calculated and saved in these variables for the successive procedures. Similarly, CD11, CD12, ..., CD23 are used in calculating Eqs. (8.118) and (8.116) for Δ_w>1/2.

```
C
C
      IF(      ITREECYL .EQ. 1 ) THEN
C                    +++ (1) BOUNCE-BACK  +++++++++++++++++
        F(I,J,K) = FTILD(I,J,KANTI)
C
      ELSE IF( ITREECYL .EQ. 2 ) THEN
C                    +++ (2A) YMLS METHOD (Quadratic) +++
        F(I,J,K) = CA21*FWALL + CA22*F(I21,J21,K)
     &                        + CA23*F(I22,J22,K)
C
      ELSE IF( ITREECYL .EQ. 3 ) THEN
C                    +++ (2B) YMLS METHOD (Liner) +++++++
        F(I,J,K) = CA11*F(I11,J11,K) + CA12*FWALL
C
      ELSE IF( ITREECYL .EQ. 4 ) THEN
C                    +++ (3A) BFL METHOD (Quadratic) +++++
        IF( CDW .LE. 0.5D0 ) THEN
C
          F(I,J,K) = CB21*FTILD(I,J,KANTI)
     &                         + CB22*FTILD(I21,J21,KANTI)
     &                         + CB23*FTILD(I22,J22,KANTI)
        ELSE
```

• The bounce-back rule.

• The quadratic YMLS method.

• The linear YMLS method.

• The quadratic BFL method.

• Eq. (8.112) is evaluated.

• Eq. (8.116) is evaluated.

```
C
                    F(I,J,K) = CD21*FTILD(I,J,KANTI)
     &                                    + CD22*FTILD(I  ,J  ,K)
     &                                    + CD23*FTILD(I21,J21,K)
                  END IF
C
```

• The linear BFL method.

```
                ELSE IF( ITREECYL .EQ. 5 ) THEN
C                            +++ (3B) BFL METHOD (Linear) ++++++++
                  IF( CDW .LE. 0.5D0 ) THEN
C
```

• Eq. (8.117) is evaluated.

```
                    F(I,J,K) = CB11*FTILD(I11,J11,KANTI)
     &                                     + CB12*FTILD(I,J,KANTI)
                  ELSE
C
```

• Eq. (8.118) is evaluated.

```
                    F(I,J,K) = CD11*FTILD(I,J,K)
     &                                     + CD12*FTILD(I,J,KANTI)
                  END IF
C
                END IF
C
C
                IF(  (K.EQ.1) .OR. (K.EQ.5) .OR. (K.EQ.7)  ) THEN
                  CDFORCE(NSMPLCD) = CDFORCE(NSMPLCD) - F(I,J,K)
                ELSE IF(  (K.EQ.2) .OR. (K.EQ.6) .OR. (K.EQ.8)  ) THEN
                  CDFORCE(NSMPLCD) = CDFORCE(NSMPLCD) + F(I,J,K)
                END IF
C
                GOTO 20
C
```

• (II) For IALPHA=K.

```
              ELSE IF( IALPHA .EQ. K     ) THEN
C
                IF(  (K.EQ.1) .OR. (K.EQ.5) .OR. (K.EQ.7)  ) THEN
                  CDFORCE(NSMPLCD) = CDFORCE(NSMPLCD) + FTILD(I,J,K)
                END IF
C
                F(I,J,K) = FTILD(I00, J00, K)
                GOTO 20
C
              END IF
C
   10       CONTINUE
          END IF
C                                              ----- FOR USUAL -----
          F(I,J,K) = FTILD(I00, J00, K)
C
   20   CONTINUE
   40   CONTINUE
C
  100 CONTINUE
                                                               RETURN
                                                               END
C**** SUB BCPROC *****
      SUBROUTINE BCPROC( PX , PY , DNS0 , ALPHAMX , ITREESID ,
     &                                             ITREEDWN )
C                                ---------- BOUNDARY CONDITION PROC. ---
      IMPLICIT REAL*8 (A-H,O-Z), INTEGER (I-N)
C
```

• A subroutine for treating the boundary surfaces.

```
      COMMON /BLOCK1/  F     , FTILD
      COMMON /BLOCK3/  RHO   , RX , RY   , VX   , VY
      COMMON /BLOCK6/  UVELX , UVELY
C
      INTEGER     PP , QQ , KK
      PARAMETER( PP=300 , QQ=400 , KK=8 , PI=3.141592653589793D0 )
C
      REAL*8     F(0:PP,0:QQ,0:KK), FTILD(0:PP,0:QQ,0:KK)
      REAL*8     RHO(0:PP,0:QQ)
      REAL*8     RX( 0:PP,0:QQ)   , RY(0:PP,0:QQ)
      REAL*8     VX( 0:PP,0:QQ)   , VY(0:PP,0:QQ)
      INTEGER    PX , PY , ALPHAMX
C
      REAL*8     FEQ , CDNS0 , UVELX0 , UVELY0
C
      CDNS0 = DNS0
C
C     ------------------------------------------ BC for Bupstream ---
      DO 100 J=0, PY
      DO 80 K=0, ALPHAMX
C                                           +++ UNIFORM FLOW +++
        F(0,J,K) = FEQ( UVELX, UVELY, K, CDNS0 )
C
```

• I. The treatment at the upstream boundary surface.

• An equilibrium distribution is assigned.

```
   80 CONTINUE
  100 CONTINUE
C
```

• II. The treatment at the side boundary surfaces.

```
C     --------------------------- BC for Bupper_side & Blower_side ---
      DO 300 I=1, PX-1
      DO 280 K=0, ALPHAMX
        IF( ITREESID .EQ. 1 ) THEN
C                                           +++ (1) EXTRAPOLATION +++
          F(I,PY,K) = 2.D0*F(I,PY-1,K) - F(I,PY-2,K)
          F(I, 0,K) = 2.D0*F(I,1,K)    - F(I,2,K)
C
```

• The extrapolation condition in Eq. (7.7) is applied.

```
        ELSE IF( ITREESID .EQ. 2 ) THEN
C                                           +++ (2) DEF=0         +++
          F(I,PY,K) = F(I,PY-1,K)
          F(I, 0,K) = F(I,1,K)
C
```

• The zero-gradient condition in Eq. (7.8) is applied.

```
        ELSE IF(  (ITREESID.EQ.3) .OR. (ITREESID.EQ.4) .OR.
     &                                 (ITREESID.EQ.5)  ) THEN
C                                           +++ (3) UNIFORM FLOW  +++
          UVELX0 = VX( I,PY)
          UVELY0 = VY( I,PY)
          CDNS0  = RHO(I,PY)
          F(I,PY,K) = FEQ( UVELX0, UVELY0, K, CDNS0 )
C
          UVELX0 = VX( I,0)
          UVELY0 = VY( I,0)
          CDNS0  = RHO(I,0)
          F(I, 0,K) = FEQ( UVELX0, UVELY0, K, CDNS0 )
C
        END IF
C
  280 CONTINUE
  300 CONTINUE
C
```

• An equilibrium distribution with each local velocity is assigned.

• III. The treatment at the downstream surface.

```
C     --------------------------------------- BC for Bdownstream ---
      DO 500 J=0, PY
      DO 480 K=0, ALPHAMX
        IF( ITREEDWN .EQ. 1 ) THEN
C                                           +++ (1) EXTRAPOLATION +++
          F(PX,J,K) = 2.D0*F(PX-1,J,K) - F(PX-2,J,K)
C
```

• The extrapolation condition in Eq. (7.7) is applied.

```
        ELSE IF( ITREEDWN .EQ. 2 ) THEN
C                                           +++ (2) DEF=0         +++
          F(PX,J,K) = F(PX-1,J,K)
C
```

• The zero-gradient condition in Eq. (7.8) is applied.

```
        ELSE IF(  (ITREEDWN.EQ.3) .OR. (ITREEDWN.EQ.4) .OR.
     &                                 (ITREEDWN.EQ.5)  ) THEN
C                                           +++ (3) UNIFORM FLOW  +++
          UVELX0 = VX( PX,J)
          UVELY0 = VY( PX,J)
          CDNS0  = RHO(PX,J)
          F(PX,J,K) = FEQ( UVELX0, UVELY0, K, CDNS0 )
C
        END IF
  480 CONTINUE
  500 CONTINUE
C
```

• An equilibrium dist. with each local velocity is assigned.

• IV. The treatment at both corner sites of the downstream surface.

```
C     -------------------------------- TWO Corners for Bdownstream ---
      DO 530 K=0, ALPHAMX
        IF( ITREEDWN .EQ. 1 ) THEN
          F(PX,PY,K) = 2.D0*F(PX-1,PY-1,K) - F(PX-2,PY-2,K)
          F(PX, 0,K) = 2.D0*F(PX-1,    1,K) - F(PX-2,    2,K)
        ELSE IF( ITREEDWN .EQ. 2 ) THEN
          F(PX,PY,K) =      F(PX-1,PY-1,K)
          F(PX, 0,K) =      F(PX-1,    1,K)
        END IF
  530 CONTINUE
                                                        RETURN
                                                        END
```

• The extrapolation condition in Eq. (7.7) is applied.

• The zero-gradient condition in Eq. (7.8) is applied.

```
C**** SUB GRAPHVEL ****
      SUBROUTINE  GRAPHVEL( NANMCTR )
C
CCC   IMPLICIT REAL*8 (A-H,O-Z), INTEGER (I-N)
      IMPLICIT REAL (A-H,O-Z), INTEGER (I-N)
C
      COMMON /BLOCK3/  RHO  , RX , RY  , VX  , VY
      COMMON /BLOCK5/  XL  , YL  , XL1 , YL1 , XL2 , YL2 , PX , PY , PXY
      COMMON /BLOCK6/  UVELX , UVELY
      COMMON /BLOCK14/ RXCYL , RYCYL , ICYL , JCYL , DCYL
C
```

• A subroutine for writing out the data used for making an animation based on the commercial software MicroAVS.

```
      INTEGER    PP , QQ , KK
      REAL*8     PI
      PARAMETER( PP=300 , QQ=400 , KK=8 , PI=3.141592653589793D0 )
C
C     --------------------------------
      REAL*8    RHO(0:PP,0:QQ)
      REAL*8    RX( 0:PP,0:QQ)    , RY(0:PP,0:QQ)
      REAL*8    VX( 0:PP,0:QQ)    , VY(0:PP,0:QQ)
      INTEGER   PX , PY ,  PXY
C     --------------------------------
      REAL*8    XL , YL , XL1 , YL1 , XL2 , YL2 , UVELX , UVELY
      REAL*8    RXCYL , RYCYL , DCYL
C     --------------------------------
      INTEGER    QQSQ , NNDUM
      PARAMETER( QQSQ=150000 , NNDUM=150000 )
C
      REAL    DUMRX(NNDUM) ,DUMRY(NNDUM) , DUMVX(NNDUM) ,DUMVY(NNDUM)
      REAL    VEL
      INTEGER NDUM, ISKIP
C
C     ------------------------- DATA OUTPUT FOR VEL-FIELD MicroAVS ---
C
      VEL  = REAL(  DSQRT( UVELX**2 + UVELY**2 )  )
C                                      +++ MAKE MicroAVS data FILE +++
      WRITE(42,83) NANMCTR
C
      II   = 0
      DO 100 J=0, PY, 1
      DO  90 I=0, PX, 1
        II = II + 1
        DUMRX(II) = REAL( RX(I,J) )
        DUMRY(II) = REAL( RY(I,J) )
        DUMVX(II) = REAL( VX(I,J) ) / VEL
        DUMVY(II) = REAL( VY(I,J) ) / VEL
        WRITE(42,85) DUMRX(II), DUMRY(II), DUMVX(II), DUMVY(II)
   90 CONTINUE
  100 CONTINUE
C
      NDUM = II
C                                       +++ MAKE MicroAVS fld FILE +++
      IF( NANMCTR .EQ. 1 ) THEN
        WRITE(41,181)
        WRITE(41,183)
        WRITE(41,185) (PX+1), (PY+1)
        WRITE(41,187)
      END IF
C
      ISKIP = (NDUM+1)*( NANMCTR-1) + 1
C
      WRITE(41,188) ISKIP-1
      WRITE(41,189) ISKIP
      WRITE(41,191) ISKIP
      WRITE(41,197)
C
C
   83 FORMAT( I5 )
   85 FORMAT( 4F8.3 )
  181 FORMAT('# AVS field file'/ '#')
  183 FORMAT( 'ndim=2' )
  185 FORMAT( 'dim1=',I4/ 'dim2=',I4)
  187 FORMAT( 'nspace= 2'/ 'veclen= 2'/ 'data= float'
     &       / 'field= uniform'/ )
  188 FORMAT( 'time file=./avsvel1.dat filetype=ascii '
     &         'skip=',I7,' close=1')
  189 FORMAT( 'variable 1 file=./avsvel1.dat filetype=ascii '
     &         'skip=',I7,' offset=2 stride=4')
  191 FORMAT( 'variable 2 file=./avsvel1.dat filetype=ascii '
     &         'skip=',I7,' offset=3 stride=4')
  197 FORMAT( 'EOT')
                                                              RETURN
                                                              END
C#### FUN FEQ ####
      DOUBLE PRECISION FUNCTION FEQ( UVELX, UVELY, ALPHA, CDNS0 )
C                         ----------------------------------------
C                             EQUILIBRIUM DISTRIBUTION FUNCTION F^(eq)
C                         ----------------------------------------
      IMPLICIT REAL*8 (A-H,O-Z), INTEGER (I-N)
C
      COMMON /BLOCK2/  CVEL , W  , IINC , ANTIALPH, ALPHAMX
C
      INTEGER    PP , QQ , KK
```

• The equilibrium distribution function.

```
      PARAMETER( PP=300 , QQ=400 , KK=8 , PI=3.141592653589793D0 )
C
      REAL*8     CVEL(2,0:KK) , W(0:KK)
      INTEGER    ALPHAMX , IINC(2,0:KK) , ANTIALPH(0:KK)
      INTEGER    ALPHA
C
      REAL*8      C0, C1, C2, C3
C
      K   = ALPHA
      C0  = W(K)*CDNS0
      C1  = CVEL(1,K)*UVELX + CVEL(2,K)*UVELY
      C2  = C1*C1
      C3  = UVELX**2 + UVELY**2
      FEQ = C0*( 1.D0 + 3.D0*C1 + (9.D0/2.D0)*C2 - (3.D0/2.D0)*C3 )
                                                             RETURN
                                                             END
```

• An equilibrium distribution is assigned according to Eq. (7.18).

8 Theoretical Background of Lattice Boltzmann Method

The lattice Boltzmann method [9–12] is a useful simulation technique for numerically solving flow problems. This method is also feasible as a simulation technique for systems such as a suspension of solid particles or a polymeric liquid. In a multicomponent system, the motion of the suspended particles or polymers must be solved together with the flow field of the solvent molecules. In a molecular simulation of a suspension composed of solid particles in a liquid, it is very difficult to treat the multibody hydrodynamic interactions among the suspended particles. Hence, it is usual to model the flow field as a simple shear flow, and under this approach only the motion of the suspended particles will be solved during the simulation. A typical simulation technique employing this concept is the Stokesian dynamics method. On the other hand, the lattice Boltzmann method enables us to solve the motion of suspended particles and the ambient flow field simultaneously, so there is much of interest in this method.

In the present chapter, we turn from the practice of molecular simulations to the theoretical background of the lattice Boltzmann method. The key equations are almost all indicated for the successive derivation procedure such that the reader will be able to derive all the important equations from the key expressions. Understanding the theoretical background is essential if, for example, the reader needs to employ a new boundary condition or develop a new version of the lattice Boltzmann method that can take into account the random motion of the suspended particles. For a clear, logical development, the fundamental equations for the following derivation may be found in Appendix A1. Note that we focus here on the BGK lattice Boltzmann method, which is the simplest and provides a solid foundation for application to various flow problems.

8.1 Equilibrium Distribution

The lattice Boltzmann method treats the particle distribution function of virtual fluid particles, which are able to move from site to site on a lattice system. A macroscopic quantity of interest, such as the fluid velocity, can be obtained from the solution of the particle distribution function. In the case of a two-dimensional system, such as the D2Q9 model shown in Figure 8.1, fluid particles at lattice site 0 have a possibility of moving to the neighboring lattice sites 1,. . .,8. If the quiescent

Introduction to Practice of Molecular Simulation. DOI: 10.1016/B978-0-12-385148-2.00008-2

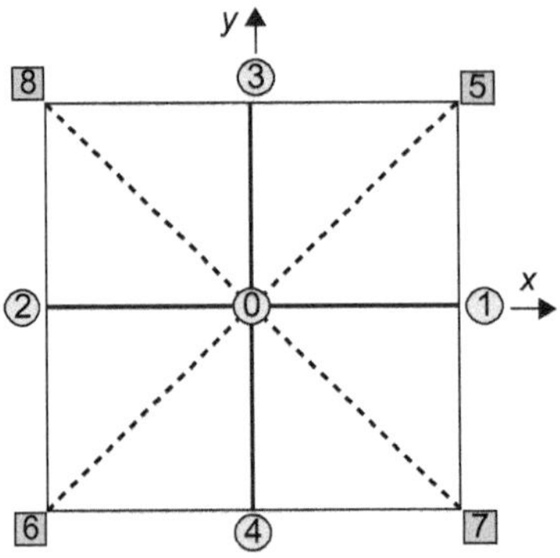

Figure 8.1 Lattice model for the D2Q9.

state is included, there are nine velocities for the fluid particles moving (or not moving) to a neighboring site; a fluid particle will arrive at its neighboring site with a given microscopic velocity during a given time interval. We use the notation $\mathbf{c}_\alpha$ for the velocity for the transfer in the α-direction ($\alpha = 0,1,2,\ldots,8$). The particle distribution function $f_\alpha(\mathbf{r},t)$ in the α-direction at the lattice site $\mathbf{r}$ at time t can be obtained by treating the collision of the fluid particles at $\mathbf{r}$ and evaluating the inflow and the outflow of fluid particles from and to the lattice site $\mathbf{r}$. In the BGK lattice Boltzmann method, the particle distribution function $f_\alpha(\mathbf{r} + \mathbf{c}_\alpha \Delta t,\ t + \Delta t)$ is obtained from the following equation:

$$f_\alpha(\mathbf{r} + \mathbf{c}_\alpha \Delta t, t + \Delta t) = \tilde{f}_\alpha(\mathbf{r}, t) \tag{8.1}$$

$$\tilde{f}_\alpha(\mathbf{r}, t) = f_\alpha(\mathbf{r}, t) + \frac{1}{\tau}\{f_\alpha^{(0)}(\mathbf{r}, t) - f_\alpha(\mathbf{r}, t)\} \tag{8.2}$$

The $\tilde{f}_\alpha$ in Eq. (8.2) is the particle distribution function after the collision at the site $\mathbf{r}$. Eq. (8.1) implies that this distribution moves to the neighboring site ($\mathbf{r} + \mathbf{c}_\alpha \Delta t$) in the α-direction. The second term on the right-hand side in Eq. (8.2) is the collision term, frequently denoted by the notation $\Omega_\alpha(\mathbf{r},t)$:

$$\Omega_\alpha(\mathbf{r}, t) = \frac{1}{\tau}\{f_\alpha^{(0)}(\mathbf{r}, t) - f_\alpha(\mathbf{r}, t)\} \tag{8.3}$$

With the above particle distribution, the macroscopic fluid density $\rho(\mathbf{r},t)$ and momentum $\rho(\mathbf{r},t)\mathbf{u}(\mathbf{r},t)$ can be evaluated as

$$\rho(\mathbf{r}, t) = \sum_\alpha f_\alpha(\mathbf{r}, t) \tag{8.4}$$

$$\rho(\mathbf{r}, t)\mathbf{u}(\mathbf{r}, t) = \sum_\alpha f_\alpha(\mathbf{r}, t)\mathbf{c}_\alpha \tag{8.5}$$

Additionally, if a system is in thermodynamic equilibrium with constant temperature T, the following equi-partition law of energies must be satisfied:

$$\frac{D}{2}kT = \sum_\alpha \frac{m}{2}(\mathbf{c}_\alpha - \mathbf{u})^2 \frac{f_\alpha}{\rho} \tag{8.6}$$

in which D is a constant for describing the dimension with the value 2 or 3 for a two- or three-dimensional space, respectively, and m is the mass of a fluid particle.

The thermodynamic equilibrium velocity distribution in the lattice Boltzmann method differs from that in the MD method. This is because virtual fluid particles in the lattice Boltzmann method are not allowed to move freely in a simulation region, but are restricted to move only from site to site. The velocity $\mathbf{c}$ of a molecule (a fluid particle), which moves freely in a three-dimensional space with a uniform flow velocity $\mathbf{u}$ of the system, is specified by the Maxwellian distribution $f^{(eq)}(\mathbf{c})$ [25]:

$$f^{(eq)}(\mathbf{c}) = \rho\left(\frac{m}{2\pi kT}\right)^{3/2} \exp\left\{-\frac{m}{2kT}(\mathbf{c}-\mathbf{u})^2\right\} \tag{8.7}$$

Note that this definition includes the density ρ, whereas the usual Maxwellian distribution does not include the density in its expression. The equilibrium distribution in the lattice Boltzmann method $f_\alpha^{(0)}$ may be expressed by expanding the exponential function in Eq. (8.7) in a Taylor series expansion as

$$f_\alpha^{(0)} = \rho w_\alpha \left\{1 + b\frac{\mathbf{c}_\alpha \cdot \mathbf{u}}{c^2} + e\frac{u^2}{c^2} + h\frac{(\mathbf{c}_\alpha \cdot \mathbf{u})^2}{c^4}\right\} \tag{8.8}$$

in which w_α, b, e, and h are unknown constants to be determined later, w_α is a weighting constant, and c is the lattice speed for fluid particles moving from site to site, expressed as $c = \Delta x/\Delta t$.

In the lattice Boltzmann method, the whole system space is divided into a fine mesh that acts as the lattice system, and the fluid particles are only able to move from lattice site to lattice site. However, any physical phenomenon should not depend on the setting of the lattice system, and Eqs. (8.4)–(8.6) are required to remain valid for an arbitrary rotation of the lattice. This requirement will determine the above-mentioned unknown constants and, because the values of these unknown constants depend on the model used, we discuss the derivation for determining the unknown constants for the D2Q9 and D3Q19 models separately.

8.1.1 D2Q9 Model

The xy-coordinate system and the α-direction are specified as shown in Figure 8.1. As already pointed out, the equilibrium distribution can be obtained explicitly by determining the unknown constants w_α, b, e, and h such that the terms on the right-hand side in Eqs. (8.4)–(8.6) remain unchanged by a rotation of the whole lattice system by an angle ϕ. Before we start the procedure of determining the unknown constants, we show preliminary expressions that are useful in the following discussion. Note that the relationship of the momentum flux is necessary for determining these constants.

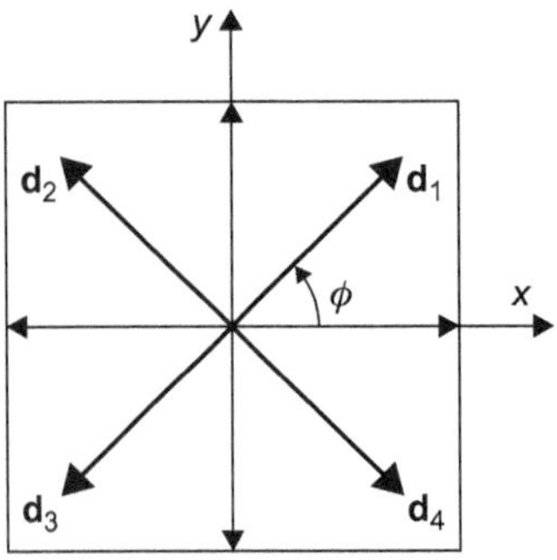

Figure 8.2 Rotation of the unit vectors.

As shown in Figure 8.2, the four unit vectors, which are along the plus and minus x- and y-axes of the orthogonal coordinate system, are rotated about the z-axis, and the new unit vectors are denoted by $\mathbf{d}_1$, $\mathbf{d}_2$, $\mathbf{d}_3$, and $\mathbf{d}_4$. These vectors are written in component expressions as

$$\left.\begin{aligned}
\mathbf{d}_1 &= (d_{1x}, d_{1y}) = (\cos\phi, \sin\phi) \\
\mathbf{d}_2 &= (d_{2x}, d_{2y}) = \left(\cos\left(\phi + \frac{\pi}{2}\right), \sin\left(\phi + \frac{\pi}{2}\right)\right) \\
\mathbf{d}_3 &= (d_{3x}, d_{3y}) = (\cos(\phi + \pi), \sin(\phi + \pi)) \\
\mathbf{d}_4 &= (d_{4x}, d_{4y}) = \left(\cos\left(\phi + \frac{3\pi}{2}\right), \sin\left(\phi + \frac{3\pi}{2}\right)\right)
\end{aligned}\right\} \tag{8.9}$$

Using these expressions, we derive several useful equations for the successive derivation. Although these equations can be derived from a simple transformation, as will be shown in the next subsection for the D3Q19 model, we here show a more sophisticated derivation based on the concept of imaginary numbers.

With the Euler formula $e^{i\theta} = \cos\theta + i\sin\theta$ for imaginary numbers, the following relationships can be obtained:

$$\sum_{k=1}^{4}(d_{kx} + id_{ky})^4 = \sum_{k=0}^{3}\left\{e^{i\left(k\frac{\pi}{2}+\phi\right)}\right\}^4 = \sum_{k=0}^{3} e^{i(2\pi k + 4\phi)} = e^{i4\phi}\sum_{k=0}^{3} e^{i2\pi k} = 4e^{i4\phi} \tag{8.10}$$

Similarly,

$$\left.\begin{aligned}
&\sum_{k=1}^{4}(d_{kx} + \mathrm{i}d_{ky})^3(d_{kx} - \mathrm{i}d_{ky}) = 0 \\
&\sum_{k=1}^{4}(d_{kx} + \mathrm{i}d_{ky})^2(d_{kx} - \mathrm{i}d_{ky})^2 = 4
\end{aligned}\right\} \tag{8.11}$$

The corresponding real and imaginary parts on the left- and right-hand sides in Eq. (8.10) are equal, which leads to the following equation:

$$\left.\begin{aligned}&\sum_{k=1}^{4}(d_{kx}^4 + d_{ky}^4 - 6d_{kx}^2 d_{ky}^2) = 4\cos 4\phi\\&\sum_{k=1}^{4}4(d_{kx}^3 d_{ky} - d_{kx} d_{ky}^3) = 4\sin 4\phi\end{aligned}\right\} \tag{8.12}$$

These relationships have been derived by expanding the left-hand side in Eq. (8.10). Similarly, from Eq. (8.11),

$$\left.\begin{aligned}&\sum_{k=1}^{4}(d_{kx}^4 - d_{ky}^4) = 0\\&\sum_{k=1}^{4}(d_{kx}^3 d_{ky} + d_{kx} d_{ky}^3) = 0\\&\sum_{k=1}^{4}(d_{kx}^4 + d_{ky}^4 + 2d_{kx}^2 d_{ky}^2) = 4\end{aligned}\right\} \tag{8.13}$$

Further preliminary relationships can be derived from Eqs. (8.12) and (8.13). From the first equation in Eq. (8.12) and the third equation in Eq. (8.13),

$$\sum_{k=1}^{4} d_{kx}^2 d_{ky}^2 = \frac{1}{2}(1 - \cos 4\phi) \tag{8.14}$$

From the second equation in Eqs. (8.12) and (8.13),

$$\left.\begin{aligned}&\sum_{k=1}^{4} d_{kx}^3 d_{ky} = \frac{1}{2}\sin 4\phi\\&\sum_{k=1}^{4} d_{kx} d_{ky}^3 = -\frac{1}{2}\sin 4\phi\end{aligned}\right\} \tag{8.15}$$

From the first and third equations in Eq. (8.13),

$$\sum_{k=1}^{4} d_{kx}^4 = \sum_{k=1}^{4} d_{ky}^4 = 2 - \sum_{k=1}^{4} d_{kx}^2 d_{ky}^2 = \frac{3}{2} + \frac{1}{2}\cos 4\phi \tag{8.16}$$

From a similar derivation procedure, the terms concerning d_{kx} or d_{ky} to the first, second, and third powers are obtained as

$$\left.\begin{aligned}&\sum_{k=1}^{4} d_{kx} = \sum_{k=1}^{4} d_{ky} = 0\\ &\sum_{k=1}^{4} d_{kx}^2 = \sum_{k=1}^{4} d_{ky}^2 = 2, \quad \sum_{k=1}^{4} d_{kx}d_{ky} = 0\\ &\sum_{k=1}^{4} d_{kx}^3 = \sum_{k=1}^{4} d_{ky}^3 = \sum_{k=1}^{4} d_{kx}^2 d_{ky} = \sum_{k=1}^{4} d_{kx}d_{ky}^2 = 0\end{aligned}\right\} \tag{8.17}$$

We have now obtained all the preliminary equations and will proceed to the determination procedures for the unknown constants w_α, b, e, and h.

As shown in Figure 8.1, we consider the rotation of the D2Q9 lattice system about the z-axis by the angle ϕ. We first evaluate the following quantity:

$$\begin{aligned}\sum_{\alpha=0}^{8} w_\alpha c_{\alpha x}^2 c_{\alpha y}^2 &= w_1 \sum_{\alpha=1}^{4} c_{\alpha x}^2 c_{\alpha y}^2 + w_5 \sum_{\alpha=5}^{8} c_{\alpha x}^2 c_{\alpha y}^2 = w_1 c^4 \frac{1}{2}(1 - \cos 4\phi)\\ &\quad + w_5\left(\sqrt{2}c\right)^4 \frac{1}{2}\left\{1 - \cos 4\left(\phi + \frac{\pi}{4}\right)\right\}\end{aligned} \tag{8.18}$$

With the assumption of

$$w_1 = 4w_5 \tag{8.19}$$

Eq. (8.18) comes to be independent of ϕ. That is,

$$\sum_{\alpha=0}^{8} w_\alpha c_{\alpha x}^2 c_{\alpha y}^2 = w_1 c^4 \tag{8.20}$$

Similar manipulation gives rise to

$$\left.\begin{aligned}&\sum_{\alpha=0}^{8} w_\alpha c_{\alpha x}^3 c_{\alpha y} = \sum_{\alpha=0}^{8} w_\alpha c_{\alpha x} c_{\alpha y}^3 = 0\\ &\sum_{\alpha=0}^{8} w_\alpha c_{\alpha x}^4 = \sum_{\alpha=0}^{8} w_\alpha c_{\alpha y}^4 = 3w_1 c^4\end{aligned}\right\} \tag{8.21}$$

The above results can be written in one expression by using the Kronecker delta δ_{ij}:

$$\sum_{\alpha=0}^{8} w_\alpha c_{\alpha i} c_{\alpha j} c_{\alpha k} c_{\alpha l} = w_1 c^4 (\delta_{ij}\delta_{kl} + \delta_{ik}\delta_{jl} + \delta_{il}\delta_{jk}) \tag{8.22}$$

Similarly,

$$\left.\begin{aligned}&\sum_{\alpha=0}^{8} w_\alpha c_{\alpha x}=\sum_{\alpha=0}^{8} w_\alpha c_{\alpha y}=0, \quad \sum_{\alpha=0}^{8} w_\alpha c_{\alpha i} c_{\alpha j}=3w_1 c^2 \delta_{ij}\\&\sum_{\alpha=0}^{8} w_\alpha c_{\alpha x}^2=\sum_{\alpha=0}^{8} w_\alpha c_{\alpha y}^2=3w_1 c^2, \quad \sum_{\alpha=0}^{8} w_\alpha c_{\alpha x} c_{\alpha y}=0\\&\sum_{\alpha=0}^{8} w_\alpha c_{\alpha x}^3=\sum_{\alpha=0}^{8} w_\alpha c_{\alpha y}^3=\sum_{\alpha=0}^{8} w_\alpha c_{\alpha x}^2 c_{\alpha y}=\sum_{\alpha=0}^{8} w_\alpha c_{\alpha x} c_{\alpha y}^2=0\end{aligned}\right\} \tag{8.23}$$

We now determine the appropriate values of the constants b, e, h, and w_α for an equilibrium distribution in Eq. (8.8). The relationships that must be satisfied for an equilibrium state are the equation of mass in Eq. (8.4), the equation of momentum in Eq. (8.5), and the equi-partition law of energies in Eq. (8.6). In these equations, $f_\alpha^{(0)}$ must be used as f_α. Substitution of Eq. (8.8) into the right-hand side of Eq. (8.4) leads to

$$\sum_{\alpha=0}^{8} f_\alpha^{(0)}=\sum_{\alpha=0}^{8} \rho w_\alpha\left\{1+b\frac{\mathbf{c}_\alpha\cdot\mathbf{u}}{c^2}+e\frac{u^2}{c^2}+h\frac{(\mathbf{c}_\alpha\cdot\mathbf{u})^2}{c^4}\right\}=\rho\left\{w_{\text{sum}}+w_{\text{sum}}\frac{u^2}{c^2}e+3w_1\frac{u^2}{c^2}h\right\} \tag{8.24}$$

In deriving this equation, the following relationships have been used:

$$\left.\begin{aligned}&\sum_{\alpha=0}^{8} w_\alpha(\mathbf{c}_\alpha\cdot\mathbf{u})=\sum_{\alpha=0}^{8} w_\alpha(c_{\alpha x}u_x+c_{\alpha y}u_y)=0\\&\sum_{\alpha=0}^{8} w_\alpha(\mathbf{c}_\alpha\cdot\mathbf{u})^2=\sum_{\alpha=0}^{8} w_\alpha(c_{\alpha x}^2u_x^2+c_{\alpha y}^2u_y^2+2u_xu_yc_{\alpha x}c_{\alpha y})=3w_1c^2u^2\end{aligned}\right\} \tag{8.25}$$

Equation (8.4) says that the quantity in Eq. (8.24) must equal the density ρ, so that the following relationships are obtained:

$$w_{\text{sum}}=1, \quad w_{\text{sum}}e+3w_1h=0 \tag{8.26}$$

in which $w_{\text{sum}}=w_0+4w_1+4w_5=w_0+5w_1$.

Similarly, we obtain the following equation:

$$\sum_{\alpha=0}^{8} c_{\alpha i} f_\alpha^{(0)}=\sum_{\alpha=0}^{8} \rho w_\alpha c_{\alpha i}\left\{1+b\frac{\mathbf{c}_\alpha\cdot\mathbf{u}}{c^2}+e\frac{u^2}{c^2}+h\frac{(\mathbf{c}_\alpha\cdot\mathbf{u})^2}{c^4}\right\}=3\rho w_1 u_i b \tag{8.27}$$

in which the following relationships have been used for the derivation.

$$\left.\begin{aligned} &\sum_{\alpha=0}^{8} w_\alpha c_{\alpha x}(\mathbf{c}_\alpha \cdot \mathbf{u}) = \sum_{\alpha=0}^{8} w_\alpha (c_{\alpha x}^2 u_x + c_{\alpha x} c_{\alpha y} u_y) = 3 w_1 c^2 u_x \\ &\sum_{\alpha=0}^{8} w_\alpha c_{\alpha x}(\mathbf{c}_\alpha \cdot \mathbf{u})^2 = \sum_{\alpha=0}^{8} w_\alpha (c_{\alpha x}^3 u_x^2 + 2 c_{\alpha x}^2 c_{\alpha y} u_x u_y + c_{\alpha x} c_{\alpha y}^2 u_y^2) = 0 \end{aligned}\right\} \tag{8.28}$$

Since the momentum equation in Eq. (8.5) must be satisfied, b is obtained as

$$b = \frac{1}{3w_1} \tag{8.29}$$

Then, we evaluate the momentum flux $\Pi_{ij}^{(0)}$ by substituting the equilibrium distribution $f_\alpha^{(0)}$ in Eq. (8.8) into this momentum flux expression:

$$\begin{aligned} \Pi_{ij}^{(0)} &= \sum_{\alpha=0}^{8} c_{\alpha i} c_{\alpha j} f_\alpha^{(0)} = \sum_{\alpha=0}^{8} \rho w_\alpha c_{\alpha i} c_{\alpha j} \left\{ 1 + b\frac{\mathbf{c}_\alpha \cdot \mathbf{u}}{c^2} + e\frac{u^2}{c^2} + h\frac{(\mathbf{c}_\alpha \cdot \mathbf{u})^2}{c^4} \right\} \\ &= \rho w_1 \left\{ 3c^2 \left(1 + \frac{u^2}{c^2} e \right) \delta_{ij} + u^2 h \delta_{ij} \right\} + 2\rho w_1 u_i u_j h \end{aligned} \tag{8.30}$$

in which the following relationships have been used for deriving this equation:

$$\left.\begin{aligned} &\sum_{\alpha=0}^{8} w_\alpha c_{\alpha x} c_{\alpha y} (\mathbf{c}_\alpha \cdot \mathbf{u}) = \sum_{\alpha=0}^{8} w_\alpha c_{\alpha x}^2 (\mathbf{c}_\alpha \cdot \mathbf{u}) = \sum_{\alpha=0}^{8} w_\alpha c_{\alpha y}^2 (\mathbf{c}_\alpha \cdot \mathbf{u}) = 0 \\ &\sum_{\alpha=0}^{8} w_\alpha c_{\alpha x} c_{\alpha y} (\mathbf{c}_\alpha \cdot \mathbf{u})^2 = 2 w_1 c^4 u_x u_y \\ &\sum_{\alpha=0}^{8} w_\alpha c_{\alpha x}^2 (\mathbf{c}_\alpha \cdot \mathbf{u})^2 = 2 w_1 c^4 u_x^2 + w_1 c^4 u^2 \end{aligned}\right\} \tag{8.31}$$

For the case of an equilibrium state, $\Pi_{ij}^{(0)}$ can be related to the pressure p as

$$\Pi_{ij}^{(0)} = p\delta_{ij} + \rho u_i u_j \tag{8.32}$$

Hence, the comparison of Eq. (8.30) with Eq. (8.32) yields the following relationships:

$$h = \frac{1}{2w_1}, \quad p = 3\rho w_1 c^2 \tag{8.33}$$

$$3e + h = 0 \tag{8.34}$$

The pressure p is related to the speed of sound c_s as $p = \rho c_s^2$, so that c_s can be written as

$$c_s = \sqrt{3w_1}c \tag{8.35}$$

Finally, we evaluate the kinetic energy. Preliminary relationships can be derived from Eq. (8.23) as

$$\left.\begin{aligned}
&\sum_{\alpha=0}^{8} w_\alpha c_\alpha^2 = 6w_1c^2\\
&\sum_{\alpha=0}^{8} w_\alpha c_{\alpha x}^2(\mathbf{c}_\alpha\cdot\mathbf{u}) = \sum_{\alpha=0}^{8} w_\alpha c_{\alpha y}^2(\mathbf{c}_\alpha\cdot\mathbf{u}) = \sum_{\alpha=0}^{8} w_\alpha c_\alpha^2(\mathbf{c}_\alpha\cdot\mathbf{u}) = 0\\
&\sum_{\alpha=0}^{8} w_\alpha c_\alpha^2(\mathbf{c}_\alpha\cdot\mathbf{u})^2 = 4w_1c^4u^2, \quad \sum_{\alpha=0}^{8} w_\alpha(\mathbf{c}_\alpha\cdot\mathbf{u})^3 = 0
\end{aligned}\right\} \tag{8.36}$$

Using these relationships, the right-hand side in Eq. (8.6) may be calculated as

$$\begin{aligned}
\sum_{\alpha=0}^{8}\frac{m}{2}(\mathbf{c}_\alpha-\mathbf{u})^2\frac{f_\alpha^{(0)}}{\rho} &= \frac{m}{2}\sum_{\alpha=0}^{8} w_\alpha({c_\alpha}^2+u^2-2\mathbf{c}_\alpha\cdot\mathbf{u})\\
&\times\left\{1+b\frac{\mathbf{c}_\alpha\cdot\mathbf{u}}{c^2}+e\frac{u^2}{c^2}+h\frac{(\mathbf{c}_\alpha\cdot\mathbf{u})^2}{c^4}\right\}\\
&= \frac{m}{2}\left\{6w_1\left(1+e\frac{u^2}{c^2}\right)c^2+4w_1hu^2+w_{\mathrm{sum}}\left(1+e\frac{u^2}{c^2}\right)u^2+3w_1h\frac{u^4}{c^2}-6w_1bu^2\right\}
\end{aligned} \tag{8.37}$$

By taking into account Eqs. (8.26), (8.29), and (8.33), the above equation is simplified as

$$\sum_{\alpha=0}^{8}\frac{m}{2}(\mathbf{c}_\alpha-\mathbf{u})^2\frac{f_\alpha^{(0)}}{\rho} = \frac{m}{2}(6w_1c^2+6w_1u^2e+w_{\mathrm{sum}}u^2) \tag{8.38}$$

Hence, Eq. (8.6) reduces to

$$\frac{2}{2}kT = \frac{m}{2}(6w_1c^2+6w_1u^2e+w_{\mathrm{sum}}u^2) \tag{8.39}$$

Since the temperature T is independent of the macroscopic velocity u, this equation yields the final relationships:

$$6w_1e + w_{\mathrm{sum}} = 0 \tag{8.40}$$

$$3mw_1c^2 = kT \tag{8.41}$$

We now have the same number of equations as the unknown constants, so that the solutions required can be obtained in a straightforward way as

$$b = 3, \quad e = -\frac{3}{2}, \quad h = \frac{9}{2} \tag{8.42}$$

$$w_{\mathrm{sum}} = 1, \quad w_0 = \frac{4}{9}, \quad w_1 = \frac{1}{9}, \quad w_5 = \frac{1}{36} \tag{8.43}$$

We summarize the final results as

$$f_\alpha^{(0)} = \rho w_\alpha \left\{ 1 + 3\frac{\mathbf{c}_\alpha \cdot \mathbf{u}}{c^2} - \frac{3}{2} \cdot \frac{u^2}{c^2} + \frac{9}{2} \cdot \frac{(\mathbf{c}_\alpha \cdot \mathbf{u})^2}{c^4} \right\} \tag{8.44}$$

$$w_\alpha = \begin{cases} 4/9 & \text{for} \quad \alpha = 0 \\ 1/9 & \text{for} \quad \alpha = 1,2,3,4\,, \\ 1/36 & \text{for} \quad \alpha = 5,6,7,8 \end{cases} \quad |\mathbf{c}_\alpha| = \begin{cases} 0 & \text{for} \quad \alpha = 0 \\ c & \text{for} \quad \alpha = 1,2,3,4 \\ \sqrt{2}c & \text{for} \quad \alpha = 5,6,7,8 \end{cases} \tag{8.45}$$

The speed of sound c_s is expressed as

$$c_s = c/\sqrt{3} \tag{8.46}$$

8.1.2 D3Q19 Model

In the case of the D3Q19 lattice model, the thermodynamic equilibrium distribution can be assumed to have the form of Eq. (8.8), and therefore the unknown constants can be derived through similar procedures to the previous D2Q9 model. Only in this present subsection, we use the notation $\tilde{c}(= \Delta x/\Delta t)$ for the lattice speed instead of c, since the notation c will be used for the abbreviated symbol of the cosine function.

In order to satisfy the isotropy condition, the lattice system has to be adopted such that it is independent of an arbitrary rotation of the lattice. In Figure 8.3, for a rotation of the lattice system about the z-axis by an angle ϕ and a rotation about the y-axis by an angle θ, the rotation matrix $\mathbf{R}$ is written as

$$\mathbf{R} = \begin{pmatrix} \cos\theta & 0 & \sin\theta \\ 0 & 1 & 0 \\ -\sin\theta & 0 & \cos\theta \end{pmatrix} \begin{pmatrix} \cos\phi & -\sin\phi & 0 \\ \sin\phi & \cos\phi & 0 \\ 0 & 0 & 1 \end{pmatrix} = \begin{pmatrix} Cc & -Cs & S \\ s & c & 0 \\ -Sc & Ss & C \end{pmatrix} \tag{8.47}$$

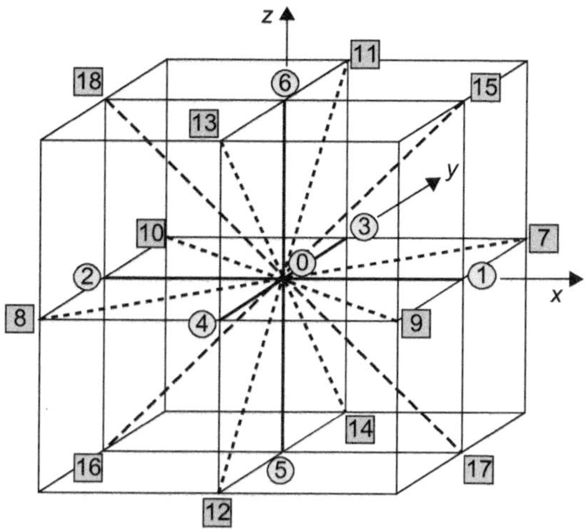

Figure 8.3 Lattice model for the D3Q19.

in which the abbreviations $C = \cos\theta$, $S = \sin\theta$, $c = \cos\phi$, and $s = \sin\phi$ are used for simplification of the equations. An arbitrary component $\mathbf{X}$ is related to the corresponding rotated component $\mathbf{X}'$ by the expression $\mathbf{X}' = \mathbf{R}\cdot\mathbf{X}$. The transferred component $\mathbf{d}_k$ ($k = 1,2,\ldots,18$) of each lattice point in Figure 8.3 is obtained as

$$\left.\begin{aligned}
&\mathbf{d}_1 = \mathbf{R}\begin{pmatrix}1\\0\\0\end{pmatrix} = \begin{pmatrix}Cc\\s\\-Sc\end{pmatrix}, \quad \mathbf{d}_3 = \mathbf{R}\begin{pmatrix}0\\1\\0\end{pmatrix} = \begin{pmatrix}-Cs\\c\\Ss\end{pmatrix}, \quad \mathbf{d}_5 = \begin{pmatrix}-S\\0\\-C\end{pmatrix}\\
&\mathbf{d}_7 = \begin{pmatrix}C(c-s)\\s+c\\-S(c-s)\end{pmatrix}, \quad \mathbf{d}_9 = \begin{pmatrix}C(c+s)\\s-c\\-S(c+s)\end{pmatrix}, \quad \mathbf{d}_{11} = \begin{pmatrix}-Cs+S\\c\\Ss+C\end{pmatrix}\\
&\mathbf{d}_{13} = \begin{pmatrix}Cs+S\\-c\\-Ss+C\end{pmatrix}, \quad \mathbf{d}_{15} = \begin{pmatrix}Cc+S\\s\\-Sc+C\end{pmatrix}, \quad \mathbf{d}_{17} = \begin{pmatrix}Cc-S\\s\\-Sc-C\end{pmatrix}
\end{aligned}\right\} \tag{8.48}$$

From symmetric considerations, the following relationship must be satisfied:

$$\mathbf{d}_{2k} = -\mathbf{d}_{2k-1} \quad (k=1,2,\ldots,9) \tag{8.49}$$

The final expressions are summarized in Table 8.1.

The results in Table 8.1 give rise to those concerning $c_{\alpha x}$, $c_{\alpha y}$, and $c_{\alpha z}$, such as $\sum_{\alpha=0}^{18} = w_\alpha c_{\alpha i} c_{\alpha j} c_{\alpha k} c_{\alpha l}$ in Table 8.2. As seen from the result of $\sum_{\alpha=0}^{18} w_\alpha c_{\alpha i}^4$, the

Table 8.1 Results of Quantities for the Successive Derivation

$\sum_{k=1}^{6} d_{kx}^4 = -4(C^2S^2 + C^4c^2s^2) + 2$	$\sum_{k=7}^{18} d_{kx}^4 = 8(C^2S^2 + C^4c^2s^2) + 8$
$\sum_{k=1}^{6} d_{ky}^4 = 4(c^4 - c^2) + 2$	$\sum_{k=7}^{18} d_{ky}^4 = -8(c^4 - c^2) + 8$
$\sum_{k=1}^{6} d_{kz}^4 = -4(C^2S^2 + S^4c^2s^2) + 2$	$\sum_{k=7}^{18} d_{kz}^4 = 8(C^2S^2 + S^4c^2s^2) + 8$
$\sum_{k=1}^{6} d_{kx}^2 d_{ky}^2 = 4C^2c^2s^2$	$\sum_{k=7}^{18} d_{kx}^2 d_{ky}^2 = -8C^2c^2s^2 + 4$
$\sum_{k=1}^{6} d_{kx}^2 d_{kz}^2 = 4C^2S^2(1 - c^2s^2)$	$\sum_{k=7}^{18} d_{kx}^2 d_{kz}^2 = -8C^2S^2(1 - c^2s^2) + 4$
$\sum_{k=1}^{6} d_{ky}^2 d_{kz}^2 = 4S^2c^2s^2$	$\sum_{k=7}^{18} d_{ky}^2 d_{kz}^2 = -8S^2c^2s^2 + 4$
$\sum_{k=1}^{6} d_{kx}^2 d_{ky} d_{kz} = 2C^2Scs(-c^2 + s^2)$	$\sum_{k=7}^{18} d_{kx}^2 d_{ky} d_{kz} = 4C^2S(c^2 - s^2)cs$
$\sum_{k=1}^{6} d_{kx} d_{ky}^2 d_{kz} = -4CSc^2s^2$	$\sum_{k=7}^{18} d_{kx} d_{ky}^2 d_{kz} = 8CSc^2s^2$
$\sum_{k=1}^{6} d_{kx} d_{ky} d_{kz}^2 = 2CS^2cs(c^2 - s^2)$	$\sum_{k=7}^{18} d_{kx} d_{ky} d_{kz}^2 = -4CS^2(c^2 - s^2)cs$
$\sum_{k=1}^{6} d_{kx}^3 d_{ky} = 2C^3cs(c^2 - s^2)$	$\sum_{k=7}^{18} d_{kx}^3 d_{ky} = -4C^3(c^2 - s^2)cs$
$\sum_{k=1}^{6} d_{kx}^3 d_{kz} = -2C^3S(1 - 2c^2s^2) + 2CS^3$	$\sum_{k=7}^{18} d_{kx}^3 d_{kz} = 4C^3S(1 - 2c^2s^2) - 4CS^3$
$\sum_{k=1}^{6} d_{ky}^3 d_{kx} = 2Ccs(-c^2 + s^2)$	$\sum_{k=7}^{18} d_{ky}^3 d_{kx} = 4C(c^2 - s^2)cs$
$\sum_{k=1}^{6} d_{ky}^3 d_{kz} = 2Scs(c^2 - s^2)$	$\sum_{k=7}^{18} d_{ky}^3 d_{kz} = -4S(c^2 - s^2)cs$
$\sum_{k=1}^{6} d_{kz}^3 d_{kx} = -2CS^3(1 - 2c^2s^2) + 2C^3S$	$\sum_{k=7}^{18} d_{kz}^3 d_{kx} = 4CS^3(1 - 2c^2s^2) - 4C^3S$
$\sum_{k=1}^{6} d_{kz}^3 d_{ky} = -2S^3cs(c^2 - s^2)$	$\sum_{k=7}^{18} d_{kz}^3 d_{ky} = 4S^3(c^2 - s^2)cs$
$\sum_{k=1}^{6} d_{kx}^2 = \sum_{k=1}^{6} d_{ky}^2 = \sum_{k=1}^{6} d_{kz}^2 = 2$	$\sum_{k=7}^{18} d_{kx}^2 = \sum_{k=7}^{18} d_{ky}^2 = \sum_{k=7}^{18} d_{kz}^2 = 8$
$\sum_{k=1}^{6} d_{kx} d_{ky} = \sum_{k=1}^{6} d_{kx} d_{kz} = \sum_{k=1}^{6} d_{ky} d_{kz} = 0$	$\sum_{k=7}^{18} d_{kx} d_{ky} = \sum_{k=7}^{18} d_{kx} d_{kz} = \sum_{k=7}^{18} d_{ky} d_{kz} = 0$

Table 8.2 Final Results of Quantities for the Successive Derivation

$$\sum_{\alpha=0}^{18} w_\alpha c_{\alpha x}^4 = \tilde{c}^4(-4w_1 + 8w_7)(C^2S^2 + C^4c^2s^2) + \tilde{c}^4(2w_1 + 8w_7)$$

$$\sum_{\alpha=0}^{18} w_\alpha c_{\alpha y}^4 = \tilde{c}^4(4w_1 - 8w_7)(c^4 - c^2) + \tilde{c}^4(2w_1 + 8w_7)$$

$$\sum_{\alpha=0}^{18} w_\alpha c_{\alpha z}^4 = \tilde{c}^4(-4w_1 + 8w_7)(C^2S^2 + S^4c^2s^2) + \tilde{c}^4(2w_1 + 8w_7)$$

$$\Downarrow \; w_1 = 2w_7$$

$$\sum_{\alpha=0}^{18} w_\alpha c_{\alpha x}^4 = \sum_{\alpha=0}^{18} w_\alpha c_{\alpha y}^4 = \sum_{\alpha=0}^{18} w_\alpha c_{\alpha z}^4 = 6w_1\tilde{c}^4$$

$$\sum_{\alpha=0}^{18} w_\alpha c_{\alpha x}^2 c_{\alpha y}^2 = \sum_{\alpha=0}^{18} w_\alpha c_{\alpha x}^2 c_{\alpha z}^2 = \sum_{\alpha=0}^{18} w_\alpha c_{\alpha y}^2 c_{\alpha z}^2 = 2w_1\tilde{c}^4$$

$$\sum_{\alpha=0}^{18} w_\alpha c_{\alpha x}^3 c_{\alpha y} = \sum_{\alpha=0}^{18} w_\alpha c_{\alpha x}^3 c_{\alpha z} = \sum_{\alpha=0}^{18} w_\alpha c_{\alpha y}^3 c_{\alpha x} = \sum_{\alpha=0}^{18} w_\alpha c_{\alpha y}^3 c_{\alpha z} = \sum_{\alpha=0}^{18} w_\alpha c_{\alpha z}^3 c_{\alpha x} = \sum_{\alpha=0}^{18} w_\alpha c_{\alpha z}^3 c_{\alpha y} = 0$$

$$\sum_{\alpha=0}^{18} w_\alpha c_{\alpha x}^2 c_{\alpha y} c_{\alpha z} = \sum_{\alpha=0}^{18} w_\alpha c_{\alpha x} c_{\alpha y}^2 c_{\alpha z} = \sum_{\alpha=0}^{18} w_\alpha c_{\alpha x} c_{\alpha y} c_{\alpha z}^2 = 0$$

$$\sum_{\alpha=0}^{18} w_\alpha c_{\alpha x}^3 = \sum_{\alpha=0}^{18} w_\alpha c_{\alpha y}^3 = \sum_{\alpha=0}^{18} w_\alpha c_{\alpha z}^3 = 0$$

$$\sum_{\alpha=0}^{18} w_\alpha c_{\alpha x}^2 c_{\alpha y} = \sum_{\alpha=0}^{18} w_\alpha c_{\alpha x}^2 c_{\alpha z} = \sum_{\alpha=0}^{18} w_\alpha c_{\alpha y}^2 c_{\alpha x} = \sum_{\alpha=0}^{18} w_\alpha c_{\alpha y}^2 c_{\alpha z} = \sum_{\alpha=0}^{18} w_\alpha c_{\alpha z}^2 c_{\alpha x} = \sum_{\alpha=0}^{18} w_\alpha c_{\alpha z}^2 c_{\alpha y} = 0$$

$$\sum_{\alpha=0}^{18} w_\alpha c_{\alpha x} c_{\alpha y} c_{\alpha z} = 0$$

$$\sum_{\alpha=0}^{18} w_\alpha c_{\alpha x}^2 = \sum_{\alpha=0}^{18} w_\alpha c_{\alpha y}^2 = \sum_{\alpha=0}^{18} w_\alpha c_{\alpha z}^2 = 6w_1\tilde{c}^2$$

$$\sum_{\alpha=0}^{18} w_\alpha c_{\alpha x} c_{\alpha y} = \sum_{\alpha=0}^{18} w_\alpha c_{\alpha x} c_{\alpha z} = \sum_{\alpha=0}^{18} w_\alpha c_{\alpha y} c_{\alpha z} = 0$$

$$\sum_{\alpha=0}^{18} w_\alpha c_{\alpha x} = \sum_{\alpha=0}^{18} w_\alpha c_{\alpha y} = \sum_{\alpha=0}^{18} w_\alpha c_{\alpha z} = 0$$

relationship of $w_1 = 2w_7$ has to be satisfied in order for this result to be independent of the rotational angle. Hence, the results after the arrow in Table 8.2 have taken into account this relationship. The expressions in Table 8.2 are written in simply unified equations as

$$\left.\begin{aligned}&\sum_{\alpha=0}^{18} w_\alpha c_{\alpha i} c_{\alpha j} c_{\alpha k} c_{\alpha l} = 2w_1 \tilde{c}^4 (\delta_{ij}\delta_{kl} + \delta_{ik}\delta_{jl} + \delta_{il}\delta_{jk})\\&\sum_{\alpha=0}^{18} w_\alpha c_{\alpha i} c_{\alpha j} c_{\alpha k} = 0\\&\sum_{\alpha=0}^{18} w_\alpha c_{\alpha i} c_{\alpha j} = 6w_1 \tilde{c}^2 \delta_{ij}, \quad \sum_{\alpha=0}^{18} w_\alpha c_i = 0\end{aligned}\right\} \tag{8.50}$$

We are now ready to derive the equilibrium distribution for the D3Q19 lattice model.

As mentioned before, the relationships that the equilibrium distribution $f_\alpha^{(0)}$ shown in Eq. (8.8) must satisfy are the mass, momentum, kinetic energy, and momentum flux equations. The former three equations are written as

$$\sum_{\alpha=0}^{18} f_\alpha^{(0)} = \rho \tag{8.51}$$

$$\sum_{\alpha=0}^{18} \mathbf{c}_\alpha f_\alpha^{(0)} = \rho \mathbf{u} \tag{8.52}$$

$$\sum_{\alpha=0}^{18} \frac{m}{2} (\mathbf{c}_\alpha - \mathbf{u})^2 \frac{f_\alpha^{(0)}}{\rho} = \frac{3}{2} kT \tag{8.53}$$

With the results shown in Table 8.2, the following equation is obtained:

$$\begin{aligned}\sum_{\alpha=0}^{18} f_\alpha^{(0)} &= \sum_{\alpha=0}^{18} w_\alpha \rho \left\{ 1 + b\frac{\mathbf{c}_\alpha \cdot \mathbf{u}}{\tilde{c}^2} + e\frac{u^2}{\tilde{c}^2} + h\frac{(\mathbf{c}_\alpha \cdot \mathbf{u})^2}{\tilde{c}^4} \right\}\\&= \rho \left\{ w_{\mathrm{sum}} + \frac{u^2}{\tilde{c}^2} (w_{\mathrm{sum}} e + 6w_1 h) \right\}\end{aligned} \tag{8.54}$$

The comparison of this equation with Eq. (8.51) leads to

$$w_{\mathrm{sum}} = 1, \quad w_{\mathrm{sum}} e + 6w_1 h = 0 \tag{8.55}$$

in which $w_{\mathrm{sum}} = w_0 + 6w_1 + 12w_7 = w_0 + 12w_1$.

In order to compare this result with the right-hand side of Eq. (8.52), the left-hand side is evaluated using the equilibrium distribution as

$$\sum_{\alpha=0}^{18} c_{\alpha i} f_\alpha^{(0)} = \sum_{\alpha=0}^{18} \rho w_\alpha c_{\alpha i}\left\{1 + b\frac{\mathbf{c}_\alpha \cdot \mathbf{u}}{\tilde{c}^2} + e\frac{u^2}{\tilde{c}^2} + h\frac{(\mathbf{c}_\alpha \cdot \mathbf{u})^2}{\tilde{c}^4}\right\} = 6bw_1\rho u_i \tag{8.56}$$

Hence, the comparison of the right-hand sides in Eqs. (8.56) and (8.52) gives rise to

$$6w_1 b = 1 \tag{8.57}$$

Similar to the D2Q9 model, the momentum flux $\Pi_{ij}^{(0)}$ is calculated as

$$\Pi_{ij}^{(0)} = \sum_{\alpha=0}^{18} c_{\alpha i} c_{\alpha j} f_\alpha^{(0)} = \left\{6\rho w_1 \tilde{c}^2\left(1 + e\frac{u^2}{\tilde{c}^2}\right) + 2\rho w_1 u^2 h\right\}\delta_{ij} + 4\rho w_1 u_i u_j h \tag{8.58}$$

In deriving this equation, the following relationships are used:

$$\left.\begin{aligned}
&\sum_{\alpha=0}^{18} w_\alpha c_{\alpha x} c_{\alpha y}\left\{1 + b\frac{\mathbf{c}_\alpha \cdot \mathbf{u}}{\tilde{c}^2} + e\frac{u^2}{\tilde{c}^2} + h\frac{(\mathbf{c}_\alpha \cdot \mathbf{u})^2}{\tilde{c}^4}\right\} = 4w_1 u_x u_y h \\
&\sum_{\alpha=0}^{18} w_\alpha c_{\alpha x}^2\left\{1 + b\frac{\mathbf{c}_\alpha \cdot \mathbf{u}}{\tilde{c}^2} + e\frac{u^2}{\tilde{c}^2} + h\frac{(\mathbf{c}_\alpha \cdot \mathbf{u})^2}{\tilde{c}^4}\right\} \\
&\qquad = 6w_1 \tilde{c}^2\left(1 + e\frac{u^2}{\tilde{c}^2}\right) + 2w_1 u^2 h + 4w_1 u_x^2 h
\end{aligned}\right\} \tag{8.59}$$

The expression for $\Pi_{ij}^{(0)}$ as defined by Eq. (8.32) is also valid for a three-dimensional system. The following relationships can therefore be obtained by comparison with Eq. (8.58) as

$$p = 6\rho w_1 \tilde{c}^2 \tag{8.60}$$

$$4w_1 h = 1, \quad 3e + h = 0 \tag{8.61}$$

Hence, the speed of sound c_s is expressed from $p = \rho c_s^2$ as

$$c_s = \sqrt{6w_1} \cdot \tilde{c} \tag{8.62}$$

Finally, in order to compare with Eq. (8.53), the following reformation is performed:

$$
\begin{aligned}
\sum_{\alpha=0}^{18}(\mathbf{c}_\alpha-\mathbf{u})^2\frac{f_\alpha^{(0)}}{\rho} &= \sum_{\alpha=0}^{18} w_\alpha(\mathbf{c}_\alpha-\mathbf{u})^2\left\{1+b\frac{\mathbf{c}_\alpha\cdot\mathbf{u}}{\tilde{c}^2}+e\frac{u^2}{\tilde{c}^2}+h\frac{(\mathbf{c}_\alpha\cdot\mathbf{u})^2}{\tilde{c}^4}\right\} \\
&= \sum_{\alpha=0}^{18} w_\alpha(c_\alpha^2-2\mathbf{c}_\alpha\cdot\mathbf{u}+u^2)\left\{1+b\frac{\mathbf{c}_\alpha\cdot\mathbf{u}}{\tilde{c}^2}+e\frac{u^2}{\tilde{c}^2}+h\frac{(\mathbf{c}_\alpha\cdot\mathbf{u})^2}{\tilde{c}^4}\right\} \\
&= \sum_{\alpha=0}^{18} w_\alpha(c_\alpha^2-2\mathbf{c}_\alpha\cdot\mathbf{u}+u^2)\left(1+e\frac{u^2}{\tilde{c}^2}\right) \\
&\quad +\sum_{\alpha=0}^{18} w_\alpha(c_\alpha^2-2\mathbf{c}_\alpha\cdot\mathbf{u}+u^2)(\mathbf{c}_\alpha\cdot\mathbf{u})\frac{b}{\tilde{c}^2} \\
&\quad +\sum_{\alpha=0}^{18} w_\alpha(c_\alpha^2-2\mathbf{c}_\alpha\cdot\mathbf{u}+u^2)(\mathbf{c}_\alpha\cdot\mathbf{u})^2\frac{h}{\tilde{c}^4} \\
&= \left(1+e\frac{u^2}{\tilde{c}^2}\right)(18w_1\tilde{c}^2+w_{\mathrm{sum}}u^2)-\frac{2b}{\tilde{c}^2}\sum_{\alpha=0}^{18} w_\alpha(\mathbf{c}_\alpha\cdot\mathbf{u})^2 \\
&\quad +\frac{h}{\tilde{c}^4}\sum_{\alpha=0}^{18} w_\alpha\left\{c_\alpha^2(\mathbf{c}_\alpha\cdot\mathbf{u})^2+u^2(\mathbf{c}_\alpha\cdot\mathbf{u})^2\right\} \\
&= 18w_1\tilde{c}^2+(w_{\mathrm{sum}}-12w_1b+18w_1e+10w_1h)u^2
\end{aligned}
\tag{8.63}
$$

in which Eq. (8.55) and the following relationship have been used for the derivation:

$$
\left.
\begin{aligned}
&\sum_{\alpha=0}^{18} w_\alpha(\mathbf{c}_\alpha\cdot\mathbf{u})^2 = 6w_1\tilde{c}^2u^2 \\
&\sum_{\alpha=0}^{18} w_\alpha c_\alpha^2(\mathbf{c}_\alpha\cdot\mathbf{u})^2 = 10w_1\tilde{c}^4u^2
\end{aligned}
\right\}
\tag{8.64}
$$

Since the temperature is independent of the macroscopic velocity u, the second term on the right-hand side in Eq. (8.63) must vanish:

$$
w_{\mathrm{sum}}-12w_1b+18w_1e+10w_1h=0 \tag{8.65}
$$

The comparison of Eq. (8.63) with Eq. (8.6) yields the following equation:

$$
\frac{m}{2}18w_1\tilde{c}^2=\frac{3}{2}kT \tag{8.66}
$$

We now have the same number of equations as the unknown constants. The final results for the unknown constants can be obtained from Eqs. (8.55), (8.57), (8.61), and (8.65) and the relationships of $w_1 = 2w_7$ and $w_{sum} = w_0 + 12w_1$, as

$$\left.\begin{aligned} &w_0 = \frac{1}{3}, \quad w_1 = \frac{1}{18}, \quad w_7 = \frac{1}{36}, \quad w_{sum} = 1 \\ &b = 3, \quad e = -\frac{3}{2}, \quad h = \frac{9}{2} \end{aligned}\right\} \tag{8.67}$$

With the original notation c for the lattice velocity, the equilibrium distribution for the D3Q19 model is finally written as

$$f_a^{(0)} = \rho w_\alpha \left\{ 1 + 3\frac{\mathbf{c}_\alpha \cdot \mathbf{u}}{c^2} - \frac{3}{2}\cdot\frac{u^2}{c^2} + \frac{9}{2}\cdot\frac{(\mathbf{c}_\alpha \cdot \mathbf{u})^2}{c^4} \right\} \tag{8.68}$$

$$w_\alpha = \begin{cases} 1/3 & \text{for} \quad \alpha=0 \\ 1/18 & \text{for} \quad \alpha = 1,2,\ldots,6 \\ 1/36 & \text{for} \quad \alpha = 7,8,\ldots,18 \end{cases}, \quad |\mathbf{c}_\alpha| = \begin{cases} 0 & \text{for} \quad \alpha=0 \\ c & \text{for} \quad \alpha = 1,2,\ldots,6 \\ \sqrt{2}c & \text{for} \quad \alpha = 7,8,\ldots,18 \end{cases} \tag{8.69}$$

8.2 Navier–Stokes Equation

In this section, we derive the Navier–Stokes equation from the preliminary equations derived in Appendix A1, which is the basic macroscopic equation for flow problems. The following derivation procedure is valid for both D2Q9 and D3Q19 models, with the exception that α is taken as $\alpha = 0,1,\ldots,8$, the axis index i is x or y for the former model, α is taken as $\alpha = 0,1,\ldots,18$, and i is x, y, or z for the latter model.

The starting equation for the derivation procedure is Eq. (A1.27), rewritten as

$$\frac{\partial}{\partial t}(\rho u_i) + \sum_j \frac{\partial}{\partial r_j}\Pi_{ij} + \frac{\Delta t}{2}\varepsilon \sum_j \frac{\partial}{\partial r_j}\left\{\frac{\partial}{\partial t_1}\Pi_{ij}^{(0)}\right\} + \frac{\Delta t}{2}\sum_j\sum_k \frac{\partial}{\partial r_j}\left\{\frac{\partial}{\partial r_k}S_{ijk}^{(0)}\right\} = 0 \tag{8.70}$$

Another starting equation is Eq. (A1.31), rewritten as

$$-\frac{1}{\tau\Delta t}f_\alpha^{(1)} = -\frac{\partial f_\alpha^{(0)}}{\partial \rho}\cdot\frac{\partial}{\partial \mathbf{r}_1}\cdot(\rho\mathbf{u}) - \sum_i\sum_j \frac{\partial f_\alpha^{(0)}}{\partial(\rho u_i)}\cdot\frac{\partial}{\partial r_{1j}}\Pi_{ij}^{(0)} + \sum_i \frac{\partial}{\partial r_{1i}}(c_{\alpha i}f_\alpha^{(0)}) \tag{8.71}$$

in which

$$f_\alpha = f_\alpha^{(0)} + \varepsilon f_\alpha^{(1)} + \varepsilon^2 f_\alpha^{(2)} + \cdots \tag{8.72}$$

$$\Pi_{ij}^{(0)} = \sum_\alpha c_{\alpha i} c_{\alpha j} f_\alpha^{(0)} = p\delta_{ij} + \rho u_i u_j = \frac{\rho}{3} c^2 \delta_{ij} + \rho u_i u_j \tag{8.73}$$

$$S_{ijk}^{(0)} = \sum_\alpha c_{\alpha i} c_{\alpha j} c_{\alpha k} f_\alpha^{(0)} \tag{8.74}$$

$$\frac{\partial}{\partial t} = \varepsilon \frac{\partial}{\partial t_1} + \varepsilon^2 \frac{\partial}{\partial t_2}, \qquad \frac{\partial}{\partial r_i} = \varepsilon \frac{\partial}{\partial r_{1i}} \tag{8.75}$$

We now begin the derivation procedure for the Navier–Stokes equation by deriving the solution for $f_\alpha^{(1)}$ from the basic equation in Eq. (8.71). If the terms higher than the order of $(u/c)^2$ are neglected, the following relationships are obtained:

$$f_\alpha^{(0)} = \rho w_\alpha \left\{ 1 + 3 \frac{\mathbf{c}_\alpha \cdot \mathbf{u}}{c^2} \right\} \tag{8.76}$$

$$\Pi_{ij}^{(0)} = \frac{\rho}{3} c^2 \delta_{ij} + \rho u_i u_j \tag{8.77}$$

$$\frac{\partial f_\alpha^{(0)}}{\partial \rho} = w_\alpha \tag{8.78}$$

$$\frac{\partial f_\alpha^{(0)}}{\partial(\rho u_i)} = w_\alpha \frac{\partial}{\partial(\rho u_i)} \left\{ \frac{3}{c^2} \sum_j c_{\alpha j} (\rho u_j) \right\} = w_\alpha \frac{3}{c^2} c_{\alpha i} \tag{8.79}$$

$$\frac{\partial}{\partial r_{1j}} \Pi_{ij}^{(0)} = \frac{c^2}{3} \cdot \frac{\partial \rho}{\partial r_{1j}} \delta_{ij} + \frac{\partial}{\partial r_{1j}} (\rho u_i u_j) \tag{8.80}$$

$$\frac{\partial}{\partial r_{1i}} (c_{\alpha i} f_\alpha^{(0)}) = w_\alpha \frac{\partial}{\partial r_{1i}} (\rho c_{\alpha i}) + 3 w_\alpha \frac{1}{c^2} \cdot \frac{\partial}{\partial r_{1i}} \left\{ \sum_j \rho c_{\alpha i} c_{\alpha j} u_j \right\} \tag{8.81}$$

By substituting these relationships into Eq. (8.71), the solution of $f_\alpha^{(1)}$ can finally be obtained as

$$f_\alpha^{(1)} = -3 w_\alpha \Delta t \tau \frac{1}{c^2} \sum_k \sum_l \left(c_{\alpha k} c_{\alpha l} - \frac{1}{3} c^2 \delta_{kl} \right) \frac{\partial}{\partial r_{1l}} (\rho u_k) \tag{8.82}$$

With this solution, the next quantity can be evaluated:

$$\begin{aligned}\varepsilon\Pi_{ij}^{(1)} &= \varepsilon\sum_{\alpha} c_{\alpha i}c_{\alpha j}f_{\alpha}^{(1)} \\ &= -3\Delta t\tau\frac{1}{c^2}\sum_{k}\sum_{l}\left(\sum_{\alpha} w_{\alpha}c_{\alpha i}c_{\alpha j}c_{\alpha k}c_{\alpha l}\right)\frac{\partial}{\partial r_l}(\rho u_k) \\ &\quad + \Delta t\tau\left(\sum_{\alpha} w_{\alpha}c_{\alpha i}c_{\alpha j}\right)\frac{\partial}{\partial \mathbf{r}}\cdot(\rho\mathbf{u}) \\ &= -\frac{\Delta t\tau c^2}{3}\left\{\frac{\partial}{\partial r_j}(\rho u_i) + \frac{\partial}{\partial r_i}(\rho u_j)\right\}\end{aligned} \tag{8.83}$$

In deriving this equation, Eqs. (8.50), (8.22), and (8.23) are used. This equation leads to

$$\sum_j \frac{\partial}{\partial r_j}(\varepsilon\Pi_{ij}^{(1)}) = -\frac{\Delta t\tau c^2}{3}\left\{\frac{\partial^2}{\partial \mathbf{r}^2}(\rho u_i) + \frac{\partial}{\partial r_i}\left(\frac{\partial}{\partial \mathbf{r}}\cdot(\rho\mathbf{u})\right)\right\} \tag{8.84}$$

On the other hand, $\Pi_{ij}^{(0)}$ is written as

$$\sum_j \frac{\partial}{\partial r_j}(\Pi_{ij}^{(0)}) = \sum_j \frac{\partial}{\partial r_j}(p\delta_{ij} + \rho u_i u_j) \tag{8.85}$$

With the approximation of $\Pi_{ij} \approx \Pi_{ij}^{(0)} + \varepsilon\Pi_{ij}^{(1)}$, the following equation is obtained from Eqs. (8.84) and (8.85):

$$\sum_j \frac{\partial}{\partial r_j}\Pi_{ij} = \sum_j \frac{\partial}{\partial r_j}(p\delta_{ij} + \rho u_i u_j) - \frac{\Delta t\tau c^2}{3}\left\{\frac{\partial^2}{\partial \mathbf{r}^2}(\rho u_i) + \frac{\partial}{\partial r_i}\left(\frac{\partial}{\partial \mathbf{r}}\cdot(\rho\mathbf{u})\right)\right\} \tag{8.86}$$

Moreover, Eqs. (A1.11) and (A1.17) give rise to

$$\frac{\Delta t}{2}\varepsilon\frac{\partial}{\partial t_1}\Pi_{ij}^{(0)} = \frac{1}{6}\Delta t c^2\delta_{ij}\varepsilon\frac{\partial \rho}{\partial t_1} = -\frac{1}{6}\Delta t c^2\delta_{ij}\frac{\partial}{\partial \mathbf{r}}\cdot(\rho\mathbf{u}) \tag{8.87}$$

From this equation, the following equation can be obtained:

$$\frac{\Delta t}{2}\varepsilon\sum_j \frac{\partial}{\partial r_j}\left(\frac{\partial}{\partial t_1}\Pi_{ij}^{(0)}\right) = -\frac{1}{6}\Delta t c^2\frac{\partial}{\partial r_i}\left\{\frac{\partial}{\partial \mathbf{r}}\cdot(\rho\mathbf{u})\right\} \tag{8.88}$$

We next evaluate the quantity $S_{ijk}^{(0)}$. First,

$$S_{ijk}^{(0)} = \sum_{\alpha} c_{\alpha i}c_{\alpha j}c_{\alpha k}f_{\alpha}^{(0)} = \sum_{\alpha} w_{\alpha}c_{\alpha i}c_{\alpha j}c_{\alpha k}\rho\left(1 + 3\frac{\mathbf{c}_{\alpha}\cdot\mathbf{u}}{c^2}\right) \tag{8.89}$$

With this equation, the partial derivative in Eq. (8.70) is obtained as

$$\begin{aligned}\sum_j\sum_k \frac{\partial}{\partial r_j}\cdot\frac{\partial}{\partial r_k}\left(S_{ijk}^{(0)}\right) &= \frac{1}{3}c^2\sum_j\sum_k \frac{\partial}{\partial r_j}\cdot\frac{\partial}{\partial r_k}(\rho u_k\delta_{ij}+\rho u_j\delta_{ik}+\rho u_i\delta_{jk})\\ &= \frac{2}{3}c^2\left\{\frac{\partial}{\partial r_i}\left(\frac{\partial}{\partial \mathbf{r}}\cdot(\rho\mathbf{u})\right)+\frac{1}{2}\cdot\frac{\partial^2}{\partial \mathbf{r}^2}(\rho u_i)\right\}\end{aligned} \tag{8.90}$$

We have now finished the preparation for deriving the Navier–Stokes equation. If the summation of the first and second terms, and the summation of the third and fourth terms on the left-hand side in Eq. (8.70), are denoted by A and B, respectively, these quantities are evaluated as

$$\begin{aligned}A &= \frac{\partial}{\partial t}(\rho u_i)+\sum_j\frac{\partial}{\partial r_j}(p\delta_{ij}+\rho u_i u_j)-\frac{\Delta t\tau c^2}{3}\left\{\frac{\partial^2}{\partial \mathbf{r}^2}(\rho u_i)+\frac{\partial}{\partial r_i}\left(\frac{\partial}{\partial \mathbf{r}}\cdot(\rho\mathbf{u})\right)\right\}\\ &= \frac{\partial}{\partial t}(\rho u_i)+\sum_j\frac{\partial}{\partial r_j}(\rho u_i u_j)+\frac{\partial p}{\partial r_i}-\frac{\Delta t\tau c^2}{3}\left\{\frac{\partial^2}{\partial \mathbf{r}^2}(\rho u_i)+\frac{\partial}{\partial r_i}\left(\frac{\partial}{\partial \mathbf{r}}\cdot(\rho\mathbf{u})\right)\right\}\end{aligned} \tag{8.91}$$

$$\begin{aligned}B &= \frac{\Delta t}{2}\varepsilon\sum_j\frac{\partial}{\partial r_j}\left(\frac{\partial}{\partial t_1}\Pi_{ij}^{(0)}\right)+\frac{\Delta t}{2}\sum_j\sum_k\frac{\partial}{\partial r_j}\left(\frac{\partial}{\partial r_k}S_{ijk}^{(0)}\right)\\ &= \frac{1}{6}\Delta tc^2\left\{\frac{\partial^2}{\partial \mathbf{r}^2}(\rho u_i)+\frac{\partial}{\partial r_i}\left(\frac{\partial}{\partial \mathbf{r}}\cdot(\rho\mathbf{u})\right)\right\}\end{aligned} \tag{8.92}$$

By substituting Eqs. (8.91) and (8.92) into Eq. (8.70), together with the relationship of $\frac{\partial}{\partial \mathbf{r}}\cdot(\rho\mathbf{u})=0$ for noncompressible fluids, the Navier–Stokes equation is finally obtained as

$$\rho\left\{\frac{\partial \mathbf{u}}{\partial t}+(\mathbf{u}\cdot\nabla)\mathbf{u}\right\} = -\nabla p+\mu^{\mathrm{LB}}\nabla^2\mathbf{u} \tag{8.93}$$

in which μ^{LB} is the viscosity, expressed as

$$\mu^{\mathrm{LB}} = \frac{\rho\Delta tc^2}{3}\left(\tau-\frac{1}{2}\right),\quad \nu^{\mathrm{LB}} = \frac{\mu^{\mathrm{LB}}}{\rho} = \frac{\Delta tc^2}{3}\left(\tau-\frac{1}{2}\right) \tag{8.94}$$

In this equation ν^{LB} is the kinematic viscosity.

8.3 Body Force

If a body force, such as the gravitational force, acts on a fluid, how do we incorporate it into the lattice Boltzmann equation? The method can be seen in the following:

$$\left.\begin{aligned} & f_\alpha(\mathbf{r}+\mathbf{c}_\alpha \Delta t, t+\Delta t)=\tilde{f}_\alpha(\mathbf{r}, t) \\ & \tilde{f}_\alpha(\mathbf{r}, t)=f_\alpha(\mathbf{r}, t)+\Omega_\alpha+g_\alpha \end{aligned}\right\} \tag{8.95}$$

in which

$$\Omega_\alpha=\frac{1}{\tau}\left\{f_\alpha^{(0)}(\mathbf{r}, t)-f_\alpha(\mathbf{r}, t)\right\} \tag{8.96}$$

$$g_\alpha=\left\{\begin{array}{ll} 0 & \text{for} \quad \alpha=0 \\ \dfrac{3 \Delta t}{c^2} w_\alpha \mathbf{c}_\alpha \cdot \mathbf{F} & \text{for} \quad \alpha \neq 0 \end{array}\right\} \tag{8.97}$$

g_α is a quantity that is due to the body force $\mathbf{F}$ per unit volume and has the following characteristics:

$$\left.\begin{aligned} & \sum_\alpha g_\alpha=0 \\ & \sum_\alpha \mathbf{c}_\alpha g_\alpha=\sum_\alpha \frac{3 \Delta t}{c^2} w_\alpha(\mathbf{c}_\alpha \mathbf{c}_\alpha) \cdot \mathbf{F}=\Delta t \mathbf{F} \end{aligned}\right\} \tag{8.98}$$

In this reformation, the following relationship has been used:

$$\sum_\alpha w_\alpha \mathbf{c}_\alpha \mathbf{c}_\alpha=(c^2/3) \mathbf{I} \tag{8.99}$$

in which $\mathbf{I}$ is the unit tensor, and Eq. (8.99) is valid for both D2Q9 and D3Q19 models. It is quite clear that the quantity g_α can be expressed in the form of Eq. (8.97), because the particle distribution tends to move in the direction of the body force $\mathbf{F}$ acting. Therefore, the relationship of the form $g_\alpha \propto \mathbf{c}_\alpha \cdot \mathbf{F}$ can be expected.

We will now confirm that the g_α, expressed in Eq. (8.97), appears in a reasonable form in the Navier–Stokes equation by deriving these equations starting from Eq. (8.95), as conducted in Section 8.2. There is no new concept applied here except for the inclusion of the new term g_α into the derivation procedure shown in Section 8.2; therefore we show only the important expressions.

From Appendix A1, the relationships of the orders ε and ε^2 are written as

$$\left.\begin{aligned}
&\frac{\partial \rho}{\partial t_1} + \nabla_1 \cdot (\rho \mathbf{u}) = 0 \\
&\frac{\partial}{\partial t_1}(\rho u_i) + \sum_j \frac{\partial}{\partial r_{1j}} \Pi_{ij}^{(0)} = \frac{1}{\varepsilon} F_i \\
&\frac{\partial \rho}{\partial t_2} + \frac{\Delta t}{2} \cdot \frac{\partial^2 \rho}{\partial t_1^2} + \frac{\Delta t}{2} \sum_i \sum_j \frac{\partial}{\partial r_{1i}} \cdot \frac{\partial}{\partial r_{1j}} \Pi_{ij}^{(0)} + \Delta t \sum_i \frac{\partial}{\partial t_1} \cdot \frac{\partial}{\partial r_{1i}} (\rho u_i) = 0
\end{aligned}\right\} \tag{8.100}$$

$$\begin{aligned}
&\frac{\partial}{\partial t_2}(\rho u_i) + \frac{\Delta t}{2} \cdot \frac{\partial^2}{\partial t_1^2}(\rho u_i) + \sum_j \frac{\partial}{\partial r_{1j}} \Pi_{ij}^{(1)} + \frac{\Delta t}{2} \sum_j \sum_k \frac{\partial}{\partial r_{1j}} \cdot \frac{\partial}{\partial r_{1k}} S_{ijk}^{(0)} \\
&\quad + \Delta t \sum_j \frac{\partial}{\partial t_1} \cdot \frac{\partial}{\partial r_{1j}} \Pi_{ij}^{(0)} = 0
\end{aligned} \tag{8.101}$$

These expressions lead to the following basic equations:

$$\frac{\partial \rho}{\partial t} + \nabla \cdot (\rho \mathbf{u}) = 0 \tag{8.102}$$

$$\frac{\partial}{\partial t}(\rho u_i) + \sum_j \frac{\partial}{\partial r_j} \Pi_{ij} + \sum_j \frac{\Delta t}{2} \cdot \frac{\partial}{\partial r_j} \left\{ \varepsilon \frac{\partial}{\partial t_1} \Pi_{ij}^{(0)} + \sum_k \frac{\partial}{\partial r_k} S_{ijk}^{(0)} \right\} = F_i \tag{8.103}$$

Also, $f_\alpha^{(1)}$ is written as

$$\begin{aligned}
f_\alpha^{(1)} &= -t\tau\Delta t \left\{ \frac{\partial f_\alpha^{(0)}}{\partial t_1} + \sum_i \frac{\partial}{\partial r_{1i}} \left(c_{\alpha i} f_\alpha^{(0)} \right) \right\} + \frac{\tau}{\varepsilon} g_\alpha \\
&= -3 w_\alpha \Delta t \tau \frac{1}{c^2} \sum_k \sum_l \left(c_{\alpha k} c_{\alpha l} - \frac{1}{3} c^2 \delta_{kl} \right) \frac{\partial}{\partial r_{1l}} (\rho u_k)
\end{aligned} \tag{8.104}$$

Finally, using these expressions in a derivation procedure similar to that used previously, the Navier–Stokes equation is obtained as

$$\rho \left\{ \frac{\partial \mathbf{u}}{\partial t} + (\mathbf{u} \cdot \nabla) \mathbf{u} \right\} = -\nabla p + \mu^{\mathrm{LB}} \nabla^2 \mathbf{u} + \mathbf{F} \tag{8.105}$$

in which μ^{LB} has already been shown in Eq. (8.94). Eq. (8.105) clearly shows that g_α defined in Eq. (8.97) gives rise to the body force $\mathbf{F}$ appearing in the appropriate form in the Navier–Stokes equation.

8.4 Boundary Conditions

In simulations by the lattice Boltzmann method, it is very important to treat the boundary conditions in an appropriate manner at all the simulation boundary surfaces. Hence, there is a lot of current interest in developing more accurate boundary conditions, and papers addressing this problem have been appearing in academic journals. For example, if we consider a flow inside a tube or around an obstacle, or a suspension composed of solid particles, the treatment of the boundary condition at the wall or particle surface is very important for obtaining reliable solutions of the flow field. In this section, we first explain the historical bounce-back boundary condition, and then focus on several alternative boundary conditions that have a clearer physical and mathematical background.

8.4.1 Bounce-back Rule

We explain the historical bounce-back rule [35,36] using Figure 8.4. The lattice position of interest in a fluid is denoted by $\mathbf{r}_l$, its neighboring site inside the material by $\mathbf{r}_p$, and the point at the material surface on a straight line between these two points by $\mathbf{r}_w$, as shown in Figure 8.4. According to the BGK lattice Boltzmann method, the particle distribution $\tilde{f}_\alpha$ after the collision at time t becomes that at the neighboring site in the α-direction at time $(t + \Delta t)$. The bounce-back rule employs the following treatment of the collision at the material surface:

$$f_{\overline{\alpha}}(\mathbf{r}_l, t + \Delta t) = \tilde{f}_\alpha(\mathbf{r}_l, t) - 2\rho w_\alpha \frac{\mathbf{u}_w \cdot \mathbf{c}_\alpha}{c_s^2} \tag{8.106}$$

$$f_\alpha(\mathbf{r}_p, t + \Delta t) = \tilde{f}_{\overline{\alpha}}(\mathbf{r}_p, t) + 2\rho w_\alpha \frac{\mathbf{u}_w \cdot \mathbf{c}_\alpha}{c_s^2} \tag{8.107}$$

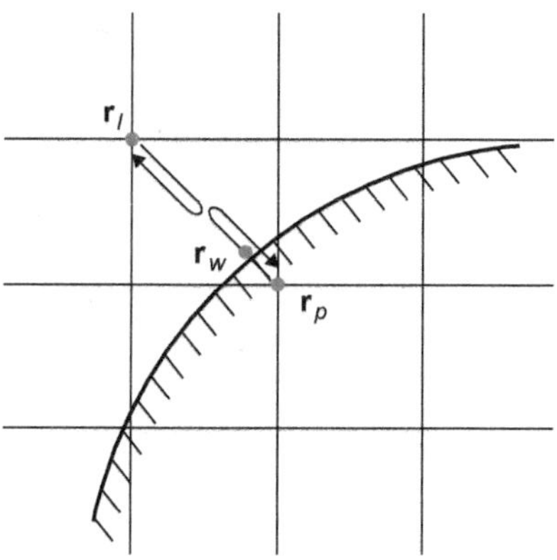

Figure 8.4 Bounce-back rule for the treatment at the material surface.

in which $\overline{\alpha}$ implies the opposite direction of α; α is the direction toward the object. Eq. (8.106) means that the fluid particles at $\mathbf{r}=\mathbf{r}_l$ move in the α-direction, collide with the obstacle at the middle point of the two lattice points, and return to the original lattice point during t and $t+\Delta t$. In this movement, if the solid surface moves in the α-direction, the number of the particles returning after the collision decreases, so that the second term on the right-hand side in Eq. (8.106) is necessary to make this modification. The following consideration makes clear that the bounce-back rule does not offer sufficient accuracy. That is, in the treatment of Eq. (8.106), the fluid particles starting from the point $\mathbf{r}_l$ do not collide with the actual solid surface $\mathbf{r}_w$, but at the exact middle point between $\mathbf{r}_l$ and $\mathbf{r}_p$, before returning to the original site. In other words, the collision procedure is conducted under the assumption that the surface of the obstacle is at the middle point between two neighboring lattice sites. In order to improve this approach, various boundary conditions have been developed. Research in this area is still a topic of interest.

Here we consider the validity of the second term on the right-hand side in Eq. (8.106). The consideration of Eqs. (8.82), (8.72), and (A1.11) leads to

$$\tilde{f}_\alpha(\mathbf{r}_l,t)=f_\alpha(\mathbf{r}_l,t)+\frac{1}{\tau}\{f_\alpha^{(0)}(\mathbf{r}_l,t)-f_\alpha(\mathbf{r}_l,t)\}=f_\alpha(\mathbf{r}_l,t)+\rho\frac{w_\alpha}{c_s^2}\left(\mathbf{c}_\alpha\cdot\frac{\partial\mathbf{u}(\mathbf{r}_l)}{\partial\mathbf{r}}\right)\cdot\mathbf{c}_\alpha\Delta t \tag{8.108}$$

in which a fluid has been assumed to be noncompressive. Substituting this equation into Eq. (8.106) yields

$$f_{\overline{\alpha}}(\mathbf{r}_l,t+\Delta t)=f_\alpha(\mathbf{r}_l,t)+\rho\frac{w_\alpha}{c_s^2}\left(\mathbf{c}_\alpha\cdot\frac{\partial\mathbf{u}(\mathbf{r}_l)}{\partial\mathbf{r}}\right)\cdot\mathbf{c}_\alpha\Delta t-2\rho w_\alpha\frac{\mathbf{u}_w\cdot\mathbf{c}_\alpha}{c_s^2} \tag{8.109}$$

If $f_\alpha(\mathbf{r}_l,t)$ in Eq. (8.109) is assumed to be not far from an equilibrium state, it is approximated from Eq. (8.68) as

$$f_\alpha(\mathbf{r}_l,t)\approx f_{\overline{\alpha}}(\mathbf{r}_l,t)+2\rho w_\alpha\frac{\mathbf{u}(\mathbf{r}_l)\cdot\mathbf{c}_\alpha}{c_s^2} \tag{8.110}$$

Substituting this equation into Eq. (8.109) gives rise to

$$\begin{aligned} f_{\overline{\alpha}}(\mathbf{r}_l,t+\Delta t)&=f_{\overline{\alpha}}(\mathbf{r}_l,t)+\frac{2\rho w_\alpha}{c_s^2}\left[\left\{\mathbf{u}(\mathbf{r}_l)+\frac{\partial\mathbf{u}(\mathbf{r}_l)}{\partial\mathbf{r}}\cdot\frac{\Delta t}{2}\mathbf{c}_\alpha\right\}-\mathbf{u}_w\right]\cdot\mathbf{c}_\alpha \\ &\approx f_{\overline{\alpha}}(\mathbf{r}_l,t)+\frac{2\rho w_\alpha}{c_s^2}\{\mathbf{u}(\mathbf{r}_l+\mathbf{c}_\alpha\Delta t/2)-\mathbf{u}_w\}\cdot\mathbf{c}_\alpha \end{aligned} \tag{8.111}$$

It is seen from Eq. (8.111) that $\mathbf{u}(\mathbf{r}_l+(1/2)\mathbf{c}_\alpha\Delta t)$ is equal to $\mathbf{u}_w$, if the medium point $(\mathbf{r}_l+(1/2)\mathbf{c}_\alpha\Delta t)$ is sufficiently near to the solid surface. Hence, we obtain the result $f_{\overline{\alpha}}(\mathbf{r}_l,t+\Delta t)=f_{\overline{\alpha}}(\mathbf{r}_l,t)$. That is, the particle distribution in the direction away from the solid surface approximates to the equilibrium distribution and is independent of time.

8.4.2 BFL Method

In this subsection we explain the BFL method [37]. In the bounce-back rule, the solid surface is regarded as being at the middle point of two lattice sites, and virtual fluid particles are reflected at this point. Hence, the exact position of the solid surface is not employed in the bounce-back method. The BFL method attempts to improve this drawback by taking into account the exact position of the solid surface in the procedure of the collision process between virtual fluid particles and the material. As shown in Figure 8.5, $\mathbf{r}_l$ is the point of interest in a fluid, $\mathbf{r}_p$ is the neighboring point inside the particle, $\mathbf{r}_w$ is the point at the solid surface on a line between these two points, and $\mathbf{r}_{l'}$ is the next neighboring point in the direction away from the solid surface. The exact position of the solid surface can be expressed using the quantity $\Delta_w = |\mathbf{r}_l - \mathbf{r}_w|/|\mathbf{r}_l - \mathbf{r}_p|$; although the lattice separation is defined to be Δx, we regarded Δx as unity in Sections 8.4.2 and 8.4.3 for simplicity's sake, because the final results derived here are unaffected even if Δx is not unity. The solid surface is at the position which is away from $\mathbf{r}_l$ in the direction toward $\mathbf{r}_p$ determined by Δ_w, as shown in Figure 8.5. The BFL method is based on an interpolation but, so as not to lose the accuracy of the interpolation, two different procedures are adopted for $\Delta_w \leq 1/2$ and $\Delta_w > 1/2$, although the concept of the treatment is the same for both cases. The fundamental concept is that fluid particles move, collide with the solid material, and return to a certain point during the time interval Δt. Since the unit lattice length is assumed in this analysis, the transportation distance is unity. In this collision process, the exact position of the solid surface is necessary. In the following text, the treatment for $\Delta_w \leq 1/2$ is discussed first.

As shown in Figure 8.5A, in the case of $\Delta_w \leq 1/2$, the particle distribution function $\tilde{f}_\alpha(\mathbf{r}_m, t)$ at $\mathbf{r}_m$ becomes that at $\mathbf{r}_l$, $f_{\bar{\alpha}}(\mathbf{r}_l, t + \Delta t)$, in which the point $\mathbf{r}_m$ is evaluated such that fluid particles move in the α-direction, collide with the solid surface, and arrive at the lattice point $\mathbf{r}_l$; the distance of travel for a fluid particle is just unity. The point $\mathbf{r}_m$ can be obtained straightforwardly as the position away from $\mathbf{r}_l$

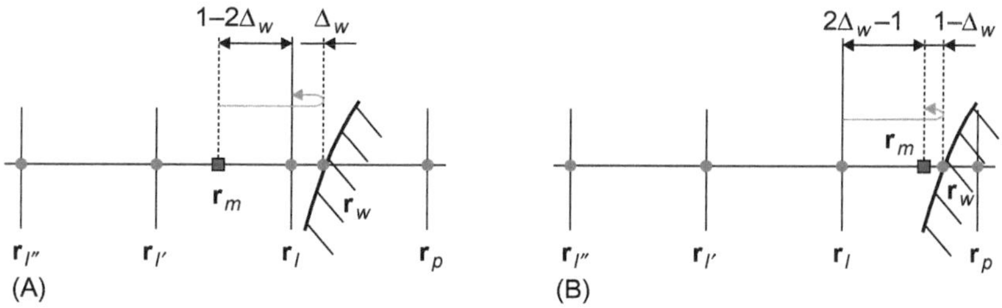

Figure 8.5 BFL method for the treatment at the material surface.

at the distance of $(1-2\Delta_w)$, shown in Figure 8.5A. Hence, the particle distribution function $\tilde{f}_\alpha(\mathbf{r}_m, t)$ is easily obtained from the quadratic extrapolation procedure as

$$\tilde{f}_\alpha(\mathbf{r}_m, t) = \Delta_w(1+2\Delta_w)\tilde{f}_\alpha(\mathbf{r}_l, t) + (1-4\Delta_w^2)\tilde{f}_\alpha(\mathbf{r}_{l'}, t) - \Delta_w(1-2\Delta_w)\tilde{f}_\alpha(\mathbf{r}_{l''}, t) \tag{8.112}$$

Since fluid particles collide with the solid surface, $f_{\bar{\alpha}}(\mathbf{r}_l, t+\Delta t)$ can finally be obtained as

$$f_{\bar{\alpha}}(\mathbf{r}_l, t+\Delta t) = \tilde{f}_\alpha(\mathbf{r}_m, t) - 2\rho w_\alpha \frac{\mathbf{u}_w \cdot \mathbf{c}_\alpha}{c_s^2} \tag{8.113}$$

Equation (8.112) has been obtained from the following formula of the quadratic interpolation method. If an arbitrary function $h(x)$ has values $h(x_1)$, $h(x_2)$, and $h(x_3)$ for $x = x_1$, x_2, and x_3, respectively, $h(x)$ at an arbitrary position x between x_1 and x_3 can be given from the quadratic interpolation as

$$h(x) = \frac{(x-x_2)(x-x_3)}{(x_1-x_2)(x_1-x_3)}h(x_1) + \frac{(x-x_1)(x-x_3)}{(x_2-x_1)(x_2-x_3)}h(x_2) + \frac{(x-x_1)(x-x_2)}{(x_3-x_1)(x_3-x_2)}h(x_3) \tag{8.114}$$

We now consider the treatment for $\Delta_w > 1/2$. As shown in Figure 8.5B, fluid particles leaving the lattice point $\mathbf{r}_l$ collide with the object, and return to the position $\mathbf{r}_m$ between $\mathbf{r}_l$ and $\mathbf{r}_w$. Hence, the following relationship is satisfied:

$$f_{\bar{\alpha}}(\mathbf{r}_m, t+\Delta t) = \tilde{f}_\alpha(\mathbf{r}_l, t) - 2\rho w_\alpha \frac{\mathbf{u}_w \cdot \mathbf{c}_\alpha}{c_s^2} \tag{8.115}$$

With this expression, the particle distribution function $f_{\bar{\alpha}}(\mathbf{r}_l, t+\Delta t)$ can be evaluated from the interpolation procedure based on a quadratic curve using values at $\mathbf{r}_m$, $\mathbf{r}_{l'}$, and $\mathbf{r}_{l''}$:

$$\begin{aligned} f_{\bar{\alpha}}(\mathbf{r}_l, t+\Delta t) = &\frac{1}{\Delta_w(2\Delta_w+1)}\tilde{f}_\alpha(\mathbf{r}_l, t) + \frac{2\Delta_w - 1}{\Delta_w}\tilde{f}_{\bar{\alpha}}(\mathbf{r}_l, t) \\ &+ \frac{1-2\Delta_w}{1+2\Delta_w}\tilde{f}_{\bar{\alpha}}(\mathbf{r}_{l'}, t) - \frac{1}{\Delta_w(2\Delta_w+1)}2\rho w_\alpha \frac{\mathbf{u}_w \cdot \mathbf{c}_\alpha}{c_s^2} \end{aligned} \tag{8.116}$$

We call the method using Eqs. (8.112), (8.113), (8.115), and (8.116) the "quadratic BFL method."

Instead of the quadratic interpolation procedure, the linear interpolation method is also applicable, and in this case each procedure for $\Delta_w \leq 1/2$ and $\Delta_w > 1/2$ can be expressed as

$\Delta_w \leq 1/2$:

$$f_{\bar{\alpha}}(\mathbf{r}_l, t+\Delta t) = (1-2\Delta_w)\tilde{f}_\alpha(\mathbf{r}_{l'}, t) + 2\Delta_w\tilde{f}_\alpha(\mathbf{r}_l, t) - 2\rho w_\alpha \frac{\mathbf{u}_w \cdot \mathbf{c}_\alpha}{c_s^2} \tag{8.117}$$

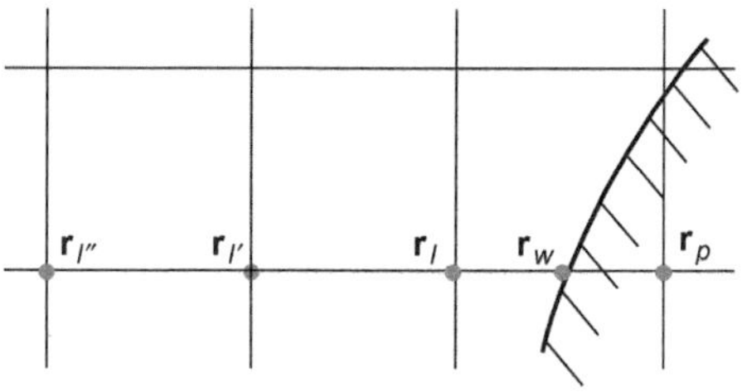

Figure 8.6 YMLS method.

$\Delta_w > 1/2$:

$$f_{\overline{\alpha}}(\mathbf{r}_l, t + \Delta t) = \frac{2\Delta_w - 1}{2\Delta_w}\tilde{f}_{\overline{\alpha}}(\mathbf{r}_l, t) + \frac{1}{2\Delta_w}\tilde{f}_{\alpha}(\mathbf{r}_l, t) - \frac{1}{2\Delta_w} 2\rho w_\alpha \frac{\mathbf{u}_w \cdot \mathbf{c}_\alpha}{c_s^2} \quad (8.118)$$

We call this scheme the "linear BFL method."

8.4.3 YMLS Method

In this subsection, we explain the YMLS method [34] using Figure 8.6. This method is also based on an interpolation scheme. The distribution function $\tilde{f}_\alpha(\mathbf{r}_m, t)$ at the position $\mathbf{r}_m$, from which fluid particles start and arrive at $\mathbf{r}_w$ after the time interval Δt, is used for the interpolation procedure. The particle distribution $f_{\overline{\alpha}}(\mathbf{r}_l, t + \Delta t)$ in the $\overline{\alpha}$-direction away from the material surface can be obtained from the interpolation using the distribution $\tilde{f}_\alpha(\mathbf{r}_m, t)$. As shown in Figure 8.6, with the notation of the point $\mathbf{r}_p$ inside the material, and the points $\mathbf{r}_l$, $\mathbf{r}_{l'}$, and $\mathbf{r}_{l''}$ on the fluid side away from the solid surface, the particle distribution functions at the solid surface in the α- and $\overline{\alpha}$-directions are written, respectively, as

$$f_\alpha(\mathbf{r}_w, t + \Delta t) = (1 - \Delta_w)\tilde{f}_\alpha(\mathbf{r}_{l'}, t) + \Delta_w \tilde{f}_\alpha(\mathbf{r}_l, t) \quad (8.119)$$

$$f_{\overline{\alpha}}(\mathbf{r}_w, t + \Delta t) = f_\alpha(\mathbf{r}_w, t + \Delta t) - 2\rho w_\alpha \frac{\mathbf{u}_w \cdot \mathbf{c}_\alpha}{c_s^2} \quad (8.120)$$

in which $\Delta_w = |\mathbf{r}_l - \mathbf{r}_w|/|\mathbf{r}_l - \mathbf{r}_p|$, as previously defined. Eq. (8.119) implies that $f_\alpha(\mathbf{r}_w, t + \Delta t)$ is obtained from the interpolation procedure using $\tilde{f}_\alpha(\mathbf{r}_{l'}, t)$ and $\tilde{f}_\alpha(\mathbf{r}_l, t)$, and Eq. (8.120) means that fluid particles are reflected at the solid surface. If $f_{\overline{\alpha}}(\mathbf{r}_w, t + \Delta t)$, $f_{\overline{\alpha}}(\mathbf{r}_{l'}, t + \Delta t)$, and $f_{\overline{\alpha}}(\mathbf{r}_{l''}, t + \Delta t)$ are used, then $f_{\overline{\alpha}}(\mathbf{r}_l, t + \Delta t)$ can be obtained from the quadratic interpolation procedure as

$$f_{\overline{\alpha}}(\mathbf{r}_l, t + \Delta t) = \frac{2}{(1 + \Delta_w)(2 + \Delta_w)} f_{\overline{\alpha}}(\mathbf{r}_w, t + \Delta t) + \frac{2\Delta_w}{1 + \Delta_w} f_{\overline{\alpha}}(\mathbf{r}_{l'}, t + \Delta t) - \frac{\Delta_w}{2 + \Delta_w} f_{\overline{\alpha}}(\mathbf{r}_{l''}, t + \Delta t) \quad (8.121)$$

This is known as the quadratic YMLS method.

The linear interpolation procedure yields the following linear YMLS method instead of Eq. (8.121):

$$f_{\overline{\alpha}}(\mathbf{r}_l, t + \Delta t) = \frac{\Delta_w}{1 + \Delta_w} f_{\overline{\alpha}}(\mathbf{r}_{l'}, t + \Delta t) + \frac{1}{1 + \Delta_w} f_{\overline{\alpha}}(\mathbf{r}_w, t + \Delta t) \tag{8.122}$$

in which $f_{\overline{\alpha}}(\mathbf{r}_w, t + \Delta t)$ is evaluated from Eq. (8.120). In this method, $f_{\overline{\alpha}}(\mathbf{r}_l, t + \Delta t)$ can be obtained from the interpolation scheme using $f_{\overline{\alpha}}(\mathbf{r}_{l'}, t + \Delta t)$ in the fluid region and $f_{\overline{\alpha}}(\mathbf{r}_w, t + \Delta t)$ at the solid surface. In the linear YMLS method, only two lattice points are used for the interpolation procedure, so that it may be suitable for particle dispersions in which a near-contact situation of dispersed particles frequently arises.

8.4.4 Other Methods

As in the MD or the MC simulations, the periodic boundary condition is applicable for the thermodynamic equilibrium case. For this case, the particle distribution function at the point $\mathbf{r}_{\text{out}}$ of the fluid particles outgoing from the simulation box, $f_\alpha(\mathbf{r}_{\text{out}}, t + \Delta t)$, is made to equal to that at the point $\mathbf{r}_{\text{in}}$ of the incoming fluid particles, $f_\alpha(\mathbf{r}_{\text{in}}, t + \Delta t)$.

Finally, we explain the extrapolation boundary condition, which is usually used for numerical simulations based on the finite difference or finite element method for a flow past an obstacle. The extrapolation boundary condition is also applicable to lattice Boltzmann simulations, for which the distribution functions at the points $\mathbf{r}_N$, $\mathbf{r}_{N-1}$, $\mathbf{r}_{N-2}$, which are taken from the boundary surface into the simulation region, are assumed to be in the linear relationship

$$f_{\overline{\alpha}}(\mathbf{r}_N, t + \Delta t) = 2 f_{\overline{\alpha}}(\mathbf{r}_{N-1}, t + \Delta t) - f_{\overline{\alpha}}(\mathbf{r}_{N-2}, t + \Delta t) \tag{8.123}$$

in which $\overline{\alpha}$ is in the direction leaving the outer boundary toward the simulation region. If the zero-gradient condition is applicable, then the differential away from the boundary is regarded as zero: that is, $f_{\overline{\alpha}}(\mathbf{r}_N, t + \Delta t) = f_{\overline{\alpha}}(\mathbf{r}_{N-1}, t + \Delta t)$. This boundary condition can be used for lattice points that are physically symmetric. If a simulation region is taken to be sufficiently large, the zero-gradient condition may be expected to give rise to results that are reasonably accurate.

8.5 Force and Torque Acting on Particles

In the case of a suspension composed of spherical or rod-like particles, the forces and torques acting on the suspended particles need to be evaluated in order to solve the particle motion and the flow field around the suspended particles simultaneously. The momentum change of the fluid particles that collide with the particle surface and are reflected during the time interval Δt is equal to the impulse acting on the particle. Hence, the force $\mathbf{F}_\alpha$ acting on the particle in the α-direction is obtained as

$$\mathbf{F}_\alpha(t+\Delta t/2) = \mathbf{c}_\alpha \left\{ f_{\bar{\alpha}}(\mathbf{r}_l^{(\text{int})}, t+\Delta t) + \tilde{f}_\alpha(\mathbf{r}_l^{(\text{int})}, t) \right\} \frac{\Delta V}{\Delta t} \tag{8.124}$$

in which ΔV is the volume occupied by one lattice site. Hence, the force $\mathbf{F}_p$ and torque $\mathbf{T}_p$ acting on the mass center of the particle are obtained from summing the contributions from the neighboring lattice sites of the particle as

$$\mathbf{F}_p(t+\Delta t/2) = \sum_{\text{all } \mathbf{r}_l^{(\text{int})}} \sum_\alpha \frac{\Delta V}{\Delta t} \left\{ f_{\bar{\alpha}}(\mathbf{r}_l^{(\text{int})}, t+\Delta t) + \tilde{f}_\alpha(\mathbf{r}_l^{(\text{int})}, t) \right\} \mathbf{c}_\alpha \tag{8.125}$$

$$\mathbf{T}_p(t+\Delta t/2) = \sum_{\text{all } \mathbf{r}_l^{(int)}} \sum_\alpha (\mathbf{r}_w - \mathbf{r}_c) \times \frac{\Delta V}{\Delta t} \left\{ f_{\bar{\alpha}}(\mathbf{r}_l^{(\text{int})}, t+\Delta t) + \tilde{f}_\alpha(\mathbf{r}_l^{(\text{int})}, t) \right\} \mathbf{c}_\alpha \tag{8.126}$$

in which $\mathbf{r}_c$ is the position vector of the particle mass center, and $\mathbf{r}_w$ is the position vector at the particle surface on a line drawn in the α-direction from the lattice point $\mathbf{r}_l^{(\text{int})}$ in the liquid region. The summation concerning α is only performed for the directions along which the above-mentioned line crosses the particle surface. Given the force and the torque from Eqs. (8.125) and (8.126), the translational and angular velocities $\mathbf{u}_p$ and $\boldsymbol{\Omega}_p$ of an arbitrary particle p with mass M_p and inertia moment I_p can be evaluated as

$$\left. \begin{aligned} \mathbf{u}_p(t+\Delta t) &= \mathbf{u}_p(t) + \frac{\Delta t}{M_p} \mathbf{F}_p(t+\Delta t/2) \\ \boldsymbol{\Omega}_p(t+\Delta t) &= \boldsymbol{\Omega}_p(t) + \frac{\Delta t}{I_p} \mathbf{T}_p(t+\Delta t/2) \end{aligned} \right\} \tag{8.127}$$

Note that here we have treated the case of the axisymmetric particle; therefore, only the inertia moment appears in the equation and not the inertia tensor.

8.6 Nondimensionalization

Finally, we show the usual nondimensionalization method used in lattice Boltzmann simulations. The following representative quantities are used in nondimensionalizing each quantity: Δx for distances, Δt for time, c ($=\Delta x/\Delta t$) for velocities, ρ_0 for the particle distribution, $\rho_0(\Delta x)^2\Delta x/(\Delta t)^2$ for forces, $(\Delta x)^2/\Delta t$ for kinematic viscosity, and $\rho_0(\Delta x/\Delta t)^2$ for pressures in the case of a two-dimensional system. Nondimensional equations are obtained by expressing a dimensional quantity as the product of the corresponding representative and nondimensional quantity

—for example, $f_\alpha = \rho_0 \times f_\alpha^*$—and substituting such quantities into the dimensional equations. Since the derivation procedure is quite straightforward, only the final results are shown in the following equations:

$$f_\alpha^*(\mathbf{r}^* + \mathbf{c}_\alpha^*, t^* + 1) = \tilde{f}_\alpha^*(\mathbf{r}^*, t^*) \tag{8.128}$$

$$\tilde{f}_\alpha^*(\mathbf{r}^*, t^*) = f_\alpha^*(\mathbf{r}^*, t^*) + \frac{1}{\tau}\{f_\alpha^{(0)*}(\mathbf{r}^*, t^*) - f_\alpha^*(\mathbf{r}^*, t^*)\} \tag{8.129}$$

in which

$$f_\alpha^{(0)*}(\mathbf{r}^*, t^*) = w_\alpha \rho^* \left\{ 1 + 3\mathbf{c}_\alpha^* \cdot \mathbf{u}^* + \frac{9}{2}(\mathbf{c}_\alpha^* \cdot \mathbf{u}^*)^2 - \frac{3}{2}u^{*2} \right\} \tag{8.130}$$

$$c^* = 1, \quad c_s^* = 1/\sqrt{3}, \quad \nu^* = (2\tau - 1)/6, \quad p^* = \rho^* c_s^{*2} \tag{8.131}$$

In these equations, the superscript * indicates the nondimensional quantities.

Appendix 1: Chapman–Enskog Expansion

In this appendix, we derive the important equations which are the starting expressions for deriving the Navier–Stokes equation, by means of the Chapman–Enskog expansion [38].

The basic equations required in the derivation are as follows:

$$\rho(\mathbf{r},t) = \sum_{\alpha} f_\alpha(\mathbf{r},t) \tag{A1.1}$$

$$\rho(\mathbf{r},t)\mathbf{u}(\mathbf{r},t) = \sum_{\alpha} \mathbf{c}_\alpha f_\alpha(\mathbf{r},t) \tag{A1.2}$$

$$\Pi_{ij} = \sum_{\alpha} c_{\alpha i} c_{\alpha j} f_\alpha(\mathbf{r},t) \tag{A1.3}$$

$$f_\alpha(\mathbf{r} + \mathbf{c}_\alpha \Delta t, t + \Delta t) = f_\alpha(\mathbf{r},t) + \Omega_\alpha(\mathbf{r},t) \tag{A1.4}$$

$$\Omega_\alpha(\mathbf{r},t) = \frac{1}{\tau}\{f_\alpha^{(0)}(\mathbf{r},t) - f_\alpha(\mathbf{r},t)\} \tag{A1.5}$$

Note that the following derivation is valid for both D2Q9 and D3Q19 models, except that α has to be taken as $\alpha = 0, 1, \ldots, 8$ and $\alpha = 0, 1, \ldots, 16$, respectively.

A Taylor series expansion of the left-hand side of Eq. (A1.4) gives rise to

$$\begin{aligned} &\Delta t \frac{\partial f_\alpha}{\partial t} + \frac{(\Delta t)^2}{2} \cdot \frac{\partial^2 f_\alpha}{\partial t^2} + \Delta t(\mathbf{c}_\alpha \cdot \nabla) f_\alpha + \frac{(\Delta t)^2}{2}(\mathbf{c}_\alpha \cdot \nabla)(\mathbf{c}_\alpha \cdot \nabla) f_\alpha \\ &\quad + (\Delta t)^2(\mathbf{c}_\alpha \cdot \nabla)\frac{\partial f_\alpha}{\partial t} = \frac{1}{\tau}(f_\alpha^{(0)} - f_\alpha) \end{aligned} \tag{A1.6}$$

The particle distribution function is expanded using the infinitesimal small quantity ε as

$$f_\alpha = f_\alpha^{(0)} + \varepsilon f_\alpha^{(1)} + \varepsilon^2 f_\alpha^{(2)} + \cdots \tag{A1.7}$$

By substituting Eq. (A1.7) into Eqs. (A1.1) and (A1.2), the following relationships are obtained:

$$\sum_{\alpha} f_{\alpha}^{(0)} = \rho, \quad \sum_{\alpha} \mathbf{c}_{\alpha} f_{\alpha}^{(0)} = \rho \mathbf{u} \tag{A1.8}$$

$$\sum_{\alpha} f_{\alpha}^{(n)} = 0, \quad \sum_{\alpha} \mathbf{c}_{\alpha} f_{\alpha}^{(n)} = 0 \quad \text{for } n = 1, 2, \ldots \tag{A1.9}$$

Next, we consider the Chapman–Enskog expansion. There are two characteristic times employed in characterizing fluid problems: T_1 relating to the fluid velocity, and T_2 relating to the viscous dissipation. It is generally satisfied that T_2 is much longer than T_1 (i.e., $T_2 \gg T_1$). Hence, if the infinitesimal quantities ε and Δt are taken as $\Delta t/T_1 = \mathrm{O}(\varepsilon)$, T_2 satisfies the relationship of $\Delta t/T_2 = \mathrm{O}(\varepsilon^2)$. On the other hand, if the representative distance is denoted by L_1, the distance Δx is generally taken such that $\Delta x/L_1 = \mathrm{O}(\varepsilon)$. With these assumptions, the time derivative is regarded as the summation of the time derivations due to the characteristics of T_1 and T_2. That is,

$$\frac{\partial}{\partial t} = \varepsilon \frac{\partial}{\partial t_1} + \varepsilon^2 \frac{\partial}{\partial t_2} \tag{A1.10}$$

Similarly, the position derivative $\partial/\partial \mathbf{r}$ is expressed, for the three-dimensional position $\mathbf{r} = (r_x, r_y, r_z)$, as

$$\frac{\partial}{\partial r_i} = \varepsilon \frac{\partial}{\partial r_{1i}} \quad (i = x, y, z) \tag{A1.11}$$

The expressions in Eqs. (A1.10) and (A1.11) imply that the original variables $(t, \mathbf{r})$ can be transformed into the new ones $(t_1, t_2, \mathbf{r}_1)$. In the usual approach, the differentiated quantities are used for comparing the magnitudes of all the terms in an equation. That is, the magnitudes of, for example, $\partial g_1/\partial t$ and $\partial g_2/\partial t$ are evaluated in such a way that $\partial g_1/\partial t = \mathrm{O}(\varepsilon)$ and $\partial g_2/\partial t = \mathrm{O}(\varepsilon^2)$, and they are compared with each other to neglect the smaller terms such as $\partial g_2/\partial t = \mathrm{O}(\varepsilon^2)$. In contrast, according to the Chapman–Enskog expansion, $\partial g_1/\partial t$ and $\partial g_2/\partial t$ are of the same order of magnitude but are moderated by the infinitesimal parameter ε and written as $\varepsilon\, \partial g_1/\partial t$ and $\varepsilon^2\, \partial g_2/\partial t$ in an equation.

We are now ready to proceed to the important equations in the derivation of the Navier–Stokes equation by means of the Chapman–Enskog expansion. The collision term in Eq. (A1.5) has the following characteristics:

$$\sum_{\alpha} \Omega_{\alpha} = 0, \quad \sum_{\alpha} \mathbf{c}_{\alpha} \Omega_{\alpha} = 0 \tag{A1.12}$$

From Eqs. (A1.6), (A1.10) and (A1.11),

$$\Delta t\left\{\varepsilon\frac{\partial f_\alpha}{\partial t_1}+\varepsilon^2\frac{\partial f_\alpha}{\partial t_2}\right\}+(\Delta t)^2\frac{\varepsilon^2}{2}\cdot\frac{\partial^2 f_\alpha}{\partial t_1^2}+\Delta t\varepsilon(\mathbf{c}_\alpha\cdot\nabla_1)f_\alpha$$
$$+(\Delta t)^2\frac{\varepsilon^2}{2}(\mathbf{c}_\alpha\cdot\nabla_1)(\mathbf{c}_\alpha\cdot\nabla_1)f_\alpha+(\Delta t)^2\varepsilon^2(\mathbf{c}_\alpha\cdot\nabla_1)\frac{\partial f_\alpha}{\partial t_1}+O(\varepsilon^3)=\Omega_\alpha \tag{A1.13}$$

By multiplying $\mathbf{c}_\alpha$ on both sides of this equation,

$$\Delta t\left\{\varepsilon\,\mathbf{c}_\alpha\frac{\partial f_\alpha}{\partial t_1}+\varepsilon^2\mathbf{c}_\alpha\frac{\partial f_\alpha}{\partial t_2}\right\}+(\Delta t)^2\frac{\varepsilon^2}{2}\mathbf{c}_\alpha\frac{\partial^2 f_\alpha}{\partial t_1^2}+\Delta t\varepsilon\mathbf{c}_\alpha(\mathbf{c}_\alpha\cdot\nabla_1)f_\alpha$$
$$+(\Delta t)^2\frac{\varepsilon^2}{2}\mathbf{c}_\alpha(\mathbf{c}_\alpha\cdot\nabla_1)(\mathbf{c}_\alpha\cdot\nabla_1)f_\alpha+(\Delta t)^2\varepsilon^2\mathbf{c}_\alpha(\mathbf{c}_\alpha\cdot\nabla_1)\frac{\partial f_\alpha}{\partial t_1}+O(\varepsilon^3)=\mathbf{c}_\alpha\Omega_\alpha \tag{A1.14}$$

Equation (A1.7) is substituted into Eq. (A1.13), the summation of α is conducted on the both sides, and the terms of the order ε are collected. Then, taking these collected terms equal to zero finally yields

$$\sum_\alpha\left\{\Delta t\frac{\partial f_\alpha^{(0)}}{\partial t_1}+\Delta t(\mathbf{c}_\alpha\cdot\nabla_1)f_\alpha^{(0)}\right\}=0 \tag{A1.15}$$

Similarly, from Eq. (A1.14),

$$\sum_\alpha\left\{\Delta t\frac{\partial}{\partial t_1}(c_{\alpha i}f_\alpha^{(0)})+\Delta t\sum_j c_{\alpha i}c_{\alpha j}\frac{\partial}{\partial r_{1j}}f_\alpha^{(0)}\right\}=0 \tag{A1.16}$$

With Eqs. (A1.8) and (A1.3), Eqs. (A1.15) and (A1.16) become

$$\frac{\partial}{\partial t_1}\rho+\nabla_1\cdot(\rho\mathbf{u})=0 \tag{A1.17}$$

$$\frac{\partial}{\partial t_1}(\rho u_i)+\sum_j\frac{\partial}{\partial r_{1j}}(\Pi_{ij}^{(0)})=0 \tag{A1.18}$$

in which $\Pi_{ij}^{(0)}=\sum_\alpha c_{\alpha i}c_{\alpha j}f_\alpha^{(0)}$.

Returning to the substitution of Eq. (A1.7) into Eq. (A1.13), but now taking the collected terms of the order ε^2 equal to zero, the following expression is derived:

$$\frac{\partial\rho}{\partial t_2}+\frac{\Delta t}{2}\cdot\frac{\partial^2\rho}{\partial t_1^2}+\frac{\Delta t}{2}\sum_i\sum_j\frac{\partial}{\partial r_{1i}}\cdot\frac{\partial}{\partial r_{1j}}\Pi_{ij}^{(0)}+\Delta t\sum_i\frac{\partial}{\partial t_1}\cdot\frac{\partial}{\partial r_{1i}}(\rho u_i)=0 \tag{A1.19}$$

Similarly, returning to the derivation of Eq. (A1.16) and taking the collected terms of the order ε^2 equal to zero yields

$$\frac{\partial}{\partial t_2}(\rho u_i) + \frac{\Delta t}{2}\cdot\frac{\partial^2}{\partial t_1^2}(\rho u_i) + \sum_j \frac{\partial}{\partial r_{1j}}\Pi_{ij}^{(1)} + \frac{\Delta t}{2}\sum_j\sum_k \frac{\partial}{\partial r_{1j}}\cdot\frac{\partial}{\partial r_{1k}} S_{ijk}^{(0)} + \Delta t\sum_j \frac{\partial}{\partial t_1}\cdot\frac{\partial}{\partial r_{1j}}\Pi_{ij}^{(0)} = 0 \tag{A1.20}$$

in which $\Pi_{ij}^{(1)} = \sum_\alpha c_{\alpha i}c_{\alpha j}f_\alpha^{(1)}$ and $S_{ijk}^{(0)} = \sum_\alpha c_{\alpha i}c_{\alpha j}c_{\alpha k}f_\alpha^{(0)}$.

Next, we reform Eqs. (A1.19) and (A1.20). Differentiating Eq. (A1.17) with respect to t_1 yields

$$\frac{\partial^2 \rho}{\partial t_1^2} = \frac{\partial}{\partial t_1}\{-\nabla_1\cdot(\rho\mathbf{u})\} = -\frac{\partial}{\partial t_1}\left\{\sum_i \frac{\partial}{\partial r_{1i}}(\rho u_i)\right\} \tag{A1.21}$$

With this result, Eq. (A1.19) is reformed and finally obtained as

$$\frac{\partial \rho}{\partial t_2} + \frac{\Delta t}{2}\sum_i \frac{\partial}{\partial r_{1i}}\left\{\frac{\partial}{\partial t_1}(\rho u_i) + \sum_j \frac{\partial}{\partial r_{1j}}\Pi_{ij}^{(0)}\right\} = 0 \tag{A1.22}$$

With Eq. (A1.18), Eq. (A1.22) reduces to

$$\frac{\partial \rho}{\partial t_2} = 0 \tag{A1.23}$$

Differentiating Eq. (A1.18) with respect to t_1 gives rise to

$$\frac{\partial^2}{\partial t_1^2}(\rho u_i) = \frac{\partial}{\partial t_1}\left\{-\sum_j \frac{\partial}{\partial r_{1j}}\Pi_{ij}^{(0)}\right\} = -\sum_j \frac{\partial}{\partial t_1}\cdot\frac{\partial}{\partial r_{1j}}\Pi_{ij}^{(0)} \tag{A1.24}$$

By substituting this result into Eq. (A1.20), the following equation is obtained:

$$\frac{\partial}{\partial t_2}(\rho u_i) + \sum_j \frac{\partial}{\partial r_{1j}}\left[\Pi_{ij}^{(1)} + \frac{\Delta t}{2}\left\{\frac{\partial}{\partial t_1}\Pi_{ij}^{(0)} + \sum_k \frac{\partial}{\partial r_{1k}} S_{ijk}^{(0)}\right\}\right] = 0 \tag{A1.25}$$

If Eqs. (A1.17) and (A1.23) are multiplied by ε and ε^2, respectively, summing each side of these equations, and taking into account Eq. (A1.10), the following mass conversation law is obtained:

$$\frac{\partial \rho}{\partial t} + \nabla\cdot(\rho\mathbf{u}) = 0 \tag{A1.26}$$

From a similar manipulation of Eqs. (A1.18) and (A1.25), the momentum conversation law is obtained as

$$\frac{\partial}{\partial t}(\rho u_i) + \sum_j \frac{\partial}{\partial r_j}\Pi_{ij} + \sum_j \frac{\Delta t}{2}\cdot\frac{\partial}{\partial r_j}\left\{\varepsilon\frac{\partial}{\partial t_1}\Pi_{ij}^{(0)} + \sum_k \frac{\partial}{\partial r_k}S_{ijk}^{(0)}\right\} = 0 \qquad \text{(A1.27)}$$

in which $\Pi_{ij} \approx \Pi_{ij}^{(0)} + \varepsilon\Pi_{ij}^{(1)}$.

Finally, we derive another important equation. The variable transformation of Eqs. (A1.10) and (A1.11) is conducted for Eq. (A1.6) to give

$$\Delta t\left\{\varepsilon\frac{\partial f_\alpha}{\partial t_1} + \varepsilon^2\frac{\partial f_\alpha}{\partial t_2}\right\} + \frac{(\Delta t)^2}{2}\varepsilon^2\frac{\partial^2 f_\alpha}{\partial t_1^2} + \Delta t\varepsilon(\mathbf{c}_\alpha\cdot\nabla_1)f_\alpha$$
$$+ \frac{(\Delta t)^2}{2}\varepsilon^2(\mathbf{c}_\alpha\cdot\nabla_1)(\mathbf{c}_\alpha\cdot\nabla_1)f_\alpha + (\Delta t)^2\varepsilon^2(\mathbf{c}_\alpha\cdot\nabla_1)\frac{\partial f_\alpha}{\partial t_1} = \frac{1}{\tau}(f_\alpha^{(0)} - f_\alpha)$$
$$\text{(A1.28)}$$

Substituting Eq. (A1.7) into this equation, collecting the terms of the order ε, and taking these collected terms equal to zero then yields

$$-\frac{1}{\tau\Delta t}f_\alpha^{(1)} = \frac{\partial f_\alpha^{(0)}}{\partial t_1} + \sum_i \frac{\partial}{\partial r_{1i}}(c_{\alpha i}f_\alpha^{(0)}) \qquad \text{(A1.29)}$$

Since $f_\alpha^{(0)}$ can be regarded as a function of the macroscopic quantities ρ and ρu_i, $\partial f_\alpha^{(0)}/\partial t$ can be reformed using Eqs. (A1.17) and (A1.18) as

$$\begin{aligned}\frac{\partial f_\alpha^{(0)}}{\partial t_1} &= \frac{\partial f_\alpha^{(0)}}{\partial \rho}\cdot\frac{\partial \rho}{\partial t_1} + \sum_i \frac{\partial f_\alpha^{(0)}}{\partial(\rho u_i)}\cdot\frac{\partial(\rho u_i)}{\partial t_1}\\ &= -\frac{\partial f_\alpha^{(0)}}{\partial \rho}\cdot\frac{\partial}{\partial \mathbf{r}_1}\cdot(\rho\mathbf{u}) - \sum_i\sum_j \frac{\partial f_\alpha^{(0)}}{\partial(\rho u_i)}\cdot\frac{\partial}{\partial r_{1j}}\Pi_{ij}^{(0)}\end{aligned} \qquad \text{(A1.30)}$$

Substituting this equation into Eq. (A1.29) yields the required equation:

$$-\frac{1}{\tau\Delta t}f_\alpha^{(1)} = -\frac{\partial f_\alpha^{(0)}}{\partial \rho}\cdot\frac{\partial}{\partial \mathbf{r}_1}\cdot(\rho\mathbf{u}) - \sum_i\sum_j \frac{\partial f_\alpha^{(0)}}{\partial(\rho u_i)}\cdot\frac{\partial}{\partial r_{1j}}\Pi_{ij}^{(0)} + \sum_i \frac{\partial}{\partial r_{1i}}(c_{\alpha i}f_\alpha^{(0)})$$
$$\text{(A1.31)}$$

Equations (A1.27) and (A1.31) are the basic equations for deriving the important relationships in Chapter 8.

Appendix 2: Generation of Random Numbers According to Gaussian Distribution

In order to set the initial velocities of particles in MD simulations, or to generate random displacements in BD and DPD simulations, it is necessary to generate random numbers according to a particular probability distribution. The probability distributions of interest here are the Gaussian distribution (also known as the normal distribution), and the Maxwell–Boltzmann distribution (or Maxwellian distribution). For example, since the velocity of particles theoretically has the Maxwellian velocity distribution for thermodynamic equilibrium, as explained in Section 2.2, the initial velocity of particles in simulations must have such a velocity distribution. We show the method of setting the initial velocity of particles according to the Maxwellian distribution in the following paragraphs.

We assume that the stochastic variable x, such as the particle velocity or a random displacement, obeys the following normal distribution $\rho(x)$:

$$\rho(x) = \frac{1}{(2\pi)^{1/2}\sigma} \exp\left\{-\frac{(x-\overline{x})^2}{2\sigma^2}\right\} \tag{A2.1}$$

in which σ^2 is the variance and $\overline{x}$ is the average of the stochastic variable x. In order to generate the stochastic variable x according to this normal distribution, the following equations are used together with a uniform random number sequence ranging from zero to unity:

$$x = \overline{x} + (-2\sigma^2 \ln R_1)^{1/2} \cos(2\pi R_2) \quad \text{or} \quad x = \overline{x} + (-2\sigma^2 \ln R_1)^{1/2} \sin(2\pi R_2) \tag{A2.2}$$

According to either equation of Eq. (A2.2), the required number of values of the stochastic variable are generated using a series of random numbers, such as R_1 and R_2, taken from a uniform random number sequence. In this way, the initial velocities of particles and random displacements can be assigned. The technique in Eq. (A2.2) is called the Box–Müller method [26].

For generating a uniform random number sequence, there is an arithmetic method and a machine-generated method; the former is shown in the last subroutine of the sample simulation program in Section 3.1.6. The arithmetic method is reproducible, and the same random number sequence can be obtained at any time in the simulations. In contrast, the machine-generated method is generally not a reproducible sequence, and a different sequence of random numbers is generated each time a simulation is run.

For the case of the Maxwellian velocity distribution, the velocity components of particle i can be assigned using the random numbers $R_1, R_2, \ldots, R_6$ taken from a uniform random number sequence as

$$\left.\begin{aligned} v_{ix} &= \{-2(kT/m)\ln R_1\}^{1/2}\cos(2\pi R_2) \\ v_{iy} &= \{-2(kT/m)\ln R_3\}^{1/2}\cos(2\pi R_4) \\ v_{iz} &= \{-2(kT/m)\ln R_5\}^{1/2}\cos(2\pi R_6) \end{aligned}\right\} \tag{A2.3}$$

In this way, all the initial velocity components can be assigned using random numbers.

Appendix 3: Outline of Basic Grammars of FORTRAN and C Languages

We here do not aim to explain the entire grammar of the programming languages; indeed, there is not sufficient space to do so. In each section of programming language, the main structure of a program is first explained in order to understand the logical framework of a program. Then, such important grammar as control statements is explained. Finally, several points of interest that may be outside of the main body of a program will be addressed. This will be followed by a short sample program that demonstrates the essence of a research simulation program and explains the grammar used in the program in detail. This approach is most effective, because the grammar is explained in relation to the logical flow of a simulation program. The skill to develop a simulation program has a strong relationship with the ability for embodying a logical flow using a programming language.

A3.1 FORTRAN Language

The general structure of a program written in the FORTRAN language is composed of a main program together with subroutine subprograms or function subprograms, as shown below.

```
123456789...                                                    ...72

      IMPLICIT REAL*8 (A-H, O-Z), INTEGER (I-N)
      PARAMETER( NN=8)
      REAL*8  RX(NN), RY(NN)
      REAK*8  VX(NN), VY(NN)
      INTEGER N

          Description of calculation procedures

      STOP
      END
```

Main program

```
      SUBROUTINE INIPOSIT(N,H)              Subroutine subprogram

          Description of calculation procedures

      RETURN
      END
      SUBROUTINE INIVEL(N)                  Subroutine subprogram

          Description of calculation procedures

      RETURN
      END
      REAL*8 FUNCTION PRESSURE(H)           Function subprogram

          Description of calculation procedures

      RETURN
      END
```

A main program first needs to be constructed, and then subroutine or/and function subprograms necessarily follow the main program. The main program begins with the definition of the variables and finishes with the STOP and END statements that are placed at the end to halt the execution of the program. A subroutine or function subprogram begins with a SUBROUTINE (name of a subroutine) or (precision) FUNCTION (name of a function) statement, respectively, and finishes with the RETURN and END statements that signal the return to the task of the main program. A main program must be written in such a way that the logical flow is clear, and calculations that disturb this logical flow should be treated in subroutine or function subprograms. In other words, when a program is constructed in such a way that a reader is able to grasp the logical flow in a straightforward manner, it becomes more than a hobby program—it becomes a common useful tool. This is an important consideration for developing a simulation program with contributions from and used by different persons in a successive research project.

In a subroutine subprogram, routine calculations are carried out. The calculation task moves from a main program to a subroutine subprogram by calling the name of the subroutine (the CALL statement) and returns to the main program on meeting the RETURN statement in the subroutine. A function subprogram is quite similar to a subroutine subprogram in that routine calculations are carried out in an area (the subprogram) aside from the main program. The difference between the two is that in a function subprogram the name of the function subprogram itself assumes a calculated value, and this value is passed to the main program by simply referring to the name of the function subprogram in the main program. In other words, the name of a function subprogram is treated as a variable in a main program: the calculation task moves to a function subprogram at the time of meeting its name, and returns to the main program with a value calculated there on meeting the RETURN statement. Hence, a CALL statement is unnecessary in order to move to a subprogram area. These are the main points of the program structure and flow of the calculation procedures. There is an important point concerning the data transfer between a main program and a subprogram. In the FORTRAN language, information regarding the value of variables cannot be transferred between a main program and a subprogram unless definite descriptions are written for that purpose. There is a significant difference between the FORTRAN and the C language in this respect. We explain the method of transferring data between a main program and a subprogram in detail later.

As shown in the preceding example, main sentences generally have to be written between the 7th and 72nd columns in a FORTRAN77 program. The first column is used for defining whether or not that line is a comment line (that does not influence calculations) by employing a C character or a blank; the sixth column is for defining whether or not the line is regarded as a continuation line from the previous line by the "&" character or blank; and the area between the 2nd and 5th columns is used for writing figures (or labels) of the end statements, such as the CONTINUE statement or of an indication of the destination of the GOTO statement. Various examples of this type of use can be seen in the sample simulation programs, and therefore we omit such explanations here.

```
IF(X.GT.0.D0)A=B+10.D0
```

```
IF(X.GT.0.D0)THEN
 ...
END IF
```

- This is the simplest IF statement, and THEN is unnecessary in this case.
- Execution only for $X>0$.

```
IF(X.GT.0.D0)THEN
 ...
ELSE
 ...
END IF
```

- One of separate procedures is chosen for $X>0$ or $X\leq 0$.

```
IF(X.GT.0.D0)THEN
 ...
ELSE IF(X.LE.-10.D0) THEN
 ...
ELSE
 ...
END IF
```

- One of three separate procedures is chosen for $X>0$ or $X\leq -10$ or the other cases.

```
IF(X.GT.0.D0)THEN
 ...
ELSE IF(X.LE.-10.D0) THEN
 ...
ELSE IF(X.EQ.-5.D0) THEN
 ...
END IF
```

- One of three separate procedures is chosen for $X>0$ or $X\leq -10$ or $X=-5$.

We now explain the IF and DO statements, which are perhaps the most important control statements for developing a calculation program. The IF statement is a control statement to select a calculation procedure by assessing the condition. The DO statement is a control statement to repeat a certain procedure a prescribed number of times. Typical examples of the IF statement are shown above. The IF statement implies the execution of a certain procedure if a condition is satisfied; another procedure is carried out if it is not satisfied. In the above examples, the first IF statement is the simplest and the following examples become more complex. Several IF statements can be combined to make a complex assessment, and such examples may be found in the sample simulation programs. In the IF statement, LT and LE imply $<$ and $\leq$, respectively; GT and GE imply $>$ and $\geq$, respectively; and EQ and NE imply $=$ and $\neq$, respectively. The statement for repeating procedures is the DO statement. Several representative examples of the DO statement are shown in the following.

```
      DO 20 I=1,N
       ...
   20 CONTINUE
```

```
      DO 30 I=N,1,-2
       ...
   30 CONTINUE
```

```
      DO 90 I=-N,N+1,5
       ...
   90 CONTINUE
```

- The procedure starts at I=1, then is conducted at I=2 and continued until I=N.
- The procedure starts at I=N, then is conducted at I=N-2 and continued at I=N-4, N-6, ...
- The procedure starts at I=-N, then is conducted at I=-N+5, I=-N+10, ..., until I becomes over N+1.

The DO statement implies that the procedure written between DO and CONTINUE is executed until the index arrives at the required end value. In the above example, I is the index and N is the end value of the loop. In the first example, the index I changes in the sequence I = 1, 2,. . ., N. In the second example, the index I changes in the sequence I = N, N − 2, N − 4,. . .; if N is even, the procedures are repeated until N = 2, and if N is odd, they are repeated until N = 1. The last example shows that a negative value, − N, is possible as a starting value of the index I. Either specific numbers or variables are possible for the starting and ending values and the increment interval value of the DO loop statement. Be aware that although REAL variables can be used as an index of the DO loop, INTEGER variables are desirable in order to remove any ambiguity in relation to the assessment concerning the termination of the DO loop. In order to move out of the DO loop at any time before the designed end, the GOTO statement employed with the previous IF statement may be used. A final point to be noted relating to the DO loop is that in the first above example, the index I does not have the figure N but (N + 1) for the end of the procedure; thus, care should be taken in using the variable I in the next task. Using variables in this way should be avoided in order to prevent causing an unexpected error.

Next, we explain several types of grammar that are relatively difficult to understand when learning the FORTRAN language. First, we explain how to transfer the values of variables between the main program and a subprogram. In FORTRAN, there are two methods for the data transfer: (1) the values of variables are transferred to a subprogram through the arguments of the subprogram, and (2) the variables to be transferred between a main program and a subprogram are declared with the COMMON statements so that they can be accessed from both the main program and the subprograms. An example of the former method is as follows:

```
CALL INIVEL(N,H,T)
...
SUBROUTINE INIVEL(N,H,T)
...
```

In this case, the values of the variables N and H are transferred from a main program to a subprogram, and the procedure returns to the main program with a value of T, which was calculated in the subprogram. A big difference between FORTRAN and the C language is that in the former language new values of N and H, which were changed in the subprogram, are reflected in the main program, but in the latter language this never arises unless a specific direction is given to do so. This will be explained in detail later in the grammar of the C language.

The second method for the data transfer is to use the COMMON statement: the variables declared in the COMMON statements can be accessed from both the main program and all the subprograms without any need for specific statements for the data transfer. An example is as follows:

```
PARAMETER (NN=100)
COMMON /BLOCK1/ N, RX, RY
REAL*8 RX(NN), RY(NN), H
INTEGER N
...
CALL INIVEL(H)
...
STOP
END
...
SUBROUTINE INIVEL(H)
PARAMETER (NN=100)
COMMON /BLOCK1/ N, RX, RY
REAL*8 RX(NN), RY(NN), H
INTEGER N
```

```
...
RETURN
END
```

In a main program, the variables, which are used in subprograms, can be defined in the COMMON statements before the definition of other standard variables. By defining the same variables in the COMMON statements in a subprogram, the values saved on the variables can be referred to; also, new values may be saved on these variables. In the above example, the values of N, RX(*), and RY(*) are transferred using the COMMON statement, and a value of H is transferred as an argument. Note that the names of the variables in the COMMON statements are not necessarily the same, but we recommend that the beginner use the same names until they obtain a more complete understanding of the language.

Another feature that the beginner may find difficult is the WRITE and FORMAT statements. These statements are used for outputting results to a data file and have no relation to the execution of the calculations. The following example is for outputting the data for the purpose of confirming either the final or intervening results of the calculation:

```
      I=3
      XI=5.D0
      YI=2.D0
      PRESS=XI*YI
      WRITE(NP,20) I, XI, YI,PRESS
   20 FORMAT('I=',I3,3X,'XI=',F7.3,2X,'YI=',F7.3,2X,
     &           'PRESSURE AT (XI,YI)=', F10.3)
```

The result of the output from this FORMAT statement is as follows:

```
I=  3 XI=  5.000  YI=  2.000  PRESSURE AT (XI,YI)=    10.000
```

The above example is a part of the program for outputting the data of the variable PRESS, which is obtained by multiplying XI by YI. For the case of NP = 6, the results are output to the display of the computer, and if the OPEN statement relates the device number (or device unit number) NP with a data file, the result is output to the data file. For example, if "OPEN (11, FILE = 'faa1.data', STATUS = 'UNKNOWN')" is declared and NP is set as NP = 11, the data is output to the file faa1.data. Since the results shown on a display can be seen only once, data is usually output to a data file. Inside the parentheses of the FORMAT statement, I3 means that the output is an integer and is output up to three spaces (columns) to the right of the space, F7.3 means that the output is a real number and is output using 7 spaces (columns), in which the number is rounded to three decimal places and is written to the right of the space; 3X means that three blank spaces are to be inserted. The reader can see many examples of FORMAT statements in the sample simulation programs in Chapters 3–7.

A long run of the execution of a simulation program is sometimes divided into several short runs. For this case, the intervening results are output to a data file, and the following run is carried out to continue from the previous run using the data saved in the file. This data may also be used for visualizing a particle configuration in a form such as a snapshot. To do so, only numerical data is suitable for the output to a data file—that is, without the specification of the names of the variables. A typical example is as follows:

```
          ...
      WRITE(NP,50) N
   50 FORMAT( I8 )
      WRITE(NP,55) (RX(I),I=1,N), (RY(I),I=1,N)
   55 FORMAT( (5E16.8) )
```

In the above example, the data saved in the array-type variables RX(*) and RY(*) are output using a simple specification without using the DO statement. The specification of (5E16.8) in the FORMAT statement means that five data are output in one line. The outer bracket ( ) is used for the repetition of the output specification 5E16.8, which means that the output data is real and is output using 16 spaces (columns) in which the data is written to the right of the space with 8 decimal places. In order to continue a separate successive run using the data which is output in the above example, we need to use the following READ statement for reading the necessary data:

```
      READ(NP1,60) N
   60 FORMAT( I8 )
      READ(NP1,65) (RX(I),I=1,N), (RY(I),I=1,N)
   65 FORMAT( (5E16.8) )
```

An important point is that the same FORMAT statement must be used for the WRITE and READ statements; otherwise, the exact numerical values cannot be read by the READ statement.

Finally, to assist the reader in understanding the grammar in more detail, we have added explanatory remarks to the following sample program, which was made by compressing a full simulation program.

• The line number is added for convenience and the first column starts from the position of C character. The C in the first column implies that the line is just a comment.

```
C*****************************
C*          diffuse_sample.f
C*
C*    MOLECULAR
C*
C*
C**************
```

• `IMPLICIT` is the implicit data type declaration. In this case, the variables with their name starting with one of `A~H` and `O~Z` are regarded to be a double-precision real, and similarly those with one of `I~N` are regarded to be an integer.

```
      IMPLICIT REAL*8 (A-H,O-Z), INTEGER (I-N)
C
      COMMON /BLOCK1/ RX0 , RY0 , RX , RY
      COMMON /BLOCK2/ FX  , FY
C
      PARAMETER( NN=80, NRANMX=50000 )
      PARAMETER( PI=3.141592653589793D0 )
C
      REAL*8  RX0(NN), RY0(NN), RX(NN)  , RY(N
      REAL*8  FX(NN)  , FY(NN)  , VELX(NN), VEL
      REAL*8  H , RC , L , T , K , NDENS
C
      REAL
      INTEGER
C
      REAL*8
      INTEGER N, NA, NB
```

• The variable defined in the `COMMON` statement can be accessed from everywhere without transferring them into subprograms as arguments. In the case of array variables, the dimension must be defined in the data type statement.

• The `PARAMETER` statement is frequently used for defining the variables used for specifying the dimensions of array variables; the change of these values in `PARAMETER` enables us to change the dimensions of the related array variables.

• `REAL*8`, `REAL`, and `INTEGER` are the data type declaration statements for double-precision reals, single-precision reals, and integers, respectively. Although a computer can usually treat integers only between ± several ten billions, the `INTEGER*8` statement enables one to use a much wider range of integers. Double-precision reals may be sufficient in scientific computations, but quad-precision reals are appropriate in some cases.

```
C
            OPEN( 9,FILE= '@aaa1.data',STATUS='UNKNOWN')
            OPEN(21,FILE='aaa001.data',STATUS='UNKNOWN')
            OPEN(22,FILE='aaa011.data',STATUS='UNKNOWN')
```

• `OPEN` statements can relate data files to the input/output devices; `CLOSE` statements must be used together. The number 5 is the keyboard, 6 is the display, and other numbers are used for data files (numbers larger than 8 may be desirable). `OLD` in the `STATUS` statement implies an already-existed file.

```
                                              NP=9
C                            ----- PARAMETER (1) -----
      T   = 5.0D0
      K   = 10.D0
      NA  = 20
      L   = DSQRT( DBLE(N)/NDENS )
      HSQ = H*H
C
      IX     = 0
      CALL RANCAL( NRANMX, IX, RAN )
      NRAN   = 1
C
C     ---------------------------
C     ---------------    INITIAL
C     ---------------------------
C
C
      CALL POSITR1( N, NA, H, K )
C
C                          --- PRINT OUT CONSTANTS ---
      WRITE(NP,5) T , K , NDENS ,
C                          --- PRIN
C     ---------------------------
C     ---------------    START OF
```

• Double-precision reals are expressed for example `5.2D0` or `0.052D2`; single-precision reals are expressed such as `5.2` or `0.052E2`. `DSQRT` means the square root, and * means the multiplication.

• The subroutine `POSITR1` is called by the `CALL` statement. The variables necessary in the subroutine are passed as arguments (`N, NA, H, K`); the description of these variables has to be described in this order in the subroutine subprogram.

• The data are written out in the format expressed in the `5 FORMAT` statement; these statements should be placed before the STOP statement.

```
C     --------------------------
C
      DO 100 NTIME=1, NTIMEMX
C
        DO 50 I=1,N
C
          IF ( I .EQ. NA+1 ) CC1
          RXI    = 2.D0*RX(I)
          RX0(I) = RX(I)
          RX(I)  = RXI
C
   50   CONTINUE
C                                  --- PRINT OUT DATA ---
        IF ( MOD(NTIME,NPRINT) .EQ. 0 ) THEN
          TIME = H*DBLE(NTIME)
          CALL PRINTOUT( N, NA,
        END IF
C
          DO 60 I =1,N
            IF( I .LE. NA ) THEN
              R = 1.D0
            ELSE
              R = 1.5D0
            END IF
            WRITE(NOPT,58) I, R, R
   60     CONTINUE
C
  100 CONTINUE
C
C     ------------------------------------
C     ------------------  END OF MAIN LOOP
C     ------------------------------------
      CLOSE(NP, STATUS='KEEP')
C
C     ---------------------- FORMAT ------
    5 FORMAT(/1H ,'-----------------------------------------'
     &       /1H ,'        MOLECULAR DYNAMICS SIMULATION      '
     &       /1H ,'FOR TWO-DIMENSI
     &       /1H ,'---------------
     &       /1H ,'TEMPERATURE=',F
     &            'NDENS=',F6.3
     &       /1H ,'NUMBER OF MOLEC
     &       /1H ,'NUMBER OF MOLEC
     &       /1H ,'MAGNITUDE OF CA
     &            F8.5 ,2X, 'CUTOFF
   56 FORMAT( 3I6, 2E13.8 )
   58 FORMAT( I5, F8.3 , 2E26.18 )
C
                                                        STOP
                                                        END
C*****************************************************************
C*********************      SUBROUTINE    *********************
C*******************************
C
C**** SUB POSITR1 ****
      SUBROUTINE POSITR1( N, NA, H, K )
C
      IMPLICIT REAL*8 (A-H,O-Z), INTEGER (I
C
      COMMON /BLOCK1/ RX0 , RY0, RX, RY
      COMMON /BLOCK2/ FX , FY
      COMMON /BLOCK3/ VELX, VELY
C
      PARAMETER( NN=80 )
C
      REAL*8  RX0(NN), RY0(NN), RX(NN)  , R
      REAL*8  FX(NN) , FY(NN) , VELX(NN), V
      REAL*8  H , K
      REAL*8  HSQ2, CC0, CC1
```

• The DO loop implies the iteration calculation; the statements between DO and CONTINUE are repeatedly carried out. The procedure starts at NTIME=1, and then is conducted at NTIME=2, 3,⋯, until I=NTIMEMX. DO loops are possible inside the DO loop.

• MOD(NTIME, NPRINT) returns the remainder after NTIME is divided by NPRINT. DMOD is used for such an operator of double-precision reals. As in this example, operators have a slightly different name depending on the data type of variables.

• DBLE(*) is used for transforming an integer into a double-precision real. For developing a universal program, it is desirable that the data types be the same between the left and right-hand sides in the equation. INT(*) is used for transforming a double-precision real into an integer.

•One of the procedures is chosen after assessing the IF statement. In this example, R=1.D0 if I≤NA, and R=1.5D0 if I>NA.

• The data file opened by the OPEN statement must be closed using the CLOSE statement. NP is the device number (name) of the I/O device, which is used to open the data file. KEEP is used in the STATUS statement in almost all cases.

• The collection of FORMAT statements before the STOP statement makes the logical structure of calculations clearer.
• / on the first line means the insertion of one blank line; / in the later lines mean starting a new line. "1H ," means one blank space indent in each line. "56 FORMAT" and "58 FORMAT" are for writing out only numerical data (or figures).

• The variables are described in the same order in which they have been written in CALL POSITR1.

• IMPLICIT, PARAMETER, and REAL*8 statements are described in the same way as in the main program. The subroutine can access the variables in the COMMON statements, as well as the arguments of N, NA, H, and K; note that the change in these variables in the subroutine is reflected in the main program. The other variables are valid only in this subroutine, and never affect the main program.

```
      INTEGER NA  , N
C
      HSQ2  = H*H/2.D0
      CC0   = 1.D0/K
      CC1   = 1.D0
C
      DO 10 I=1,N
        IF( I .EQ. NA+1 ) CC1 = CC0
        RX(I) = RX0(I) + H*VELX(I) + HSQ2*FX(I)*CC1
        RY(I) = RY0(I) + H*VELY(I) + HSQ2*FY(I)*CC1
   10 CONTINUE
                                                    RETURN
                                                    END
```

The use of the RETURN statement arbitrary times is possible in the subroutine. The END statement is necessary for specifying the end of the descriptions of the subroutine.

A3.2 C Language

We will explain the grammar of the C language in a way similar to our discussion of the FORTRAN language. The main structure of a program written in the C language is made up of the function **main** and a set of functions that correspond to subprograms in FORTRAN. The C language has considerable flexibility in writing a program in comparison to FORTRAN. However, the logical structure concerning the arrangement of **main** and functions has similarities to FORTRAN, and therefore it may be beneficial for the reader to write a simulation program in a similar structure to one in FORTRAN.

We show a typical structure of a C program in the following. Note that there is no requirement that the statements be written between 7th and 72nd columns in the C language.

```
#include <stdio.h>
#include <math.h>
#define PI    3.1415926535
#define NN    20
    double  RX[NN], RY[NN] ;
main()
  {
    double rxi, ryi ;
    int    n, i, j ;

        Description of calculation procedures

  }
```

The function **main**

```
     setiniposit(n,temp)
       double temp ;
       int     n ;
       {
        double rxi, ryi ;
          int i, j ;

        Description of calculation procedures

  }
```

Function (setiniposit)

```
setinivel(n,press)
  double press ;
  int     n ;
  {
   double vxi, vyi ;
   int i, j ;

    [Description of calculation procedures]

}
```

Function (setinivel)

The function **main** is placed first, and other functions, corresponding to subprograms in FORTRAN, follow **main**. The main body of the statements in each function begins with the notation { and ends with the notation }. Since there is no limit on the number of characters in one line, the notation **;** is used for terminating a line of statement—that is, it means the end of the line. Except for special statements, every line must end with such a notation **;**. Because mathematical functions such as **sin** and **sqrt** are necessarily used in scientific calculations, the statement of **#include <math.h>** needs to be declared in the first description area. Also, the statement of **#include <stdio.h>** is indispensable to any program to facilitate the input or output of data to a display or data file, and for reading data from a keyboard or data file. The statements of **#define PI 3.14**... and **#define NN 20** correspond to the PARAMETER statement in FORTRAN. A statement beginning with the notation **#** is a preprocessor directive, which is a command to the compiler's preprocessor that treats instructions before the compilation procedure starts. The preprocessor directive of **#define NN 20** implies that the value of 20 is assigned to the variable **NN**. The next statement, corresponding to the COMMON statement in FORTRAN, is the declaration of the array-type variables RX[NN] and RY[NN] being used as global variables (which can be used in the other function programs with no further definitions). The C language typically uses lower-case characters, but it may be best that the names of global variables are declared using upper case, so a programmer can be more aware of treating the global variables. In a way similar to FORTRAN, the execution of a program starts with the function **main**; the procedures move to a function when the function name is met and return to the main program (**main**) after completion of the procedures in the function. As seen in this explanation, the C language does not need the CALL statement used in the FORTRAN language for transferring the task to another function. It employs only the name of the function. The function **main** is written in a way to clarify the flow of calculations, whilst any complex calculation procedure is recommended as a separate function.

Next, we explain the most important statements for developing a program: the **if**, **for**, **do while**, and **switch** statements. We first explain the **if** statement, which is used for choosing tasks according to certain conditions specified by the instructions. Some typical examples are as follows:

```
if(i= =3) x=a ;
```

• This is the simplest **if** statement. "i=3" is expressed as "i==3" in the C language.

```
if( x>=0.)            z=b ;
else if( x<-10. ) z=c ;
else  z=d ;
```

• If x≥0, z=b is set, if x<-10, z=c, and z=d for the other cases.

```
if( x>=5. ) {
  z=a1 ;
}
```

• This is a block-type **if** statement.

```
if( x>5. ) {
  z=a1 ;
}else if( x<=-10. ) {
  z=b1 ;
}else {
  z=c1 ;
}
```

• This is also a block-type **if** statement. One procedure is chosen depending on the condition; there are three cases x>5, x≤-10, and the other cases.

```
if( (x>=-10.) && (X<=10.) ) {
  z=a1 ;
} else if( (x>=50.)||(x<=-50.) ){
  z=b1 ;
}
```

• This is also a block-type **if** statement. "&&" means that if both the conditions are satisfied, z=a1 is assigned and "| |" means that if one of the conditions at least is satisfied, z=b1 is assigned.

The **if** statement implies that the procedure is carried out if a certain condition specified in the **if** statement is satisfied; otherwise, another assessment or another procedure (including the end of the execution of the **if** directive) is conducted. The specification "$<\ =$" in the condition statements represents the mathematical meaning $\leq$, "$>\ =$" means $\geq$, "$=\ \ =$" means $=$, and "$!=$" means $\neq$. We next explain the statements of **for**, **while**, and **do while**, which are used for specifying the repeating procedures. Several typical examples follow.

```
for(i=1; i<=n; i++) {
     ... ;
}
```

• The procedure starts at i=1, then is conducted at i=2 and continued until i=n.

```
for(i=100; i>=0; i-=2) {
     ... ;
}
```

• The procedure starts at i=100, and is conducted at i=98, 96,..., while i≥0. "i-=2" means"i=i-2."

```
 i=3 ;
 do {
   xnew = xold + xdef ;
   i++ ;
 } while( i<=n ) ;
```

• The procedure starts at i=3 and is conducted at i=4,5, ..., while i≤n. "i++" means "i=i+1" and "i--" means " i=i-1."

```
 i=3 ;
 while( i<=n ) {
   xnew = xold + xdef ;
   i++ ;
 }
```

• The procedure is the same as in the previous case, but terminating the procedure is assessed in a different position.

The above statements correspond to the **DO** statement in FORTRAN. The procedures specified between { and } are repeated, with the value of the index variable i increasing or decreasing after the execution of each cycle step. The way of changing the index value is specified by the statement between (and) in the **for** statement, such as "i++" or "i−=2". In the case of the **do while** and **while** statements, the way of changing the index value is specified by "i++." If the statement of "i+=3" is used, the index i will change so that "i=i+3." A difference between the **do while** and the **while** statements is the position for assessing the termination of the procedures. The procedure specified between { and } is repeatedly carried out, whilst the condition indicated in the **while** statement is satisfied.

A statement with characteristics similar to **if** is the **switch** statement. This statement is quite simple to use; an example follows:

```
switch (itree ) {
  case 2 ;
    x = a1 ;
    y = b1 ;
    break ;
  case 3 ;
    x = a2 ;
    y = b2 ;
    break ;
  default ;
    x = a4 ;
    y = b4 ;
    break ;
}
```

• When `itree=2`, a series of statements defined in "`case 2`" are executed, and **`break`** means the exit from the **`switch`** statement. A similar procedure is carried out for "`case 3`." In the other cases, a series of statements defined in **`default`** are executed; the **`break`** statement is possibly unnecessary in the default area.

As already pointed out, the function **main** and other functions correspond to a main program and subprograms in the FORTRAN language, respectively. There are two types of functions in the C language. That is, the first type of function corresponds to a function subprogram in the FORTRAN language, and therefore a value calculated in the function is transferred through the variable (i.e., the name of the function) in the function **main**. The second type of function corresponds to a subroutine subprogram, and a value calculated there is not transferred through the name of the function. Several examples that explain these two types of functions are shown here:

```
setinivel(n,press)
   double press ;
   int     n ;
   {
      double vxi, vyi ;
      int i, j ;
        ...
   }
```

• This is a function that returns no calculated values to the main function. It corresponds to the subroutine subprogram in FORTRAN.

```
double press(x, y)
   double x, y;
   {
      double c1, c2, cans ;
      c1=1. ; c2=2. ;
      cans = c1*x + c2*y ;
      return( cans ) ;
   }
```

• This corresponds to the function subprogram in FORTRAN; the calculated value "cans" is substituted into the double-precision variable "`press`," and the value of "`press`" is returned to **main**.

```
int press(x,y)
   double x, y ;
   {
      int ic, jc, ians ;
      ians = ic*(int)x
           + jc*(int)y ;
      return( ians ) ;
   }
```

• This also corresponds to the integer function subprogram in FORTRAN; the calculated value "`ians`" is substituted into the integer variable "`press`," and the value of "`press`" is returned to **main**.

In the second and third examples, a value calculated in the function is transferred to the main program **main** through the function name. The descriptor of the function type, such as **double** and **integer**, is, therefore, attached before the function name. In the first example, the function does not return a calculated value to the main program, but certain procedures are carried out in this function, so that the declaration of the function type is unnecessary and not attached to the function name. Note that **int**, **float**, and **double** imply that a variable (or data) is integer-type, single-precision-real-type, and double-precision-real-type, respectively.

Next, we explain several important points that seem to be relatively difficult or may be misunderstood by the beginner who is learning the grammar of the C language. Array-type variables are defined in the declaration statements of the data type in such a way as **double a[100]** or **rx[20][20]**. For example, in the case of a one-dimensional array such as **double a[100]**, it is noted that **a[0]**, **a[1]**, . . ., **a[99]** storage spaces are prepared, but **a[100]** is not available. The second example of **double rx[20][20]** means the declaration of a two-dimensional array variable, and **rx[0]][0]**, **rx[0][1]**, **rx[0][2]**, . . ., **rx[19][19]** storage spaces are prepared.

A significant difference between FORTRAN and the C language concerns the data transfer between the function **main** (main program) and other functions (subprograms). In FORTRAN, when one transfers data to a subprogram as arguments, one does not take the values themselves saved in the variables but rather takes the positions or addresses of the variables in which the data are saved. This means that the values saved in the variables can freely be accessed from the subprogram, and also that new data can be assigned to such variables; these new values are reflected in the main program. This data transfer type is the "call by reference." In the case of the C language, the specification of variables as arguments, as in the FORTRAN language, does not mean the transfer of the address of the variables; rather, the values themselves saved in the variables are transferred to the function; therefore the assignment of new values to the variables in the function is never reflected in the main program. This type of data transfer is the "call by value." This means that in respect to data transfer, the C language is much safer than FORTRAN. If the data transfer is carried out by "call by reference" in a similar way to FORTRAN, then the variables of the "pointer" class must be used in the C language. A pointer variable saves the position or address of a standard variable, and therefore it is important to declare what type of data is saved at the position. For example, if an integer value is to be saved in a variable, the address of which a pointer variable “pa” saves, then the asterisk * must be attached to the pointer variable like “*pa,” and the data type must be declared like “int *pa.” In the body of the program, the variable “*pa” is treated as a standard integer variable. If “int *pb, ix” is declared in the definition statement of the data types, the statement “pb = &ix” is used in order to save the address of the integer variable “ix” in the pointer variable “pb.” If “&” is attached to a standard variable, for example, “&ix,” it will return the value of the address of the variable ix. Therefore, since a pointer variable—for example, “pa”—has the information about the address of a standard variable, a value (data) saved at the address of the standard variable can be extracted using the notation “*pa.” We are now ready to begin the explanation of “call by reference.”

In order to return from a function with the calculated values, the information of the addresses of the variables, in which the calculated values are saved, need to be transferred to the function by using arguments of the pointer type. For example, consider a sample program in which a calculation is carried out using a value saved in the variable “h” in the function “anscal,” and the calculated data is returned to the main program through the variable “ans.” One has to call the function using the statement “anscal (h, &ans),” in which a value (i.e., not the pointer information) saved in the variable “h” is transferred to the function “anscal,” and the address of the variable “ans” can be transferred to the function using the pointer information “&ans.” It is important that the data type of the variable “*ans” is declared in the function “anscal,” so that the variable “*ans” can be treated as a standard variable in the function. Several typical examples (including a bad example) follow.

```
double h, ans ;
   ...
x = anscal(h, &ans) ;
   ...
anscal(h, pans)
  double h, *pans ;
  {
  *pans = h*h ;
  }
```

• The address of “ans” is transferred to the function; “&ans” is the address of the variable “ans.” In the function, the pointer variable “pans” is used for receiving the value of “ans” in the main function. Since “ans” is a double-precision real, “*pans” has to be defined as a double-precision-real variable.

```
double h, ans ;
   ...
x = anscal(h, ans) ;
   ...
anscal(h, ans)
  double h, ans ;
  {
    ans = h*h ;
  }
```

• This is a bad example. In this case the values saved in “h” and “ans” are transferred to the function “anscal,” but the values calculated in the function can never be returned (reflected) to the main function.

```
double h, ans[100] ;
   ...
x = anscal(h, ans) ;
   ...
anscal(h, ans)
  double h, ans[100] ;
  {
     for( i=0, i<=99; i++) {
        ans[i] = h*(dble)i ;
     }
  }
```

• For array variables, the data transfer to the function is quite similar to FORTRAN; the pointer variables are unnecessary for the data transfer for the case of array variables. The variable name itself is used as an argument in calling the function and also in the definition of the function name.

In the first example, the address of the variable "ans" in the main program is transferred as an argument "&ans." This value is saved in the pointer variable "pans" in the function; the data type of the variable "ans" is recognized in the function by declaring "double *pans" there. Through these statements, the original value saved in the variable "ans" in the main program is changed into a new value after this new value is substituted into the variable "*pans" in the function. Clearly identifying pointer variables from standard variables by attaching the asterisk * may significantly assist the programmer by removing the danger of mistakes arising from substituting new values to those variables in other functions.

The second example demonstrates a bad example of programming, where new values calculated in the function "anscal" are not transferred to the variables "h" and "ans" in the main program, since the connection of the variables between the main function and the function "anscal" can never be made using a statement of the type "anscal (h, ans)." If the arguments are defined without pointer variables, then a function that returns a calculated value to the main program may be used, as already explained; in this case, "anscal (h, ans)" has to be changed into "double anscal (h, ans)," and "return (ans);" is added to the line after "ans = h*h;," which corresponds to a function subprogram in the FORTRAN language.

The third example demonstrates how to transfer values saved in array-type variables such as "ans." The data transfer of array-type variables can be conducted in the same way as for the FORTRAN language, and therefore pointer variables are unnecessary. That is, calling a function with the arguments that are array-type variables will have a direct type of connection, so that new values assigned to the array-type variables in the function are reflected in **main** without the need for pointer variables.

We have shown the three methods of returning calculated values from a function back to the main function. The first method is to use pointer variables; the second is to use array-type variables; and the third is to use a function that returns a calculated value through the name of the function itself. In addition to these three methods is another method that uses global variables that correspond to variables declared in the COMMON statements in FORTRAN. The global variables have to be declared before the "main( )" statement, and for these variables we recommend the use of capital characters in their names, to help the programmer recognize them. An example of using global variables is in the sample simulation program shown in Section 5.6 as the array-type variables such as RX[NN], RY[NN], and RZ[NN].

Next, we explain the statements for inputting data, **scanf** and **fscanf** statements, and for outputting data, **printf** and **fprintf** statements. The **scanf** and **printf** statements correspond to READ(5,*) and WRITE(6,*) statements in FORTRAN; in these statements, data are input from a keyboard and results shown on a display. In the case of the **fscanf** and **fprintf** statements, data files are used for reading and writing the data. If the reader understands the latter reading and writing statements, the former statements are quite straightforward to understand, so we only focus on the explanation of the **fscanf** and **fprintf** statements. In order to use data files, pointer variables must be connected to the data files used in a program. To do so, the **fopen** statement is used, and **fclose** must be used to disconnect the data file used before the end of the main program; this means that a data file connected by the **fopen** statement should always be disconnected in a program. Some examples follow.

```
main()
    {
      double a, b, c;
      int i;
      FILE *fopen(),*np1,*np2,*np[4];
      np1 = fopen("aaa0.data","r");
      np2 = fopen("aaa1.data","w");
      np[1] = fopen("bbb1.data","w");
      np[2] = fopen("bbb2.data","w");
      np[3] = fopen("bbb3.data","w");
        ...
      fscanf(np1,"%lf", &c );
      fprintf(np2,"a = %10.3f b = %10.3f\n", a, b);
      i = 2;
      fprintf(np[i],"a = %10.3f b = %10.3f c = %10.3f\n",
      a,b,c);
        ...
      fclose(np[1]);
      fclose(np[2]);
      fclose(np[3]);
      fclose(np1);
      fclose(np2);
```

As shown in the above example, a data file must be connected to the file pointer variable, which is declared in the **FILE** statement, by using the **fopen** statement. After a data file is opened (connected), data can be input from the data file by using the **fscanf (np1,. . .)** statement, and also can be output by the **fprintf (np2,. . .)** or **fprintf (np[2],. . .)**. The latter example for **fprintf** is quite useful for outputting the particle positions at given time step intervals, which may be used for making an animation of the particle motion. In this case, the index "i" in "np[i]" is made to change in such a way as i=1,2,3,. . ., with advancing time for the output. The arguments "r" and "w" in the **fopen** statement indicate the reading and the writing mode, respectively. A data file opened by the **fopen** statement must be closed (disconnected) using the **fclose** statement before the end of the program. If a data is read and saved in a standard variable "c," the pointer information (address) of "c" is necessary as an argument in the **fscanf** statement. In contrast, when a data saved in the variable "a" is output to a data file, only a value is necessary, so that the name itself is used as an argument in the **fprintf** statement; the pointer information is unnecessary in this case.

Next, we explain how to describe the format to output data, using the following example:

```
i = 3;
xi = 5.;
```

```
yi = 2.;
press = xi*yi;
fprintf(np2,"i = %3d  xi = %7.3f  yi = %7.3f  pressure =
%10.3f\n", i, xi, yi, press);
```

The output result of the above **fprintf** statement is as follows:

```
i=  3 xi=   5.000 yi=   2.000 pressure=     10.000
```

The C language does not have a statement corresponding to the FORMAT statement in FORTRAN. Instead, the output format for the data is specified in the **fprintf** statement. In the above example, "%3d" is used for integer-type data and is written using 3 columns (spaces) from the right. Similarly, "%7.3f" is for real-type data and is written using 7 columns from the right with three decimal places, and "\n" means the start of a new paragraph. If a data is output in exponential form, for example, using "%10.2e," this implies that a value is written using 10 columns with 2 decimals. The reader sees many examples in the sample simulation program shown in Section 5.6.

In order to make a visualization, such as an animation or snapshot, using the data of the particle positions, it is necessary to write out only data (figures) in a data file without any characters for explaining the data such as the names of variables. An example for this output is as follows:

```
   ...
for (i = 1; i < 100; i ++ ) {
  fprintf(np2,"%10.4f%10.4f%10.4f\n", rx[i], ry[i],
  rz[i]);
}
   ...
```

In this example, the components of the particle position vector, rx[*], ry[*], and rz[*] are output at each time step using the **for** loop statement; the position data of particle 1, particle 2, and particle i are written in the first, second, and ith lines, respectively, of the data file. In order to conduct another run using the data saved in the above-mentioned manner, one needs to read such data from the data file in the following way:

```
   ...
for (i = 1; i < 100; i ++ ) {
  fscanf(np1,"%lf%lf%lf\n", &rx[i], &ry[i], &rz[i]);
}
   ...
```

In the above example, "rx[*], ry[*], and rz[*]" are assumed to be defined as double-precision-real-type variables. As this example demonstrates, in the C language, data does not need to be read using the same format description that was used in the **fprintf** statement, but only described as "%lf%lf%lf/n" in the **fscanf** statement; this is in contrast to FORTRAN. As already pointed out, the address of the variables—not the name itself—is necessary in reading the data by the **fscanf** statement.

The C language has several characteristic concepts for using variables, such as structure variables, which are not contained in the FORTRAN77 language. We do not explain them in this book, because these characteristic statements are not used in the sample simulation programs. Since imaginary variables may be useful in certain cases, the reader may find them in a textbook on the C language, if necessary.

Finally, we show some additional features of the grammar using a short sample simulation program.

• The statements enclosed by "/*" and "*/" are regarded as comment lines and therefore have no influence on the calculation. Comment lines are placed at any positions, which is dissimilar to FORTRAN.

```
/*----------------------------
/*                        alde
/*
/*           ----- Hard Sphere Molecular Dynamics  -----           */
/*              Simulation of phase transition for a two            */
/*              dimensional system.                                 */
/*-----------------------------------------------------------------*/
#include <stdio.h>
#include <math.h>
#define PI     3.141592653589793
#define NN     201
#define NNCOLMX 2001
#define NRANMX  100001
      double  RX[NN] , RY[NN] ;
      double  VX[NN] , VY[NN] ;
      double  XL, YL ;
      float   RAN[NRANMX] ;
      int     NRAN, IX ;

/*--------------------------------------------- main function ---*/
main()
  {
      int     n, partnr[NN] ;
      int     n, partnr[NN] ;
      double  coltim[NN] , tstep, tij , tim, timbig ;
      float   rx0[NN][NNCOLMX], ry0[NN][NNCOLMX] ;
      int     i, j, k, ii, ncol , ncolmx , nbump ;
      FILE    *fopen(), *np[10], *np1, *np2 ;

           np1    = fopen("@baa1.data", "w");
           np[1]  =
           np[2]  =


      n      = 36 ;
      vdens  = 0.1 ;
      ndens  = vdens*(4.
      dsq    = d*d ;
      timbig = 1.e10 ;

      IX     = 0 ;
```

• "#include <stdio.h>" is necessary for the input/output of data, and "#include <math.h>" is necessary for the use of mathematical calculations.
• The "define" statement corresponds to the PARAMETER statement in FORTAN, which is useful for defining the size of the array-type variables.
• The variables defined using "double," "float," and "int" are regarded as global variables that can be accessed from any functions without any definition in each function.

• In order to output the calculated data on a data file, the file has to be related to the pointer variable (device number) using the "fopen" statement; the opened file has to be closed using the "fclose" statement before the end of the main function. "w" and "r" are used for writing and reading the data, respectively.

• This is calling the function rancal(*), in which arguments are unnecessary because of the use of the global variables. This is a void function of returning no calculated results, which corresponds to a subroutine subprogram in FORTRAN.

```
rancal() ;
NRAN   = 1 ;

/*-------------------------------------------------------------*/
/*-----------------   initial configuration   -----------------*/
/*-------------------------------------------------------------*/

iniposit( n, ndens ) ;

inivel( n, temp ) ;
```

- The given parameters are written out in the data file `@baal.data`.
- There is no statement that corresponds to the FORMAT statement in FORTRAN.

```
fprintf(np1,"----------------------------------------------------¥n") ;
fprintf(np1,"       Molecular dynamics of hard spheres     ¥n") ;
fprintf(np1,"                                              ¥n") ;
fprintf(np1," n=%4d ndens=%8.3f vdens=%6.3f temp=%7.3f¥n",
             n, ndens, vdens, temp) ;
fprintf(np1," XL=%6.3f YL=%6.3f¥n", XL, YL) ;
fprintf(np1," ncolmx=%8d¥n", ncolmx) ;
fprintf(np1,"-----------------------------------------------------¥n") ;

/*---------------------------
/*----------------------   eq
/*---------------------------

for ( ncol=1 ; ncol<=ncolmx ;
```

- The "for" loop implies the iteration calculation. The procedure starts at `ncol=1` and then is conducted at `ncol=2,3,⋯`, until `ncol=ncolmx`. Another "for" loop is possible inside the "for" loop.

```
                     /
  tim   += tstep ;
  nbump += 1 ;
```

- The "`tim+=tstep`" implies "`tim=tim+tstep`." "`rint`" is a round-up function.

```
                             /*--- advance particle position ---*/
  for ( k=1 ; k<=n ; k++ ) {
    coltim[k] += - tstep ;
    RX[k]     += VX[k]*tstep ;
    RY[k]     += VY[k]*tstep ;
    RX[k]     += - rint( RX[k]/XL - 0.5 )*XL ;
    RY[k]     += - rint( RY[k]/YL - 0.5 )*YL ;
  }
```

- In the "if" statement, "`==`" means "`=`," "`<=`" means "≤," and "`>=`" means "≥." Also, "| |" means "OR," and "`&&`" means "AND."

```
  for( k=1 ; k<=n ; k++ ) {
    if( (partnr[k] == i) || (partnr[k] == j) )
                             collist( n, dsq, k, coltim, partnr ) ;
  }
                                   /*--- for data output ---*/
  for( k=1 ; k<=n ; k++ ) {
    rx0[k][ncol] = (float)RX[k] ;
    ry0[k][ncol] = (float)RY[k] ;
  }
}
```

- The "(float)" is added just before the variable in order to change a double-precision to a single-precision data.

```
/*-------------------------------------------------------------*/
/*--------------------   end of main loop   -------------------*/
/*-------------------------------------------------------------*/
```

- This is not for the postprocessing analysis, just for reconfirming the validity of results.

```
fprintf(np1,"time=%11.3e num.of coll.=%7d nbump/ncolmx=%6.3f¥n",
             tim, nbump, (float)nbump / (float)ncolmx ) ;
                                   /*--- data output ---*/
fprintf(np2,"%4d%8.4f%8.4f%8.3f%9.3f%9.3f%8d¥n",
             n, ndens, vdens, temp, XL, YL , ncolmx ) ;
                                   /*--- data output ---*/
```

- This "fprintf" statement is for the postprocessing analysis, such as making snapshots and analyzing data; thus, only numerical values are written out.

```
for ( ii=1 ; ii<=7 ; ii++ ) {
  op = 5*(ii-1) ; inp += 1 ;
  for( k=1 ; k<=ncolmx ; k++ ) {
    fprintf(np[inp],
```

```
              "%6.2f%6.2f%6.2f%6.2f%6.2f%6.2f%6.2f%6.2f%6.2f%6.2f¥n",
              rx0[op+1][k],ry0[op+1][k], rx0[op+2][k],ry0[op+2][k],
              rx0[op+3][k],ry0[op+3][k], rx0[op+4][k],ry0[op+4][k],
              rx0[op+5][k],ry0[op+5][k] ) ;
      }
      fclose(np[inp]) ;
    }
              fclose (np1) ;
  }

/*---------------------------
/*------------------------- fu
/*---------------------------
/*+++ fun iniposit +++*/
    iniposit( n, ndens )

    double  ndens ;
    int     n ;
    {
      double  rxi, ryi, rx0, ry0, a , ax , ay ,
      int     i , j , kx , ky , k  , p , iface ;

                    /*--- set mol. at close-pack
      a  = sqrt(  (2./sqrt(3.))/ndens  ) ;
      p  = rint(  sqrt( (double)(n/4) )  ) ;
      XL = sqrt(3.)*a*(double)p ;

      ax = sqrt(3.)*a ;  ay   =  2.*a ;
      kx = p ; ky = p ;

      for ( iface=1 ; iface<=4 ; iface++ ) {
        if( iface == 1 ) {
          rx0 = c1 ; ry0 = c1 ;
        } else if( iface == 2 ) {
          rx0 = c1 ; ry0 = a + c1 ;
        } else if( iface == 3 ) {
          rx0 = ax/2. + c1  ;
          ry0 = a/2.  + c1  ;
        } else {
          rx0 = ax/2. + c1  ;
          ry0 = a*3./2. + c1 ;
        }
        for( j=0 ; j<=ky-1 ; j++ ) {
          ryi = (double)j*ay + ry0 ;
          if ( ryi >= YL )   break ;
          for ( i=0 ; i<=kx-1 ; i++ ) {
            rxi = (double)i*ax + rx0 ;
            if ( rxi >= XL )  break ;

            k += 1 ;
            RX[k]   =  rxi ;
          }
        }
      }
    }
/*+++ fun inivel +++*/
    inivel( n, temp )

    int     n ;
    double  temp ;
    {
      int     i ;
      double  c0 , c1 , c2 , c3 , t , vxi , vyi ;

      c0 = 2.*PI ;
      for ( i=1 ; i<=n ; i++ ) {

L5:     NRAN += 1  ;
```

• The file opened by the "fopen" statement is necessarily closed by the "fclose" statement.

• The arguments have to be described in the same order as being called in the main function.
• Even if any values are saved in "n" and "ndens," these values are not reflected in the main function.

• The variables defined are valid only in this function, and these values have no influence on the main function.
• The results calculated here are returned to the main program through the global variables.

• "sqrt" means a mathematical function that calculates the square root of a value.

• The "break" statement enables the procedure to be terminated and to leave the calculation in the "for" or the "if" statement unit.

• "log(x)" means a natural logarithm, "cos(x)" and "sin(x)" mean a cosine and a sine function, "fabs(x)" returns the absolute value of x, "pow(x,y)" means x^y, and "m%n" returns the reminder. Also, "floor(x)" means truncation, "rint(x)" means rounding-up, and "exp(x)" is an exponential function. In the above functions, x is regarded to be a double-precision real, and m and n are integer variables.

```
        c1    = sqrt( -t*log( (double)(RAN[NRAN]) ) ) ;
        NRAN += 1 ;
        c2    = c0*(double)( RAN[NRAN] ) ;
        vxi   = c1*cos(c2) ;

        if( (vxi*vxi+vyi*vyi) >= c3 )   goto L5 ;
        VX[i] = vxi ;
      }
    }
```

• The "goto" statement tends to make the logical flow complex, so this statement should be used limitedly.

In this example, the line numbers are attached for the sake of convenience—they are not necessary in writing a program. In the C language, all variables used in a program must be defined using the data type statement such as **int**, **float**, and **double**.

A3.3 Execution Procedures of FORTRAN and C Programs

The execution of a program in the FORTRAN or the C language involves two procedures: one to make an executive-type program by compiling the program, and another to conduct a command for running the executive-type program. When error messages appear in compiling a program, one has to modify the program so as to completely remove those errors. Error messages are quite useful for the beginner in the process to learn how to develop a program, so that the reader is recommended to spend sufficient time on tackling such problems. Note that if there are no error messages, it does not mean that there are no bugs in the simulation program, but just implies there are no grammatical errors. Hence, after error messages disappear in compiling, one should check a program another 5 times. Since this kind of careful verification procedure is necessary to remove fatal bugs, programmers have to avoid employing complex logical structures in writing a program.

The sample simulation programs shown in each chapter of this book are almost directly portable to free FORTRAN and C compilers, for example, in a free Linux system. However, if the reader intends to conduct a large-scale simulation, it is desirable to introduce a commercial compiler, which may offer higher performance for the computer.

If a Linux system is installed with GNU family compilers in the FORTRAN and C languages, typical execution procedures are as follows:

```
> f77 sample1.f
> ./a.out
```

```
> cc sample1.c -lm
> ./a.out
```

```
> f77 -o sample1.out sample1.f
> ./sample1.out
```

```
> cc -o sample1.out sample1.c -lm
> ./sample1.out
```

The "a.out" is a default name of an executive-type program, but in the second example, the name of an executive-type program is assigned to a chosen name and the execution is carried out using this name. Since mathematical functions are usually used in a program, the compile option "-lm" is necessary for a C program.

If you use a commercial compiler, offered by Intel or other companies, installed on a Linux system, a typical example for the execution is as follows:

```
> ifort -o sample1.out sample1.f
> ./sample1.out
```

in which "ifort" is the command for starting the FORTRAN compiler. If the reader is using a freeware, the required command may be "g77," "f90," "f95," "gfortran," "ifc," or "fort."

If the reader wants more information on the compile options, "man ifort" or "man ifc" can be used to access to the manual of the compiler. Note that since the grammar is slightly different among different compilers, one compiler may output error messages in compiling, but another does not. Hence, we recommend that the reader devise a program in a general form, otherwise, a large amount of tuning tasks may be necessary to apply it to a compiler on another computer.

If error messages are output in compiling the same programs in this book, the following data type statement may be a reason; in this case, the reader is advised to replace "REAL*8" with "DOUBLE PRECISION." Also, error messages may be resolved by reducing the size of array-type variables.

Appendix 4: Unit Systems of Magnetic Materials

The CGS unit system and the SI unit system, which was developed from the MKSA unit system, are generally used in the field of magnetic materials. Although the CGS unit system is commonly used in the commercial world, the SI unit system is invariably used in textbooks on magnetic materials. Using quantities expressed in different unit systems at the same time will lead to wrong expressions for physical quantities, so one must adhere to the same unit system for handling equations or physical values of magnetic materials. Many textbooks on magnetic materials provide tables to transform values from one unit system to another. We here summarize the two unit systems based on the MKSA system. In the first unit system, the magnetization **M** corresponds to the magnetic field **H** in units. In the second unit system, **M** corresponds to the magnetic flux density **B**. Some typical quantities used for magnetic materials are tabulated below.

Note that in this book we use the first unit system of **M** corresponding to **H** in units.

	$\mathbf{B} = \mu_0(\mathbf{H} + \mathbf{M})$	$\mathbf{B} = \mu_0\mathbf{H} + \mathbf{M}$
Magnetic field strength, **H**	[A/m]	[A/m]
Magnetization strength, **M**	[A/m]	[Wb/m^2]
Magnetic flux density, **B**	[T] (=[Wb/m^2])	[T] (=[Wb/m^2])
Permeability of free space, μ_0	$\mu_0 = 4\pi \times 10^{-7}$ [H/m] (=[Wb/(A · m)]	$\mu_0 = 4\pi \times 10^{-7}$ [H/m] (=[Wb/(A · m)]
Magnetic charge, q	[A · m]	[Wb] (=[N · m/A])
Magnetic moment, **m**	[A · m^2]	[Wb · m] (=[N · m^2/A])
Potential energy, U	$U = -\mu_0 \mathbf{m} \cdot \mathbf{H}$ [J] (=[Wb · A])	$U = -\mathbf{m} \cdot \mathbf{H}$ [J] (=[Wb · A])
Torque, **T**	$\mathbf{T} = \mu_0 \mathbf{m} \times \mathbf{H}$ [N · m] (=[Wb · A])	$\mathbf{T} = \mathbf{m} \times \mathbf{H}$ [N · m] (=[Wb · A])
Magnetic field induced by magnetic charge, **H**	$\mathbf{H} = \frac{q}{4\pi r^2} \cdot \frac{\mathbf{r}}{r}$ [A/m]	$\mathbf{H} = \frac{q}{4\pi\mu_0 r^2} \cdot \frac{\mathbf{r}}{r}$ [A/m]
Magnetic force acting between two magnetic charges, **F**	$\mathbf{F} = \frac{\mu_0 q q'}{4\pi r^2} \cdot \frac{\mathbf{r}}{r}$ [N] (=[Wb · A/m])	$\mathbf{F} = \frac{q q'}{4\pi\mu_0 r^2} \cdot \frac{\mathbf{r}}{r}$ [N] (=[Wb · A/m])
Magnetic interaction between two magnetic moments, U	$U = \frac{\mu_0}{4\pi r^3}\{\mathbf{m}_1 \cdot \mathbf{m}_2 - \frac{3}{r^2} \times (\mathbf{m}_1 \cdot \mathbf{r})(\mathbf{m}_2 \cdot \mathbf{r})\}$ [J](=[Wb · A])	$U = \frac{1}{4\pi\mu_0 r^3}\{\mathbf{m}_1 \cdot \mathbf{m}_2 - \frac{3}{r^2} \times (\mathbf{m}_1 \cdot \mathbf{r})(\mathbf{m}_2 \cdot \mathbf{r})\}$ [J] (=[Wb · A])

Combined units: [H] = [Wb/A], [T] = [Wb/m^2], [J] = [N · m]
Equivalent units: [N] = [Wb · A/m]

How to Acquire Simulation Programs

A copy of the sample simulation programs that are shown in this book can be requested directly from the author via e-mail:

asatoh_book2010@excite.co.jp

Please note that the following information is required:

1. the purchase date,
2. the number of purchased copies,
3. the profession of the purchaser.

The sample simulation programs in this book can be used free of charge for educational purposes in an academic environment such as a university, but are not permitted to be used for commercial purposes. In addition, the user takes responsibility for all results obtained from using the sample simulation programs.

The author would deeply appreciate the report of any bugs in the programs, but regrets that he is unable to accept any inquiries concerning the content of the simulation programs.

References

[1] M.P. Allen, D.J. Tildesley, Computer Simulation of Liquids, Clarendon Press, Oxford, 1987.

[2] D.C. Rapaport, The Art of Molecular Dynamics Simulation, Cambridge University Press, Cambridge, 1995.

[3] J.M. Haile, Molecular Dynamics Simulation, Elementary Methods, John Wiley & Sons, New York, 1992.

[4] A. Satoh, Introduction to Molecular-Microsimulation of Colloidal Dispersions, Elsevier Science, Amsterdam, 2003.

[5] P.J. Hoogerbrugge, J.M.V.A. Koelman, Simulating microscopic hydrodynamic phenomena with dissipative particle dynamics, Europhys. Lett. 19 (1992) 155−160.

[6] J.M.V.A. Koelman, P.J. Hoogerbrugge, Dynamic simulations of hard-sphere suspensions under steady shear, Europhys. Lett. 21 (1993) 363−368.

[7] P. Espanol, Hydrodynamics from dissipative particle dynamics, Phys. Rev. E 52 (1995) 1734−1742.

[8] C.A. Marsh, G. Backx, M.H. Ernst, Static and dynamic properties of dissipative particle dynamics, Phys. Rev. E 56 (1997) 1676−1691.

[9] S. Succi, The Lattice Boltzmann Equation for Fluid Dynamics and Beyond, Clarendon Press, Oxford, 2001.

[10] D.H. Rothman, S. Zaleski, Lattice-Gas Cellular Automata, Simple Models of Complex Hydrodynamics, Cambridge University Press, Cambridge, 1997.

[11] J.-P. Rivet, J.P. Boon, Lattice Gas Hydrodynamics, Cambridge University Press, Cambridge, 2001.

[12] B. Chopard, M. Droz, Cellular Automata Modeling of Physical Systems, Cambridge University Press, Cambridge, 1998.

[13] L. Verlet, Computer experiments on classical fluids. I. Thermodynamical properties of Lennard-Jones molecules, Phys. Rev. 159 (1967) 98−103.

[14] W.C. Swope, H.C. Andersen, P.H. Berens, K.R. Wilson, A computer simulation method for the calculation of equilibrium constants for the formation of physical clusters of molecules: application to small water clusters, J. Chem. Phys. 76 (1982) 637−649.

[15] R.W. Hockney, The potential calculation and some applications, Methods Comput. Phys. 9 (1970) 136−211.

[16] S. Kim, S.J. Karrila, Microhydrodynamics: Principles and Selected Applications, Butterworth-Heinemann, Stoneham, 1991.

[17] H. Brenner, Rheology of a dilute suspension of axisymmetric Brownian particles, Int. J. Multiphase Flow 1 (1974) 195−341.

[18] S. Kim, R.T. Mifflin, The resistance and mobility functions of two equal spheres in low-Reynolds-number flow, Phys. Fluids 28 (1985) 2033−2045.

[19] D.A. McQuarrie, Statistical Mechanics, Harper & Row, New York, 1976.

[20] J.P. Hansen, I.R. McDonald, Theory of Simple Liquids, second ed., Academic Press, London, 1986.

[21] N. Metropolis, A.W. Rosenbluth, M.N. Rosenbluth, A.H. Teller, E. Teller, Equation of state calculations by fast computing machines, J. Chem. Phys. 21 (1953) 1087–1092.
[22] H.C. Tuckwekk, Elementary Applications of Probability Theory, second ed., Chapman & Hall, London, 1995.
[23] A. Jeffrey, Mathematics for Engineers and Scientists, fifth ed., Chapman & Hall, London, 1996.
[24] D.L. Ermak, J.A. McCammon, Brownian dynamics with hydrodynamic interactions, J. Chem. Phys. 69 (1978) 1352–1360.
[25] G.A. Bird, Molecular Gas Dynamics and the Direct Simulation of Gas Flows, second ed., Oxford University Press, Oxford, 1994.
[26] G.E.P Box, M.E. Müller, A note on the generation of random normal deviates, Ann. Math. Stat. 29 (1958) 610–611.
[27] B. Quentrec, C. Brot, New method for searching for neighbors in molecular dynamics computations, J. Comput. Phys. 13 (1975) 430–432.
[28] R.W. Hockney, J.W. Eastwood, Computer Simulation Using Particles, McGraw-Hill, New York, 1981.
[29] S.M. Thompson, Use of neighbor lists in molecular dynamics, CCP5 Quarterly 8 (1983) 20–28.
[30] A.W. Lees, S.F. Edwards, The computer study of transport processes under extreme conditions, J. Phys. C 5 (1972) 1921–1929.
[31] R.E. Rosensweig, Ferrohydrodynamics, Cambridge University Press, Cambridge, 1985.
[32] R.E. Rosensweig, J.W. Nestor, R.S. Timminins, Ferrohydrodynamics fluids for direct conversion of heat energy, Symp. AIChE-I Chem. Eng. 5 (1965) 104–118.
[33] C. Pozrikidis, Introduction to Theoretical and Computational Fluid Dynamics, Oxford University Press, Oxford, 1997.
[34] D. Yu, R. Mei, L.-S. Luo, W. Shyy, Viscous flow computations with the method of lattice Boltzmann equation, Prog. Aerospace Sci. 39 (2003) 329–367.
[35] A.J.C. Ladd, Numerical simulations of particulate suspensions via a discretized Boltzmann equation. Part 1. Theoretical foundation, J. Fluid Mech. 271 (1994) 285–309.
[36] A.J.C. Ladd, Short-time motion of colloidal particles: numerical simulation via a fluctuating lattice-Boltzmann equation, Phys. Rev. Lett. 70 (1993) 1339–1342.
[37] M. Bouzidi, M. Firdaouss, P. Lallemand, Momentum transfer of a Boltzmann-lattice fluid with boundaries, Phys. Fluids 13 (2001) 3452–3459.
[38] S. Chapman, T.G. Cowling, The Mathematical Theory of Non-Uniform Gases, Cambridge University Press, Cambridge, 1960.

FSC
www.fsc.org
MIX
Papier | Fördert gute Waldnutzung
FSC® C083411

Kolibri 360 GmbH
Ferdinand-Jühlke-Straße 7
99095 Erfurt, Deutschland
produktsicherheit@kolibri360.de